METHODS IN MOLECULAR BIOLOGY™

Series Editor
John M. Walker
School of Life Sciences
University of Hertfordshire
Hatfield, Hertfordshire, AL10 9AB, UK

For further volumes:
http://www.springer.com/series/7651

Genome-Wide Association Studies and Genomic Prediction

Edited by

Cedric Gondro

*The Center for Genetic Analysis and Applications,
University of New England, Armidale, NSW, Australia*

Julius van der Werf

*School of Environmental and Rural Science,
University of New England, Armidale, NSW, Australia*

Ben Hayes

*Biosciences Research Division,
Department of Primary Industries, Bundoora, VIC, Australia*

Humana Press

Editors
Cedric Gondro
The Center for Genetic Analysis
  and Applications
University of New England
Armidale, NSW, Australia

Ben Hayes
Biosciences Research Division
Department of Primary Industries
Bundoora, VIC, Australia

La Trobe University, Bundoora
VIC, Australia

Julius van der Werf
School of Environmental and Rural Science
University of New England
Armidale, NSW, Australia

Additional material to this book can be downloaded from http://extras.springer.com

ISSN 1064-3745   ISSN 1940-6029 (electronic)
ISBN 978-1-62703-446-3   ISBN 978-1-62703-447-0 (eBook)
DOI 10.1007/978-1-62703-447-0
Springer New York Heidelberg Dordrecht London

Library of Congress Control Number: 2013937825

© Springer Science+Business Media, LLC 2013
This work is subject to copyright. All rights are reserved by the Publisher, whether the whole or part of the material is concerned, specifically the rights of translation, reprinting, reuse of illustrations, recitation, broadcasting, reproduction on microfilms or in any other physical way, and transmission or information storage and retrieval, electronic adaptation, computer software, or by similar or dissimilar methodology now known or hereafter developed. Exempted from this legal reservation are brief excerpts in connection with reviews or scholarly analysis or material supplied specifically for the purpose of being entered and executed on a computer system, for exclusive use by the purchaser of the work. Duplication of this publication or parts thereof is permitted only under the provisions of the Copyright Law of the Publisher's location, in its current version, and permission for use must always be obtained from Springer. Permissions for use may be obtained through RightsLink at the Copyright Clearance Center. Violations are liable to prosecution under the respective Copyright Law.
The use of general descriptive names, registered names, trademarks, service marks, etc. in this publication does not imply, even in the absence of a specific statement, that such names are exempt from the relevant protective laws and regulations and therefore free for general use.
While the advice and information in this book are believed to be true and accurate at the date of publication, neither the authors nor the editors nor the publisher can accept any legal responsibility for any errors or omissions that may be made. The publisher makes no warranty, express or implied, with respect to the material contained herein.

Printed on acid-free paper

Humana Press is a brand of Springer
Springer is part of Springer Science+Business Media (www.springer.com)

Preface

Genome-wide association studies (GWAS) have rapidly spread across the globe over the last few years becoming the de facto approach to identify candidate regions associated with complex diseases in human medicine. GWAS and genome analysis are also powerful tools to provide a handle on genetic variation in a wide range of traits important for human health, agriculture, conservation, and the evolutionary history of life. As the technology matured and results from the various studies came to light, a clear picture emerged that, as often seems to be the case, biological processes prove to be much more complex than we would want them to be. While many new insights were and are being attained, maybe just as many new questions emerged. In many cases, GWAS successfully identified variants that were unquestionably associated with variation in traits and, in some cases, culminated in the discovery of the causal variant(s) and its mechanism of action. In other cases, these studies evidenced that a large number of traits are highly polygenic, and, instead of identifying a few genomic region *culprits*, we saw a scattering of effects across the whole genome. While in these latter cases the underlying biology largely remains elusive, it still allows making predictions of phenotypes based on the genomic information of an individual. And this in the short term might be even more important for translational outcomes. This brings us to the focus of this volume in the series *Methods in Molecular Biology: Genome-Wide Association Studies and Genomic Prediction*.

This volume in the series covers from the preliminary stages of understanding the phenotypes of interest and design issues for GWAS, passing through efficient computational methods to store and handle large datasets, quality control measures, phasing, haplotype inference and imputation, and moving on to then discuss the various statistical approaches to data analysis where the experimental objective is either to *nail down* the biology by identifying genomic regions associated to a trait or to use the data to make genomic predictions about a future phenotypic outcome (e.g. predict onset of disease). One of the chapters even merges the two approaches into a unified framework.

Other chapters discuss how to validate results from GWAS, how to incorporate prior biological knowledge to improve the power of GWAS, how the wealth of data generated from GWAS can be used to understand the evolutionary history and diversity of populations, or how genomic selection can be used in real world scenarios to improve industry outcomes. The results from GWAS have brought to the fore the fact that biological processes are highly connected—traits are polygenic and genes do not act on their own; epistasis and interactions are prevalent, systems are complex. Later chapters address methods to *tame* this complexity with computational heuristic and machine learning techniques which attempt to reduce the dimensionality of the problem and make it more tractable, or from another angle, propose statistical methods to explore the genomic connectivity or make better use of other sources of information.

Authors of the chapters in this book come from the various *walks of life*. We have tried to find a balance between biomedical, plant, and animal researchers. Each *habitat* has its own questions and ways of solving problems; we have tried with this volume to bring these sometimes contradictory approaches together. Frequently, a glance into other fields can be very revealing.

Finally, this volume purposely covers both theoretical and practical aspects of GWAS and genomic prediction. Some chapters are more focussed on principles and theoretical aspects while others are motivated by the *hands on/how to do it* approach. We wanted this balance in the book: robust theoretical foundations but also step by step instructions on how to use the available tools and implement them in a real world setting.

We want to thank John Walker (series editor) and David Casey (editor at Springer) for their continuous support and assistance throughout the project. And of course, special thanks to the authors of the chapters who have generously contributed their time and knowledge to this book.

Armidale, NSW, Australia   *Cedric Gondro*
Armidale, NSW, Australia   *Julius van der Werf*
Bundoora, VIC, Australia   *Ben Hayes*

Contents

Preface		v
Contributors		ix

1. R for Genome-Wide Association Studies 1
   Cedric Gondro, Laercio R. Porto-Neto, and Seung Hwan Lee

2. Descriptive Statistics of Data: Understanding the Data Set
   and Phenotypes of Interest 19
   Sonja Dominik

3. Designing a GWAS: Power, Sample Size, and Data Structure 37
   Roderick D. Ball

4. Managing Large SNP Datasets with SNPpy 99
   Faheem Mitha

5. Quality Control for Genome-Wide Association Studies 129
   Cedric Gondro, Seung Hwan Lee, Hak Kyo Lee, and Laercio R. Porto-Neto

6. Overview of Statistical Methods for Genome-Wide Association
   Studies (GWAS) 149
   Ben Hayes

7. Statistical Analysis of Genomic Data 171
   Roderick D. Ball

8. Using PLINK for Genome-Wide Association Studies (GWAS)
   and Data Analysis 193
   Miguel E. Rentería, Adrian Cortes, and Sarah E. Medland

9. Genome-Wide Complex Trait Analysis (GCTA): Methods, Data Analyses,
   and Interpretations 215
   Jian Yang, Sang Hong Lee, Michael E. Goddard, and Peter M. Visscher

10. Bayesian Methods Applied to GWAS 237
    Rohan L. Fernando and Dorian Garrick

11. Implementing a QTL Detection Study (GWAS) Using Genomic
    Prediction Methodology 275
    Dorian J. Garrick and Rohan L. Fernando

12. Genome-Enabled Prediction Using the BLR (Bayesian Linear Regression)
    R-Package 299
    Gustavo de los Campos, Paulino Pérez, Ana I. Vazquez, and José Crossa

13. Genomic Best Linear Unbiased Prediction (gBLUP) for the Estimation
    of Genomic Breeding Values 321
    Samuel A. Clark and Julius van der Werf

14. Detecting Regions of Homozygosity to Map the Cause of Recessively
    Inherited Disease 331
    James W. Kijas

15. Use of Ancestral Haplotypes in Genome-Wide Association Studies .. 347
    Tom Druet and Frédéric Farnir

16	Genotype Phasing in Populations of Closely Related Individuals	381
	John M. Hickey	
17	Genotype Imputation to Increase Sample Size in Pedigreed Populations	395
	John M. Hickey, Matthew A. Cleveland, Christian Maltecca,	
	Gregor Gorjanc, Birgit Gredler, and Andreas Kranis	
18	Validation of Genome-Wide Association Studies (GWAS) Results	411
	John M. Henshall	
19	Detection of Signatures of Selection Using F_{ST}	423
	Laercio R. Porto-Neto, Seung Hwan Lee, Hak Kyo Lee, and Cedric Gondro	
20	Association Weight Matrix: A Network-Based Approach Towards Functional Genome-Wide Association Studies	437
	Antonio Reverter and Marina R.S. Fortes	
21	Mixed Effects Structural Equation Models and Phenotypic Causal Networks	449
	Bruno Dourado Valente and Guilherme Jordão de Magalhães Rosa	
22	Epistasis, Complexity, and Multifactor Dimensionality Reduction	465
	Qinxin Pan, Ting Hu, and Jason H. Moore	
23	Applications of Multifactor Dimensionality Reduction to Genome-Wide Data Using the R Package 'MDR'	479
	Stacey Winham	
24	Higher Order Interactions: Detection of Epistasis Using Machine Learning and Evolutionary Computation	499
	Ronald M. Nelson, Marcin Kierczak, and Örjan Carlborg	
25	Incorporating Prior Knowledge to Increase the Power of Genome-Wide Association Studies	519
	Ashley Petersen, Justin Spratt, and Nathan L. Tintle	
26	Genomic Selection in Animal Breeding Programs	543
	Julius van der Werf	
Index		563

Contributors

Roderick D. Ball • *Scion (New Zealand Forest Research Institute Limited), Rotorua, New Zealand*

Gustavo de los Campos • *Department of Biostatistics, University of Alabama at Birmingham, Birmingham, AL, USA*

Örjan Carlborg • *Department of Clinical Sciences, Swedish University of Agricultural Sciences, Uppsala, Sweden*

Samuel A. Clark • *School of Environmental and Rural Science, University of New England, Armidale, NSW, Australia*

Matthew A. Cleveland • *Genus plc, Hendersonville, TN, USA*

Adrian Cortes • *University of Queensland Diamantina Institute, Princess Alexandra Hospital, University of Queensland, Brisbane, QLD, Australia*

José Crossa • *Biometrics and Statistics Unit, Crop Research Informatics Lab, International Maize and Wheat Improvement Center (CIMMYT), México, D.F., Mexico*

Sonja Dominik • *CSIRO Animal, Health and Food Sciences, Armidale, NSW, Australia*

Tom Druet • *Fonds National de la Recherche Scientifique, Brussels, Belgium; Unit of Animal Genomics, Faculty of Veterinary Medicine and Centre for Biomedical Integrative Genoproteomics, University of Liège, Liège, Belgium*

Frédéric Farnir • *Department of Animal Productions, Faculty of Veterinary Medicine, University of Liège, Liège, Belgium*

Rohan L. Fernando • *Department of Animal Science, Iowa State University, Ames, IA, USA*

Marina R.S. Fortes • *School of Veterinary Science, University of Queensland, Brisbane, QLD, Australia*

Dorian J. Garrick • *Department of Animal Science, Iowa State University, Ames, IA, USA*

Michael E. Goddard • *Department of Food and Agricultural Systems, University of Melbourne, Melbourne, VIC, Australia; Biosciences Research Division, Department of Primary Industries, Bundoora, VIC, Australia*

Cedric Gondro • *The Centre for Genetic Analysis and Applications, University of New England, Armidale, NSW, Australia*

Gregor Gorjanc • *Department of Animal Science, University of Ljubljana, Domžale, Slovenia*

Birgit Gredler • *Qualitas AG, Zug, Switzerland*

Ben Hayes • *Biosciences Research Division, Department of Primary Industries, Bundoora, VIC, Australia; La Trobe University, Bundoora, VIC, Australia*

John M. Henshall • *CSIRO Livestock Industries, Armidale, NSW, Australia*

JOHN M. HICKEY • *School of Environmental and Rural Science, University of New England, Armidale, NSW, Australia; Biometrics and Statistics Unit, International Maize and Wheat Improvement Centre (CIMMYT), México, D.F., Mexico*

TING HU • *Computational Genetics Laboratory, Dartmouth Medical School, Dartmouth College, Lebanon, NH, USA*

MARCIN KIERCZAK • *Department of Animal Breeding and Genetics, Swedish University of Agricultural Sciences, Uppsala, Sweden*

JAMES W. KIJAS • *Division of Animal, Food, and Health Sciences, CSIRO, Queensland Bioscience Precinct, Brisbane, QLD, Australia*

ANDREAS KRANIS • *Aviagen, Edinburgh, UK*

SANG HONG LEE • *The Queensland Brain Institute, The University of Queensland, Brisbane, QLD, Australia*

SEUNG HWAN LEE • *Hanwoo Experimental Station, National Institute of Animal Science, RDA, Pyongchang, Republic of Korea*

HAK KYO LEE • *Department of Biotechnology, Hankyong National University, Ansung-City, Kyonggi-Do, Republic of Korea*

GUILHERME JORDÃO DE MAGALHÃES ROSA • *Department of Animal Science, University of Wisconsin-Madison, Madison, WI, USA; Department of Biostatistics & Medical Informatics, University of Wisconsin-Madison, Madison, WI, USA*

CHRISTIAN MALTECCA • *Department of Animal Science, North Carolina State University, Raleigh, NC, USA*

SARAH E. MEDLAND • *Quantitative Genetics Laboratory, Queensland Institute of Medical Research, Brisbane, QLD, Australia*

FAHEEM MITHA • *Duke University, Durham, NC, USA*

JASON H. MOORE • *Department of Genetics, Dartmouth Medical School, Dartmouth College, Lebanon, NH, USA*

RONALD M. NELSON • *Department of Clinical Sciences, Swedish University of Agricultural Sciences, Uppsala, Sweden*

QINXIN PAN • *Computational Genetics Laboratory, Dartmouth Medical School, Dartmouth College, Lebanon, NH, USA*

PAULINO PÉREZ • *Statistics and Computer Science Departments, Colegio de Postgraduados, Mexico*

ASHLEY PETERSEN • *Department of Biostatistics, University of Washington, Seattle, WA, USA*

LAERCIO R. PORTO-NETO • *School of Environmental and Rural Science, University of New England, Armidale, NSW, Australia; School of Veterinary Science, The University of Queensland, St. Lucia, Brisbane, QLD, Australia*

MIGUEL E. RENTERÍA • *Queensland Institute of Medical Research, Brisbane, QLD, Australia; School of Psychology, University of Queensland, Brisbane, QLD, Australia*

ANTONIO REVERTER • *CSIRO Livestock Industries, Queensland Bioscience Precinct, Brisbane, QLD, Australia*

JUSTIN SPRATT • *Department of Mathematics, Statistics, and Computer Science, Dordt College, Sioux Center, IA, USA*

NATHAN L. TINTLE • *Department of Mathematics, Statistics, and Computer Science, Dordt College, Sioux Center, IA, USA*

BRUNO DOURADO VALENTE • *Department of Animal Science, University of Wisconsin-Madison, Madison, WI, USA; Department of Biostatistics & Medical Informatics, University of Wisconsin-Madison, Madison, WI, USA*
ANA I. VAZQUEZ • *Department of Biostatistics, University of Alabama at Birmingham, Birmingham, AL, USA*
PETER M. VISSCHER • *University of Queensland Diamantina Institute, Princess Alexandra Hospital, University of Queensland, Brisbane, QLD, Australia; The Queensland Brain Institute, The University of Queensland, Brisbane, QLD, Australia*
JULIUS VAN DER WERF • *School of Environmental and Rural Science, University of New England, Armidale, NSW, Australia*
STACEY WINHAM • *Division of Biomedical Statistics and Informatics, Department of Health Sciences Research, Mayo Clinic, Rochester, MN, USA*
JIAN YANG • *University of Queensland Diamantina Institute, Princess Alexandra Hospital, University of Queensland, Brisbane, QLD, Australia*

Chapter 1

R for Genome-Wide Association Studies

Cedric Gondro, Laercio R. Porto-Neto, and Seung Hwan Lee

Abstract

In recent years R has become de facto statistical programming language of choice for statisticians and it is also arguably the most widely used generic environment for analysis of high-throughput genomic data. In this chapter we discuss some approaches to improve performance of R when working with large SNP datasets.

Key words Genome-wide association studies, R programming, High-throughput analysis, Parallel computation

1 Introduction

In recent years R [1] has become de facto statistical programming language of choice for statisticians and it is also arguably the most widely used generic environment for analysis of high-throughput genomic data. R's core strength lies in the literally thousands of packages freely available from repositories such as CRAN and Bioconductor [2] which build up on the base installation. These packages (somewhat akin to programming libraries) are developed independently from the core program and can be downloaded from central repositories or, less frequently, directly from the developers' websites. There is a good chance that for any given task there already is a package that can do the job at hand, which will save hours or days of programming it yourself. This wide adoption of R by researchers can be seen throughout this book with many chapters using R to directly implement the methods described or to illustrate the concepts.

More specifically for genome-wide association studies (GWAS), there are hundreds of packages available for the various analytical steps. There are packages for importing a wide range of data formats, preprocessing data, to perform quality control tests, to run the analysis per se, downstream integration with biological databases, and so on. A large number of new algorithms and methods are published and at the same time released as an R package, thus providing quick access to the most current methodologies without the slow turnover time of commercial software. While this chapter will not review R packages for GWAS, good starting points to find useful packages are the CRAN task view webpage found at http://cran.r-project.org/web/views/Genetics.html and the Bioconductor (http://www.bioconductor.org/) packages for SNP found in BiocViews (under Software/AssayDomains).

R is free (released under the GNU public license). It is free to use, free to modify, and an open source. R is also platform-independent. Scripts will generally run on any operating system without changes, but a small number of packages are not available on all platforms. Since R is a scripted language it is very easy to essentially *assemble* various packages (e.g., use the output of a function in one package as the input for another one), add some personalized routines, and chain-link it all into a full analysis pipeline all the way from raw data to final report. This of course dramatically reduces development and deployment times for complex analyses. The downside is that the development speed and ease come along with a certain compromise in computational times because R is a scripted language and not a compiled one. But there are some tricks for writing R code which will improve performance, and this will be the focus of this chapter. Herein we assume that the reader is reasonably familiar with R and its syntax. For those who are unfamiliar with it, two excellent texts more focused on the programming aspects of the language are Chambers [3] and Jones et al. [4].

In this chapter we will compare different programming approaches in R and illustrate how small changes in programming practices can lead to substantial gains in speed. Advanced R programmers/users will probably be familiar with most of these *tricks* (and maybe have even better ones); this tutorial is more targeted at users in an earlier stage of the learning curve.

2 Reading Data into R

One of the first bottlenecks commonly encountered when working with genomic data is getting it into R. Datasets tend to be rather large and it can take a significant amount of time just to load it in. Ideally data will be in binary format and there will be a specific serialization/deserialization available for storing and retrieving the data, but this involves custom-tailored coding and forces rigid data formats (*see* **Note 1**). More commonly, data will be stored as plain text files in somewhat variable formats, requiring flexibility in how data is read into R. We will focus on this latter scenario.

Let's illustrate by using a small example file *square.txt* consisting of 1,000 samples (rows) × 50,000 SNPs (columns) with SNPs coded as 0/1/2, plus a first alphanumeric column with *status* information (e.g., A—affected and N—normal) and a second column with a continuous phenotypic measurement (data and scripts for this chapter are available from the book publisher's website, http://extras.springer.com/). This is a common data format for GWAS. An easy way to get the data into R is as a *dataframe* using

```
myData = read.table("square.txt", sep="\t")
```

this takes around 1.09 min to load the ~96 MB file. The first thing we can do to speed things up is to tell R what kind of data to expect in each column using *colClasses* to define the first column as a factor and the others as numeric.

```
myData=read.table("square.txt",sep="\t",
colClasses=c("factor", rep("numeric",50001)))
```

and we are already down to 45.4 s, roughly 35 % faster but, of course, this only works if you know the exact structure of the data file. A couple of extra parameters that can, albeit usually marginally, boost performance is to define the number of rows in the table beforehand to avoid reallocating memory on the fly (*ncols=50000* in this case) and if there are no comments in the data, set comment to empty with *comment.char = ""*. This is still rather inefficient. The *read.table* function works much better when there are more rows than columns. If the data is stored as *SNP X sample* instead of *sample X SNP*, the read time comes down to only 12.4 s which is over 5 times faster than the first example. Of course, if the data was a transpose of the original *square.txt* it would not be possible to define the classes of the columns anymore since the first line would be a factor (the *status* information). Generally a good approach is to keep genotypic and phenotypic data in separate files; this allows reading in the genotypic data faster by defining the columns with *colClasses*.

If the file has only genotypes, the *scan* function can be used:

```
myData=matrix(scan(file="genotypes.txt",what=
integer(),sep="\t"),50000,1000,byrow=T)
```

which takes 9.6 s to run and is about 23 % faster than using *read.table*. Here the actual order of SNPs and samples makes no difference. We can do even better with *readLines*:

```
myData=readLines(con="genotypes.txt")
```

and this brings it down to 4.35 s. But the downside is that *readLines* does not provide a useful matrix format structure as the previous functions; it simply returns a vector of character lines. Even faster is to use *readChar*. This loads the whole file as a raw unparsed string. This is extremely fast and takes only 0.43 s to run, but again without any structure to the data and you need to know beforehand the size of the data file (can use the argument *size* from the *file.info* function for this). The additional parsing steps required to convert the data to matrix format can make the whole process even slower than simply using *read.table*.

```
myData=readChar(con="genotypes.txt",nchars=file.
info("genotypes.txt")$size,useBytes=T)
```

Another common data format for SNP data is as *long* tables. A typical Illumina Genome Studio output file will have columns for sample identifiers, SNP identifiers, the alleles themselves, and other information. For our example the data file would increase more than 6 times in size to 604 MB and take 1.02 min to load with *read.table* and 2.47 s with *readChar*.

If the plan is to load the genotypes more than once into R, it is worthwhile to load them once using one of the previously described approaches and then save it in binary format. This makes the file much smaller and is quick to read in. A simple solution is to use the *save* function

```
save(myData,file="genotypes.bin")
```

The binary file is a quarter of the original file (22.5 MB) and it takes only 1.47 s to load, in the correct matrix format, with

```
load("genotypes.bin")
```

Apart from R's built-in possibilities, a few packages that are worth exploring for GWAS data manipulation are sqldf, ncdf, and GWASTools.

In a nutshell, we started with a load time of 69 s and ended with 0.43 s. That's a 160-fold improvement in performance. We've achieved this by using different functions and preparing the dataset in a format that is more *digestible* to R; the compromise is that the faster the data gets imported, the less useful the format becomes. The trade-off is speed to read in the data vs. the additional overhead in later stages to parse it. For repeated use of the data, binary formats can save a lot of time.

3 Loops and Vectorization

Let's start this section with a rather obvious comment: R is not C or FORTRAN. Usual programming practices do not always translate well into R. The first point to remember is that R is an interpreted language and not a compiled one. Things will always be slower in R than in, e.g., C or FORTRAN; at best the speed will be the same. On the positive side, most of R itself is written in C or FORTRAN and many functions have been implemented in these languages. This naturally suggests that the fewer times the interpreter has to be called upon to then send a command to run a compiled function, the faster the code will execute.

But before looking into this, let's look at something that most languages do not like: memory reallocation. Since R does not force us to declare variable types or their size before we use them, we tend to forget that as we grow the contents of a variable we have to allocate memory for it. This is time-consuming (and fragments the memory). Typical villains are *c*, *cbind*, or *rbind*, very handy commands but also very slow. Let's illustrate with a vector of one million random numbers that are generated and added to a matrix of undefined size

```
nums=rnorm(1000000)
numsMat=nums
for (i in 1:100) nums2=cbind(nums2,nums)
```

This takes 28.3 s to complete. If instead we pre-allocated the size of the matrix and then filled in the data it would take only 1.7 s.

```
numsMat=matrix(NA,1000000,101)
for (i in 1:100) nums2[,i]=nums
```

Even better is to simply create the whole matrix in a single line in just 0.51 s.

```
numMat=matrix(nums,1000000,101)
```

Very significant time savings (~55×). This is one of the easiest ways of improving performance in R and probably also one of the most common pitfalls.

The last snippet of code above returns to the initial point of this section that it is better to avoid repeated calls to the R interpreter. To illustrate let's revisit the initial genotypes dataset and after you load the file, calculate the allele frequencies.

```
freqA=numeric(50000)
freqB=numeric(50000)
for (i in 1:50000)
{
hold=0
for (j in 1:1000) hold=hold+myData[i,j]
freqB[i]=hold/2000
freqA[i]=1-freqB[i]
}
```

This is a rather literal translation from C or FORTRAN and quite inefficient in R (takes 45.7 s). We can make some good progress by using vectorization. Generally R functions are vectorized—they will perform the same operation on a single value or a vector of values using a lower level language. Of course this still implies looping over the vector but if this is done from within a compiled function, the speed gains can be large. For example, replacing the inner loop with the *sum* function which adds all elements of a vector we come down to 1.33 s.

```
freqA=numeric(50000)
freqB=numeric(50000)
for (i in 1:50000)
{
freqB[i]=sum(myData[i,])/2000
freqA[i]=1-freqB[i]
}
```

And we can do even better with the compiled function *rowSums* and make use of vectorization in the whole process:

```
freqB=rowSums(myData)/2000
freqA=1-freqB
```

Here, as the name suggests, *rowSums* calculates the sum for each row in a compiled function, then each element is divided by the number of alleles (2× number of individuals), finally we calculate the frequency for the other allele by subtracting the frequencies in *freqB* from one for all SNPs. Now we are down to 0.15 s and only two lines of code. If there is no function for your particular needs consider matrix operations; they can be nearly as effective (0.21 s), particularly for GWAS.

```
freqB = as.vector((myData %*% rep(1, 1000))/2000)
freqA=1-freqB
```

In summary, R is slow with loops. With a loop you are each time going back and forth through the interpreter. Try to send everything as much as possible as a vector which can then be iterated through using compiled functions. Recall that most of R is written in C or FORTRAN; it's just a matter of presenting the data more efficiently to these functions. This chapter clearly cannot cover all the different functions available in R but as a rule of thumb whenever the code involves double loops it's worthwhile to spend some time investigating if there is a low-level language function already written that will perform the operation.

At this point it is worth mentioning the *apply* family of functions. Using these functions it is possible to avoid loops and vectorize the task. Briefly, these functions iterate through a vector (*sapply*), list (*lapply*), matrix (*apply*), or some other object applying a function on their elements and return the results. For example *apply* runs a function over the lines or columns of a matrix. These functions are convenient and make for concise coding (albeit not intuitive until you are familiar with the syntax) but they are not always very useful to improve performance. To some extent it is just *loop hiding*, the speed gains are more related to the function that is being applied rather than the *apply* function itself. To illustrate let's run a single SNP regression on a phenotype (additive model) and store the p-values. Some housekeeping first:

```
load("genotypes.bin")
pheno=rnorm(1000,mean=100,sd=10)
pvals=numeric(50000)
```

First we loaded the genotypes from the binary file created in the previous section, then invented a trait value (the *pheno* vector) sampled from a normal distribution with mean 100 and a standard deviation of 10, and finally created a vector to store the p-values for the single SNP regressions (*see* **Note 2**). Now we can run the analysis using a loop and store the p-values with

```
for (i in 1:50000) pvals[i]=coef(summary(lm(phe-
no~myData[i,])))[2,4]
```

and this takes 2.11 min. The downside with this approach is that we make a call to the *lm* function (which is compiled code), return to R, call the function again, and repeat this 50,000 times. Intuition would suggest that *apply* would do a better job.

```
pvals=apply(myData,1, function(geno) coef(summary
(lm(pheno~geno)))[2,4])
```

but this takes even slightly longer (2.13 min). In reality we did not vectorize anything; all we did was call *lm* 50,000 times again. The take-home message is that the *apply* functions will not automatically vectorize a task, if the code has reached a stage where the only option is to iterate, it will seldom make any difference in terms of speed if a *for loop* or an *apply* function is used.

A last point is that not everything is vectorizable. Some computations may depend on the result from a previous stage (e. g., Bayesian approaches). In these instances it can be worthwhile to write the function in a low-level language instead. If you are going to stick with R, try not to repeat any calculations within loops and if memory is not an issue, sometimes it's slightly more efficient to use suboptimal code (e.g., doubling up entire objects with some change) rather than trying to modify specific elements in an object.

4 Byte-Code Compiler

A byte-code compiler developed by Luke Tierney has been available for R since version 2.14 and most packages are already making use of it. The package itself (*compiler*) is part of the R base installation. In some scenarios, particularly *for loops*, significant speed gains are achieved just by adding a couple of extra lines of code. Essentially the compiler converts the R commands from a *sourceable* file into instructions that are executed by a virtual machine (rather loosely think in terms of Java and its runtime environment). The byte-code *compiler* is well documented in [5]. The actual speed gains using the *compiler* cannot be generalized; they will be null or marginal for functions written in low-level languages and for straight R functions in packages that already make use of it.

The package itself has quite a few functions; here we will focus only on a single one, but further details can be found in the R help files for the package (and it's a highly recommended read). The easiest way to use it is by adding the following lines at the beginning of the code:

```
library(compiler)
enableJIT(3)
```

The command *enableJIT* enables/disables just-in-time compilation. Use argument 0 to disable and 3 to compile loops before usage (*see* **Note 3**). With the single SNP regression example from the previous session there will be no difference in runtimes (*see* **Note 4**). On the other hand, if we run the double loop example to calculate allele frequencies, without the compiler we had a runtime of 45.7 s and with the compiler this comes down to 11.4 s; a fourfold improvement. Of course this is far from our optimal solution of 0.15 s but can be handy for tough-to-vectorize problems. Just to illustrate, a toy genomic prediction run using a Bayesian approach (250 samples, 2,000 markers during 5,000 iterations) took 12.58 min to run without the compiler and 9.7 min with it; this improvement (~30 %) while not dramatic could mean the difference between R being a viable platform to solve the problem instead of having to resort to other languages.

5 Managing Memory

R was not designed for large datasets and it is quite a memory hog. Datasets are getting larger and larger and R not only stores everything in RAM but tends to duplicate objects unnecessarily as well. The footprint for the same operations in C can take less than a quarter of the memory used by R. The most efficient solution is to simply get a bigger machine and more memory—think about how much you have spent on your SNP data; is it really sensible to be willing to analyze it on a $1,000.00 computer? Ok, it might be doable, but performance will take a hit. First, forget about 32 bit operating systems; they cannot make use of all available memory, e.g., in Windows the maximum for a single process is ~3 GB (*see* **Note 5**).

There is no specific allocation, deallocation, and garbage collection in R, or not at least what a programmer would expect. Garbage collection is automatic but memory is not released when objects are deleted, they will only be excluded when needed. For example:

```
a=matrix(123,10000,15000)
memory.size()
```

returns ~1.15 GB. After the variable is *deleted* with rm(a), a call to *memory.size* still returns 1.15 GB. An explicit call to the garbage collector (with the *gc* function) will release the memory.

```
gc()
memory.size()
```

This now returns 8.8 M which, for this particular case, is just the footprint of R itself. Generally non-release of memory does not cause any problems because as soon as R needs the memory the garbage collector is called. If the OS requests memory the collector is also triggered, but if a very large object is removed it can be worthwhile to manually call gc(), even if only for the *psychological comfort* of other users who might notice that your R process is eating up 200 GB of RAM.

Of more concern is that, if we return to the matrix created before and try to modify it, R will double the memory used. For example

```
a=matrix(123,10000,15000)
b=matrix(123,10000,15000)
memory.size()
```

is now using 2.35 GB, we would not be able to create another matrix of this size in 32 bit R for Windows (~3.5 GB which is more than the 3 GB limit) without running out of memory. But we would expect to be able to change the values of these matrices without any problems. This is not the case:

```
a=a+1
Error: cannot allocate vector of size 1.1 Gb
```

R does not replace values, it makes a copy of the entire object with the modifications. This effectively means that to be able to manipulate the matrix you'll need at least the size of the object in free memory (once finished memory is released, but it has to be available for the operation). On the positive side this only happens once and repeated changes will not blow out the memory usage any further (Chambers [3] discusses the programming logic behind this in detail). Trying to split the problem with *apply* will not work either.

```
a=apply(a,1,function(x) x+1)
```

This will take up 4.6 GB and as discussed above, is much slower. The last resort is a *for* loop and it will do better at 1.99 GB, but it is exceedingly slow.

```
for (i in 1:length(a)) a[i]=a[i]+1
```

An easy workaround is to split objects into *chunks*, work on these one at a time, and then modify the object.

```
a=matrix(123,10000,15000)
b=a[1:1000,]+1
a[1:1000,]=b
```

This takes up 1.3 GB—much more manageable and easy to scale between performance (number of *chunks*) and memory (size of *chunks*) given the hardware restrictions.

The limitations of working with large datasets in R are recognized; a recent package that aims to address this is *bigmemory*. For the example we have been using, the matrix will occupy only a fraction of the R matrix—1.46 M which is ~8,000 times smaller. If this still does not work, the package allows using disk storage as well.

```
library(bigmemory)
a=big.matrix(init=123,nrow=10000,ncol=15000,type=
"integer")
```

The downside is that most R functions will not work on the *big.matrix* object (our simple a = a + 1 will return an error) and it can be rather convoluted to make use of it. Additional packages are available to facilitate usual tasks and worth exploring (e.g., *bianalytics*, *bigtabulate*, *bigalgebra*).

One additional point about memory is that R does not copy an object immediately.

```
a=matrix(123,10000,15000)
b=a
memory.size()
> 1166.11
```

This will still use the same 1.15 GB of memory. But as soon as an element is changed the matrix is instantiated.

```
b[1,1]=124
memory.size()
> 2305.58
```

This seldom has any practical effect but it can be confusing why a single change to one element of a matrix suddenly causes an out of memory error (and of course the matrix has to be copied as well, which makes the seemingly small change seem very slow).

6 Parallel Computation

Parallel computation became a hot topic with the shift from single-core to multicore architectures but programs and programming practices are lagging a little behind the technology; mainly because there is still a reasonable amount of work involved in developing a program that runs in parallel. Currently, applications written in R are easier to parallelize than in C or Fortran, but this should change in the near future as parallel development tools mature. Some R packages/functions already run in parallel and typically transparent to the user, but it is also easy to parallelize iterative routines in R (albeit this can be rather problem-specific of course).

Interest in parallel computation in R has grown rapidly in the last few years. This is mainly driven by the rapid increase in the size of datasets, particularly data derived from genomic projects which are outstripping hardware performance. And also, there is a strong drive towards wide adoption of Bayesian approaches which are computationally demanding. A variety of packages and technologies have been developed for parallel computing with R. This is a very active area of research so the landscape is changing quite rapidly; a good starting point for the current state of affairs is the CRAN Task View: High-Performance and Parallel Computing with R (http://cran.r-project.org/). In a nutshell, there are packages to target different platforms (multicore machines, clusters, grid computing, GPUs) and these provide either low-level functionality (mainly communication layers) or easier-to-use wrappers that make the low-level packages more accessible for rapid development/deployment. Currently, the most widely used approach for parallel computing is the MPI (Message Passing Interface) which is supported in R through the *Rmpi* package. At a higher level the package snow[6] provides easy-to-use functions that hide details of the communication layer and allows communication using different methods (PVM, MPI, NWS, or simple sockets). An even higher level wrapper is snowfall which wraps the snow package. Herein we will illustrate how to parallelize single SNP regressions (as detailed in Subheading 3 above) using snowfall and simple socket communication; a detailed overview of parallel methods/packages in R is given in [7] and a good introductory tutorial is [8].

```
library(snowfall)
# set maximum number of cpus available - default is 32
sfSetMaxCPUs(number=128)
sfInit(parallel=TRUE,cpus=128, type="SOCK", sock-
etHosts=rep("localhost",128))
```

The first line loads the snowfall library and then the parameter *sfSetMaxCPUs* is used to change the default maximum number of

CPUs that can be accessed (in this example we want to use 128 cores). If less than 32 nodes/cores will be used there's no need to change the parameter—this is simply a safeguard to avoid reckless overloading of shared resources. The function *sfInit* initializes the cluster. The parameter *cpus* stores the number of processors to be used. Here we are using simple socket connections (type="SOCK") but MPI, PVM, or NWS could be used for other cluster modes. Note that for these to work the packages *Rmpi*, *rpvm*, or *nws* would also have to be installed (and the cluster setup in these modes). Sockets do not require any additional setup but they are slower than other modes. The argument *socketHosts* is a list of cpus/computers in the cluster; for processors on the same machine *localhost* (here repeated 128 times—once for each node) can be used but IP addresses or network computer ids can be used for distributed systems (*see* **Note 6**).

With this the cluster is up and running. After the job has finished, the cluster can be stopped with

```
sfStop()
```

This is all that is needed is to set up a simple cluster in R. Now let's illustrate how to parallelize single SNP regressions. All that is needed is a small change from the sequential code previously described in Subheading 3.

```
load("genotypes.bin")
pheno=rnorm(1000,mean=100,sd=10)
pvals=apply(myData,1, function(geno) coef(summary(lm(pheno~geno)))[2,4])
```

Recall that previously we ran a simple SNP by SNP regression using random numbers as a proxy for the phenotypes. This took 2.13 min to run. To parallelize the job all that's needed after setting up the cluster is to use the *sfApply* function which is a parallel version of *apply*. It is simply a matter of replacing *apply* with *sfApply* in the code above. The full code on four *cpus* is

```
library(snowfall)
sfInit(parallel=TRUE,cpus=4, type="SOCK", socketHosts=rep("localhost",4))
load("genotypes.bin")
pheno=rnorm(1000,mean=100,sd=10)
sfExport(list=list("pheno"))
pvals=sfApply(myData,1, function(geno) coef(summary(lm(pheno~geno)))[2,4])
sfStop()
```

This took 36 s to run, around 3.55 times faster. Runtimes do not improve linearly with the number of processors used (some of the reasons for this are detailed below) but the speed gains are still considerable, particularly with long runs.

In the parallel code an additional function was used

```
sfExport(list=list("pheno"))
```

Each node in the cluster needs to have access to the data structures it has to work with. In this example the nodes had to use the phenotypic (*pheno*) data to run the regressions. Objects that will be used by the cluster have to be exported with *sfExport* (*see* **Note 7**).

When considering parallel work in R a couple of key points are worth taking into account. The first point is that computation is much faster than communication. With parallel computing it is important to minimize the transfer of data between nodes. For example, under the exact same conditions, consider a run that takes 50.6 s and the function returns a single value (as example, the above single SNP regressions). If instead, the function each time returned a vector of 10,000 numbers, the runtime increases to 91.8 s and with 100,000 numbers the runtime is 112.2 s. Considerable speed gains can be achieved by minimizing the shuffling of data between nodes.

The second point is that the number of calls made to the nodes also impacts performance. In practice, the more time the workers independently spend on the computation itself, the more efficient the use of the computational resources will be. Ideally calculations themselves will be the slowest aspect of the computation and not the number of times they have to be performed. But this is not necessarily always the case and frequently computations on the nodes are trivial and the workers spend hardly any time computing but do spend a lot of time communicating and transferring data (particularly if networks are involved). In these cases it is worthwhile to partition the problem into larger subsets but there will be some additional coding overhead.

Figure 1 illustrates this last point quite clearly, using varying numbers of workers (between 1 and 40) and with six different computing times (0, 0.01, 0.1, 1, 2, and 5 s); that is, the number of seconds that each node spends computing before returning results to the master. Results are shown as a ratio between sequential vs. parallel runs. When the node spends 0 s on the computation (just a call to the node and returns) performance is poor due to the communication overhead incurred. As more time is spent on the computation, the gains obtained through additional cores become more evident but the compromise between computation and communication is still clear.

For times 0.01 and 0.1 s the ideal numbers of cores are, respectively, 4 and 8 (1.6 and 3.7 times faster). Adding more workers after this reduces the performance. For calls of 1 and 2 s the performance

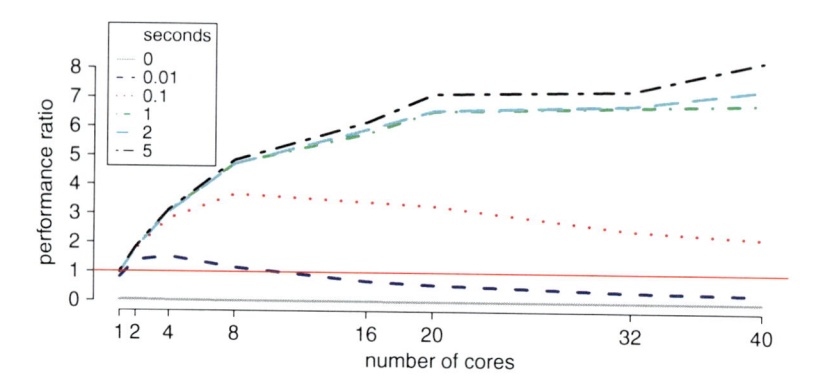

Fig. 1 Comparison of performance of parallel runs for varying number of nodes and various time lengths spent on independent computations within nodes. The labels on the *x* axis represent the actual number of cores used. The straight line through 1 (*y* axis) represents the sequential runtime (baseline)

starts to taper off after 20 workers. Take the 1 s scenario—with 20 workers the performance is 6.6 times better than the sequential job but this only increases to 6.8 with 40 workers. Computations of 5 s show the best results in terms of speed gains but while the performance has not yet plateaued there is evidence of *diminishing returns*. Performance is a balance between computational times and the number of cores used. Once performance starts to taper off there is little value in adding more workers. Either performance will decline due to additional communication overhead or there will be unnecessary strain placed on computational resources, which becomes quite relevant in shared environments.

There are a few additional points worth noting. First, there is a fixed time lag to initialize the communication protocols of around 1.5 s. That is why the parallel run on a single worker performs worse than the sequential one (Fig. 1). Second, performance gains through parallelization are not linear (e.g., 4 cores perform fourfold better than 1 core) but rather incremental. For example, consider the 5 s computation time in Fig. 1—with 40 cores the performance is 8.3 times better (instead of the wishful 40×). While this may not sound optimal, if actual runtimes are taken into consideration the picture becomes much more interesting: the sequential run takes 36.7 min while the 40 core run takes 4.4 min. Returning to the example we have been working with, the sequential run took 127.8 s to run and parallel runs with 2, 4, 8, 16, and 32 cores took, respectively, 64.04 (1.95 times faster), 36.95 (3.45), 24.91 (5.13), 24.6 (5.19), and 27.24 (4.69) seconds. For this example adding more than 8 cores starts to become of limited value. With 16 nodes performance is only marginally improved and with 32 nodes performance starts to drop back. Note that the runtime for 4 cores is higher than before because now the overhead of setting up the nodes for the run is included.

In summary, the key points to consider when setting up parallel tasks in R are: (1) the more time spent on computations, the more useful it becomes to parallelize the process and more cores can be used, (2) communication is more expensive than computation—minimize data transfer and calls to nodes while maximizing computation within workers, (3) performance gains are not linear to the number of cores but actual computing times can be significantly lower.

7 Running Software from R

If still none of this does the trick, R can be dynamically linked to compiled code in C or FORTRAN (and also other languages to various degrees); this opens the possibility of using prior code or developing code specifically tailored for solving a computationally intensive task and then sending the results back into R for further downstream analyses. A lot can be done in R, but sometimes we have code or other applications that we need or want to use.

The simplest solution is to run applications in another language straight from within R. All data preprocessing and setup of files could be done in R, then call the program from within R, and finally read the results back into R for further analyses. The R function for this is *system* (*see* **Note 8**). By default *system* will wait for the job to finish before continuing, but you can also work asynchronously using the argument *wait=FALSE* or redirect the output to R (provided it's a console program) with *show.output.on.console = TRUE* and *intern=TRUE*. Arguments can also be passed using *system*. To illustrate, a call to perform a multiple sequence alignment in *Clustal* would be

```
system("clustalw2 -INFILE=myDataFile.txt -OUTFILE=
myPhylo.fasta  -OUTPUT=FASTA  -TYPE=DNA  -PWDNAMA-
TRIX=IUB -PWGAPOPEN=10 -OTHER_PARAMETERS =FOO")
```

A much more refined way of interfacing is by linking directly to the external source and calling/running routines straight from R. This is beyond the scope of our discussion, but it's quite simple to work with C (a fair chunk of R is written in C) and almost as easy with FORTRAN (search for *Foreign* in the help files to get started). Other languages are also supported to some extent or other (see packages *RSJava*, *RSPerl*, and *RSPython*). The reader interested in the topic could start with Chapter 6 of Gentleman's *R Programming for Bioinformatics* [9] and then move on to Chamber's book [3]. Rizzo's [10] text while it does not discuss language interfacing might also be appealing due to its more algorithmic approach.

8 Notes

1. For various genomic-platforms or data-types efficient read/write functions are available in various R packages. It is worthwhile exploring what is already out there before using the more generic R functions.

2. Single SNP regressions are far from being an ideal approach to GWAS as discussed in other chapters throughout the book. Nonetheless it is extremely easy to perform and interpret (and also comparatively quick). It still has its place for a first quick look at the data.

3. The JIT compiler can also be enabled/disabled for all R jobs by setting the start-up environment variable R_ENABLE_JIT to a value between 0 and 3 (see details of flags in the R help files).

4. The JIT compiler might make the run even slower if there's nothing to optimize because of the overhead in converting the source code to byte code. This is usually not significant though.

5. A further limitation in 32 bit R for Windows is the size of a single object. The maximum contiguous memory allocation is ~2 GB, so if your object is larger than that, which often will be the case with GWAS data, it cannot be stored as a single variable.

6. One point of note and common pitfall when working with clusters across different physical machines is that socket connection is tunneled through SSH. This has to be set up and accessible to R for between-computer communication to work. In Windows firewall blocking rules may be triggered the first time the parallel nodes are initialized, make sure to allow traffic if the firewall popup appears.

7. Objects exported for use by the cluster are duplicated on each node; this can take up sizable quantities of memory if exported objects are large. Once the run is completed memory is released, but some thought has to be given as to whether there is enough memory on each node of the cluster (in distributed systems) or in total (for jobs parallelized on the same machine).

8. Command line arguments cannot be passed using *system* in Windows. Use *shell* instead or the newer implementation *system2*.

Acknowledgments

This work was supported by a grant from the Next-Generation BioGreen 21 Program (No. PJ008196), Rural Development Administration, Republic of Korea.

References

1. R Development Core Team (2012) R: a language and environment for statistical computing. R Foundation for Statistical Computing, Vienna, Austria
2. Gentleman RC, Carey VJ, Bates DM, Bolstad B, Dettling M, Dudoit S, Ellis B, Gautier L, Ge YC, Gentry J, Hornik K, Hothorn T, Huber W, Iacus S, Irizarry R, Leisch F, Li C, Maechler M, Rossini AJ, Sawitzki G, Smith C, Smyth G, Tierney L, Yang JYH, Zhang JH (2004) Bioconductor: open software development for computational biology and bioinformatics. Genome Biol 5(10):R80
3. Chambers JM (2008) Software for data analysis—programming with R. Springer, New York
4. Jones O, Maillardet R, Robinson A (2009) Introduction to scientific programming and simulation using R chapman & Hall/CRC, Boca Raton
5. Tierney L, Rossini AJ, Li N (2009) Snow: a parallel computing framework for the R system. Int J Parallel Program 37:78–90
6. Rossini AJ, Tierney L, Li NM (2007) Simple parallel statistical computing in R. J Comput Graph Stat 16(2):399–420
7. Schmidberger M, Morgan M, Eddelbuettel D, Yu H, Tierney L, Mansmann U (2009) State of the art in parallel computing with R. J Stat Softw 31(1):27p
8. Eugster MJA, Knaus J, Porzelius C, Schmidberger M, Vicedo E (2011) Hands-on tutorial for parallel computing with R. Comput Stat 26:219–239
9. Gentleman RC (2009) R programming for bioinformatics. Chapman & Hall/CRC, Boca Raton
10. Rizzo ML (2007) Statistical computing with R. Chapman & Hall/CRC, Boca Raton

Chapter 2

Descriptive Statistics of Data: Understanding the Data Set and Phenotypes of Interest

Sonja Dominik

Abstract

A good understanding of the design of an experiment and the observational data that have been collected as part of the experiment is a key pre-requisite for correct and meaningful preparation of field data for further analysis. In this chapter, I provide a guideline of how an understanding of the field data can be gained, preparation steps that arise as a consequence of the experimental or data structure, and how to fit a linear model to extract data for further analysis.

Key words Phenotype, Response variable, Binary data, Continuous data, Descriptive statistics, Diagnostics

1 Introduction

The aim of genome-wide association studies is to determine relationships between phenotypes and genotypes. Therefore, rare phenotypes and their associated genotypes are of great interest; however, apparently rare phenotypes are often a result of technical errors or poor models rather than true outliers. Consequently, equally rigorous preparation should be applied to the analysis of phenotypic and genotypic data. To prepare phenotype data for analysis, linear models are fitted to strip down observations as much as possible to their genetic components. In order to do this, researchers need to spend some time exploring and understanding the structure of their data set and the phenotype(s) of interest if they are to fit sensible models. However, because of the scale of modern genetic experiments, the individual researcher often has little involvement in the experimental design and empirical data collection.

The purpose of this chapter is to guide the reader through a series of steps aimed at summarizing and displaying data to gain an understanding of the data set and in particular the phenotype(s) of interest.

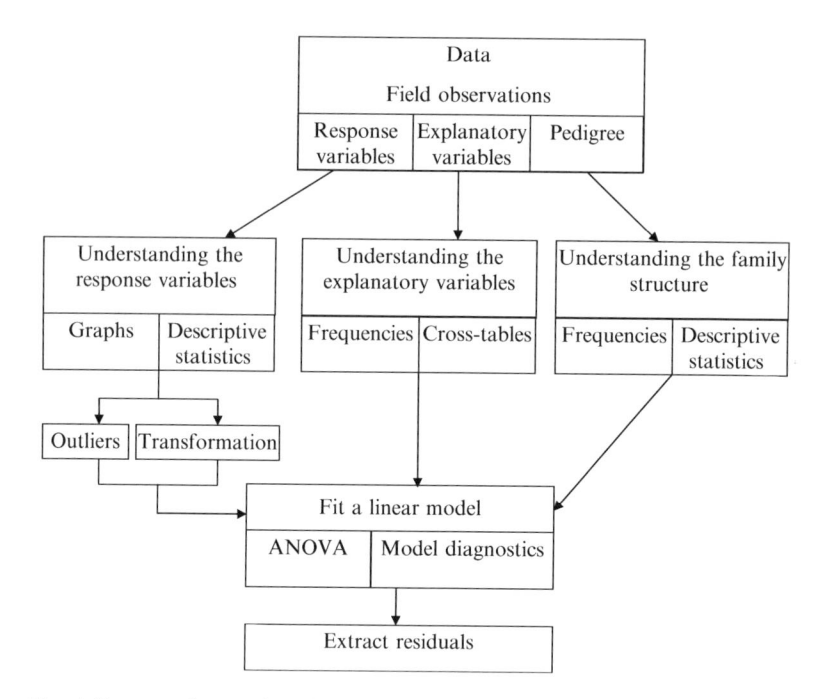

Fig. 1 Process of preparing observed field data for analysis

It outlines simple procedures that explore the data, facilitate error detection, and describes, in its simplest form, how to fit a linear model. An overview of the process is provided in Fig. 1.

This chapter does not provide an exhaustive description of statistical tools. Further reading is recommended. Suggested reading is compiled at the end of this chapter [1–22]. Rather, the aim is to encourage the reader to explore their data prior to detailed analysis of a data set. An example data set, representing data from an experiment investigating internal parasites in sheep, is used to illustrate some of the issues (the data are available on the publisher's website, http://extras.springer.com/). Even though this is a livestock example, the problems discussed here are applicable to data collected for genetic analysis in other disciplines. Practically, the individual steps are demonstrated in the Notes section using code for R (10). The R software was chosen, because it has a high level of flexibility and is freely available. The instructions in this chapter and in the Notes should also provide a useful guideline for analysis in other software applications.

2 Tools

2.1 The Example Data Set

The example data set described in Tables 1 and 2 has been used for illustration purposes. The data represents records from an experiment on internal parasites in sheep. The level of parasite infection in the sheep was estimated using fecal worm egg count (FWEC),

Table 1
First lines of the example data set—explanatory variables that describe the experimental unit (see additional material on publisher's website (http://extras.springer.com/) for full example data set)

Indiv	Sire	Dam	Birth year	Birth type	Rear type	Birth_rear type	Sex	Birth date	Birth weight	Weaning date	Weaning weight
4	101	1,011	2009	Single	Single	11	Female	285	4.2	22/01/2004	29
6	101	1,012	2009	Single	Single	11	Female	285	4.9	22/01/2004	32
13	102	1,015	2009	Single	Single	11	Male	286	NA	22/01/2010	23
14	102	1,016	2009	Multiple	Single	21	Female	287	3.5	22/01/2010	18.5
16	102	1,016	2009	Multiple	Multiple	22	Female	286	4.1	22/01/2010	24
17	103	1,017	2009	Multiple	Multiple	22	Male	287	4.1	22/01/2010	24.5
18	103	1,017	2009	Multiple	Single	21	Female	287	2.1	22/01/2010	20
27	104	1,021	2010	Multiple	Dead	NA	Female	287	2.9	NA	NA
28	104	1,021	2010	Multiple	Multiple	22	Male	287	3.5	22/01/2010	26

Table 2
First lines of the example data set—explanatory variables that describe the data collection event and response variables (see additional material on publisher's website (http://extras.springer.com/) for full example data set)

Indiv	Sire	Dam	Collection date	Worm egg counter	WEC	Wormy status
4	101	1,011	24/06/2010	1	3,700	1
6	101	1,012	24/06/2010	1	9,000	1
13	102	1,015	24/06/2010	2	5,500	1
14	102	1,016	24/06/2010	2	6,900	1
16	102	1,016	24/06/2010	2	7,400	1
17	103	1,017	24/06/2010	2	6,100	1
18	103	1,017	24/06/2010	2	0	0
27	104	1,021	NA	3	NA	NA
28	104	1,021	24/06/2010	3	5,600	1

which is the phenotype to be analyzed. Similar data on disease trials are collected in human, plant, or aquaculture studies. The data set contains records for each experimental unit, which is the individual animal in the example, in multiple columns with one row per experimental unit. Variables are generally of two types of data, explanatory and response. Type "explanatory" describes the circumstances under which the data were collected and provides detail on the experimental unit (e.g., sire, birth year, body weight, sample collector), while "response variables" and the second type comprise measurements of phenotypes of interest for analysis (e.g., FWEC, wormy status). In the example data set, we have two response variables: "fecal worm egg count (FWEC)" and "worm status". "Worm status" describes FWEC information as a binary variable with animals classified as "wormy" = 1 if FWEC > 2,999 and "not wormy" = 0, if FWEC < 3,000.

2.2 Descriptive Statistics

Descriptive statistics summarize data to facilitate the exploration of data characteristics. Descriptive statistics include measures of centrality and dispersion. Measures of centrality include the mean, mode, and median. Measures of dispersion include the standard deviation and the range. In the following paragraphs, "x" describes a data point and "n" describes the total number of data points.

2.2.1 Measures of Centrality

The mean or average ($\overline{X}$) is the sum of all data points (x) divided by the total number of data points (n):

$$\overline{X} = \frac{\sum x}{n}$$

The mode is a measure of which value occurs most frequently in the data. The data values are sorted and the occurrences of each value counted. The value that occurs most frequently is the mode.

e.g., Assume the following data points 1,1,1,2,2,3, 4,5,6 Mode=1

The median describes the value that divides the data set in half. An equal number of records are above it as are below it. Data points are sorted and the value of the record that divides the data set equally is the median. Note that for an even number of records, the median is between the two middle records.

For uneven number of records Median = $(n/2)$th case

For even number of records Median = $(n + 1/2)$th case

2.2.2 Measures of Dispersion

The range describes the difference between the maximum and the minimum values in the data set.

$$\text{Range} = \text{Maximum} - \text{Minimum}$$

The standard deviation (σ) is a measure that illustrates the dispersion of the data points. It is the square root of the average of the sum of the squared data points:

$$\sigma = \sqrt{\frac{\sum x^2}{n}}$$

The variance of a measure (σ^2) is the standard deviation squared.

2.2.3 Graphical Display

Displaying data graphically is a powerful tool to detect error and the need for data transformation. It also provides a visual tool to explore relationships between variables. In the data preparation step, the most useful types of graphs for data visualization are histograms, *xy* scatter plots, and box plots.

An *xy* scatter plot simply plots two variables against each other. It can be informative about their relationship to each other. Figure 2 shows that there are interestingly animals that have low FWEC at one time point and high FWEC at the other time point.

To prepare a histogram, sorted data are grouped into bins and the frequencies of the data in each bin are plotted in a bar chart. Figure 3 shows that FWEC data are skewed positively, with the long tail in the positive direction.

Box plots can display differences in the variation of a variable and centrality and dispersion across experiments or across years. In such box plots, the bottom line indicates the 25th and top line the 75th percentile, the median within the box is indicated by a line, and whiskers on the box describe the range of values. Figure 4 shows that there were no differences in FWEC between male and female animals.

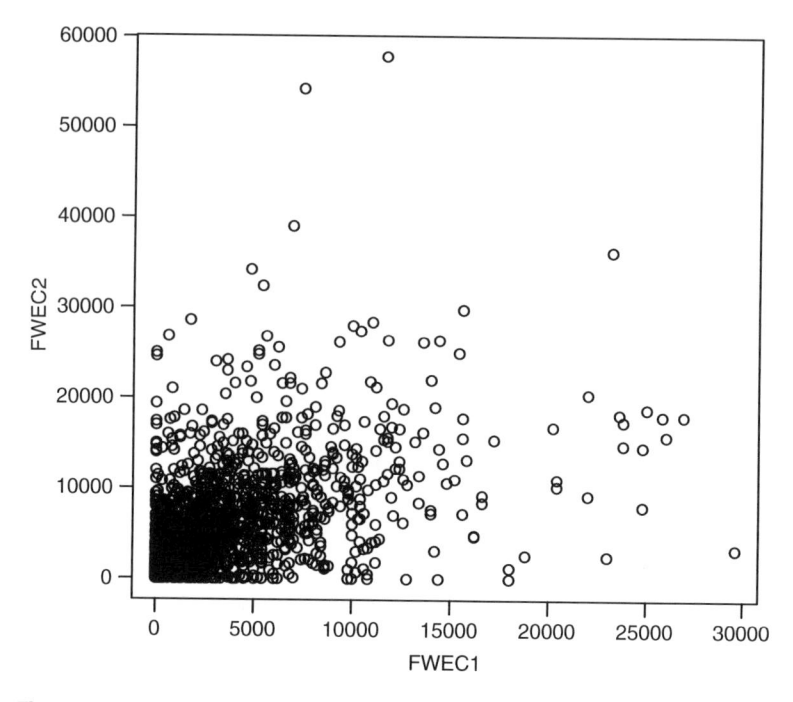

Fig. 2 *xy* scatter plot of the FWEC1 and FWEC2, fecal worm egg counts taken at different time points

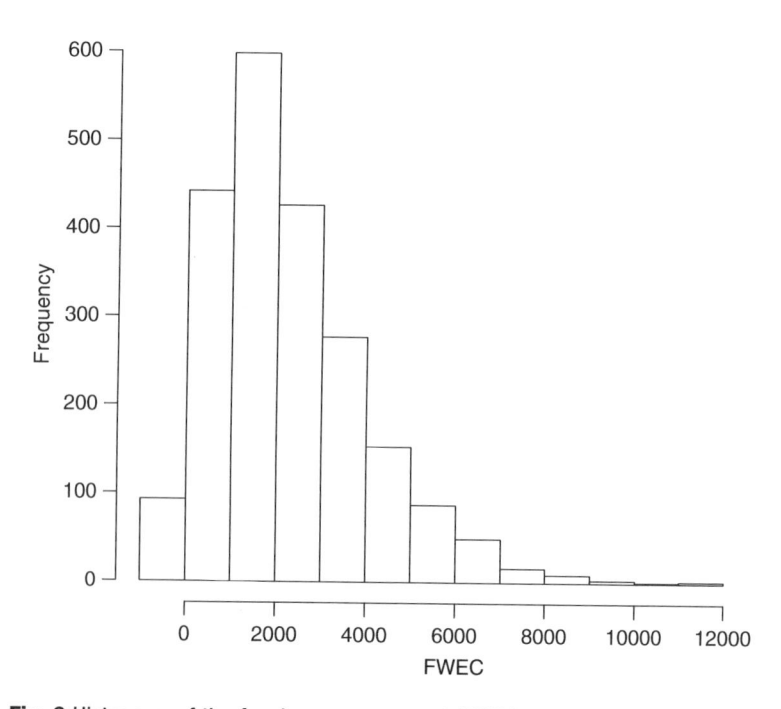

Fig. 3 Histogram of the fecal worm egg count (FWEC)

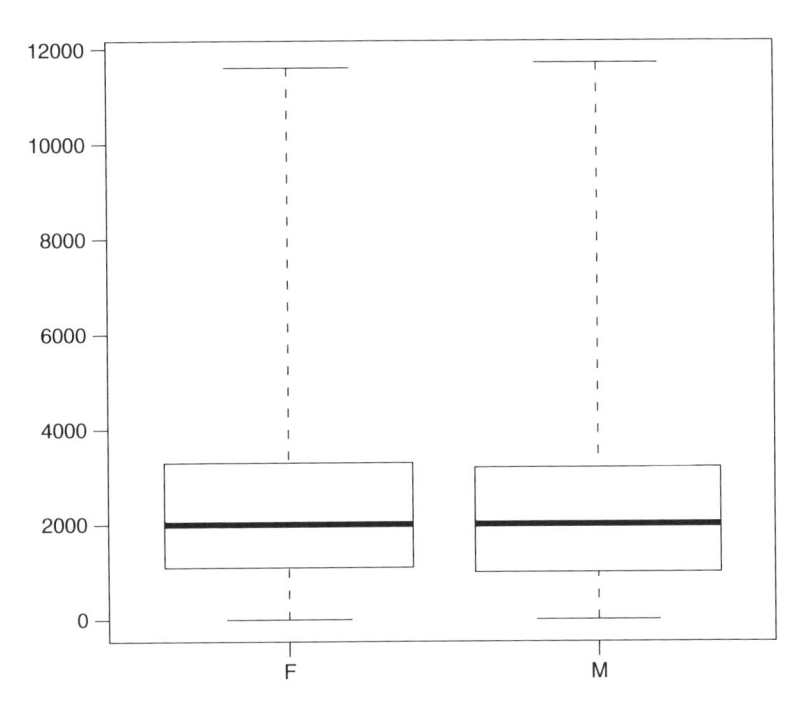

Fig. 4 Box plot of FWEC for female (*F*) and male (*M*) sheep

2.2.4 Model Comparisons

Linear models and ANOVA

Within the context of this chapter, an analysis of variance (ANOVA) is used to determine the linear model that best fits the data. Remember that the goal is to adjust the phenotype for environmental effects. The remaining variation which is not accounted for is called the "error component" or "residual". It contains the genetic component and other unaccounted effects. Environmental effects, so-called fixed effects, are explanatory variables. These can consist of a number of categories (e.g., sex with "male" and "female", birth type in the example data set with "multiple", "singles", or "dead") or can be continuous like, e.g., weight. We are focusing on a fixed effect model and on the practical aspects of an ANOVA.

Assume a fixed effect model

Response = mean + fixed effects + residual/error component

In statistical terms, it is generally written as

$$y_{ij} = \mu + \beta_i + e_{ij}$$

where y_{ij} is the response variable of the *j*th experimental unit on the *i*th explanatory variable. β_i is the effect of the *i*th treatment and e_{ij} is the random error $\sim(0, I\sigma^2)$.

The aim is to build a biologically meaningful model that minimizes the residual sums of squares in the ANOVA. Using an ANOVA, it can be determined if the explanatory variable should be retained in the model or not. The ANOVA partitions the total sums of squares attributed to variables in the linear model. In very simple terms, it is separating the overall variance observed in the response variable into baskets with each basket containing the variance attributed to a particular variable. For each explanatory variable fitted in the model, the sums of squares and mean sums of squares are computed and presented in an ANOVA table. The residual sums of squares are the difference between the sums of squares attributed to a variable and the total sums of squares. An F statistic and P-value are reported for each variable. A significant F-value indicates that the variable should be retained in the model.

Explanatory variables are often not independent from each other. If the variance of a dependent variable explained by two explanatory variables is somewhat confounded, the sums of squares for each explanatory variable will be affected, depending on the order that the variables are fitted in the model. The effect of the order that variables are fitted in the model can be investigated by computing Type I, II, and III sums of squares. Since we are not interested in the explanatory variables we are fitting, the order is not our focus and the easiest way of testing significance of explanatory variables is to use Type III sums of squares. The Type III sums of squares for each variable are obtained after all other variables are fitted. Therefore, the order that variables are fitted with becomes irrelevant. Statistically, Type III sums of squares are frowned upon, but they are a useful tool in this case. In contrast, Type I sums of squares are obtained after fitting explanatory variables sequentially in the order that they appear in the model. To illustrate the difference between Type I and Type III sums of squares, Tables 3 and 4 outline the output of two ANOVAs. Type I sums of squares fit the fixed effects sequentially and produce an F-value after fitting each effect, therefore the order is important. With Type III sums of squares, F-values are produced for each variable after all other effects are fitted. As a result, the F-value for the final fixed effect (sex) in the example is the same for Type I and Type III (Tables 3 and 4).

Table 3
R output of ANOVA table with Type I sums of squares

Fixed effect	Df	Sums of squares	Means sums of squares	F-value	Pr(>F)
Factor (birthtype)	1	59,973,998	59,973,998	12.38	0.0004***
Factor (age)	2	81,457,529	40,728,765	8.41	0.0002***
Factor (sex)	1	49,959,527	49,959,527	10.32	0.0013**
Residuals	1,194	5,782,456,539	4,842,928		

Significance levels: *Pr < 0.05, ** Pr < 0.01, *** Pr < 0.001

Table 4
R output of ANOVA table with Type III sums of squares

Fixed effect	Sums of squares	Df	F-value	Pr(>F)
Mean	3,230,479,603	1	667.05	<2.20E-16***
Factor (birthtype)	45,706,395	1	9.44	0.0022**
Factor (age)	81,021,186	2	8.37	0.0002***
Factor (sex)	49,959,527	1	10.32	0.0014**
Residuals	5,782,456,539	1,194		

Significance levels: *Pr < 0.05, ** Pr < 0.01, *** Pr < 0.001

Logistic regression and deviance tables
Often response variables are binary with values of "0" or "1" expressing a successful or unsuccessful outcome. Binary data are analyzed with logistic regression models. These have non-normal error terms and error variances that are not constant. However, the response function can be linearized with a so-called logit transformation. Models with different explanatory variables are compared using a Chi-square test. The results of the test are presented in an analysis-of-deviance table. Fixed effects are fitted sequentially and the difference in deviance tested against the Chi-squared statistic. Significance can easily be assessed using the Chi-squared statistics or *P*-value that is associated with each effect. However, because the effects are fitted sequentially, rearrangement can change the significance level. It is recommended to test fixed effects in various positions. The overall fit of the model compared to others can be assessed using the residual deviance; the lower the deviance, the higher the likelihood.

3 Methods

3.1 Understanding the Family Structure

The pedigree of each experimental unit, i.e., sire and dam or mother and father, contains information on the family structure of the data set. In human studies, the family structure is generally obvious and simple; however, in livestock, aquaculture species, or plants, family structures can vary widely and be complex. Guiding questions to understand the family structure in a data set are

1. How many individuals are in the data set?
2. How many mates does each male or female have?
3. Are parents used across years?
4. How many full-sib and/or half-sib families are in the data set?
5. How many offspring does each full- and/or half-sib family have?

These questions can simply be answered by computing frequencies and descriptive statistics of number of offspring per full- and/or half-sib family (*see* **Notes 1–6**). If the aim is to conduct a linkage analysis, families with low number of offspring should be excluded to avoid potential bias. In addition, the information on family structure also needs to be compared with the explanatory variables (e.g., year) to investigate potential bias.

3.2 Understanding the Explanatory Variables

To understand the circumstances of the data collection and to fit an appropriate model, it is important to understand the explanatory variables and how they relate to each other. Only with this knowledge, can explanatory variables be fitted meaningfully in a model. Key questions for this task are

1. How many categories does each variable have?
2. How many observations per category?
3. How many observations are missing?
4. Are variables completely confounded with each other?
5. Are any of the categories of variables confounded with others?

The computation of frequencies of categories and missing data points per category are a useful starting point. Cross tables are useful to identify complete confounding of variables or of categories of multiple variables. If some categories of variables are lowly populated, it is advisable to either remove the records or to recode the categories (*see* **Notes 1–6** for relevant R code).

The example data set demonstrates these issues of recoding in the explanatory variables "birth type" and "birth rearing type". The variable "birth type" contains "singles" and "multiples". Sheep can have more than two lambs, even though it is not very common. Therefore, the actual data recorded would have been "single", "twins", "triplets", and "quadruplets". Since the latter two categories are not very common, they could be removed if the category averages are not representative due to low number of records. Alternatively, they can be regrouped together with "twins" as "multiples". Another alternative would be to create a "multiples" category that includes "triplets" and "quadruplets", but "twins" are a separate category.

The issue of confounding variables can also be explained with the example data set. The variables "birth type" and "rearing type" are confounded, because a single can only be raised as a single or not at all, but it can never be raised as a twin. This problem can be addressed by creating a new variable, which is a concatenation of the two. For example, an experimental unit which was born as a twin (2) and raised as a single (1) receives a birth rearing type category of 21. This way, the variable can be fitted appropriately. Another option is discussed in Subheading 3.4.1.

3.3 Understanding the Response Variable

3.3.1 Error and Outlier Diagnostics

It is an advantage to have expert knowledge of the phenotypic trait (response variable) of interest; however, even with limited knowledge errors can be detected. Outliers can influence the fit of a model; therefore, time needs to be spent exploring outlying data points. It is also important therefore to distinguish between erroneous data and outliers. Erroneous data need to be removed if the error cannot be corrected, whereas outliers can be particularly interesting for genome-wide association studies.

An initial visualization of the data in graphical form in combination with descriptive statistics provides information on the distribution and potential errors and outliers (*see* **Notes 7–9**). Errors, like, e.g., incorrect data entries, and outliers can easily be spotted. In some cases, it can be difficult to distinguish between an outlier and an incorrect data point. Obvious erroneous data should be removed unless original data sheets are available to track down the error to correct it. As a rule of thumb, outlying values should be removed; in particular, if they are extreme values, they are more than likely to be an error.

Error checking is an even more straightforward process for binary or ordinal traits. If the response variable is binary or ordinal, it can only take a limited number of values or it is missing. A histogram or a count of number of records in each category quickly identifies incorrect values.

3.3.2 Transformation

The type of data collected for genetics studies are groups of unrelated observations. Therefore, the distribution of the phenotype of interest is expected to be normally distributed. However, that might not always be the case. Normality of the distribution and also homogeneity of variances need to be adhered to when fitting a linear model. Note that this is not the case for binary or binomial traits that characteristically do not conform to the normality and homogeneity of variances requirements. Therefore, the normality properties of the data distribution and the homogeneity of variances need to be checked prior to analysis. The data might need to be transformed to correct potential violations. Histograms provide an indication of potential skewness (*see* **Note 7**) and a Shapiro–Wilk test (*see* **Note 11**) provides a formal test of the normality properties of the data distribution. A QQ plot also provides information on the distribution of data by comparing two probability distributions (*see* **Note 11**).

The second step is to investigate homogeneity of variances. For example in the example data set, it needs to be tested if variances are constant across years. This can be visualized with a scatter plot or a box plot of the phenotype of interest across years (*see* **Notes 8** and **9**). Also, calculating the standard deviation of FWEC for each year shows the level of consistency of the variances (*see* **Note 5**) or conducting a formal Leven's test (*see* **Note 14**).

If non-normality or heterogeneous variances are detected, the data need to be transformed. The appropriate transformation depends on the direction of skewness. Positively skewed distributions, with the long tail into the positive direction, can be corrected with a logarithmic or square root transformation. Cubing or squaring of the data can correct negatively skewed distributions that have a long tail in the negative direction (*see* **Note 12**). Plotting transformed data as a histogram allows visualization of the effect of the transformation. We revisit the transformation again after discussing the fit of a linear model. If the appropriate transformation is not evident from diagnostic plots, the Box–Cox procedure automatically identifies the best transformation, but can only be applied to positive values (*see* **Note 13**). Note that it is necessary to remove outliers before applying the Box–Cox transformation. Homogeneity of variances can be investigated using Leven's test (*see* **Note 14**).

3.4 Data Analysis

3.4.1 Model Comparison

Subheading 3.2 provided a brief overview on how to build a sensible model using an ANOVA or deviance tables. Practically, there are two ways a model can be built (*see* **Notes 15–17**). Variables can either be included step wise one at a time and an ANOVA or deviance table is generated every time a new variable is fitted. Variables with significant F-value or Chi-squared statistics, as indicated by the P-value, are retained in the model and nonsignificant ones are excluded. The second way of building the model works in reverse. All explanatory variables are included in the model and step wise nonsignificant variables are removed starting with the least significant variable.

When building a model, it is important to only include sensible explanatory variables. One needs to be able to defend the variables that have been fitted in the model. To illustrate this issue, assume an investigation of human height where gender, age, and also the color of the clothing of the jumper of each test individual was recorded. Most likely the color will show an effect on height, but only because there will be some confounding effects with gender because males preferably wear blues, blacks, and greens, whereas females tend to wear brighter colors. Even though phenotypic variation is seemingly associated with the color of the jumper, it would be inappropriate to fit this variable in a model for human height.

The other important issue to keep in mind when building a model is the confounding of exploratory variables, which was outlined earlier in Subheading 3.2. If variables are partially confounded, they are most appropriately fitted by nesting one within the other. If they are fully confounded, test each of them and fit the variable with the higher sums of squares. As an example, we can refer back to the two explanatory variables birth type and rearing type in the example data set. Previously, it was suggested to create a new variable by forming new categories, i.e., birth rearing type, and the new variable can be fitted in the model. Another option is to nest birth type within rearing type in the model. Both approaches lead to an appropriate fit of the variables in a linear model (*see* **Notes 16–18**). Be careful in over-fitting a model in relation to the number of observations in the data set. The more variables, the more complex the model becomes. In particular, interactions can be problematic when they only describe a small number of observations. Think very carefully about fitting interactions because even though they might show significance in the model, they can be difficult to interpret.

Experimental units that have missing values in an explanatory variable that has been fitted in the model will be excluded from the analysis. Therefore, knowledge on the extent of missing data in the explanatory variables is important and needs to be considered when fitting the variables.

3.4.2 Model Diagnostics

After the model for the data has been identified, the residuals need to be checked for influential outliers. Model diagnostics plots include plots of residual vs. fitted values (no particular relationship should be obvious), QQ plots (checking normality), and scale-location plots (checking changes in variance of residual with predicted value) (*see* **Note 19**). In R, influential outliers are marked with their sequential index number in the data set. The relevant records of the outlying record, including response and explanatory variables in the model, should be checked for obvious reasons of the outlying characteristics. If no problems are obvious, outliers may be removed if they are outside the range of three standard deviations from the mean.

3.4.3 Extraction of Residuals

Once the most appropriate linear model has been fitted and influential outliers have been removed, the last step is to extract the residuals for further analysis (*see* **Note 20**). Ideally, the residuals contain only the component of the phenotype that is influenced by the genetic makeup of the individual. Since this is virtually impossible, the best efforts should be made to prepare the data as best as possible and to fit the best possible model prior to the actual association test.

4 Notes

1. Reading data into R—the data file includes a header and "NA" is a missing value.

    ```
    data_in <- read.table(file="example.dat", header
        =T, na.string="NA")
    ```

2. Getting access to the data frame (all variables will relate to this data frame).

    ```
    attach(data_in)
    ```

3. Overview of data (mean, median, maximum, minimum, first and third quartiles, number of missing values).

   ```
   summary(data_in)
   ```

4. Calculating means and standard deviations (2 indicates that the function is applied to all columns, na.rm=T means that missing values are removed).

   ```
   apply(data_in,2,mean,na.rm=T)
   apply(data_in,2,sd,na.rm=T)
   ```

5. Means or standard deviations can be calculated for data points grouped by a factor (e.g. year).

   ```
   aggregate(data_in,list(data_in$year),FUN=mean)
   aggregate(data_in,list(data_in$year),FUN=sd)
   ```

6. Frequencies for a single variable and across two variables.

   ```
   table(sex)
   table(wormy)
   table(wormy,sex)
   ```

7. Histogram.

   ```
   hist(FWEC)
   ```

8. *xy* scatter plots.

   ```
   plot(FWEC)
   ```

9. Box plot.

   ```
   #the whiskers of the box plot can be set at various ranges, range=0 includes all data points in the range of the whiskers
   boxplot(FWEC~sex,data=data_in, range=0)
   ```

10. Shapiro–Wilk's tests to check normality of data distribution.

    ```
    shapiro.test(FWEC)
    ```

11. Checking data distribution with QQ plot—if data are normally distributed, the plotted data and the line are well aligned.

    ```
    qqplot(FWEC)
    qqline(FWEC)
    ```

12. Data transformation—log, square root, and cube root transformation.

    ```
    log_FWEC <- log(FWEC)
    sqrt_FWEC <- sqrt(FWEC+1)
    cbrt_FWEC <- (FWEC)^(1/3)
    ```

13. Box–Cox transformation.

    ```
    #Code to find suitable lambda for Y to the power lambda
    #download the MASS library
    ```

```
library(MASS)
#seq(min value, max value, step) defines the range
    from which lambda is drawn
boxcox(FWEC~factor(sex)+factor(birth_rearing_
    type), lambda = seq(0,1.0,0.01)
savePlot("boxcox","jpeg")
lambda = "insert maximum lambda value in graph here"
trans(FWEC) <- ((FWEC^lambda)-1)/lambda
```

14. Checking homogeneity of variances.

    ```
    #download library (Rcmdr)
    library(Rcmdr)
    #run the Leven's test, specifying the vector of
        data y and group, the factor across which the var-
        iances are tested (e.g., year)
    leveneTest(y,group)
    ```

15. Fitting a linear model and ANOVA.

    ```
    #need to load the "car" package for Type III ANOVA
    library(car)
    lmod <- lm(cbrt_FWEC~factor(sex))
    #Type I ANOVA
    anova(lm)
    #Type III ANOVA—Note that the first letter in the
        command below has to be a capital "A" (ensure that
        you loaded the "car" package as shown above)
    Anova(lmod, type="III")
    ```

16. Addressing confounding of explanatory variables in a linear model.

    ```
    lmod1 <- lm(cbrt_FWEC~factor(sex)+factor(birth_
        type)*factor(rearing_type))
    lmod2 <-lm(cbrt_FWEC~factor(sex)+factor(birth_
        rearing_type))
    ```

17. Check the difference with an ANOVA.

    ```
    Anova(lmod1,type="III")
    Anova(lmod2,type="III")
    ```

18. Model comparison using logistic regression for binary data.

    ```
    logres <- glm(formula=wormy_status~factor(sex) +
        factor(birth_rearing_type), family = binomial
        (link="logit"))
    #producing an analysis-of-deviance table to test
        fixed effects
    anova(logres,test="Chisq")
    #produces the deviance of the model (the lower the
        better the fit)
    ```

```
summary(glm(formula=wormy_status~factor(sex) +
   factor(birth_rearing_type),family=binomial
   (link="logit"))$deviance))
#the difference in deviance can be formally tested
   with a loglikelihood ratio test
#install library(lme4)
library(lme4)
#comparing two nested models ("nested" means that
   one has one more factor than the other)
logres1 <- lmer(wormy_status~factor(sex)),family
   ="binomial",method="Laplace")
logres2 <- lmer(wormy_status~factor(sex) +factor
   (birth_rearing_type),family="binomial",
   method="Laplace")
anova(logres1,logres2)
#to assess the model, plot predicted probability
   against observed proportion
#install library(languageR)
library(languageR)
plot.logistics.fit.fnc(logres1,logres2)
```

19. Model diagnostics.

```
#the following produces plot of residual vs. fitted
   value, QQ plot, and scale-location plot of the
   previously tested model 1 (lmod1)
plot(lmod1)
#assessing a logit model for binary data by plotting
   the predicted probability against observed
   proportions
#download library(languageR)
library(languageR)
plot.logistic.fit.fnc(logres1,data_in)
```

20. Extracting residuals and writing them to a file—assuming lmod2 is the model of choice.

```
res_lmod2 <-residuals(lmod2)
write.table(res_lmod2,file="res_FWEC")
```

References

1. Aitkin M, Francis B, Hinde J (2009) Statistical modelling in R. Oxford University Press, New York
2. Bowman AW, Robinson DR (1990) Introduction to regression and analysis of variance. Briston, Cambridge
3. Crawley MJ (2005) Statistics: an introduction using R. Wiley, Chichester, U.K
4. Fisher L, McDonald J (1978) Fixed effects analysis of variance. Academic, New York
5. Cox DR (1977) The analysis of binary data. Chapman and Hall, London

6. Fox J (2002) An R and S-Plus companion to applied regression. Sage, Thousand Oaks, California
7. Hocking RR (1985) The analysis of linear models. Books/Cole, Monterey, California
8. Hocking RR (2003) Methods and applications of linear models: regression and the analysis of variance, 2nd ed. Wiley-Interscience, Hoboken, New Jersey
9. Hoaglin DC, Mosteller F, Tukey JW (1991) Fundamentals of exploratory analysis of variance. Wiley, New York
10. Ihaka R, Gentleman R (1996) R: a language for data analysis and graphics. J Comput and Graph Stat 5(3):299–314
11. Lynch M, Walsh B (1998) Genetics and analysis of quantitative traits. Sinauer, Sunderland, MA, USA
12. Mendenhall W (1968) Introduction to linear models and the design and analysis of experiment. Wadsworth, Belmont, California
13. Mittal HV (2011) R graphs cookbook. Packt, Birmingham
14. Murrell P (2011) R graphics, 2nd edn. Chapman & Hall, London
15. Naylor GFK, Enticknap LE (1981) Statistics simplified. An introductory course for social scientist and others. Harcourt Brace & Company, Australia
16. Neter J, Kutner MG, Nachtsheim CJ, Wasserman W (1996) Applied linear statistical models, 4th edn. The McGraw-Hill, USA
17. Rencher AC, Schaalje GB (2008) Linear models in statistics, 2nd ed. Wiley-Interscience, Hoboken, New Jersey
18. Shababa B (2012) Biostatistics with R: an introduction to statistics through biological data. Springer, New York
19. Spector P (2008) Data manipulation with R. Springer, New York
20. The comprehensive CRAN network. http://cran.r-project.org/ (last viewed 31 May 2012)
21. Verzani J (2005) Using R for introductory statistics. John Verzani. Chapman & Hall, London
22. Zuur AF, Ieno EN, Meesters EHWG (2009) A beginner's guide to R. Springer, London, New York

Chapter 3

Designing a GWAS: Power, Sample Size, and Data Structure

Roderick D. Ball

Abstract

In this chapter we describe a novel Bayesian approach to designing GWAS studies with the goal of ensuring robust detection of effects of genomic loci associated with trait variation.

The goal of GWAS is to detect loci associated with variation in traits of interest. Finding which of 500,000—1,000,000 loci has a practically significant effect is a difficult statistical problem, like finding a needle in a haystack. We address this problem by designing experiments to detect effects with a given Bayes factor, where the Bayes factor is chosen sufficiently large to overcome the low prior odds for genomic associations. Methods are given for various possible data structures including random population samples, case–control designs, transmission disequilibrium tests, sib-based transmission disequilibrium tests, and other family-based designs including designs for plants with clonal replication. We also consider the problem of eliciting prior information from experts, which is necessary to quantify prior odds for loci. We advocate a "subjective" Bayesian approach, where the prior distribution is considered as a mathematical representation of our prior knowledge, while also giving generic formulae that allow conservative computations based on low prior information, e.g., equivalent to the information in a single sample point. Examples using R and the R packages ldDesign are given throughout.

Key words Experimental design, Frequentist statistics, Bayesian statistics, Bayes factors, BIC criterion, Posterior probabilities, Power calculations, Bayesian power calculations, ldDesign R package, Robust detection of effects, Genome-wide association studies (GWAS), Quantitative trait loci (QTL), Quantitative traits, Case–control studies, TDT test, S-TDT test, SDT test, Family-based association studies, Linkage, Linkage disequilibrium, Prior elicitation, Genome scans, Candidate gene association studies, Replication and validation

1 Introduction

The goal of genome-wide association studies (GWAS) is to understand the variation in complex traits and diseases by relating genotypes of large numbers of markers (typically SNPs) to observed phenotypes. To do this, it is necessary to detect which markers are associated with variation in the traits. In this section, we introduce the fundamental statistical concepts needed for experimental design of GWAS with sufficient power to robustly detect genomic effects. To motivate the choice of methods, we first discuss the history of spurious associations and the need for better evidence; followed by subsections on frequentist and Bayesian measures of evidence; the debate between frequentists and Bayesians, with particular interest in testing scientific hypotheses, and the inference problem in genomics; indicating why the Bayesian approach, using probability theory to quantify statistical evidence, is particularly important for GWAS that are typically testing for small effects with low prior odds and large sample sizes.

Spurious Associations and the Need for Better Evidence:
Early attempts to find associations for complex diseases or quantitative traits have lead to many published associations which are likely to be spurious [1–4]. Altshuler et al. (2000) discussed in [5] retested 13 published associations of SNPs with type II diabetes in an independent population. Only one was significant. These results are summed up by Altshuler [6]:

> The lack of replication of the others points to the need for larger samples, controls for population differences, and stronger statistical evidence prior to claiming an association. *(emphasis added)*

Figure 4 in [2] shows the distribution of around 260 reported p-values from association studies in two journals, and note that there is no evidence of departure from the uniform distribution (i.e., no evidence of any real effect):

> "… investigators are too frequently gambling on and publishing results in situations where the evidence is not at all compelling."

Emahazion et al. [3] re-tested a number of associations from 13 genes putatively associated with Alzheimer's disease. Apart from the APOE ϵ_4 allele used as a positive control, only one was validated with $p < 0.01$ rising to $p = 0.33$ after adjusting for multiple comparisons.

Even the stringent threshold of $P < 5 \times 10^{-7}$ may not be sufficient. Our calculations showed that around half of the published associations with $P < 5 \times 10^{-7}$ from recent large case–control studies [7, 8] with samples of 2,000 cases and 3,000 controls had posterior probabilities less than 0.5. Subsequently 5×10^{-8} has been adopted as a de facto standard [9].

Effects detected with low power to detect the true size of effect are not robustly detected, and are likely to be spurious and/or subject to selection bias, a phenomenon where those effects detected tend to be overestimated, because overestimated effects have higher probability to be detected.

The key to robustly detecting effects is to have good power to detect them with sufficiently strong evidence. The need for stronger evidence for detection has been recognized. The critical problem is: what measure of evidence and how strong? To answer that question it is important to understand the fundamentals of statistical inference. We first review the measures of evidence used by the two major schools of statistics, frequentist, and Bayesian.

1.1 Frequentist Measures of Evidence

Frequentist, or non-Bayesian statisticians, view probabilities as limiting frequencies of repeated events. A frequentist would consider an unknown parameter as having a true value (only, and not a probability distribution) and make inference based on frequency distributions of statistics under hypothetical repeated samples. Frequentists dislike the subjectivity of prior probability distributions used in Bayesian statistics. However most practicing frequentist statisticians do in fact use prior probability distributions for unknown parameters (e.g., random effects in mixed models), and do use prior information (e.g., if they are more sceptical about an effect they may require a lower threshold before accepting it).

Frequentists use the p-value, defined as the probability that the test statistic is more extreme than the observed value under hypothetical replications under the null hypothesis H_0, e.g.,

$$p = \Pr(T \geq t_{\text{obs}} \mid H_0). \tag{1}$$

Small values of the p-value, e.g., $p < \alpha$ for some previously chosen threshold α are considered to be "significant." A small p-value says that the event $\{T \geq t_{\text{obs}}\}$ in Eq. 1 has low probability. This suggests that H_0 may not be the perfect model for the data. However there is no reason that the corresponding event $\{T \geq t_{\text{obs}} \mid H_1\}$ under the alternative hypothesis, H_1, is any more likely.

Note: the uppercase and lowercase P, p when referring to p-values refer to the random variable P and an observed value, p, respectively.

The advantages of the p-value are that it can be readily calculated, and that under H_0 the p-value has a uniform distribution. The p-value effectively standardizes the test statistic, whether F, t, z, χ^2, or other. However the p-value does not successfully standardize the strength of evidence for a real effect across different sample sizes, experimental designs, and test setups.

A fundamental criticism of the p-value is that using a p-value to summarize the data amounts to conditioning on an unobserved event (e.g., $\{T \geq t_{\text{obs}}\}$ in Eq. 1) [10].

The main problem with p-values is, how to use and interpret the p-value? When is it good evidence, and how should we make a decision? Problems with the interpretation of p-values have been pointed out by, e.g., [10–12], and demonstrated in a genetics context by [13–16]. Unlike the Bayes factors discussed below the p-value does not have a natural interpretation as strength of evidence. The problems with spurious associations are symptomatic of the difficulties in interpreting p-values.

1.2 Bayesian Measures of Evidence

Bayesian statistics is statistics soundly based on probability theory [17]. Probability theory is the only sound approach to making decisions under uncertainty. Despite centuries of use, no better approach has been demonstrated. Probabilistic models (where often all the information is "prior" information) are widely used (e.g., in the field of artificial intelligence). Bayesian statistics can be considered as one special case of probabilistic modeling, where data together with a model for the process generating the data are available in addition to prior information.

Bayesian statisticians use probability theory to model uncertainty. Bayes' theorem is used to combine prior information, represented by a probability distribution, with data and a model for the process generating the data, represented by a likelihood function.

Note: The Bayesian notion of probability theory is often considered as distinct from the frequentist probability theory based on limiting probabilities of repeated events. However, probability theory is essentially unique—both theories satisfy the basic axioms of probability theory and are hence essentially the same. A Bayesian probability has the same limiting frequency interpretation if an infinite sequence of repeated events is available. Frequentists deny that a probabilistic model can be constructed for prior information while at the same time using a probabilistic model for the likelihood, the choice of which often has a much stronger effect. A strong argument for the Bayesian approach is that, in practice, repeated events are hypothetical and not available in practice, and only the observed data is available and relevant.

In statistics we are given observed data y and are interested in learning about unknown parameters θ. Given prior knowledge about θ represented by a probability distribution $\pi(\theta)$ and the likelihood function $f(y \mid \theta)$ defining a probability distribution for the observed data, *Bayes theorem* gives an expression for the *posterior distribution* $g(\theta \mid y)$.

Bayes theorem [17] follows from a basic property of conditional probabilities: for any events A, B:

$$\Pr(A \mid B)\Pr(B) = \Pr(B \mid A)\Pr(A). \qquad (2)$$

Substituting and rearranging we obtain

$$g(\theta \mid y) = \frac{f(y \mid \theta)\pi(\theta)}{m(y)} \qquad (3)$$

$f(y \mid \theta)$ is the *likelihood*, $\pi(\theta)$ is the prior distribution, and $g(\theta \mid y)$ is the posterior distribution where $m(y)$ is the marginal distribution for the data, i.e., the probability of observing the data.

Integrating (Eq. 3) over θ we obtain:

$$m(y) = \int f(y \mid \theta)\pi(y)\mathrm{d}\theta. \qquad (4)$$

When comparing two or more hypotheses or models the model itself becomes part of the set of unknown parameters, and each model may have different sets of parameters. The Bayesian approach to hypothesis testing is to find the probability that H_1 or H_0 is the true model.

The *Bayes factor* for comparing two hypotheses (or models) H_0, H_1 is defined as the ratio

$$B = \frac{\Pr(y \mid H_1)}{\Pr(y \mid H_0)}. \tag{5}$$

From Bayes theorem, it follows that the posterior odds are given by:

$$\frac{\Pr(H_1 \mid y)}{\Pr(H_0 \mid y)} = \frac{\Pr(y \mid H_1)}{\Pr(y \mid H_0)} \times \frac{\Pr(H_1)}{\Pr(H_0)}$$
$$= B \times \frac{\Pr(H_1)}{\Pr(H_0)}. \tag{6}$$

Thus, the Bayes factor is the factor by which prior odds, $\Pr(H_1)/\Pr(H_0)$, are increased to give posterior odds after observing the data, i.e., the Bayes factor represents *the strength of evidence* in the data.

Note:

1. The key technical difficulty in applying Bayesian statistics is evaluating the integral (Eq. 4) to obtain $m(y)$, where required. Difficulties with the integral can be avoided by use of MCMC, conjugate priors (where the integral is obtainable in closed form), or approximations.

2. The numerator and denominator in the expression for the Bayes factor (Eq. 5) are the values of $m(y)$ under H_1 and H_0 respectively.

3. We have used different symbols $f(\cdot), \pi(\cdot), m(\cdot), g(\cdot)$ for the likelihood, prior, marginal, and posterior distributions. Some authors use the same symbol $f(\cdot)$ for all distributions, e.g., $f(y \mid \theta), f(y)$ where the actual function being referred to is implicit from the context. Or, drop the symbol altogether, e.g., $[y \mid \theta], [\theta], [\theta \mid y]$.

4. Some authors define the Bayes factor as the reciprocal of Eq. 5. We prefer Eq. 5, where larger numbers represent stronger evidence.

Solving for B in Eq. 6 it follows that if we can design an experiment to be sufficiently powerful to detect effects with a given Bayes factor, where the Bayes factor is sufficiently large to overcome the prior odds, we can ensure the posterior odds are respectably high. This is critical since in GWAS prior odds are typically low, e.g., there may be 500,000 or more SNP markers but only 10 loci with detectable effects.

For further information on Bayesian philosophy and methods see the books [18–21]. Lavine [22] gives a simplified example of Bayesian computation for evaluation of evidence for an effect and comparison with a nearly significant *p*-value.

1.3 Fisher vs. Bayes

There has been much discussion between Bayesian and frequentist statisticians that has often shed more heat than light. Many observers think Bayesian and frequentist statistics are similar, and, for some problems and situations they are.

Maximum likelihood estimation. Estimation using maximum likelihood and frequentist confidence intervals is qualitatively similar to Bayesian estimation and Bayesian credible intervals with a non-informative (e.g., flat) prior distribution, or where the influence of the prior is relatively weak. The Bayesian and frequentist approaches are asymptotically equivalent.[1]

One-sided tests and tail probabilities. In a Bayesian context, some authors, e.g., [23], use tail probabilities (similar to p-values) for testing hypotheses. Again this is qualitatively similar to using p-values for inference. Tests based on tail probabilities are effectively *1-sided hypothesis tests* such as $H_a : \theta < a$ vs. $H_b : \theta > a$. These tests are appropriate when it is known the effect in question is real, and we are just interested in its size, or sign. In this situation Bayesian and frequentist inference can be reconciled [24]. These tests are not appropriate when the question of interest is whether or not there is a real effect: e.g., $H_0 : \theta = 0$ vs. $H_1 : \theta \neq 0$. For this type of "precise hypothesis," a Bayesian approach should base decisions on posterior probabilities for the hypotheses themselves.

Testing scientific or 'precise' hypotheses. When testing scientific hypotheses such as Newton's theory of gravitation vs. Einstein's theory of general relativity or that a given treatment enhances plant growth, or that a given marker is associated with a trait, we are often considering two-sided tests such as $H_0 : \theta = 0$ vs. $H_1 : \theta \neq 0$. In this case note that the null hypothesis is a lower dimensional subspace of H_1 and hence would have probability zero under the induced probability measure from H_1. It is, however, logically inconsistent to test the hypothesis unless there is some finite probability that H_0 is true. It is therefore not surprising that the p-value based on the probability measure under H_1 (the tail probability for the test statistic under H_0 is equivalent to a tail probability for the parameter under H_1 (R. Ball, unpublished)) is inadmissible. This type of hypothesis, more appropriately described as a "precise hypothesis," is qualitatively different. Bayesian and frequentist inference for testing precise hypotheses cannot be reconciled [12].

Model uncertainty. Where there are multiple possible alternative models, decisions need to be based on posterior probabilities for models as well as distributions of parameters within models. When using Bayes factors or posterior probabilities of models for inference, Bayesian and frequentist approaches are not similar. Moreover, in a Bayesian model selection or hypothesis testing context using a non-informative prior distribution does not make sense. In the limit as

---

[1] In asymptotics where $n \to \infty$ while prior information is fixed.

prior information on the parameter being tested tends to zero the Bayes factor or strength of evidence for H_1 also tends to zero.

Below, we use the Spiegelhalter and Smith [25] Bayes factor, or approximate Bayes factors based on asymptotic approximations to the posterior, where a conservative "default" prior equivalent to a single sample can be obtained. Our results indicate this is more conservative than using p-values, in fact p-values, naively interpreted, are more anti-conservative than the maximum Bayes factor.

Fisher had extensive discussions, often heated, with other statisticians, notably Karl Pearson. Fisher rejected the Pearson approach of choosing a threshold, α, and reporting (or making a decision based on) only whether or not $P < \alpha$ for this reason—the information as to whether $P = \alpha$ or $P = \alpha/10$ is lost. Fisher proposed reporting the p-value which decreases continuously as the test statistic becomes more extreme, hence does not lose information. However Fisher perhaps did not realize he was being inconsistent in conditioning on $\{T \geq t_{obs}\}$ in the definition of the p-value itself (Eq. 1) amounts to the same thing—the event that *was* observed was $\{T = t_{obs}\}$.[2]

Modern texts often merge the Fisher and Pearson approaches into one seamless treatment. However α is only a valid type I error rate if chosen pre-experimentally. This is however inefficient as Fisher pointed out. The p-value itself is not a valid error rate. Pearson was valid but not efficient. Fisher was efficient but not valid—he did not give a valid procedure for inference, or making a decision. The criterion commonly used is $p \leq 0.05$; however it is mistakenly believed that this represents strong evidence.

The p-value violates the intuitively reasonable likelihood principle, that inference should depend on the data only through the likelihood function. The likelihood principle is an important underpinning principle in Bayesian theory but not adopted by frequentists. Fisher's principles, widely adopted, include *consistency* (that we should seek estimators that converge to the true value with large samples), *sufficiency* (that an estimator should incorporate all the information, i.e., we must use all the information in the data), and *ancillarity* (that one should condition on ancillary information that is not informative for the parameter(s) in question, e.g., the marginal totals in a contingency table—hence we should use Fisher's exact test, not Pearson's chi-squared). Frequentists were shocked when it was shown that Fishers' principles essentially implied the likelihood principle, or, technically, a weak form thereof ([26, 27] and reviewed in [28]), and hence that the Fisher p-value was not consistent with Fisher's own principles. Acknowledging the p-value, and hence hypothesis testing, is a failure means that there is no method in frequentist methodology for making decisions.

---

[2] Bayesian inference does condition on exactly the observed data.

Consequences of violating the likelihood principle are:

- That p-values give a misleading impression of strength of evidence and are often misinterpreted as "the probability of being wrong," i.e., the posterior probability of H_0, which is what many people would like it to be.
- That there is no interpretation of p-values as strength of evidence independent of sample size, experimental design, and test setup.
- Inference depends on the experimenter's intentions as to what might have been tested, which are not verifiable and not of interest to end users or decision makers—the p-value depends on what the experimenter might have done, e.g., if the experimenter would have continued for further samples, if the test was not significant or would have stopped earlier, if the test was significant with a smaller number of samples (cf. [10]), the p-value changes; hence p-values are problematic in a sequential sampling application.

Regarding Bayesian methods, Fisher thought he could solve the problem of needing to specify the prior distribution through the use of *fiducial probabilities*, or "probabilities" consistent with confidence intervals. This attempt to "make the Bayesian omelette without breaking the Bayesian egg" was not, however, an unqualified success:

> Fisher's lack of clarity over the basis of the fiducial method led to considerable unease about it, which seemed justified when counterexamples showed that it need not yield unique inferences [29–31].
> —Davison [28].

Fisher thought only a uniform prior made sense, if using Bayes' theorem. Fisher thought the Bayesian approach was only relevant when prior knowledge was quantified as "probability of the mathematicians," meaning that the prior distribution was given as a more or less exact formula such as the probability of observations from tossing a coin, dice, or roulette wheel. Thus, he believed it could not be applied to real-world situations such as the probability that a horse would win a race.

> While as Bayes perceived, the concept of mathematical probability affords a means, in some cases, of expressing inferences from observational data, involving a degree of uncertainty, and of expressing them rigorously, in that the nature and degree of the uncertainty is specified with exactitude, yet it is by no means axiomatic that the appropriate inferences, though in all cases involving uncertainty, should always be rigorously expressible in terms of this same concept.—Fisher [32, p. 37]

In genetics, allele frequencies provide examples where probabilities are not exactly known. Fisher [33, pp. 18–20] considered the probability of observing heterozygous or homozygous mice in a cross. Suppose there are two alleles A, a, and suppose the AA genotypes have black coats while Aa genotypes have brown coats. The aa genotypes were not considered. Given the parental genotypes

AA, *Aa* the probability of each offspring to be heterozygous is exactly known, i.e., 0.5. However, if the parental genotypes are unknown we have to use our prior knowledge of the population from which the parents are drawn. We may have prior knowledge from previous experiments. But what if there is no prior information? In practice this is unlikely, e.g., we would have observed a certain number of black and brown mice, but suppose for arguments sake there is no prior information. Simply assuming equal probabilities of 0.5 for heterozygous and homozygous parents is not a good strategy, as Fisher rightly pointed out. This was perhaps not so rightly attributed as following from Laplace's suggestion of assigning equal probabilities to the various alternatives in the absence of prior information [34]; the mistake being to effectively specify an exact value for the allele frequency p below, giving equal probabilities of heterozygous and homozygous parents, which he knew would probably be wrong, and which is wrong in principle, because it ignores the knowledge that the parental mice come from a population. Instead, Laplace's principle could be applied to values of p. A modern Bayesian approach would be to say the population allele frequency for A is p, and, in the absence of other prior information, let p have, e.g., a Beta(0.5, *sigma*1) prior distribution. This gives a *hierarchical model*

$$p \to (g_p, g_m) \to g_{x,i}, \tag{7}$$

where g_p, g_m denote the paternal and maternal genotypes, and $g_{x,i}$ denotes the ith offspring genotype. In the hierarchical model, probabilities for each parameter depend on the values of its ancestors. For example g_p depends on p. If Hardy–Weinberg equilibrium applies, $\Pr(g_p = AA) = p^2$, $\Pr(g_p = Aa) = 2p(1-p)$, $\Pr(g_p = aa) = (1-p)^2$. The progeny genotype $g_{x,i}$ depends on g_p and on g_m. We have $\Pr(g_{x,i} = Aa \mid g_p = Aa$ and $g_m = aa) = 0.5$.

Note: A hierarchical model is similar to a family tree, where the probabilities for the genotypes of an individual depend on the genotypes of its ancestors. The Beta family of distributions Beta(a, b) gives a range of shapes, useful as prior distributions for proportions, with any given mean value $a/(a+b)$ and variance $ab/[(a+b)^2(a+b+1)]$, ranging from the "uninformative" Beta$(1/2, 1/2)$ or Beta$(1,1)$ to highly informative distributions when $a+b$ is large.

Of equal or greater importance than theoretical or foundational considerations is the experience in practice that p-values are misleading. Many people believe that a p-value of 0.05 represents strong evidence, but this is not the case, particularly with the larger sample sizes used in genome-wide association studies, where even p-values of 5×10^{-8} may not be sufficient. A common naive interpretation is that the p-value is the probability of being wrong.

If $\alpha = 0.05$ there is only a 5 % chance of being wrong if H_0 is true. However if $P = 0.05$ this does not mean there is only a 5 % chance of being wrong. We can't use $\alpha = 0.05$ even though $P \leq \alpha$ holds.

The appropriate α value (under fairly general distributional assumptions) once $P = p$ has been observed is the *conditional alpha* of Sellke et al. [34] who show that, under fairly general conditions, a lower bound for the conditional α is given by

$$\alpha_c \geq \left(1 + [-ep\log(p)]^{-1}\right)^{-1}. \tag{8}$$

So that if $P = 0.05$, $\alpha_c \geq 0.289$. So the probability of being wrong is around 0.3 or more, and the evidence is now apparently very weak.

The probability of being wrong, which is what people would like the p-value to be, is the Bayesian posterior probability $\Pr(H_0 \mid y)$. Regardless of the prior used, $\Pr(H_0 \mid y)$ can be much larger than P (e.g., 0.2 compared to 0.05) in common examples [10], and with moderately large sample sizes we can have $\Pr(H_0 \mid y) > \Pr(H_1 \mid y)$ even with equal prior odds.

Early "successes" with the use of p-values were perhaps fortuitous. Applications in agriculture, for which the p-values were initially applied, tended to have small sample sizes, the effects being tested were reasonably large, and, the hypotheses being tested were a priori quite likely, e.g., that nitrogen fertilizer promoted growth. Scientists were using their prior knowledge and getting a stamp of approval in the form of one or more *'s from a statistician and everyone was happy. Frequentist statisticians were content to be "objective" by not using prior knowledge. No one cared to investigate whether the strength of evidence from the data was in fact any stronger than the prior knowledge.

This differs from the situation for GWAS for complex traits and diseases, where the effects are quite small and the probability that any given marker is in substantial linkage disequilibrium with a detectable functional locus is very low, and the sample sizes are large. This is a radically different situation than that for which p-values were initially conceived. In the frequentist domain this is considered as a complex multiple testing problem.

1.4 The Inference Problem in Genomics

We wish to determine which loci are functional or in LD with functional loci. This is in essence a model selection problem, not a hypothesis testing problem, and not a multiple comparisons problem. Multiple comparisons/hypothesis testing is not appropriate—testing for the hypothesis that no loci are associated (genome-wise significance level) is of no interest. Nor are we testing multiple hypotheses relating to the same effect.

Consider alternative possible models, where a model specifies which set of loci are functional or in LD with (and in close proximity to) functional loci. We would like to solve the model selection problem: which model is the true model, representing the underlying genetic architecture of the trait? The traditional (frequentist) paradigm is to first choose a model then make inferences assuming the chosen model is the true model. This works well in common applications where the true model is either unequivocally determined by the data, or the choice of model is ancillary to the question of interest.

For example, either a linear or quadratic model might both fit a dataset reasonably well and use of either in estimating the time to reach a certain value may give similar answers. It is usually the case in gene mapping, the true model is not unequivocally determined by the data, and the choice of model is central to the question of interest—which loci are influencing a trait? Hence we need to consider model uncertainty. This is best done in a Bayesian framework, by computing probabilities for all possible models.

1.5 Summary

Bayesian and frequentist inference are not similar for testing scientific or "precise hypotheses." Frequentists believe they are being objective by calculating p-values and not using prior information, but do not give a procedure for making a decision. This is often misleading because p-values may be much smaller than the posterior probability of H_0, particularly in large samples. Any decision based on p-values is *ad hoc* or somehow using prior information. Bayesians acknowledge subjectivity and use prior information transparently by specifying a prior distribution which is a mathematical representation of prior uncertainty as a probability distribution. Prior distributions can be elicited from experts (cf. Subheading 5) and based on experience and data from related experiences (e.g., different species) but are inevitably subjective. There is no one method to analyze data that all statisticians will agree on, but, with a Bayesian prior distribution, other observers are free to recalculate with their own prior. It is therefore important to explain the reasoning in the determination of the prior, particularly if the prior represents comparable or greater information than the data.

In genomics, we have low prior odds requiring strong evidence, or high Bayes factors in order to obtain respectable posterior odds. Often the true model is not determined unequivocally from the data, rather a number of models (e.g., two models representing H_0 and H_1 in the case of single-marker tests, or more in the case of a multi-locus analysis) representing alternative possible genetic architectures are consistent with the data. Inference needs to be based on the posterior probability of the alternative possible models.

2 Power Calculations

2.1 Traditional Power Calculations

Power is the probability of detecting a specified effect with the criterion used. Traditional power calculations give the probability of detecting an effect with $P \leq \alpha$, for a given size of effect.

F-Test

Consider an ANOVA model with ν_1 degrees of freedom for the effects being tested, and ν_2 degrees of freedom for error. Under the null hypothesis the F-statistic has a (central) F-distribution. Under the alternative hypothesis the F-statistic has a non-central F-distribution.

Let $F_{1-\alpha;\nu_1,\nu_2}$ denote the $1 - \alpha$ quantile of the F-distribution with ν_1, ν_2 degrees of freedom, and let $F(\cdot; \nu_1\nu_2, \delta)$ denote the cumulative distribution function (c.d.f.) of the non-central F-distribution with ν_1, ν_2 degrees of freedom, and non-centrality parameter δ.

The power is given by:

$$\mathcal{P} = \Pr(F \geq F_{1-\alpha} \mid \nu_1, \nu_2, \delta) = 1 - F(F_{1-\alpha;\nu_1,\nu_2}; \nu_1\nu_2, \delta). \quad (9)$$

t-Test

Suppose we observed an independent identically distributed (i.i.d.) sample $x_1 \ldots x_n$, where $x_i \sim N(\beta, \sigma^2)$. We wish to test $H_0 : \beta = 0$ vs. $H_1 : \beta \neq 0$. The test statistic is $T = \hat{\beta}/(\hat{\sigma}/\sqrt{n})$.

Claim: T^2 has a non-central F-distribution on 1, n − 1 d.f. with non-centrality parameter $\delta = t^2$, where $t = \beta/(\sigma/\sqrt{n})$.

Proof

$$\hat{\beta} = \beta + \frac{1}{n}\sum(x_i - \beta) \quad (10)$$
$$= \beta + u$$

where $u \sim N(0, \sigma^2/n)$.

$$T^2 = \frac{\hat{\beta}^2}{\hat{\sigma}^2} = \frac{(\beta + u)^2}{\hat{\sigma}^2} = \frac{(t + z)^2}{n(\hat{\sigma}^2/\sigma^2)}, \quad (11)$$

where $z \sim N(0, 1)$.

The top line is a chi-squared distribution on 1 d.f. with non-centrality parameter $\delta = t^2$ (by definition of the non-central χ^2; [35]) under H_1 and a central chi-squared distribution under $H_0 : \beta = 0$ (since t = 0). The bottom line is a chi-squared on n − 1 d.f. (sum of squares on n − 1 d.f.). Hence the ratio is a non-central F distribution with degrees of freedom 1, n − 1 and non-centrality parameter t^2 under H_1 and a central F distribution with 1, n − 1 d.f. under H_0 (by definition of the F-distribution; [35]).

Hence the power is given by:

$$\mathcal{P} = \Pr(F \geq F_{1-\alpha} \mid \nu_1, \nu_2, \delta)$$
$$= 1 - F(F_{1-\alpha;1,n-1}; 1, n-1, t^2). \quad (12)$$

R example 5.1: Find the power of a two-sample *t*-test to detect an effect with $t = 1.96, P \leq \alpha = 0.05, n = 100$. Assume two samples of size $n/2 = 50$. Determine the equivalent Bayes factor using the Spiegelhalter and Smith Bayes factor.

```
> options(digits=3, width=65)
> library(ldDesign)
> t.value <- 1.96; alpha <- 0.05; n <- 100
```

Designing a GWAS: Power, Sample Size, and Data Structure 49

```
> power <- 1 - pf(qf(1-alpha,1,n-1),1,n-1,ncp=t.
value^2)
> power
```

[1] 0.492

```
> f.value <- t.value ^2
> f.value
```

[1] 3.84

```
> B <- SS.oneway.bf(group.sizes=c(50,50),Fstat=f.
value)
> B
```

[1] 1.12

Note:

1. The *F*-value is the square of the *t*-value.
2. The power is 0.49 to obtain $P \leq \alpha = 0.05$.
3. The corresponding Bayes factor is 1.12 representing very weak evidence. In a typical hypothesis testing example we would like to get a Bayes factor of 10 or preferably 20. The following examples illustrate how to achieve this.

R example 5.2: Find the power of a *t*-test with $t = 1.96$, $n = 100$, to detect an effect with Bayes factor $B \geq 20$. Also try a sample size of $n = 1,000$. Use the function oneway.bf.alpha.required () to find the critical alpha value corresponding to $B = 20$ for each sample size.

```
> library(ldDesign)
> source("functions.R")
> # calculations with n=100
> t.value <- 1.96; alpha <- 0.05; n <-100
> # find the value of alpha required for the given B,n
> alphac <- oneway.bf.alpha.required(n=n, group.sizes=c
(0.5,0.5)*n,
+                                                   B=20)
> alphac
      100
   0.0022
> power <- 1 - pf(qf(1-alphac,1,n-1),1,n-1,ncp=t.
value^2)
> power
      100
   0.125
> # calculations with n=1000
> n <-1000
> # calculate effect size in units of sigma=1
```

```
> b <- 1.96/sqrt(100)
> b
```

[1] 0.196

```
> # calculate t-value for the given sample size
> t.value <- b*sqrt(n)
> alphac <- oneway.bf.alpha.required(n=1000, group.
sizes=c(0.5,0.5)*n,
+                                                B=20)
> alphac
     1000
0.000565
> power <- 1 - pf(qf(1-alphac,1,n-1), 1, n-1, ncp=t.
value^2)
> power
   1000
0.997
```

Note:

1. $n = 100$ is not sufficient, with negligible power.
2. $n = 1{,}000$ is more than sufficient, with a power 0.997 to obtain $B \geq 20$.
3. The Bayesian power calculation in this case uses the frequentist power calculation with α chosen corresponding to the specified Spiegelhalter and Smith [25] Bayes factor.
4. The function oneway.bf.alpha.required() uses interpolation to find the value of α needed for given Bayes factor, n, and power. When increasing n, the value of α required for a given Bayes factor decreases.
5. We could try various values of n to solve for the sample size required for a given power. An iterative method is demonstrated in the next example.

R example 5.3: Find the sample size needed to detect the same size of effect as in R Example 5.1 with power 0.8 and Bayes factor 20.

```
> # effect size in units of sigma=1
> b <- 1.96/sqrt(100)
> # intitial guess
> n1 <- 400
> for(iter in 1:5){
+   alphac <- oneway.bf.alpha.required(n1,group.sizes=
                                       c(0.5,0.5)*n1,
                                       B=20)
+   n1 <- find.n(b=b, n.min=100, n.max=1000,
              group.proportions=c(0.5,0.5), alpha=alphac,
              n.interp=20,power=0.8)
```

Designing a GWAS: Power, Sample Size, and Data Structure

```
+    B1<-oneway.bf.alpha(n=n1,group.sizes=c(0.5,0.5)
                          *n1,alpha=alphac)
+    cat("alpha =",alphac,"n =",n1,"B =",B1,"\n")
+}
alpha = 0.000944   n = 454   B = 18.7
alpha = 0.000878   n = 459   B = 19.9
alpha = 0.000873   n = 459   B = 20
alpha = 0.000873   n = 459   B = 20
alpha = 0.000873   n = 459   B = 20

> n1

[1] 459
```

Note:

1. The helper function `find.n()` uses interpolation to find the value of n needed for given, α, and power. For a given Bayes factor the value of α changes as the sample size is increased, hence the function is called iteratively.

2. For design of association studies, the functions `ld.design()` and `cc.design()` use interpolation similarly.

The helper function `find.n()` is defined as:

```
> find.n<-function(b,n.min,n.max,group.proportions=
                   c(0.5,0.5),
+                  alpha,n.interp=20,power=0.8){
+ log10.ns   <-   seq(log10(n.min),   log10(n.max),
length=n.interp)
+ ns <- 10^log10.ns
+ t.values <- b*sqrt(ns)
+ powers <- 1 -pf(qf(1-alpha,1,ns-1),1,ns-1,ncp=t.
values^2)
+ log10.n1 <-approx(powers,log10.ns,xout=power)$y
+ n1 <- 10^log10.n1
+ n1
+ }
```

R example 5.4: Find the sample size needed for a t-test to detect an effect explaining 1 % of the variance with a Bayes factor 20 or more with power 0.8. Also find the sample size needed to detect the effect with a Bayes factor 1 million with power 0.8.

```
> # assume sigma=1
> Vb <- 0.01
> # H0: y = mu + e, e ~ N(0,1)
> # H1: y = mu + (I(group=1) - I(group=2))b + e
> b <- sqrt(Vb)
> # sample size for B=20
> n1 <- 400
> for(iter in 1:5){
```

```
+     alphac <- oneway.bf.alpha.required(n1,group.sizes=
c(0.5,0.5)*n1,
+                                          B=20)
+     n1 <- find.n(b=b, n.min=100, n.max=10000,
                   group.proportions = c(0.5,0.5),
                   alpha=alphac,n.interp=20,power=0.8)
+     B1 <- oneway.bf.alpha(n=n1, group.sizes=c(0.5,0.5)
                               *n1,
+                              alpha=alphac)
+     cat("alpha =", alphac, "n =", n1, "B =",B1,"\n")
+ }
alpha = 0.000944   n = 1736   B = 9.35
alpha = 0.000418   n = 1931   B = 18.9
alpha = 0.000394   n = 1946   B = 19.9
alpha = 0.000393   n = 1947   B = 20
alpha = 0.000392   n = 1947   B = 20

> n1

[1] 1947
> # sample size for B=1e6
> n1 <- 4000
> for(iter in 1:5){
+     alphac <- oneway.bf.alpha.required(n1,group.sizes=
                                         c(0.5,0.5)*n1,
+                                B=1e6,min.F=10, max.F=40)
+     n1 <- find.n(b=b, n.min=100,n.max=10000,
+                  group.proportions=c(0.5,0.5),
+                  alpha=alphac, n.interp=20, power=0.8)
+     B1 <- oneway.bf.alpha(n=n1, group.sizes=c(0.5,0.5)
+                              *n1,
+                              alpha=alphac)
+     cat("alpha =", alphac, "n =", n1, "B =",B1,"\n")
+ }
alpha = 3.43e-09   n = 4616   B = 929360
alpha = 3.18e-09   n = 4630   B = 998355
alpha = 3.18e-09   n = 4630   B = 1e+06
alpha = 3.18e-09   n = 4631   B = 1e+06
alpha = 3.18e-09   n = 4631   B = 1e+06

> n1

[1] 4631
```

2.1.1 The Calibration Problem for p-Values

Despite the problems discussed above it was thought that p-values could still be used if appropriately "calibrated." The calibration problem for p-values means calibrating p-values to an interpretable strength of evidence, or choosing a threshold α small enough so that we can ensure sufficiently strong evidence if $P < \alpha$. If we knew what the optimal threshold, α, to use was, in any situation, p-values would often suffice since with any other measure of evidence some threshold would have to be chosen. However the optimal threshold to use is not clear.

Multiple comparisons does not solve the problem since, after adjusting for multiple comparisons, there is no reason that $P < 0.05$ genome-wise is the optimal, or even a good, strategy. For example, using a comparison-wise error rate $\alpha = 0.05$ is clearly too permissive. But is a genome-wise error rate too conservative?

There has been a de facto standard threshold that was $\alpha = 5 \times 10^{-7}$ at the time of WTCCC [7] and DGI [8], justified using a quasi Bayesian argument:

$$B = \frac{\Pr(E \mid H_1)}{\Pr(E \mid H_0)}, \qquad (13)$$

where E is the event $\{P < \alpha\}$ (WTCCC 2007). This corresponds to a Bayes factor of 10^6 if $\alpha = 5 \times 10^{-7}$ and $\mathcal{P} = 0.5$. This only applies, however, to the average in some sense of effects with $P \leq \alpha$ and not for effects with $P \approx \alpha$. We cannot use this formula for effects with $P = \alpha$.

Subsequently, a threshold of $\alpha = 5 \times 10^{-8}$ has become the new de facto standard. This is based on $\alpha = 0.05$ adjusted for the equivalent number of independent comparisons corresponding to a dense set of SNP markers [9].

The *false discovery rate* (FDR) has been proposed as an alternative to controlling type I errors [36–39]. The FDR was proposed as a measure intermediate between using comparison-wise and genome-wise significance levels. While preferable to the *p*-value, since it does give a measure approximating the probability of being wrong, the FDR is in essence a poor man's Bayesian posterior probability. The FDR is (modulo "estimating" the prior) approximately the average posterior probability of H_{0_i} of a set of effects:

$$\text{FDR} \sim \frac{1}{k} \sum_{i=1}^{k} \Pr(H_{0_i} \mid y). \qquad (14)$$

The individual posterior probabilities are more informative, and, according to Fisher we *must* use all the information. Moreover there are no power calculations we know of that are given in terms of FDR.

3 Bayesian Power Calculations

We propose using the Bayes factor as the measure of strength of evidence in the data, and give power calculations to ensure that effects are detected with at least a given Bayes factor, i.e., $B \geq B_c$, where B_c is chosen sufficiently large to overcome the low prior odds in GWAS. We have developed methods for quantitative traits and case–control studies [14, 16], and give methods for TDT and S-TDT tests in this chapter.

Using Bayes factors solves the calibration problem because Bayes factors have a natural interpretation as strength of evidence. Calculating the corresponding p-value solves the calibration problem for p-values for a given problem, but the critical p-value depends on experimental design, test setup, and sample size, suggesting that no single value of α will be sufficient.

3.1 Criteria for Robust Detection of Effects

In GWAS, prior odds are very low. Finding loci associated with variation in a trait is like finding a needle in the haystack. The difficult problem is finding the needle, not detecting the needle once we know where it is.

For effects to be robustly detected we require good power to detect effects with a sufficiently strong evidence, i.e., high Bayes factor, to overcome the low prior odds and obtain respectable posterior odds.

A consequence of effects being robustly detected is that we would expect to detect the effects fairly consistently in replicate experiments of similar power, and the estimated effects would be stable. The consequence of effects not being robustly detected is that effects would often not be detected and/or the estimated size of effects would tend to decrease, often substantially, on replication (unless using Bayesian model averaging). As noted in Subheading 1, selection bias, where the selected effects tend to be overestimated, is a common problem unless the power is good to detect the true size of effects.

Given prior odds we can determine the Bayes factor required from Eq. 6. For example if the prior odds are 1:50,000 and the posterior odds desired are 20:1 the Bayes factor required is

$$B = \frac{20}{1/50,000} = 20 \times 50,000 = 10^6. \tag{15}$$

3.2 Alternative Forms of Bayes Factors and Closed Form Expressions for PowerCalculations

Spiegelhalter and Smith [25] consider Bayes factors for linear models with low prior information, obtaining, for a one-way analysis of variance model:

$$B = \left[\frac{1}{2}\frac{(m+1)}{n}\prod_{i=1}^{m} n_i\right]^{-1/2}\left[1 + \frac{(m-1)}{(n-m)}F\right]^{n/2}, \tag{16}$$

3.2.1 Spiegelhalter and Smith Bayes Factor

where m is the number of groups, n_i the number in each group, n is the total sample size, and F is the classical F-value. This is implemented in SS.oneway.bf() function in ldDesign.

This form of the Bayes factor is particularly convenient, because it links directly to the F-statistic used in existing power calculations. To detect an association with power $\mathcal{P}$ to achieve a given Bayes factor B_c we solve for F in Eq. 16, and select a design using the existing deterministic power calculation. The solution, F_c, for F is given by:

Designing a GWAS: Power, Sample Size, and Data Structure   55

$$F_c = \frac{n-m}{m-1}\left(\left[B_c^2 \frac{1}{2}\frac{(m+1)}{n}\prod_{i=1}^m n_i\right]^{1/n} - 1\right) \qquad (17)$$

and the power is given by Eq. 9:

$$\mathcal{P} = 1 - F(F_c; \nu_1, \nu_2, \delta), \qquad (18)$$

where $\nu_1 = 2$ is the number of degrees of freedom being tested, ν_2 is the residual error d.f., and δ is the non-centrality parameter.

In our experience, the Spiegelhalter and Smith Bayes factor is approximately equivalent to assuming a prior distribution for the effect being tested with information equivalent to a sample of size 1. The approximate Bayes factors described in the next subsection have the advantage that we can choose the prior variance.

R example 5.5: Consider the Oats split plot dataset [40] in the R nlme package [41]. From the R help page:

> The experimental units were arranged into six blocks, each with three whole-plots subdivided into four subplots. The varieties of oats were assigned randomly to the whole-plots and the concentrations of nitrogen to the subplots. All four concentrations of nitrogen were used on each whole-plot.

Find the multi-stratum ANOVA table, the *F*-statistic and *p*-value for testing for variety. Ignoring the blocking structure use the function SS.oneway.bf() to find the corresponding Bayes factor.

Note: For purposes of illustration, we assume that the blocks are subdivided into plots corresponding to levels of nitrogen, as might be the case if the experiment was designed to compare varieties. Conveniently, the effects of Variety in this model have *P* slightly less than 0.05, so we can see what sort of Bayes factor this corresponds to.

```
> library(ldDesign)
> library(nlme)
> options(digits=3,width=65,show.signif.stars=FALSE)
> data(Oats)
> Oats$N <- ordered(factor(Oats$nitro))
> Oats$plot <- factor(Oats$nitro)
> aov1 <- aov(yield ~ Variety + N + Error(Block/
plot),data=Oats)
> summary(aov1,split=list(N=list(L=1,Dev=2:3)))

Error: Block
          Df Sum Sq Mean Sq F value Pr(>F)
Residuals  5  15875    3175

Error: Block:plot
```

	Df	Sum Sq	Mean Sq	F value	Pr(>F)
N	3	20020	6673	55.98	2.2e-08
N:L	1	19536	19536	163.88	1.8e-09
N:Dev	2	484	242	2.03	0.17
Residuals	15	1788	119		

Error: Within

	Df	Sum Sq	Mean Sq	F value	Pr(>F)
Variety	2	1786	893	3.28	0.046
Residuals	46	12516	272		

```
> # means for Variety, balanced data
> with(Oats, tapply(yield, Variety, mean))
```

Golden Rain	Marvellous	Victory
104.5	109.8	97.6

```
> table(Oats$Variety)
```

Golden Rain	Marvellous	Victory
24	24	24

```
> # ignoring the blocking structure to find SS Bayes
factor
> SS.oneway.bf(group.sizes=c(24,24,24), Fstat=3.28)
[1] 1.34

> # alternative using lme
> fit1 <- lme(yield ~ Variety+nitro, random=~1/Block/
plot, data=Oats)
> summary(fit1)
```

Linear mixed-effects model fit by REML
  Data: Oats

AIC	BIC	logLik
600	613	-294

Random effects:
  Formula: ~1/Block/plot | Block

	(Intercept)	Residual
StdDev:	15.7	15.3

Fixed effects: yield ~ Variety+nitro

	Value	Std.Error	DF	t-value	p-value
(Intercept)	82.4	7.52	63	10.96	0.000
VarietyMarvellous	5.3	4.42	63	1.20	0.236
VarietyVictory	-6.9	4.42	63	-1.55	0.125
nitro	73.7	8.07	63	9.12	0.000

  Correlation:

	(Intr)	VrtyMr	VrtyVc
VarietyMarvellous	-0.294		
VarietyVictory	-0.294	0.500	
nitro	-0.322	0.000	0.000

```
Standardized Within-Group Residuals:
    Min      Q1     Med      Q3     Max
-1.8407 -0.8085  0.0402  0.7048  2.2215

Number of Observations: 72
Number of Groups: 6

> anova(fit1)
            numDF denDF F-value p-value
(Intercept)     1    63   245.1  <.0001
Variety         2    63     3.8  0.0275
nitro           1    63    83.2  <.0001

> SS.oneway.bf(group.sizes=c(24,24,24),Fstat=3.8)
[1] 2.2
```

Note:

- The Bayes factors are only 1.3 or 2.2, despite *p*-values of 0.05 and 0.03 respectively. This means that the data are only about two times more likely under H_1 than under H_0, i.e., the evidence for an effect of Variety is very weak.
- The *F*-statistic and Bayes factor from the lme analysis are slightly higher due to recovery of inter-block information.
- The evidence for an effect of nitrogen is stronger ($P = 6.5 \times 10^{-5}$). We leave it as an exercise for the reader to calculate the corresponding Bayes factor.

3.2.2 Approximate Bayes Factors

Ball [16, 42] use the Savage–Dickey density ratio and an asymptotic approximation to the posterior to derive approximate Bayes factors for the S-TDT test and a test for the log-odds ratio in genome-wide case–control studies.

The Savage–Dickey density ratio [43] estimate of the Bayes factor for comparing nested models (e.g., $H_0 : \theta = 0$ vs. $H_1 : \theta \neq 0$) is given by the ratio of prior to posterior densities at 0:

$$B = \frac{\pi(\theta = 0)}{g(\theta = 0 \mid y)}, \qquad (19)$$

where θ is the parameter being tested, and $\pi(\theta), g(\theta \mid y)$ are the prior and posterior distributions for θ. This formula is a consequence of Bayes' theorem. The formula is exact in nested models (i.e., $\theta = 0$ vs. $\theta \neq 0$) with common prior for "nuisance parameters" (i.e., parameters not being tested, not shown), if we integrate over nuisance parameters to obtain $\pi(\theta), g(\theta \mid y)$. The formula is approximate if we condition on nuisance parameters.

A generic form of the approximate Bayes factor for asymptotically normally distributed test statistics can be obtained using the Savage–Dickey density ratio as follows. Suppose Z is a test statistic with sampling distribution

$$Z \mid z \sim N\left(z, \frac{1}{n}\right) \tag{20}$$

for an i.i.d. sample of size n. Suppose we have a prior distribution

$$\pi(Z) \sim N\left(0, \frac{1}{a}\right). \tag{21}$$

Then the posterior of z is

$$g(z \mid Z) \sim N\left(\frac{n}{n+a} Z, \frac{1}{n+a}\right). \tag{22}$$

The Bayes factor for testing $H_1 : z \neq 0$ vs. $H_0 : z = 0$ is given by

$$B = \frac{\pi(\theta = 0)}{g(\theta = 0 \mid y)} \tag{23}$$

$$= \frac{\sqrt{a}}{\sqrt{n+a}} \exp\left(\frac{n^2 Z^2}{2(n+a)}\right). \tag{24}$$

Note:

- Here the prior precision parameter a represents the amount of information in the prior in sample units, e.g., $a = 1$ represents a prior equivalent to a single sample point. This often represents a good conservative default. However, we can and should use an informative prior for the parameter of interest, where prior information is available. For example if θ is the parameter of interest with prior

$$\pi(\theta) \sim N(0, \sigma_\theta^2) \tag{25}$$

and sampling distribution

$$\hat{\theta} \mid \theta \sim N\left(\theta, \frac{c}{n}\right) \tag{26}$$

then letting $Z = \hat{\theta}/\sqrt{c}, z = \theta/\sqrt{c}$ we obtain

$$\pi(Z) \sim N\left(0, \frac{\sigma_\theta^2}{c}\right) = N\left(0, \frac{1}{a}\right) \tag{27}$$

$$Z \mid z \sim N\left(z, \frac{1}{n}\right), \tag{28}$$

where $a = c/\sigma_\theta^2$.

- Approximate Bayes factors of a form similar to Eq. 23 were first derived in [42], and used for an S-TDT test (cf. Subheading 4.3), and subsequently by [44].
- Approximate Bayes factor formulae can also be derived using the method of Bayes factors based on test statistics [45, 46]).

3.2.3 The Schwarz BIC Criterion

For a linear model, the BIC criterion can be defined as:

$$\text{BIC} = n \log(1 - R^2) + p \log n, \qquad (29)$$

where p is the number of parameters in the model excluding the intercept. With this definition the marginal probability of the data is proportional to $\exp(-\text{BIC}/2)$ and the Bayes factor for comparing two models is approximately $\exp(-\Delta\,\text{BIC}/2)$ [47]. Using the BIC criterion as an approximation to the Bayes factors leads to an expression for the posterior probability of alternative models

$$\Pr(\mathcal{M}_\gamma) \propto \exp(-\text{BIC}/2) \times \text{prior} \qquad (30)$$

enabling determination of approximate probabilities $\Pr(\mathcal{M}_\gamma)$ for sets of alternative models [13, 48].

Note:

- The BIC criterion is used for comparing models but is defined only up to an arbitrary additive constant. Using Eq. 29 sets $\text{BIC} = 0$ for the null model. The choice of constant affects all models equally and hence does not effect Bayes factors.
- The first term in Eq. 29 is, up to a constant, minus twice the log likelihood.
- Often the BIC criterion is used to select models, with the implicit assumption that all models are a priori equally likely. In gene mapping models this is generally not the case, and the BIC criterion on its own does not give a good approximation to posterior probabilities of models; it is important to take into account the prior probability for models ([13, 49]; Eq. 30)
- The BIC criterion does not always give a good approximation to Bayes factors, e.g.,
  - In different asymptotic frameworks, e.g., comparing multiple groups when the group size is fixed and the number of groups, p, tends to infinity with n [50, 51]; however the Spiegelhalter and Smith Bayes factor (Eq. 16) does have the correct asymptotic behavior in this situation);
  - In non-regular models where a tubular neighborhood of a sub-model is not locally Euclidean [52];
  - For non-i.i.d. data, e.g., mixed models, the appropriate n in the BIC may not be the total sample size [53, 54].

Broman [55] and Broman and Speed [56] used an adjustment to the BIC criterion. They proposed an additional multiplicative penalty factor δ for QTL mapping:

$$\text{BIC} = n \log(1 - R^2) + \delta p \log n \qquad (31)$$

suggesting $\delta = 2$ or 3. However when taking into account prior probabilities for QTL we found the additional penalty δ was not necessary [13, 42, 57].

Using the BIC criterion is approximately equivalent to assuming a prior centered at 0 with information equivalent to a single sample point for the effect being tested. This is often a reasonable conservative default; however it is desirable to incorporate prior information where possible, which can result in higher Bayes factors and posterior probabilities. To take into account prior knowledge of the variance of QTL effects we propose a different adjustment to the BIC criterion. Comparing Eqs. 29 and 23, noting that minus twice the log likelihood ratio terms are respectively, $n \log(1 - R^2)$ (up to a constant) and $n Z_n^2 / 2$ suggests that the BIC criterion can be adjusted as follows:

$$\mathrm{BIC}_a = \frac{n^2}{(n+a)} \log(1 - R^2) + p \log \frac{n+a}{a}. \qquad (32)$$

Note that when $a = 1$ this is approximately

$$\begin{aligned}\mathrm{BIC}_1 &= \frac{n^2}{(n+1)} \log(1 - R^2) + p \log \frac{n+1}{1} \\ &= n \log(1 - R^2) + p \log n + O(1/n) \\ &\approx \mathrm{BIC} \end{aligned} \qquad (33)$$

i.e., the BIC criterion is equivalent to BIC_1 to order $O(1/n)$.

4 Power Calculations and Sample Size Determination for Various Data Structures

In this section we consider power and sample size calculations for various data structures including: independent population samples; case–control studies, where similar sizes samples or cases and controls are taken from the population; family-based tests for disease (presence/absence) including the transmission disequilibrium test (TDT), sib transmission disequilibrium test (S-TDT) test, sib disequilibrium test (SDT), and more general family-based tests for quantitative traits, which are more robust against spurious associations due to population structure.

4.1 GWAS for Quantitative Traits

We consider the power of bi-allelic marker loci to detect bi-allelic QTL affecting a quantitative trait. In general, we don't observe the QTL locus but one or more markers in LD with the QTL. Factors affecting power include marker and QTL allele frequencies p, q, QTL heritability (or percent of variation explained) h_Q^2, linkage disequilibrium coefficient D, and Bayes factor required B, and genetic model (e.g., dominant, recessive, additive).

Table 1
Expected genotypic frequencies and phenotypic values from [58] for QTL/marker genotype combinations

Marker	Frequencies (f_{ij})			Expected Value
QTL	MM	Mm	mm	
AA	$p^2 Q^2$	$2p(1-p)QR$	$(1-p)^2 R^2$	d
Aa	$2p^2 Q(1-Q)$	$2p(1-p)(Q+R-2QR)$	$2(1-p)^2 R(1-R)$	h
aa	$p^2(1-Q)^2$	$2p(1-p)(1-Q)(1-R)$	$(1-p)^2(1-R)^2$	$-d$

Marker genotypes are MM, Mm, mm;QTL genotypes are AA, Aa, aaReproduced with permission required from Genetics 2005

4.1.1 Genetic Model

We will assume a bi-allelic marker with alleles M, m and a bi-allelic QTL with alleles A, a. Following [58], let p, q be the probability of M, A respectively. In the bi-allelic case, linkage disequilibrium is specified by a single coefficient, D, such that the joint probabilities of alleles are given by:

$$\Pr(A, M) = \Pr(A)\Pr(M) + D = qp + D, \quad (34)$$

$$\Pr(A, m) = \Pr(A)\Pr(m) - D = q(1-p) - D \quad (35)$$

(cf. [59]). It follows that the conditional probabilities are given by

$$Q = \Pr(A \mid M) = q + D/p, \quad (36)$$

$$R = \Pr(A \mid m) = q - D/(1-p). \quad (37)$$

Genotypic frequencies and phenotypic expected values are as in Table 1.

4.1.2 Statistical Model

Luo [58] uses the statistical models

$$y = \mu + b_i + \omega_{ij} + \epsilon_{ijk}, \quad (38)$$

$$y = \mu + b_i + e_{ij}, \quad (39)$$

where i indexes marker genotype and j indexes QTL genotype, and k indexes units within genotype combinations. The first model parameterizes effects in terms of QTL genotypes; the second model (analysis model) parameterizes effects in terms of marker genotypes only. Since QTL genotypes are unobserved, the analysis model parameterizes effects in terms of marker genotypes. The parameters b_i and ω_{ij} are defined so that the correct expectations are obtained, e.g.,

$$\begin{aligned}\mu = E(y) &= E(y\mid AA)\Pr(AA) + E(y\mid Aa)\Pr(Aa) + E(y\mid aa)\Pr(aa)\\ &= q^2 d + 2q(1-q)h - (1-q)^2 d\\ &= (2q-1)d + 2q(1-q)h,\end{aligned} \qquad (40)$$

$$\begin{aligned}\mu + b_2 &= E(y\mid Mm)\\ &= d\Pr(AA\mid Mm) + h\Pr(Aa\mid Mm) - d\Pr(aa\mid Mm)\\ &= dQR + h(Q + R - 2QR) - d(1-Q)(1-R)\end{aligned} \qquad (41)$$

$$\therefore b_2 = \frac{[D(1-2p)]d + d[2D + (1-2p)(1-2q)]h}{p(1-p)}. \qquad (42)$$

(cf. [58, Eq. 3.2]). Solving for ω_{23} gives:

$$\mu + b_2 + \omega_{23} = E(y\mid Mm, aa) = -d \qquad (43)$$

$$\therefore \omega_{23} = -d - \mu - b_2 \qquad (44)$$

$$\begin{aligned}&= -d - (2q-1)d - 2q(1-q)h\\ &\quad - \frac{D(1-2p)d + [2D^2 + D(1-2p)(1-2q)]h}{p(1-p)}\end{aligned} \qquad (45)$$

[14, Appendix I].

The power calculation implemented in the R package ldDesign [16, 60] uses

$$B \approx \left[4n^2 p^3 (1-p)^3\right]^{-1/2} \left[1 + \frac{2}{(n-3)} F\right]^{n/2} \qquad (46)$$

obtained by substituting expected values $E(n_1) = np^3$, $E(n_2) = 2np(1-p)$, $E(n_3) = n(1-p)^2$ for the group sizes in Eq. 16. The critical F-value is given by:

$$F_c = \frac{n-3}{2}\left(\left[4n^2 p^3 (1-p)^3 B_c^2\right]^{1/n} - 1\right) \qquad (47)$$

and the power is given by

$$\mathcal{P} = 1 - F(F_c; \nu_1, \nu_2, \delta), \qquad (48)$$

where $\nu_1 = 2$ are the degrees of freedom being tested, ν_2 are the residual d.f., and δ is the non-centrality parameter.

Under the alternative hypothesis, F is distributed as a non-central F-distribution with non-centrality parameter δ given by:

$$\delta = \frac{\mathrm{EMS}_b}{\mathrm{EMS}_w} \times \frac{\nu_1(\nu_2 - 1)}{\nu_2} - \nu_1, \qquad (49)$$

Table 2
ANOVA table for single marker analysis

	d.f.	SS	MS	F
Between marker classes	$\nu_1 = 2$	SS_b	$MS_b = SS_b/\nu_1$	$F = MS_b/MS_w$
Within marker classes	$\nu_2 = n - 3$	SS_w	$MS_w = SS_w/\nu_2$	

Reproduced with permission required from Genetics 2005

Table 3
p-Values corresponding to various Bayes factors, for testing for linkage disequilibrium between a bi-allelic marker and QTL, for marker minor allele frequency p = MAF = 0.5, 0.05

n	\multicolumn{6}{c}{Bayes factor (B)}
	1	100	1,000	10^4	10^5	10^6
MAF = 0.5						
200	2.1×10^{-2}	2.3×10^{-4}	2.4×10^{-5}	2.4×10^{-6}	2.5×10^{-7}	2.6×10^{-8}
1,000	4.1×10^{-3}	4.1×10^{-5}	4.2×10^{-6}	4.2×10^{-7}	4.2×10^{-8}	4.2×10^{-9}
5,000	8.0×10^{-4}	8.1×10^{-6}	8.1×10^{-7}	8.1×10^{-8}	8.1×10^{-9}	8.1×10^{-10}
25,000	1.6×10^{-4}	1.6×10^{-6}	1.6×10^{-7}	1.6×10^{-8}	1.6×10^{-9}	1.6×10^{-10}
125,000	3.2×10^{-5}	3.2×10^{-7}	3.2×10^{-8}	3.2×10^{-9}	3.2×10^{-10}	3.2×10^{-11}
MAF = 0.05						
200	0.25	2.6×10^{-3}	2.7×10^{-4}	2.8×10^{-5}	2.9×10^{-6}	3.0×10^{-7}
1,000	0.05	4.9×10^{-4}	5.0×10^{-5}	5.0×10^{-6}	5.0×10^{-7}	5.1×10^{-8}
5,000	0.001	9.7×10^{-5}	9.7×10^{-6}	9.7×10^{-7}	9.8×10^{-8}	9.8×10^{-9}
25,000	0.002	1.9×10^{-5}	1.9×10^{-6}	1.9×10^{-7}	1.9×10^{-8}	1.9×10^{-9}
125,000	0.0004	3.9×10^{-6}	3.9×10^{-7}	3.9×10^{-8}	3.9×10^{-9}	3.9×10^{-10}

where EMS_b, EMS_w are the expected values of the mean squares between and within marker classes (cf. Table 2; [58, equation 2] and [35, p. 189]) (Tables 3 and 4). The expected mean squares are calculated in terms of expressions for b_i, ω_{ij} and genotype frequencies (cf. Eqs. 41, 45, and Table 1).

R example 5.6: Recent meta-analyses for human height with $n \approx 40{,}000$ found 54 putative associations putatively explaining 2–3 % of the variation in human height. Find the power of the study to detect the average of such effects, i.e., an effect explaining 0.05 % of the variation with a Bayes factor of 20. Find the sample sizes needed to detect the effect with Bayes factors of 10^5, 10^6. Find the Bayes factors corresponding to the recent de facto standards of $\alpha = 5 \times 10^{-7}, 5 \times 10^{-8}$. Find the α thresholds needed to obtain Bayes factors of 10^5, 10^6. Assume the most optimistic situation of bi-allelic markers and QTL with allele frequencies of 0.5 in complete linkage disequilibrium.

Table 4
Comparison with results from [58]

Pop.	n	p	q	D	h_Q^2	ϕ	$\mathcal{P}_{0.05}$	B	$n_{B_{20}}$	$n_{B_{1,000}}$	$n_{B_{10^6}}$
1	100	0.5	0.5	0.1	0.1	0.0	0.18	0.88	1521	2138	3173
2	200	0.5	0.5	0.1	0.1	0.0	0.34	0.42	1521	2138	3173
3	200	0.5	0.5	0.2	0.1	0.0	0.91	0.42	309	464	719
4	200	0.5	0.5	0.1	0.2	0.0	0.62	0.42	694	1003	1523
5	200	0.5	0.5	0.1	0.1	0.5	0.31	0.42	1693	2381	3515
6	200	0.5	0.5	0.1	0.1	1.0	0.25	0.42	2188	3047	4498
7	200	0.3	0.3	0.1	0.1	0.0	0.46	0.54	993	1433	2163
8	200	0.7	0.7	0.1	0.1	0.0	0.46	0.54	993	1433	2163
9	200	0.3	0.5	0.1	0.1	0.0	0.39	0.54	1212	1739	2605
10	200	0.5	0.3	0.1	0.1	0.0	0.39	0.42	1248	1774	2636
11	200	0.4	0.6	0.1	0.2	1.0	0.45	0.45	1033	1479	2213
12	200	0.6	0.4	0.1	0.2	1.0	0.54	0.45	818	1177	1782

Results are shown for the 12 example populations (cf. [58, Tables 2, 3]) with sample size n, marker and QTL allele frequencies p and q respectively, linkage disequilibrium D, QTL heritability h_Q^2, and dominance ratio ϕ. $\mathcal{P}_{0.05}$ is the power to detect an effect with $\alpha = 0.05$, B is the corresponding Bayes factor, and $n_{B_{20}}, n_{B_{1,000}}, n_{B_{10^6}}$ are the sample sizes required to achieve a Bayes factor of 20, 1, 000, 10^6 with power 0.8 respectively Reprinted with permission required from Genetics 2005//024752.

```
> library(ldDesign)
> options(digits=3, width=65)
> ld.power(B=20, D=0.25, p=0.5, q=0.5, n=40000,
h2= 0.0005, phi=0)
      n power
[1,] 40000 0.359 attr(,"parms")
           p            q                  D         h2
      0.5000       0.5000             0.2500     0.0005
         phi           Bf     missclass.rate
      0.0000      20.0000             0.0000

> ld.design(B=1e5, D=0.25, p=0.5, q=0.5, nmin=40000,
            nmax=120000,
+     h2=0.0005, phi=0, power=0.8)

[1] 108307
>   ld.design(B=1e6,D=0.25,p=0.5,q=0.5,   nmin=1e5,
            nmax=120000,
+     h2=0.0005, phi=0, power=0.8)

[1] 119102
```

Designing a GWAS: Power, Sample Size, and Data Structure 65

```
> oneway.bf.alpha(n=40000, group.sizes=c(0.25,0.5,
                  0.25)*40000,
+        alpha=5e-7)
[1] 200
> oneway.bf.alpha(n=40000,group.sizes=c(0.25,0.5,
                  0.25)*40000,
+        alpha=5e-8)
[1] 2003
>   oneway.bf.alpha.required(n=40000,group.sizes=c
                  (0.25,0.5,0.25)*40000,
+          B=1e5)
40000
1e-09
>   oneway.bf.alpha.required(n=40000,group.sizes=c
                  (0.25,0.5,0.25)*40000,
+          B=1e6)
40000
1e-10
```

Note:

- The thresholds $\alpha = 5 \times 10^{-7}$, 5×10^{-8} correspond to Bayes factors of around 200, 2,000 respectively. If the effects were detected at $\alpha = 5 \times 10^{-8}$ and replicated at $\alpha = 5 \times 10^{-7}$ we would obtain a Bayes factor of the order required to detect genomic effects. The sample size $n = 40,000$ has power only 0.35 to detect these effects at $B = 20$.
- The de facto standard requires, however, only the combined p-value from all detection and replication samples to be less than 5×10^{-8}.
- Clearly these effects were not robustly detected.
- It is not currently possible to specify the prior variance for quantitative traits in the ldDesign functions ld.design() and ld.power(). It is in principle possible to use the approximate Bayes factors from Subheading 2, as is done for case–control studies in the next subsection. This may be implemented in a future version of ldDesign.

4.2 Genome-Wide Case–Control Studies

Following [16], we consider the power of bi-allelic marker loci to detect bi-allelic QTL affecting a quantitative trait. In general we don't observe the QTL locus but one or more markers in LD with the QTL. Factors affecting power include marker and QTL allele frequencies p, q, QTL effect (odds ratio(s)), linkage disequilibrium coefficient D, Bayes factor required B, and genetic model (e.g., dominant, recessive, additive, general). We consider the dominant

Table 5
Counts and row probabilities for a case–control study with two genotypic classes

	Counts		Row probabilities	
	$X=0$	$X=1$	$X=0$	$X=1$
Control	n_{11}	n_{12}	p_{11}	p_{12}
Case	n_{21}	n_{22}	p_{21}	p_{22}

or recessive model with two genotypic classes, denoted by $X = 0, X = 1$. For additive and general models see [16] and the ldDesign package [16, 60].

Contingency tables for the observed counts and corresponding row probabilities are shown in Table 5, where p_{ij} are the probabilities *within rows*, e.g., $p_{11} = \Pr(X = 0 \mid \text{control})$, $p_{12} = \Pr(X = 1 \mid \text{control})$. For the dominant model: $X = 0$ for genotype aa, $X = 1$ for genotypes Aa, AA. For the recessive model: $X = 0$ for genotypes aa, Aa, $X = 1$ for genotype AA. Let n_{ij} denote the corresponding counts, and n the total sample size.

The log-odds ratio is:

$$\eta = \log p_{11} - \log p_{12} - \log p_{21} + \log p_{22} \quad (50)$$

and is estimated as:

$$\hat{\eta} = \log\left(\frac{n_{11} n_{22}}{n_{12} n_{21}}\right) \quad (51)$$

$$= \log \hat{p}_{11} - \log \hat{p}_{12} - \log \hat{p}_{21} + \log \hat{p}_{22}. \quad (52)$$

The sampling distribution of $\hat{\eta}$ is [61]

$$\hat{\eta} \sim N(\eta, \sigma_n^2) \quad \text{(sampling distn.)}, \quad (53)$$

where

$$\eta = \log p_{11} - \log p_{12} - \log p_{21} + \log p_{22} \quad (54)$$

$$\sigma_n^2 = \frac{1}{n_1 p_{11}} + \frac{1}{n_1 p_{12}} + \frac{1}{n_2 p_{21}} + \frac{1}{n_2 p_{22}}, \quad (55)$$

The normalized test statistic and variance are given by:

$$\sigma_1^2 = n\sigma_n^2 = \frac{1}{c}\left(\frac{1}{p_{11}} + \frac{1}{p_{12}}\right) + \frac{1}{1-c}\left(\frac{1}{p_{21}} + \frac{1}{p_{22}}\right), \quad (56)$$

$$Z_n = \frac{\hat{\eta}}{\sqrt{n\hat{\sigma}_n}} = \frac{\hat{\eta}}{\hat{\sigma}_1} \sim N\left(\frac{\eta}{\sigma_1}, \frac{1}{n}\right) \quad \text{(asymptotic sampling distn.)}, \quad (57)$$

where c is the proportion of controls in the sample.

Designing a GWAS: Power, Sample Size, and Data Structure

The generic form of the approximate Bayes factor is calculated as follows. Assume a prior with information equivalent to a sample points

$$\pi(z) \sim N\left(0, \frac{1}{a}\right) \quad (58)$$

the posterior density is given by:

$$g(z \mid Z_n) \cong N(\frac{n}{n+a} Z_{n'} \frac{n}{n+a}) \quad \text{(Bayes' theorem)} \quad (59)$$

and the Bayes factor is given by:

$$B = \frac{\pi(z=0)}{g(z=0 \mid Z_n)} \quad \text{(Savage–Dickey)}, \quad (60)$$

$$B \approx \frac{\sqrt{a}}{\sqrt{n+a}} \exp\left(\frac{n^2 Z_n^2}{2(n+a)}\right). \quad (61)$$

Let Z_c be the solution to Eq. 61 for Z_n:

$$Z_c^2 = \frac{n+a}{n^2} \log\left(\frac{n+a}{a} B^2\right). \quad (62)$$

The power is given by:

$$\mathcal{P} \approx 1 - \Phi\left(\sqrt{n}\left(Z_c - \frac{\eta}{\sigma_1}\right)\right) \quad (63)$$

To apply Eqs. 62 and 63 we need expressions for σ_1^2 in terms of η, a. To incorporate linkage disequilibrium between the observed and causal loci, we need expressions for the log-odds ratio (denoted by η_x) at the observed locus in terms of the log-odds ratio (η) at the causal locus. Optionally, we need expressions for a in terms of the prior variance, σ_η^2, of η. The required expressions are calculated in terms of the prior variance (σ_η^2) for η, disease prevalence, linkage disequilibrium coefficients, and allele frequencies at the causal locus in [16, Appendix I] and in the `ldDesign` R functions `cc.power()`, `cc.design()`.

R example 5.7: Find the power of a case–control design similar to [7] with 2,000 cases and 3,000 controls to detect an effect with odds ratio 2.0, marker allele frequency $p = 0.3$, and QTL risk allele frequency $q = 0.1$, in additive, dominant, recessive, and general models with a Bayes factor of 10^6 for a disease with prevalence 0.1. Assume a prior with information equivalent to a single sample point. Assume linkage disequilibrium $D' = 0.5$ or $D' = 1$. For the additive and general models assume an odds ratio of 2.0 per copy of the risk allele.

```
> library(ldDesign)
> options(width=65, digits=3)
> cc.power(B=1e6, OR=2.0, Dprime=0.5, p=0.3, q=0.1,
          prevalence=0.1,
```

```
+           n.cases=2000,n.controls=3000,model="addi-
            tive",a=1)
[1] 0.016
>cc.power(B=1e6,OR=2.0,Dprime=1,p=0.3,q=0.1,pre-
            valence=0.1,
+           n.cases=2000,   n.controls=3000,   mod-
            el="additive",a=1)
[1] 0.888
> cc.power(B=1e6, OR=2.0, Dprime=0.5, p=0.3, q=0.1,
            prevalence=0.1,
+           n.cases=2000,   n.controls=3000,   model
            ="dominant",a=1)
[1] 0.000586
>cc.power(B=1e6,OR=2.0,Dprime=1,p=0.3,q=0.1,pre-
            valence=0.1,
+           n.cases=2000,   n.controls=3000,   model
            ="dominant",a=1)
[1] 0.244
> cc.power(B=1e6, OR=2.0, Dprime=0.5, p=0.3, q=0.1,
            prevalence=0.1,
+           n.cases=2000,   n.controls=3000,   model
            ="recessive",a=1)
[1] 5.52e-08
> cc.power(B=1e6, OR=2.0, Dprime=1, p=0.3, q=0.1,
            prevalence=0.1,
+           n.cases=2000,   n.controls=3000,   model
            ="recessive",a=1)
[1] 1.03e-06
> cc.power(B=1e6, OR=c(2.0,2.0), Dprime=0.5, p=0.3,
            q=0.1,
+           prevalence=0.1,  n.cases=2000,  n.con-
            trols=3000,
+           model="general",a=1)
[1] 1.32e-05
> cc.power(B=1e6, OR=c(2.0,2.0), Dprime=1, p=0.3,
            q=0.1,
+           prevalence=0.1,n.cases=2000,n.controls
            =3000,
+           model="general",a=1)
[1] 0.629
```

Note: We see that the WTCCC [7] design can only detect effects with OR = 2, $p = 0.3$, $q = 0.1$ with $B = 10^6$ under favorable conditions: an additive or general model with D' close to 1.

R example 5.8: Assuming linkage disequilibrium given as $r^2 = 0.25$, find the sample size needed to detect the effect in each genetic model with power 0.8. What happens if $r^2 = 0.5$?

```
> options(width=65)
> r.squared <- 0.25
> p <- 0.3
> q <- 0.1
> D <- sqrt(r.squared*p*(1-p)*q*(1-q))
> c(D=D,p=p,q=q)
     D      p      q
0.0687 0.3000 0.1000
> cc.design(B=1e6, OR=2.0, D=D, p=0.3, q=0.1, pre-
          valence=0.1,
+         n.cases=2000,n.controls=3000,model="addi-
          tive",
power=0.8, +    a=1)
Power curve:
     n.controls n.cases power
 [1,]     1255    837 0.100
 [2,]     1333    889 0.125
 [3,]     1416    944 0.156
 [4,]     1503   1002 0.192
 [5,]     1597   1065 0.234
 [6,]     1696   1131 0.282
 [7,]     1801   1201 0.337
 [8,]     1913   1276 0.397
 [9,]     2032   1355 0.461
[10,]     2159   1439 0.529
[11,]     2293   1529 0.597
[12,]     2435   1624 0.665
[13,]     2586   1724 0.730
[14,]     2747   1832 0.790
[15,]     2918   1945 0.842
[16,]     3099   2066 0.886
[17,]     3292   2195 0.922
[18,]     3496   2331 0.949
[19,]     3714   2476 0.968
[20,]     3944   2630 0.982
[21,]     4190   2793 0.990
2779 controls and 1853 cases for power 0.8
     n n.controls n.cases
  4632    2779    1853

> cc.design(B=1e6, OR=2.0, D=D, p=0.3, q=0.1, preva-
          lence =0.1,
+         n.cases=2000, n.controls=3000, model =
          "dominant",power=0.8,
+         print.power.curve=FALSE, a=1)

5294 controls and 3529 cases for power 0.8
```

```
         n  n.controls  n.cases
      8823        5294     3529
```
```
> cc.design(B=1e6, OR=2.0, D=D, p=0.3, q=0.1, preva-
      lence = 0.1,
+     n.cases=2000,    n.controls=3000,   model
      ="recessive", power=0.8,
+     print.power.curve=FALSE, a=1)
```
155616 controls and 103744 cases for power 0.8
```
         n  n.controls  n.cases
    259360      155616   103744
> cc.design(B=1e6, OR=c(2.0,2.0), D=D, p=0.3, q=0.1,
      prevalence=0.1,
+     n.cases=2000,n.controls=3000,model="gen-
      eral", power=0.8,
+     print.power.curve=FALSE, a=1)
```
3644 controls and 2429 cases for power 0.8
```
         n  n.controls  n.cases
      6073        3644     2429
> r.squared <- 0.5
> D <- sqrt(r.squared*p*(1-p)*q*(1-q))
> c(D=D,p=p,q=q)
      D       p       q
 0.0972  0.3000  0.1000
> error <- try(cc.design(B=1e6, OR=2.0, D=D, p=0.3,
      q=0.1,
+     prevalence=0.1,n.cases=2000,n.controls=3000,+

      model="additive", power=0.8, a=1))
```

The last command gave the error

```
Error in function (B, OR, D, p, q, baseline.
risk, Dprime = NULL, R = NULL, : must have max(-p*q,
-(1-p)*(1-q)) = Dmin = -0.03 < D < min((1-p)*q,(1-q)
*p) = Dmax = 0.07
```

Note: The run with $r^2 = 0.5$ failed because $r^2 = 0.259$ at the maximum LD for the given allele frequencies. The functions cc.power(), cc.design() are conveniently called with the Dprime argument since any value of D' from -1 to 1 is always valid.

R example 5.9: Assuming $D' = 0.5$, and other parameters as above, find the sample size needed to detect the effect in an additive model. Assume a prior for the log odd ratio such that a 95 % c.i. from the prior contains the range (1/3,3) for the odds ratio. With this prior the standard deviation for the log odds ratio is 1/4 of the width of the interval on the log scale. Compare with the results from a prior equivalent to a single sample point (a = 1).

```
> sigma2.eta <- (1/4*(log(3) - log(1/3)))^2
> sigma2.eta
[1] 0.302

> cc.design(B=1e6, OR=2.0, Dprime=0.5, p=0.3, q=0.1,
            prevalence=0.1,
+           n.cases=2000, n.controls=3000, model="additive",
+           sigma2.eta=sigma2.eta, power=0.8, print.power.curve=FALSE)
10267 controls and 6845 cases for power 0.8
       n    n.controls   n.cases
   17112         10267      6845
> cc.design(B=1e6, OR=2.0, Dprime=0.5, p=0.3, q=0.1,
            prevalence=0.1,
+           n.cases=2000, n.controls=3000, model="additive", a=1,
+           power=0.8, print.power.curve=FALSE)
10398 controls and 6932 cases for power 0.8
       n    n.controls   n.cases
   17330         10398      6932
```

Note: The sample sizes are approximately similar, hence both priors are approximately equally non-informative, i.e., the prior for odds ratio has comparable information to a single sample point. These examples show that even quite large effects may be undetected by current GWAS if their allele frequencies are lower than the SNP marker frequencies. This is generally the case because common SNPs tend to have intermediate allele frequencies [62].

4.3 Genome-Wide Transmission Disequilibrium Tests

In this section, we discuss three forms of TDT test: the discrete TDT test based on many small families (trios) with a single affected offspring; and the S-TDT test based on *discordant sib pairs*, where one affected and one unaffected offspring are sampled from each family, and the SDT test [63]. The TDT test requires both parental and offspring genotypes. The S-TDT [64] requires only genotypes from each discordant sib pair, so can be used where parental DNA is not available. The SDT test can be used when data from more than one affected and unaffected sib are available.

The discrete transmission disequilibrium test (TDT) [65] is based on transmission of marker alleles from parents to offspring in many small families. The TDT and related tests compare probabilities of transmission of alleles for affected offspring to expectation under H_0 or between affected and unaffected offspring.

Note:

- The TDT test is an application of McNemar's test [66].
- The TDT test was based on an earlier idea [67] of presenting data related to transmission and non-transmission of alleles.

Table 6
Contingency table for transmission of alleles in a TDT test based on parent–offspring trios

Transmitted Alleles	Non-transmitted alleles 1	2
1	n_{11}	n_{12}
2	n_{21}	n_{22}

In unstructured population association studies, spurious associations can be generated by differences in allele frequencies between sub-populations. Transmission disequilibrium tests reduce the occurrence of spurious associations due to population structure, since transmission of alleles in each family is conditionally independent given the parental genotypes.

4.3.1 TDT Tests

The TDT test is based on the fact that each parental allele is transmitted randomly, with 50 % probability, to each offspring. If an allele is associated with a disease phenotype, then there will be a higher or lower proportion of the marker amongst the cases.

The TDT test statistic for a bi-allelic marker is given by

$$T = \frac{(n_{12} - n_{21})^2}{(n_{12} + n_{21})} \sim \chi_1^2 \qquad \text{under } H_0, \qquad (64)$$

where n_{ij} are as in Table 6. For a multi-allelic marker with m alleles, the TDT test statistic is given by

$$T = \frac{m-1}{m} \sum_{i=1}^{m} \frac{(n_{i\cdot} - n_{\cdot i})^2}{(n_{i\cdot} + n_{\cdot i} - 2n_{ii})} \sim \chi_{(m-1)}^2, \qquad (65)$$

where n_{ij} is the number of times allele i is transmitted and allele j is not transmitted, $n_{i\cdot} = \sum_j n_{ij}$ and $n_{\cdot i} = \sum_j n_{ji}$ [68].

Allele transmission status is known if one parent is heterozygous and the other homozygous (e.g., $Aa \times aa$) or both parents are heterozygous with different alleles.

The TDT test simultaneously tests for linkage and linkage disequilibrium:

- If there is no linkage disequilibrium, that means that the marker and trait loci are independent in the population. Hence any parental trait QTL allele will be independent of the parental marker alleles. Hence there will be no association between QTL allele and the transmitted marker allele.

- If there is linkage disequilibrium between marker and QTL alleles, but no linked QTL this means that there is an association between parental marker and QTL alleles, but because there is no linkage ($r = 0.5$), the QTL allele transmitted will be independent of the marker allele transmitted. Again, there will be no association between the transmitted QTL allele and the transmitted marker allele. This reduces spurious associations from population structure, by eliminating associations with unlinked QTL.

- If there is linkage disequilibrium and linkage, linked QTL effects with $r < 0.5$ will be reduced by a factor $(1 - 2r)$. This reduces spurious associations from population structure, by reducing associations with or weakly linked QTL

- The disadvantage of the TDT design is the need to genotype trios—each transmission requires genotyping of three markers—two parents with one parent heterozygous and the other homozygous plus one offspring. Power may be lower than for a random population sample with the equivalent amount of genotyping.

Most spurious associations between markers unlinked to the trait locus will be eliminated by the TDT test. However, it is important to realize that some spurious associations between markers linked to the trait locus may still occur. These are associations between markers linked at low recombination distances to QTL. For example a marker at recombination distance $r = 0.1, 0.2$ would be reduced by a factor $(1 - 2r) = 0.8, 0.6$, respectively. For a species with 24 chromosomes and recombination distance averaging 100cM per chromosome, approximately 2 % of loci would be within $r \leq 0.2$ or a given locus. Hence we would expect 98 % of spurious associations to be eliminated or reduced by at least 60 %. These associations are "spurious" in the sense that although the marker is linked to the trait locus, it may still be very far from the trait locus compared to the potential resolution of the association study. This should be considered in the analysis and experimental design.

Recall n_{12}, n_{21} were the numbers of times allele 1 but not allele 2 was transmitted, and the number of times allele 2 but not allele 1 was transmitted, respectively (Table 6).

We condition on $n_{12} + n_{21}$, the number of times exactly one allele was transmitted. Under the null hypothesis, alleles 1 and 2 are equally likely to be transmitted, so n_{12} has a binomial distribution with $n = n_{12} + n_{21}$, and $p = 0.5$. Under the alternative hypothesis n_{12} has a binomial distribution with $p = p_1$. We assume a Beta$(a/2, a/2)$ prior.

$$n_{12} \sim \text{Binomial}(n_{12} + n_{21}, 0.5)p_1 \quad \text{under } H_0, \quad (66)$$

$$n_{12} \sim \text{Binomial}(n_{12} + n_{21}, p_1)0.5 \quad \text{under } H_1. \quad (67)$$

Table 7
Parental and offspring genotypes for the TDT test

Parental Genotype	Offspring (case) genotype M_1M_1	M_1M_2	M_2M_2	Total
1. $M_1M_1 \times M_1M_1$	18	0	0	18
2. $M_1M_1 \times M_1M_2$	21	30	0	51
3. $M_1M_1 \times M_2M_2$	0	8	0	8
4. $M_1M_2 \times M_1M_2$	2	15	19	36
5. $M_1M_2 \times M_2M_2$	0	3	3	6
6. $M_2M_2 \times M_2M_2$	0	0	1	1

Reprinted from Association Mapping in Plants (Springer 2007)

Recall that the Beta distribution is the conjugate prior for binomial sampling—if the prior is Beta(a, b), and k successes are observed in n Bernoulli trials the posterior is Beta$(a + k, b + n - k)$. Hence, under H_1 the posterior for p_1 under H_1 is Beta$(n_{12} + 0.5, n_{21} + 0.5)$. We assume a Beta$(a/2, a/2)$ prior for Beta$(a/2, a/2)$. The Beta$(a/2, a/2)$ prior has mean 0.5 and information equivalent to a sample points.

$$\pi(p) = \frac{1}{B(a/2, a/2)} p^{a/2-1}(1-p)^{a/2-1}, \qquad (68)$$

$$g(p \mid n_{12}, n_{21}) = \frac{1}{B(a/2 + n_{12}, a/2 + n_{21})} p^{a/2+n_{12}-1}(1-p)^{a/2+n_{21}-1}, \qquad (69)$$

where $B(a, b)$ is the beta function.

H_0 and H_1 are nested models with H_0 corresponding to $p_1 = 0.5$. Therefore the Bayes factor is given by the Savage–Dickey density ratio:

$$\begin{aligned}
B &= \frac{\pi(p_1 = 0.5)}{g(p_1 = 0.5 \mid n_{12}, n_{21})} = \frac{0.5^{a-1}}{B(a/2, a/2)} \\
&\quad \times \frac{B(a/2 + n_{12}, a/2 + n_{21})}{0.5^{a+n_{12}+n_{21}-1}} = \frac{B(a/2 + n_{12}, a/2 + n_{21})}{B(a/2, a/2)} \times 2^n,
\end{aligned} \qquad (70)$$

where $n = n_{12} + n_{21}$.

Frequencies of parent–offspring genotype combinations for 120 nuclear families, each with one affected sibling, are shown in Table 7. Frequencies of transmitted and non-transmitted alleles are shown in Table 8.

Table 8
Counts of transmitted and non-transmitted alleles for a TDT test

Transmitted Allele	Non-transmitted allele M_1	M_2	Total
M_1	99	39	138
M_2	86	16	102
Total	185	55	240

Note: When both parents are homozygous (parental genotypes 1,6) the transmitted and non-transmitted alleles are the same, affecting n_{11}, n_{22} only; hence these crosses do not contribute to the test statistic. When one parent is homozygous (parental genotypes 2,5) and the other parent heterozygous only the allele transmitted from the heterozygous parent contributes to the test statistic.

From Table 8 we have $n_{12} = 39$, and $n_{21} = 86$. The TDT test statistic is

$$T = \frac{(39-86)^2}{(86+39)} = 17.7. \tag{71}$$

The p-value is $\Pr(\chi_1^2 \geq 17.7) = 2.6 \times 10^{-5}$.

R example 5.10: Bayes factor calculation for a TDT using a ratio of Beta densities.

```
> # TDT test statistic and p-value
> n12 <- 39; n21 <- 86
> T <- (n12 - n21)^2/(n12+n21)
> T
[1] 17.7
> p.value <- 1 - pchisq(T,1)
> p.value
[1] 2.62e-05
> # Savage--Dickey Bayes factor calculation (a=1)
> dbeta(0.5,0.5,0.5)/dbeta(0.5,86.5,39.5)
[1] 611
```

4.3.2 Power of the TDT Test

Using the formula 70 for the Bayes factor, the power to detect an effect with allele frequency p_1 for cases is

$$\mathcal{P} = \Pr\left(2^n \frac{B(a/2 + n_{12}, a/2 + n - n_{12})}{B(a/2, a/2)} > B_c\right)$$
$$= \sum f(k; n, p_1) I\left(\frac{2^n B(a/2 + k, a/2 + n - k)}{B(a/2, a/2)} > B_c\right), \tag{72}$$

where

$$f(k; n, p) = nkp^k(1-p)^{n-k} \tag{73}$$

is the binomial density and $I(\cdot)$ is the *indicator function*, $I(x) = 1$ if x is true or 0 otherwise.

For large values of n the posterior density can be approximated by a normal density. The Beta(A, B) density has mean $A/(A+B)$ and variance $AB/((A+B)^2(A+B+1))$ [19, Appendix A, pp. 476–477]. It follows that

$$g(p) \sim N\left(\frac{a/2 + n_{12}}{a+n}, \frac{(a/2+n_{12})(a/2+n_{21})}{(a+n)^2(a+n+1)}\right). \tag{74}$$

We rescale the test statistic as

$$Z_n = \frac{\hat{p} - \frac{1}{2}}{\sqrt{p(1-p)}}, \qquad n\hat{p} \sim \text{Binomial}(n, p) \tag{75}$$

so that

$$Z_n \sim N\left(\frac{p - 1/2}{\sqrt{p(1-p)}}, \frac{1}{n}\right). \tag{76}$$

The Beta$(a/2, a/2)$ prior does not allow a closed form solution for the Bayes factor. Suppose a normal prior adequately represents prior information:

$$\pi(Z) \sim N\left(0, \frac{1}{a^*}\right) \tag{77}$$

For $\pi(Z)$ to have the same mean and variance as the Beta$(a/2, a/2)$ we require

$$\text{var}\hat{p} = \frac{AB}{(A+B)^2(A+B+1)} \qquad \text{(where } A = B = a/2\text{)}$$
$$= \frac{1}{4(a+1)} \tag{78}$$

$$\therefore a^* = 4(a+1)\sqrt{p(1-p)}. \tag{79}$$

Equations 61, 62, and 63 hold, with a replaced by a^*, and η/σ_1 replaced by $E(Z) = (p - 1/2)/\sqrt{p(1-p)}$. Hence the power is given by:

$$\mathcal{P} \approx 1 - \Phi\left(\sqrt{n}\left(Z_c - \frac{p - 1/2}{\sqrt{p(1-p)}}\right)\right). \tag{80}$$

Designing a GWAS: Power, Sample Size, and Data Structure   77

The power to detect an effect with a given odds ratio can be found by first solving for p in

$$\text{OR} = \frac{2p\nu}{1-2p\nu} \cdot \frac{1-2(1-p)\nu}{2(1-p)\nu}, \qquad (81)$$

where $\nu = \Pr(\text{case})$ is the disease prevalence.

R example 5.11: Find the power of a TDT test with 800 families with one heterozygous parent and one affected offspring to detect an effect. Assume that the transmission frequency of the allele in cases is $p = 0.6$. Calculate the power to detect the effect with $P \leq \alpha$ for $\alpha = 0.001, 5 \times 10^{-7}, 5 \times 10^{-8}$. Calculate the power to detect the effect with Bayes factor $B > B_c$ for $B_c = 1000, 10^6$. Calculate the equivalent odds ratio.

```
> options(digits=3)
> library(ldDesign)
> tdt.frequentist.power <- function(alpha,p,n){
+ ncp <- n*(p-0.5)^2/(p*(1-p))
+ power <- 1-pchisq(qchisq(1-alpha,1),1,ncp)
+ power
+ }
> tdt.frequentist.power(alpha=0.001,p=0.6,n=800)
[1] 0.993
> tdt.frequentist.power(alpha=5e-7,p=0.6,n=800)
[1] 0.773
> tdt.frequentist.power(alpha=5e-8,p=0.6,n=800)
[1] 0.626
> tdt.power <- function(B, p, n, a){
+ sigma1 <- sqrt(p*(1-p))
+ astar <- 4*(a+1)*sigma1
+ Zc <- calc.Zc.ABF(B,n,astar)
+ delta <- abs((p - 0.5)/sigma1)
+ power <- 1-pnorm(sqrt(n)*(Zc - delta))
+
+ power
+ }
> tdt.power(B=1000,p=0.6,n=800,a=1)
[1] 0.917
> tdt.power(B=1e6,p=0.6,n=800,a=1)
[1] 0.508
> tdt.calc.OR <- function(p, prevalence){
+ nu <- prevalence
+ OR <- (1-2*nu*(1-p))*p/((1-p)*(1-2*nu*p))
+
OR + }
> tdt.calc.OR(p=0.6,prevalence=0.1)
[1] 1.57
```

Note: The prevalence here is the population prevalence. If considering only a particular set of cross types, the prevalence should, strictly speaking, be the prevalence in the subpopulation consisting of the given cross types.

4.3.3 Sib TDT-Tests

We consider two forms of sib-based TDT tests, the S-TDT test [64], and the SDT test [63]. The S-TDT test [64] compares marker allele frequencies between affected and unaffected sibs from a large number of nuclear families. The S-TDT gives a test for both linkage and linkage disequilibrium if only one affected and one unaffected sib per family are used. If data from more than one affected and/or unaffected sib is used the S-TDT gives a test for linkage [68, 69].

The SDT test bases inference on a summary statistic calculated for each family, based on whether or not the average allele frequency for affected sibs differs from the family mean. The SDT tests for both linkage and linkage disequilibrium provided there is at least one affected and unaffected sib per family.

Let y_i denote the number of M_1 alleles in affected sibs, a_i, u_i the number of affected and unaffected sibs, $t_i = a_i + u_i$ the total sibship size, and r_i, s_i the number of $M_1 M_1$ and $M_1 M_2$ genotypes in the ith nuclear family, respectively. The test statistic is

$$T_n = \frac{\Upsilon_n - A_n}{\sqrt{V_n}} \sim N(0,1) \qquad \text{asymptotically, under } H_0, \qquad (82)$$

where

$$\Upsilon_n = \sum_{i=1}^{n} y_i, \qquad (83)$$

$$A_n = \sum_{i=1}^{n} (2r_i + s_i) a_i / t_i, \qquad (84)$$

$$V_n = \sum_{i=1}^{n} \frac{a_i u_i [4 r_i (t_i - r_i - s_i) + s_i (t_i - s_i)]}{t_i^2 (t_i - 1)}. \qquad (85)$$

Υ_n is the total number of M_1 alleles in affected sibs, A_n is an estimate of the total expected value of Υ_n under H_0 (assuming the allele frequency for affected sibs is the same as for all sibs), and V_n is an estimate of the variance of $\Upsilon_n - A_n$.

Bayes factor for the S-TDT test: Let Z_n be given by:

$$Z_n = T_n / \sqrt{n} \qquad (86)$$

$$= \frac{(\Upsilon_n - A_n)/n}{\sqrt{\overline{V}_n}} \sim N(0, 1/n) \qquad \text{under } H_0, \qquad (87)$$

where $\overline{V}_n = V_n/n$. Notice that the quantities in the numerator and denominator for Z_n are stable, i.e., estimating the same quantity, independent of n.

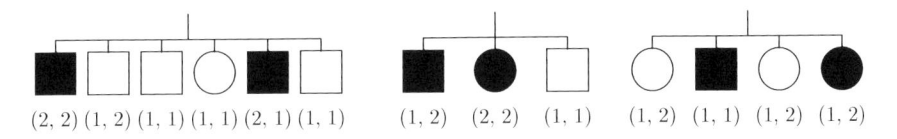

Fig. 1 Families for the S-TDT test (Example 5.1). Affected offspring are shown in *black*

Table 9
Values of t_i, a_i, u_i, r_i, s_i, and y_i for S-TDT families in Example 5.1

Family (*i*)	t_i	a_i	u_i	r_i	s_i	y_i
1	6	2	4	2	1	4
2	3	2	1	1	1	3
3	4	2	2	0	3	1

Under H_1 the sampling variance for Z_n is $1/n$, and its estimate is the value of the statistic. We take a prior for Z, the quantity that Z_n is estimating under H_1 to be

$$\pi(Z) \sim N\left(0, \frac{1}{a}\right). \tag{88}$$

By construction, this is a prior with the information equivalent to a sample points. The posterior distribution for Z under H_1 is

$$z \mid Z_{\text{obs}} = N\left(\frac{n}{n+a} Z_{\text{obs}}, \frac{1}{n+a}\right) \tag{89}$$

and the Bayes factor is given by Eq. 23.

Example 5.1 (An S-TDT Test): Three families are shown in Fig. 1. Affected sibs are shown as solid black boxes or circles; unaffected sibs are shown with blank boxes or circles. Marker alleles for each individual are indicated as, e.g., (1, 2). The summary statistics $t_i, a_i, u_i, r_i, s_i, y_i$ for marker allele 2 are summarized for each family in Table 9.

Applying Eqs. 83–85 we obtain $\Upsilon = 8, A = 5.17, V = 2.21$, and hence

$$T = \frac{8 - 5.17}{\sqrt{2.21}} = 1.90. \tag{90}$$

The *p*-value is $\Pr(|Z| \geq 1.9) = 2(1 - \Phi(1.9)) = 0.057$, where $\Phi(\cdot)$ is the standard normal c.d.f., which is not quite significant at the 5 % level.

Now let Z_n be given by

$$Z_n = T_n/\sqrt{n}, \tag{91}$$

$$= 1.90/\sqrt{3}. \tag{92}$$

We have $T = 1.90$, and $n = 3$, so $Z_{obs} = T/\sqrt{3} = 1.10$. The Savage–Dickey density ratio estimate of the Bayes factor with $a = 1$ is

$$B = \frac{\exp(\frac{n^2}{n+1} Z^2_{obs}/2)}{\sqrt{n+1}} = 1.94 \tag{93}$$

representing very weak evidence for H_1, despite the p-value close to 0.05 (0.057).

Power for the S-TDT test: We will assume a bi-allelic marker with alleles A, a and n families of cross type $Aa \times aa$, i.e., one segregating parent. We will assume complete linkage disequilibrium between the marker and QTL locus.

Let $T = 1$ (resp. $T = 0$) denote the transmission (resp. non-transmission) of allele A. We modify the test statistic to be an unbiased estimator of

$$\delta p = \Pr(T = 1 \mid \text{case}) - \Pr(T = 1 \mid \text{control}). \tag{94}$$

Let $p_T = \Pr(T = 1 \mid \text{case})$ and $p'_T = \Pr(T = 1 \mid \text{control})$.

Note that p'_T can be calculated from p_T using Bayes' theorem:

$$\begin{aligned} p'_T = \Pr(T = 1 \mid \text{control}) &= \frac{\Pr(\text{control} \mid T = 1)\Pr(T = 1)}{\Pr(\text{control})} \\ &= \frac{(1 - \Pr(\text{case} \mid T = 1))0.5}{1 - \nu} \\ &= \frac{(1 - \Pr(T = 1 \mid \text{case})\nu/\Pr(T = 1))}{2(1 - \nu)} \\ &= \frac{1 - 2p_T\nu}{2(1 - \nu)}, \end{aligned} \tag{95}$$

where $\nu = \Pr(\text{case})$ is the disease prevalence.

The odds ratio is given by:

$$\begin{aligned} \text{OR} &= \frac{\Pr(\text{case} \mid T = 1)}{\Pr(\text{control} \mid T = 1)} \cdot \frac{\Pr(\text{control} \mid T = 0)}{\Pr(\text{case} \mid T = 0)} \\ &= \frac{\Pr(T = 1 \mid \text{case})\Pr(\text{case})/\Pr(T = 1)}{\Pr(T = 1 \mid \text{control})\Pr(\text{control})/\Pr(T = 1)} \times \\ &\quad \frac{\Pr(T = 0 \mid \text{control})\Pr(\text{control})/\Pr(T = 0)}{\Pr(T = 0 \mid \text{case})\Pr(\text{case})/\Pr(T = 0)} \\ &= \frac{\Pr(T = 1 \mid \text{case})}{\Pr(T = 1 \mid \text{control})} \cdot \frac{\Pr(T = 0 \mid \text{control})}{\Pr(T = 0 \mid \text{case})} \\ &= \frac{p_T}{p'_T} \cdot \frac{1 - p'_T}{1 - p_T} \\ &= \frac{2\nu p_T^2 + (1 - 2\nu)p_T}{2\nu p_T^2 - (1 + 2\nu)p_T + 1}. \end{aligned} \tag{96}$$

In family i, δp is estimated as:

$$\widehat{\delta p}_i = \frac{y_i}{a_i} - \frac{s_i - y_i}{u_i}. \qquad (97)$$

The estimator $\widehat{\delta p}_i$ has expected value and variance:

$$E(\widehat{\delta p}_i) = \delta p \qquad (98)$$

$$\operatorname{var} \widehat{\delta p}_i = \frac{p_T(1-p_T)}{a_i} + \frac{p'_T(1-p'_T)}{u_i} \qquad (99)$$

and

$$w_i^* = (\operatorname{var} \widehat{\delta p}_i)^{-1}. \qquad (100)$$

Let $S_w = \sum w_i^*$, $w_i = w_i^*/S_w$.

Define Z as

$$Z = \sum w_i \widehat{\delta p}_i. \qquad (101)$$

Then

$$\operatorname{var} Z = \frac{\sum w_i^{*2}(1/w_i^*)}{S_w^2} = \frac{1}{S_w}. \qquad (102)$$

If we further assume $a_i = u_i = k/2$ cases and controls in each family, and use the small deviations approximation $p_T \approx 0.5$ in Eq. 99

$$w_i^* = 4\left(\frac{2}{k} + \frac{2}{k}\right)^{-1} = k. \qquad (103)$$

Then

$$Z \sim N\left(\delta p, \frac{1}{n^*}\right), \qquad (104)$$

where $n^* = nk$. More generally let $n^* = S_w$.

We calibrate the prior for Z as

$$\pi(Z) \sim N\left(0, \frac{c}{a}\right) \qquad (105)$$

so that $a = 2$ corresponds to a single minimally informative family with $a_i = u_i = 1$. In this case $S_w = 2$ so $c = 1$.

Equations 61, 62, and 63 hold, with η/σ_1 replaced by $E(Z) = \delta p$. Hence the power to detect an effect with transmission probability difference δp and the specified Bayes factor is given by:

$$\mathcal{P} \approx 1 - \Phi(\sqrt{n^*}(Z_c - \delta p)). \qquad (106)$$

We have derived the power for a given value of δp to detect the effect with a given Bayes factor, for n families of cross type $Aa \times aa$ under complete disequilibrium. Under complete disequilibrium $\delta p_i = \delta p$. Under incomplete disequilibrium δp_i has additional variance between families, and δp is reduced. We conjecture that δp is reduced approximately by

$$(\delta p')^2 = r^2 (\delta p)^2. \qquad (107)$$

Other cross types are possible, though for bi-allelic markers, only cross types $Aa \times Aa$ and $Aa \times aa$ (or $Aa \times AA$) are of interest. Results for $Aa \times AA$ are identical to those of $Aa \times aa$ under common independence assumptions. This method is currently being developed and will be incorporated in a future release of `ldDesign`.

R example 5.12: Find the number of families with one affected and one unaffected sib per family required to detect a transmission probability difference of 0.1 between cases and controls with a Bayes factor 10^6 with power 0.8.

```
> library(ldDesign)
> stdt.power <- function(B, deltap, n.affected, n.
    unaffected, a=1){
+ # power to detect a transmission probability
+ # difference deltap with Bayes factor >= B
+ # a[i], u[i] = number of affected, unaffected in
    family i
+ # n[i] = a[i] + u[i] = number of genotyped indivi-
    duals in family i
+ # assumes all cross type Aaxaa
+ wstar <- 4/(1/n.affected + 1/n.unaffected)
+ nstar <- Sw <- sum(wstar)
+ Zc <- calc.Zc.ABF(B,nstar,a)
+ power <- 1-pnorm(sqrt(nstar)*(Zc-deltap))
+ power
+ }
> # first try several guesses
>   stdt.power(B=1e6, deltap=0.1, n.affected=rep
            (1,2000),
+           n.unaffected=rep(1,2000))
[1] 0.629
>   stdt.power(B=1e6, deltap=0.1, n.affected=rep
            (1,3000),
+           n.unaffected=rep(1,3000))
[1] 0.957
> # then interpolate
> na <- exp(seq(log(2000), log(3000), length=100))
> powers <- list()
> for(ii in seq(along=na)){
+ powers[[ii]] <- stdt.power(B=1e6, deltap=0.1,
+                     n.affected=rep(1,na[ii]),
```

Designing a GWAS: Power, Sample Size, and Data Structure   83

```
+                               n.unaffected=rep(1,na[ii])))
+ }
> powers <- unlist(powers)
> approx(powers,na,xout=0.8)$y
[1] 2346
> # check result
> stdt.power(B=1e6,deltap=0.1,n.affected=rep(1,2346),
+            n.unaffected=rep(1,2346))
[1] 0.8
```

4.3.4 SDT Test

As noted above, the sib-disequilibrium test (SDT; [63]) tests for both linkage and linkage disequilibrium if multiple affected and unaffected offspring per family are used. The SDT test compares the average rate of occurrences of a given allele between affected an unaffected sibs scoring 1, 0, − 1 for a family if rate of occurrences for affected sibs is greater, equal, or less (respectively) than the rate for unaffected sibs.

The total score, S, is

$$S = \sum_{i=1}^{n} \text{sign}(d_i), \qquad (108)$$

where $d_i = y_i/a_i - [(2r_i + s_i) - y_i]/u_i$ is the difference in rates of occurrence of the allele, and $\text{sign}(d_i)$ is 1, 0, − 1 if d_i is positive, zero, or negative respectively.

The number of nonzero differences is

$$W = \sum_{i=1}^{n} \text{sign}(d_i)^2, \qquad (109)$$

and the test statistic is given by

$$T = \frac{S^2}{W} \sim \chi_1^2. \qquad (110)$$

Note:

- Summarizing family data by the sign of differences in allele frequencies serves to "balance" the data at the family level, by effectively weighting all families equally. This may be suboptimal, except for families of similar size.
- The χ^2 approximation for the test statistic is likely to be poor except for very large samples, since a χ^2 distribution is a sum of squares of normal random variables, and $d_i = \pm 1$ is binary for informative families. Since a reasonable approximation far in the tails of the distribution is needed to get high Bayes factors or low p-values, we proceed similarly to the TDT test.

Bayes factor for the SDT: Let n_{12}, n_{21} be the number of times $d_i > 0$ and $d_i < 0$ respectively, and let $n = n_{12} + n_{21}$ be the number of times d_i is nonzero. We condition on n. The Bayes factor and power are as for the TDT (Eqs. 70 and 80), i.e.,

$$B = \frac{B(a/2 + n_{12}, a/2 + n_{21})}{B(a/2, a/2)} \times 2^n, \tag{111}$$

$$\mathcal{P} \approx 1 - \Phi\left(\sqrt{n}\left(Z_c - \frac{p - 1/2}{\sqrt{p(1-p)}}\right)\right), \tag{112}$$

where p is the binomial proportion:

$$n_{12} \sim \text{Binomial}(n, p). \tag{113}$$

Note: In the case of unequal sized families or different numbers of cases and controls sampled per family, the probability that $d_i > 0$ will vary between families. Hence, strictly speaking, we can only apply the power calculation (Eq. 112) for equal sized families, with the same number of cases and controls per family. In the case of equal sized families p is given as a binomial probability in terms of the transmission probabilities p_T, p'_T.

4.4 Family-Based Tests for Quantitative Traits

For quantitative traits there are two main approaches.

1. Linear model based approaches, considering within- and between-family effects [70–76].
2. Approaches generalizing the TDT, where the continuous trait is split into high ("affected") and low ("unaffected") subsets [77–79].

See Lange [75] for a review. Allison [70] considers five variants of TDT tests for quantitative traits, called TDT Q1–Q5. TDT-Q1 assumes random sampling. TDT-Q3, Q4 assume "extreme sampling," a form of selective genotyping where trios are preselected so the offspring lie in the tails of the phenotypic distribution. TDT Q1–Q4 use only families with one heterozygous parent, while TDT-Q5 attempts to use information from all possible matings.

The QTDT (quantitative TDT; [73, 75]) is based on the model:

$$E(Y_{ij}) = \mu + \beta_b b_i + \beta_w w_{ij}, \tag{114}$$

where β_b, β_w are between and within family effects, b_i is a "family mean" genotype, and w_{ij} the within family genotype adjusted for b_i. The precise definitions depend on the available data [73, 75], e.g., if $X_{ij} = 0, 1, 2$ for genotypes aa, Aa, AA

$$b_i = \frac{1}{m_i}\sum_{j=1}^{m_i} X_{ij}, \tag{115}$$

$$w_{ij} = X_{ij} - b_i. \qquad (116)$$

The QTDT is a likelihood ratio test for $H_0 : b_w = 0$.

The pedigree disequilibrium test (PDT; [79]) generalizes the TDT test to families without size restrictions on the numbers of parent and child genotypes. Power calculations are given.

A Bayesian method generalizing the power calculations in `ldDesign` to multiple families is being developed for bi-allelic QTL and markers (R. Ball, 2012, unpublished). The method combines simulations of the number of families of each cross type at the marker ($aa \times aa$, $Aa \times aa$, $Aa \times aa$, ...) and number of progeny of each within each family of each possible genotype, with analytical calculations of an approximate Bayes factor analogous to the method for case–control studies from Subheading 4.2. An approximate Bayes factor and power calculation are derived for a statistic Z defined as a weighted average of within-family estimates. This gives a test for LD in the presence of possible linkage. In the case of linkage but not LD, Z has nonzero variance but the expected value of Z is zero.

Additionally, for increased efficiency of experimental designs for plants, multiple replicates of each genotype (ramets) can be specified. Using multiple ramets per genotype reduces environmental error and increases the effective heritability of a trait, while not increasing genotyping cost.

R example 5.13: Find the power of an experimental design with 400 families of size 10, and each genotype replicated three times to detect a QTL effect explaining 5 % of the genetic variance ($h_q^2 = 0.05$). Assume that trait heritability $h^2 = 0.6$, marker and QTL allele frequencies of 0.5, and linkage disequilibrium half the maximum possible ($D' = 0.5$). Assume that the recombination rate between marker and QTL is approximately zero. Compare with a calculation for a population sample of size 4000 using `ldDesign`.

```
> library(ldDesign)
>   fbat.power(B=1e6,h2.q=0.05,p=0.5,q=0.5,h2=0.6,
            a=1,n.families=400,
+       family.size=10,n.ramets=3,nsim=10000,Dprime=
0.5,r=0)
[1] 0.903
attr(,"parms")
        B      Dprime           p           q     nstar
1.00e+06    5.00e-01    5.00e-01    5.00e-01  3.08e+03
      ncp      alphac  n.families family.size
5.53e+01    2.32e-09    4.00e+02    1.00e+01
> # power with n=4000 genotypes, 3 ramets per genotype
> ld.power(Bf=1e6,n=4000,p=0.5,q=0.5,phi=0,D=0.125,
+      h2=0.05*1/(0.6+0.4/3))
    n power
```

```
           [1,] 4000 0.974
           attr(,"parms")
                  p             q                    D            h2
           5.00e-01      5.00e-01             1.25e-01      6.82e-02
                phi            Bf         missclass.rate
           0.00e+00      1.00e+06             0.00e+00
    > # number of genotypes for power 0.9
    > ld.design(Bf=1e6, p=0.5, q=0.5, phi=0, D=0.125,
    +    h2=0.05*1/(0.6+0.4/3), power=0.9)
    [1] 3361
```

Note: The sample size of 3,361 genotypes from a population sample required for power 0.9 compared to 4,000 for the 400 family design shows that the relative efficiency of the multi-family design per genotype is 3,361/4,000=84 %. The function `fbat.power()` or equivalent will be included in a future release of `ldDesign`.

5 Prior Distributions and Elicitation

Prior distributions both for the number and size of QTL are critical inputs into the experimental design calculations, yet this information is unknown or imperfectly known, but will be increasing as more GWAS studies are published. The number of QTL present is a factor in determining prior odds per marker, and hence the Bayes factor required, while the Bayes factors obtained depends on the prior variance for QTL effects (parameter a in Eq. 23).

In practice we need to consider the number of QTL of detectable size: large numbers of very small QTL would be indistinguishable from a polygenic non-QTL variance component. As well as the size of effects it is necessary to consider allele frequencies. Power to detect effects with low minor allele frequency decreases particularly if tagged by SNPs with higher minor allele frequencies.

Prior information may include trait heritability, which is generally known, giving an upper bound for the number of QTL exceeding a threshold.

Prior elicitation [80] where a prior distribution is "elicited" from various experts is an important but relatively neglected component of Bayesian analysis. In this paradigm, various experts are consulted and their opinions synthesized into a prior distribution by a statistician.

In genomics, such experts could include molecular geneticists, clinicians, biochemists, or statistical geneticists. Information considered could include number and size of effects detected in different traits and/or species, and bioinformatics, and knowledge of biochemical pathways. When considering previous effects, e.g., from the literature, it is important to also consider the posterior probabilities for the effects as many published effects may be

---

- genome scan with 500000 SNP markers
- 10 causal loci
- extent of LD comparable to marker spacing (*e.g.* 6*kb* in a genome of length $L = 3 \times 10^9 b$).
- ⇒ prior odds 1:50000
- Posterior odds desired: 20:1
- ⇒ Bayes factor required 10^6

Fig. 2 Example parameters leading to prior odds and Bayes factor required

---

- genome scan with 500000 SNP markers
- 200 causal loci
- extent of LD comparable to marker spacing (*e.g.* 6*kb* in a genome of length $L = 3 \times 10^9 b$).
- ⇒ prior odds 1:2500
- Posterior odds desired: 20:1
- ⇒ Bayes factor required 50000

Fig. 3 Example parameters leading to prior odds and Bayes factor required

spurious. The weight to give each of the various sources of information is subjective and would also be considered by the experts. This would include both the *reliability* of the information (e.g., posterior probabilities and robustness of the analysis to spurious associations) and the *relevance* (e.g., how similar the related species or traits are, or how well understood the relationship between a biochemical pathway and the trait of interest is).

In the following subsections we give examples of eliciting prior information, quantifying it as a probability distribution, and incorporating prior uncertainty into power calculations for genome scans, where a set of markers covering the genome is considered, and also candidate gene studies, where candidate loci are preselected according to various criteria, thought likely to increase prior odds.

5.1 Genome Scans

In this subsection we consider several scenarios for genome scans. Due to the complexity of jointly considering all the above factors in prior elicitation, it may be useful to consider several scenarios such as Figs. 2 and 3. The following scenario for genome scans (Fig. 2) is based on 10 expected causal loci of detectable size with extent of LD comparable to a Northern European population was considered [9, 16].

The scenario in Fig. 2 is of course not the only possibility. Higher extent of linkage disequilibrium would result in lower Bayes factors being needed, while higher Bayes factors, and more markers would be needed for a more diverse African population. Lower Bayes factors may be needed if there was more prior information either on the size of possible effects or on their number.

Meta-analyses for human height [81–84] have detected 54 putative associations ($P < 5 \times 10^{-7}$), explaining only a small percentage of the variance. Allen et al. [85] reported "hundreds of genetic variants, in at least 180 loci" influencing human height in an analysis using 183,727 individuals, suggesting the scenario in Fig. 3.

This scenario would not need such a high Bayes factor, but would nevertheless need very large sample sizes due to the smaller size of effects. As noted above, we should be aware that a proportion of these putative effects may not be real, e.g., only 3 of the 54 putative associations were detected in all 3 of the 2008 meta-analyses. A probable explanation is low power [80], suggesting these loci are not robustly detected and should be considered provisional until replicated at a similar level of evidence. A sceptic might be justified in considering an expected number of 10 QTL per genome as still being realistic.

Lango Allen et al. [85] also note:

> the 180 loci are not random, but instead are enriched for genes that are connected in biological pathways ($P = 0.016$) that and that underlie skeletal growth defects ($P < 0.001$)...Our data explain approximately 10% of the phenotypic variation in height, and we estimate that unidentified common variants of similar effect sizes would increase this figure to approximately 16 % of phenotypic variation (approximately 20% of heritable variation)

R example 5.14: Find the size of effect that can be detected in the [85] analysis of human height with parameters as in Eq. 3. Assume maximum LD ($D' = 1$). Repeat for several values of MAF and ratio between marker and QTL allele frequencies.

```
> # MAF=0.5, p=q
> # Initial guesses
> ld.power(B=50e3,D=0.25,p=0.5,q=0.5, n=183727, h2=0.0001,
      phi=0)
      n power
[1,] 183727 0.0144 attr(,"parms")
           p              q              D             h2
      5.0e-01        5.0e-01        2.5e-01        1.0e-04
         phi             Bf   missclass.rate
      0.0e+00        5.0e+04        0.0e+00
> ld.power(B=50e3,D=0.25,p=0.5,q=0.5, n=183727, h2=
      0.0005, phi=0)
      n power
```

```
[1,] 183727 0.999 attr(,"parms")
         p           q          D         h2
   5.0e-01     5.0e-01    2.5e-01    5.0e-04
       phi          Bf missclass.rate
   0.0e+00     5.0e+04    0.0e+00
> h2s <- seq(0.0001, 0.0005, length=20)
> # Interpolate
> powers <- numeric(20)
> res <- list()
> for(ii in 1:20){
+ res[[ii]] <- ld.power(B=50e3,D=0.25,p=0.5,q=0.5,
                        n=183727,
+                       h2=h2s[ii],phi=0) + }
> powers <- sapply(res,function(u)u[1,2])
> h2.soln <- approx(powers,h2s,xout=0.9)$y
> h2.soln
[1] 0.29
> ld.power(B=50e3,D=0.25,p=0.5,q=0.5,n=183727,h2=
         h2.soln,phi=0)
         n power
[1,] 183727 0.901 attr(,"parms")
          p           q          D         h2
   5.00e-01    5.00e-01   2.50e-01   3.29e-04
       phi          Bf missclass.rate
   0.00e+00    5.00e+04   0.00e+00
> # MAF= 0.1, p=q
> powers <- numeric(20)
> calc.D(Dprime=1,p=0.1,q=0.1)
[1] 0.09
> res <- list()
> for(ii in 1:20){
+ res[[ii]] <- ld.power(B=50e3,calc.D(Dprime=1,p=0.1,
                        q=0.1),
+                       p=0.1, q=0.1, n=183727, h2=h2s
                        [ii],phi=0) + }
> powers <- sapply(res,function(u)u[1,2])
> h2.soln2 <- approx(powers,h2s,xout=0.9)$y
> h2.soln2
[1] 0.09
> ld.power(B=50e3, D=calc.D(Dprime=1, p=0.1, q=0.1),
          p=0.1, q=0.1,
+         n=183727, h2=h2.soln2, phi=0)
     n power
[1,] 183727 0.9 attr(,"parms")
          p           q          D         h2
   1.00e-01    1.00e-01   9.00e-02   3.09e-04
       phi          Bf missclass.rate
   0.00e+00    5.00e+04   0.00e+00
```

```
> # MAF= 0.01, p=5q
> calc.D(Dprime=1,p=0.05,q=0.01)
[1] 0.0095
> ld.power(B=50e3, calc.D(Dprime=1, p=0.05, q=0.01),
p=0.05, q=0.01,
+      n=183727, h2=0.0005, phi=0)
     n power
[1,] 183727 0.0303 attr(,"parms")
            p            q            D           h2
    5.0e-02      1.0e-02      9.5e-03      5.0e-04
          phi           Bf missclass.rate
    0.0e+00      5.0e+04      0.0e+00
> ld.power(B=50e3, calc.D(Dprime=1, p=0.05, q=0.01),
          p=0.05, q=0.01,
+          n=183727, h2=0.002, phi=0)
     n power
[1,] 183727 0.989
attr(,"parms")
            p            q            D           h2
    5.0e-02      1.0e-02      9.5e-03      2.0e-03
          phi           Bf missclass.rate
    0.0e+00      5.0e+04      0.0e+00
> h2s <- exp(seq(log(0.0005), log(0.002), length=20))
> powers <- numeric(20)
> res <- list()
> for(ii in 1:20){
+   res[[ii]] <- ld.power(B=50e3, calc.D(Dprime=1, p=0.05,
                  q=0.01),+p=0.05,q=0.01,n=183727,
                  h2=h2s[ii],phi=0)
+ }
> powers <- sapply(res,function(u)u[1,2])
> h2.soln3 <- approx(powers,h2s,xout=0.9)$y
> h2.soln3
[1] 0.00155
> ld.power(B=50e3, calc.D(Dprime=1, p=0.05, q=0.01),
p=0.05, q=0.01,
+      n=183727, h2=h2.soln3, phi=0)
     n power
[1,] 183727 0.902
attr(,"parms")
             p            q            D           h2
    5.00e-02     1.00e-02     9.50e-03     1.55e-03
           phi           Bf missclass.rate
    0.00e+00     5.00e+04     0.00e+00
```

In practice, the reader will want to try a set of combinations of effect size, allele frequencies, and linkage disequilibrium, and choose the sample size large enough to obtain the required power for each combination. Budget or sample size constraints may mean it is necessary to accept lower power for some combinations.

The calculations shown here apply when the marker spacing is comparable to or greater than the extent of LD. When substantially more markers are available, a multi-locus method should be used. In this case it is not necessary to localize an effect to the nearest marker. Sufficient power will be obtained by designing for the case when marker spacing is comparable to the extent of LD. Actual power will be slightly higher.

5.2 Candidate Genes

A candidate gene approach tests loci segregating within "candidate genes." These are the genes selected by some criteria e.g.

C1. Genes mapping within a QTL region. The increase in prior probability for a candidate gene (or polymorphism) mapping into a QTL region depends on the QTL probability intensity from a previous QTL analysis on independent data, which in turn depends on the distance from the estimated QTL location, size of QTL, and power of the previous QTL experiment. This can be quantified by the multi-locus method [15].

C2. Genes differentially expressed according to the phenotype.

C3. Genes expressed or differentially expressed in a tissue related to the trait of interest, e.g., genes expressed in developing xylem.

C4. Genes obtained from literature searches, e.g., genes expressed in relevant biochemical pathways in the same or related species.

C5. Randomly sampled genes.

When candidate genes (or polymorphisms within genes) are obtained that meet one or more of the above criteria, their prior probability may be higher than for a random locus within a genome scan. In the case of (C1) this may be quantified by analysis; in other cases, the increase in prior odds could be elicited from experts.

For example, suppose there are 40,000 genes in a species of which 3,000 are differentially expressed in some test. Suppose our prior expected number of QTL is 100 but that we expect only about 50 of these could be found in candidate genes, and of these half would be expected to be differentially expressed. Then the prior probability per gene would be $50/40,000 = 1/800$. The prior probability per differentially expressed gene would be $25/3000 = 1/120$. The prior probability per non-differentially expressed gene would be $25/37,000 = 1/1,480$. For posterior odds 10:1 we would require a Bayes factor of 1,200 for differentially expressed genes or 14,800 for non-differentially expressed genes. There would be prior uncertainty in these numbers, for example an expert might say he is 90 % sure the proportion of QTL, p_c, found in candidate genes ranges from 30 % to 70 % which could be approximated by a Beta distribution with 5, and 95 percentiles 0.3 and 0.7 respectively. In this case the distribution is symmetric so is of the form Beta(n, n) which can be found by trial and error or interpolation:

R example 5.15: Find a Beta prior distribution with elicited 5 % and 95 % quantiles 0.3 and 0.7.

```
> # initial guess
> qbeta(0.05,5,5)
[1] 0.251
> qbeta(0.05,10,10)
[1] 0.32
> # interpolate
> ns <- seq(5,10,length=20)
> q5s <- numeric(20)
> for(ii in seq(along=ns)){
+ q5s[ii] <- qbeta(0.05,ns[ii],ns[ii]) + }
> nsoln <- approx(q5s,ns,xout=0.3)$y
> qbeta(0.05,nsoln,nsoln)
[1] 0.3
```

A good approximation, for practical purposes, may be found by approximating this distribution by a discrete distribution with, e.g., 3 or 5 values. Using a discrete distribution simplifies the computation by approximating the Bayesian integral (Eq. 4) by a sum—we calculate the power for each discrete value and average the powers according to the discrete distribution. We find an approximating distribution with the same mean and variance by first choosing quantiles, then simulating the means of a random sample grouped according to the quantiles, then adjusting the means of the outer quantiles such that using these means with the quantile frequencies gives the require distribution.

R example 5.16: Discrete approximation to an elicited Beta prior distribution.

```
> var.beta <-function(a,b){(a*b)/((a+b)^2*(a+b+1))}
> "the variance"
[1] "the variance"
> v2 <- var.beta(8,8)
> v2
[1] 0.0147
> # discrete distribution approximating the beta prior
> quants <- c(0,qbeta(c(0.25,0.75),8,8),1)
> u <- rbeta(10000,8,8)
> u1 <- u[u<quants[2]]
> u2 <- u[u>quants[2] & u < quants[3]]
> u3 <- u[u>quants[3]]
> mean(u1)
[1] 0.345
> mean(u2)
[1] 0.5
> mean(u3)
```

```
[1] 0.655
> # first approximation
> x <- c(mean(u1),mean(u2),mean(u3))
> x
[1] 0.345 0.500 0.655
> f <- c(0.25,0.5,0.25)
> # its variance
> s2 <- sum((x-0.5)^2*f)
> # interpolate to adjust variance
> dx <- seq(0,0.2,length=100)
> s2s <- numeric(length(dx))
> for (ii in 1:length(dx)){
+ x1 <- c(x[1] - dx[ii], x[2], x[3] + dx[ii])
+ s2s[ii] <- sum((x1-0.5)^2*f)
+ }
> dx1 <- approx(s2s,dx,xout=v2)$y
> x1 <- c(x[1] - dx1, x[2], x[3] + dx1)
> # the final approximation
> rbind(x1,f)
        [,1]   [,2]   [,3]
x1     0.329  0.5   0.672
f      0.250  0.5   0.250
> # check its variance, OK
> sum((x1-0.5)^2*f)
[1] 0.0147
```

Thus, our prior for the proportion of causal effects in candidate genes, p_c, is represented as a discrete distribution with values 0.33, 0.50, and 0.67 with probabilities 0.25, 0.5, and 0.25 respectively. Similarly, prior uncertainty in the number of QTL, n_Q, may be approximated by a discrete distribution. We will assume a discrete distribution with values $n_Q = 20, 40, 100$ and probabilities 1/3 for each value has been elicited, and calculate the power as a weighted average of the power for each of the possible combinations of p_c, n_Q in the following R example.

R example 5.17 : Power calculation for candidate genes incorporating elicited priors for the number of QTL and proportion of causal effects in candidate genes. Find the power to detect effects with $h_q^2 = 0.05$, a minor allele frequency of 0.05, with markers with a minor allele frequency 0.1, $D' = 0.5$, and a sample size of 6700 with posterior odds at least 10:1.

```
> # power calculation
> # find power of a sample of size 6700
> # do power calculations
> pc.values <- round(x1,2)
> pc.probs <- c(0.25,0.5,0.25)
> rbind(pc.values,pc.probs)
```

```
                    [,1] [,2] [,3]
      pc.values 0.33  0.5 0.67
      pc.probs  0.25  0.5 0.25
> nq.values <- c(20,40,100)
> nq.probs <- rep(1/3,3)
> rbind(nq.values,nq.probs)
                 [,1]     [,2]      [,3]
   nq.values  20.000   40.000   100.000
   nq.probs    0.333    0.333     0.333
> parms.df <- expand.grid(pc=pc.values,nq=nq.values)
> probs <-  apply(expand.grid(pc.probs,nq.probs),1,
                  function(u){ u[1]*u[2]})
> post.odds.required <- 10
> powers <- numeric(nrow(parms.df))
> bf.required <- numeric(nrow(parms.df))
> for(ii in 1:nrow(parms.df)){
+   pc <- parms.df[ii,1]
+   nq <- parms.df[ii,2]
+   prior.prob <- nq*pc/3000
+   prior.odds <- prior.prob/(1-prior.prob)
+   # Bayes factor required for differentially expressed genes
+   bf.required[ii] <- post.odds.required/prior.odds
+   powers[ii] <- ld.power(Bf=bf.required[ii],n=6700,
+       D=calc.D(0.5,0.1,0.05),p=0.1,q=0.05,h2=0.05,
+       phi=0)[2]
+ }
> cbind(parms.df,bf.required,powers,probs)
    pc    nq   bf.required   powers   probs
1  0.33   20          4535    0.857  0.0833
2  0.50   20          2990    0.874  0.1667
3  0.67   20          2229    0.885  0.0833
4  0.33   40          2263    0.885  0.0833
5  0.50   40          1490    0.900  0.1667
6  0.67   40          1109    0.910  0.0833
7  0.33  100           899    0.916  0.0833
8  0.50  100           590    0.929  0.1667
9  0.67  100           438    0.937  0.0833
> power <- sum(powers*probs)
> power
[1] 0.899
```

Note:

- The power is the weighted average of powers for the discrete parameter combinations, with weights equal to their prior probabilities.
- We have assumed that the number of QTL and proportion in candidate genes are a priori independent.

- Note the use of `expand.grid()` to generate the required combinations of parameters and the corresponding probabilities. In practice, the above calculation may need to be iterated to find the sample size giving the desired power.

5.3 Replication and Validation

Replication is an important principle in statistics. Within the frequentist paradigm, an independent replication sample may be used along with an often less stringent threshold. A replication sample may involve less genotyping, since only loci selected at the first stage need to be genotyped. In the frequentist paradigm, effects may be considered and validated if effects are "significant" in both samples. Strictly speaking, the sample sizes and thresholds need to be determined prior to seeing any results. Otherwise, as noted in Subheading 1.3, the p-value(s) may change if a sequential sampling approach is subsequently contemplated, depending on the results.

In the Bayesian paradigm, we can consider Bayes factors at each stage or use the posterior distributions from the first sample as priors for the second. The former approach is convenient for power calculations but somewhat conservative since the data has to overcome the low prior information twice.

With either a frequentist or Bayesian approach, it is important to realize that while all false positives may be eliminated with sufficient replication, just replicating once does not imply effects are true. The evidence still needs to be quantified. The replication sample needs sufficient power to achieve a Bayes factor large enough to give the desired combined strength of evidence while avoiding losing too many true associations. Bayesian decision theory [15, 18] can be used to find the optimal strategy—at each stage when making a decision. In Bayesian decision theory, a utility is defined representing net benefits of alternative options. The optimal decision is the one maximizing the expected utility, where expectation is taken over unknown parameters with respect to the posterior distribution. This should be applied, considering the posterior for effect size for each individual locus, and selecting each locus for replication where the marginal benefit is positive. This is in contrast to frequentist approaches which consider a class of loci satisfying $P < \alpha$ as being "significant" and implicitly suggest the significant effects should be accepted as if real.

References

1. Altshuler D, Hirschhorn JN, Klannemark M, Lindgren CM, Vohl M-C, Nemesh J, Lane CR, Schaffner SF, Bolk S, Brewer C, Tuomi T, Gaudet D, Hudson TJ, Daly M, Groop L, Lander ES (2000) The common PPARγ Pro12Ala polymorphism is associated with decreased risk of type 2 diabetes. Nat Genet 26:76–80

2. Terwilliger JD, Weiss KM (1998) Linkage disequilibrium mapping of complex disease: fantasy or reality? Curr Opin Biotechnol 9:578–594

3. Emahazion T, Feuk L, Jobs M, Sawyer SL, Fredman D et al (2001) SNP association studies in Alzheimer's disease highlight problems for complex disease analysis. Trends Genet 17:407–413

4. Neale DB, Savolainen O (2004) Association genetics of complex traits in conifers. Trends Plant Sci 9:325–330
5. Gura T (2000) Can SNPs deliver on susceptibility genes? Science 293:593–595
6. Hampton T (2000) Research brief, Focus September 29, 2000. Harvard University, Cambridge
7. The Wellcome Trust Case Control Consortium (2007) Genome-wide association study of 14,000 cases of seven common diseases and 3,000 shared controls. Nature 447:661–678
8. Diabetes Genetics Initiative of Broad Institute of Harvard and MIT, Lund University and Novartis Institutes for Biomedical Research (2007) Genome-wide association analysis identifies loci for type 2 diabetes and triglyceride levels. Science 316:1331–1345
9. Dudbridge F, Gusnato A (2008) Estimation of significance thresholds for genome-wide association scans. Genet Epidemiol 32:227–234
10. Berger J, Berry D (1988) Statistical analysis and the illusion of objectivity. Am Sci 76:159–165
11. Edwards W, Lindman H, Savage LJ (1963) Bayesian statistical inference for psychological research. Psychol Rev 70:193–242
12. Berger JO, Sellke T (1987) Testing a point null hypothesis: the irreconcilability of P-values and evidence (with discussion). J Am Stat Assoc 82:112–139
13. Ball RD (2001) Bayesian methods for quantitative trait loci mapping based on model selection: approximate analysis using the Bayesian Information Criterion. Genetics 159:1351–1364. http://www.genetics.org/cgi/content/abstract/159/3/1351. Accessed 29 May 2012
14. Ball RD (2005) Experimental designs for reliable detection of linkage disequilibrium in unstructured random population association studies. Genetics 170:859–873. http://www.genetics.org/cgi/content/abstract/170/2/859. Accessed 29 May 2012
15. Ball RD (2007) Quantifying evidence for candidate gene polymorphisms: Bayesian analysis combining sequence-specific and quantitative trait loci colocation information. Genetics 177:2399–2416. http://www.genetics.org/cgi/content/abstract/177/4/2399. Accessed 29 May 2012
16. Ball RD (2011) Experimental designs for robust detection of effects in genome-wide case-control studies. Genetics 189:1497–1514. http://www.genetics.org/cgi/content/abstract/189/4/1497. Accessed 29 May 2012
17. Bayes T (1763) An essay towards solving a problem in the doctrine of chances. Phil Trans Roy Soc 53:370–418
18. De Groot MH (1970) Optimal statistical decisions. McGraw-Hill, New York
19. Gelman A, Carlin JB, Stern HS, Rubin DB (1995) Bayesian data analysis. Chapman & Hall, London
20. Gilks WR, Richardson S, Spiegelhalter DJ (eds) (1996) Markov Chain Monte Carlo in practice. Chapman & Hall, London
21. Bernardo JM, Smith AFM (2000) Bayesian theory. Wiley, New York. ISBN 0 471 92416 4
22. Lavine M (1999) What is Bayesian statistics an why everything else is wrong. J Undergraduate Math Appl 20:165–174
23. Bolstad WM (2007) Introduction to Bayesian statistics, 2nd edn. Wiley, Hoboken. ISBN 978-0-470-14115-1
24. Casella G, Berger RL (1987) Reconciling Bayesian and frequentist evidence in the one-sided testing problem. J. Am. Stat. Assoc. 82:106–111
25. Spiegelhalter D, Smith AFM (1982) Bayes factors for linear and log-linear models with vague prior information. J R Stat Soc B 44:377–387
26. Birnbaum A (1962) On the foundations of statistical inference (with discussion). J Am Stat Assoc 57:269–306
27. Kalbfleisch JD (1975) Sufficiency and conditionality (with discussion). Biometrika 62:251–268
28. Davison AC (2001) Biometrika centenary: theory and general methodology. Biometrika 88:13–52
29. Anscombe FJ (1957) Dependence of the fiducial argument on the sampling rule. Biometrika 44:464–469
30. Lindley DV (1958) Fiducial distributions and Bayes' theorem. J R Stat Soc B 20:102–107
31. Fraser DAS (1964) Fiducial inference for location and scale parameters. Biometrika 51:17–24
32. Fisher RA (1959) Statistical methods and scientific inference, 2nd edn. Oliver and Boyd, Edinburgh
33. marquis de Laplace P-S (1820) Théorie analytique des probabilities, 3rd edn. Courcier, Paris
34. Sellke T, Bayarri MJ, Berger J (2001) Calibration of P-values for testing precise null hypotheses. Am Stat 55:62–71
35. Johnson NL, Kotz N (1970) Distributions in statistics: continuous univariate distributions. Houghton Mifflin, Boston
36. Benjamini Y, Hochberg Y (1995) Controlling the false discovery rate a practical and powerful

approach to multiple testing. J R Stat Soc B 57:289–300
37. Benjamini Y, Hochberg Y (2000) On the adaptive control of the false discovery rate in multiple testing with independent statistics. J Educ Behav Stat 25:60–83
38. Benjamini Y, Yekutieli D (2001) The control of the false discovery rate in multiple testing under dependency. Ann Stat 29:1165–1188
39. Storey JD (2003) The positive false discovery rate: a Bayesian interpretation and the q-value. Ann Stat 31:2013–2035
40. Yates F (1935) Complex experiments. J R Stat Soc Suppl 2:181–247
41. Pinheiro JC, Bates DM (2000) Mixed-effects models in S and S-PLUS. Springer, New York
42. Ball RD (2007) Statistical analysis and experimental design. In: Oraguzie NC et al (eds) Association mapping in plants. Springer, 69 pp. ISBN 0387358447 (Chapter 8)
43. Dickey JM (1971) The weighted likelihood ratio, linear hypothesis on normal location parameters. Ann Math Stat 42:204–223
44. Wakefield J (2007) A Bayesian measure of the probability of false discovery in genetic epidemiology studies. Am J Hum Genet 81:208–227
45. Johnson VE (2005) Bayes factors based on test statistics. J R Stat Soc B 67:689–701
46. Johnson VE (2008) Properties of Bayes factors based on test statistics. Scand J Stat 35:354–368
47. Schwarz G (1978) Estimating the dimension of a model. Ann Stat 6:461–464
48. Raftery AE (1995) Bayesian model selection in social research (with discussion). In: Marsden PV (ed) Sociological methodology. Blackwells, Cambridge, pp 111–196
49. Bogdan M, Ghosh JK, Doerge RW (2004) Modifying the Schwarz Bayesian information criterion to locate multiple interacting quantitative trait loci. Genetics 167:989–999
50. Stone M (1979) Comments on model selection criteria of Akaike and Schwarz. J R Stat Soc Ser B 41:276–278
51. Berger JO, Ghosh JK, Mukhopadhyay N (2003) Approximations and consistency of Bayes factors as model dimension grows. J Stat Plann Inference 112:241–258
52. Geiger D, Heckerman D, King H, Meek C (2001) Stratified exponential families: graphical models and model selection. Ann Stat 29:505–529
53. Kass RE, Wasserman L (1995) A reference Bayesian test for nested hypothesis and its relationship to the Schwarz criterion. J Am Stat Assoc 90:928–934
54. Pauler D (1998) The Schwarz criterion and related methods for normal linear methods. Biometrika 85:13–27
55. Broman KW (1997) Identifying quantitative trait loci in experimental crosses. PhD dissertation, Department of Statistics, University of California, Berkeley. http://www.biostat.wisc.edu/~kbroman/publications/thesis.pdf. Accessed 29 May 2012
56. Broman KW, Speed TP (2002) A model selection approach for the identification of quantitative trait loci in experimental crosses (with discussion). J R Stat Soc B 64:641–656, 731–775
57. Ball RD (2002) Discussion of: Broman K. W. and Speed T. P. 2002: A model selection approach for the identification of quantitative trait loci in experimental crosses (with discussion). J R Stat Soc B 64:749–750
58. Luo ZW (1998) Detecting linkage disequilibrium between a polymorphic marker locus and a trait locus in natural populations. Heredity 80:198–208
59. Weir BS (1996) Genetic data analysis II. Sinauer Associates, Sunderland
60. Ball RD (2004/2011) ldDesign — design of experiments for genome-wide association studies version 2 incorporating quantitative traits and case-control studies. http://cran.r-project.org/web/packages/ldDesign/index.html. Accessed 29 May 2012
61. Menashe I, Rosenberg PS, Chen BE (2008) PGA: power calculator for case-control genetic association analyses. BMC Genet 9:36. doi:10.1186/1471-2156-9-3608
62. Yang J, Beben B, McEvoy DP, Gordon S, Henders AK, Nyholt DR, Madden PA, Heath AC, Martin NG, Montgomery GW, Goddard ME, Visscher PM (2010) Common SNPs explain a large proportion of the heritability for human height. Nat Genet 42:565–569, 608
63. Horvath S, Laird NM (1998) A discordant-sibship test for disequilibrium and linkage: no need for parental data. Am J Hum Genet 63:1886–1897
64. Spielman RS, Ewens WJ (1998) A sibship test for linkage in the presence of association: the sib transmission/disequilibrium test. Am J Hum Genet 62:450–458
65. Spielman RS, McGinnis RE, Ewens WJ (1993) Transmission test for linkage disequilibrium: the insulin gene region and insulin-dependent diabetes mellitus (IDDM). Am J Hum Genet 52:506–516

66. McNemar Q (1947) Note on the sampling error of the difference between correlated proportions or percentages. Psychometrika 12:153–157
67. Terwilliger JD, Ott J (1992) A haplotype based haplotype relative risk approach to detecting allelic associations. Hum Hered 42:337–346
68. Weir BS (2001) Population genetic data analysis 2001. Southern Summer Institute of Statistical Genetics, North Carolina State University
69. Monks AA, Kaplan NL, Weir BS (1998) A comparative study of sibship tests of linkage and/or association. Am J Hum Genet 63:1507–1516
70. Allison DB (1997) Transmission-disequilibrium tests for quantitative traits. Am J Hum Genet 60:676–690
71. Allison DB, Heo M, Kaplan N, Martin ER (1999) Sibling-based tests of linkage and association for quantitative traits. Am J Hum Genet 64:1754–1763
72. Fulker DW, Cherny SS, Sham PC, Hewitt JK (1999) Combined linkage and association sib-pair analysis for quantitative traits. Am J Hum Genet 64:259–267
73. Abecasis GR, Cardon LR, Cookson WOC (2000) A general test of association for quantitative traits in nuclear families. Am J Hum Genet 66:279–292
74. Abecasis GR, Cookson WOC, Cardon LR (2001) The power to detect linkage disequilibrium with quantitative traits in selected samples. Am J Hum Genet 68:1463–1474
75. Lange C, DeMeo DL, Laird NM (2002) Power and design considerations for a general class of family-based association tests: quantitative traits. Am J Hum Genet 71:1330–1341
76. Lange C, Laird NM (2002) Analytical sample size and power calculations for a general class of family-based association tests: dichotomous traits. Am J Hum Genet 71:575–584
77. Schaid DJ (1996) General score tests for associations of genetic markers with disease using cases and their parents. Genet Epidemiol 13:423–449
78. Rabinowitz D (1997) A transmission disequilibrium test for quantitative trait loci. Hum Hered 47:342–350
79. Monks SA, Kaplan NL (2000) Removing the sampling restrictions from family-based tests of association for a quantitative-trait locus. Am J Hum Genet 66:576–592
80. O'Hagan A, Buck CE, Daneshkhah A, Eiser JR, Garthwaite PH, Jenkinson DJ, Oakley JE, Rakow T (2006) Uncertain judgements: eliciting experts' probabilities. Wiley, Hoboken, xiii + 321 pp. ISBN: 978-0-470-02999-2
81. Visscher PM (2008) Sizing up human height variation. Nat Genet 40:489–490
82. Gudbjartsson DF et al (2008) Many sequence variants affecting diversity of adult human height. Nat Genet 40:609–615
83. Lettre G et al (2008) Identification of ten loci associated with height highlights new biological pathways in human growth. Nat Genet 40:584–590
84. Altshuler D, Hirschhorn JN, Klannemark M, Lindgren CM, Vohl M-C, Nemesh J, Lane CR, Schaffner SF, Bolk S, Brewer C, Tuomi T, Gaudet D, Hudson TJ, Daly M, Groop L, Lander ES (2000) The common PPARγ Pro12Ala polymorphism is associated with decreased risk of type 2 diabetes. Nat Genet 26:76–80
85. Lango Allen H, Estrada K, Lettre G et al (2010) Hundreds of variants clustered in genomic loci and biological pathways affect human height. Nature 467:832–838

Chapter 4

Managing Large SNP Datasets with SNPpy

Faheem Mitha

Abstract

Using relational databases to manage SNP datasets is a very useful technique that has significant advantages over alternative methods, including the ability to leverage the power of relational databases to perform data validation, and the use of the powerful SQL query language to export data. SNPpy is a Python program which uses the PostgreSQL database and the SQLAlchemy Python library to automate SNP data management. This chapter shows how to use SNPpy to store and manage large datasets.

Key words SNP, Bioinformatics, Database, PostgreSQL, Data management, Genomics, GWAS

1 Introduction

Statistical analysis of SNP data from Genome-wide Association Studies (GWAS) typically involves the management and integration of patient information, including phenotypic data, with the genomic SNP data across multiple studies. Issues that have to be addressed include (1) data validation of the patient data and the SNP data, (2) performance issues with operating on large datasets, and (3) accurately updating the portions of the data that are rapidly changing, usually the patient data.

Consider the basic task of converting GWAS data, consisting of phenotype and genotype data, into standard format files like PED/MAP or TPED/TFAM for processing with statistical tools. A common approach to this task is direct code manipulation of raw data files. For example one can extract SNP and patient data from the files, perform any necessary transformations, and then write the results to another file. While this ad hoc approach is simple and therefore superficially attractive, it has major problems in practice.

For example, suppose the data is corrupt. With the flat file approach given above, problems will only be discovered when the data is being processed by statistical tools, as this will be the first time the data has undergone any validation. At this point the data has already undergone processing, and such errors are harder to trace back to the source.

Also, an approach based on manipulating these files requires the software to accommodate different source data formats like those used by Affymetrix and Illumina. Thus, a procedure that converts source data into output data files necessarily depends on both the source data format and the output data format. Therefore, in a real life system with multiple source and output data formats, a separate procedure would need to be written for each combination of source data format and output data format, resulting in the number of such procedures being the product of the number of source and data formats. For this reason, among others, such an approach quickly becomes both very complex and very fragile. There are better approaches to this problem.

One such approach uses relational databases. Relational databases are a well-known solution to parts of this problem, particularly data validation. However, they have not seen much use in this context, possibly because of performance issues caused by the typically large size of GWAS datasets, coupled with the complex data manipulations the database would need to handle. In this chapter, I show that these obstacles are surmountable, and that a usable and useful system can be built based on a relational database. This system is called SNPpy.

The functionality of SNPpy divides naturally into two parts. The first part involves *loading* data from various files into the database; a task which is commonly referred to as Extract-Transform-Load (ETL). The second part involves *exporting* data from the relational database via Structured Query Language (SQL) queries to standard format data files for use in further analyses. This process is illustrated in Fig. 1.

As justification that databases offer a better approach, consider the data validation and multiple source/data format issues described above. How is a database approach superior in the handling of these problems?

With regard to data validation, observe that unlike the naive file-based approach, databases validate data when they are loaded into the database. Thus, errors are discovered in the raw data *before* it has undergone processing, and therefore it is easy to trace the origin of the error. Clearly, it is extremely desirable to be working with validated data at the outset.

With regard to source/data formats, one can write a single procedure for the import of each source format into the database. Similarly, one can write a single procedure for the export of each output format from the database. This works because the layout of data in the database (henceforth called the data model, *see* **Note 1**) is largely independent of the source format. Thus, the total number of procedures required is approximately the sum of the number of source and data formats. As already discussed, the number of

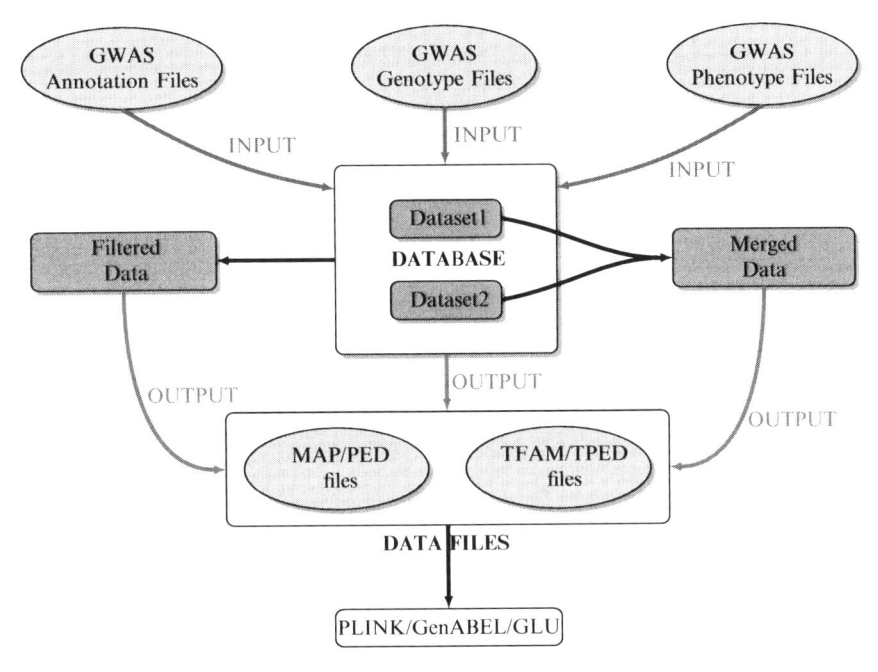

Fig. 1 Workflow chart. This figure shows the data workflow. First the genotypic and phenotypic data are loaded into the database. The data is then exported from the database as standard format files, including a possible filtering and/or merging step. Finally, the output files are further analyzed using third party tools.

procedures in the flat file approach is approximately the product of the number of source and data formats. So, the database approach is simpler in this respect.

A final advantage of relational databases is that they provide a domain-specific language dedicated to the problem of data selection and transformation, namely SQL. This is greatly preferable to using a general-purpose programming language for the task.

To expand on the validation issue, databases offer considerable and customizable machinery for low level data validation, which can be used to detect corrupt data. It is particularly important in the case of patient data, which is exceptionally mutable and corruption-prone. It is mutable because it is frequently updated as new patient information becomes available, and it is corruption-prone because patient information is typically entered manually and therefore errors can easily be introduced by human error or data entry software. More details about the kinds of validation provided by databases, along with their application to SNPpy, are provided in **Note 4**.

Statistical tools, which typically do not use a database, "hard-wire" their more limited validation checks. For example, PLINK [1] and GenABEL [2] verify that for each SNP the same number of patients have been called and that the SNPs are bi-allelic.

The heart of SNPpy is the data model illustrated in Fig. 2 (*see* **Note 1**). In addition to the data model, there are two classes

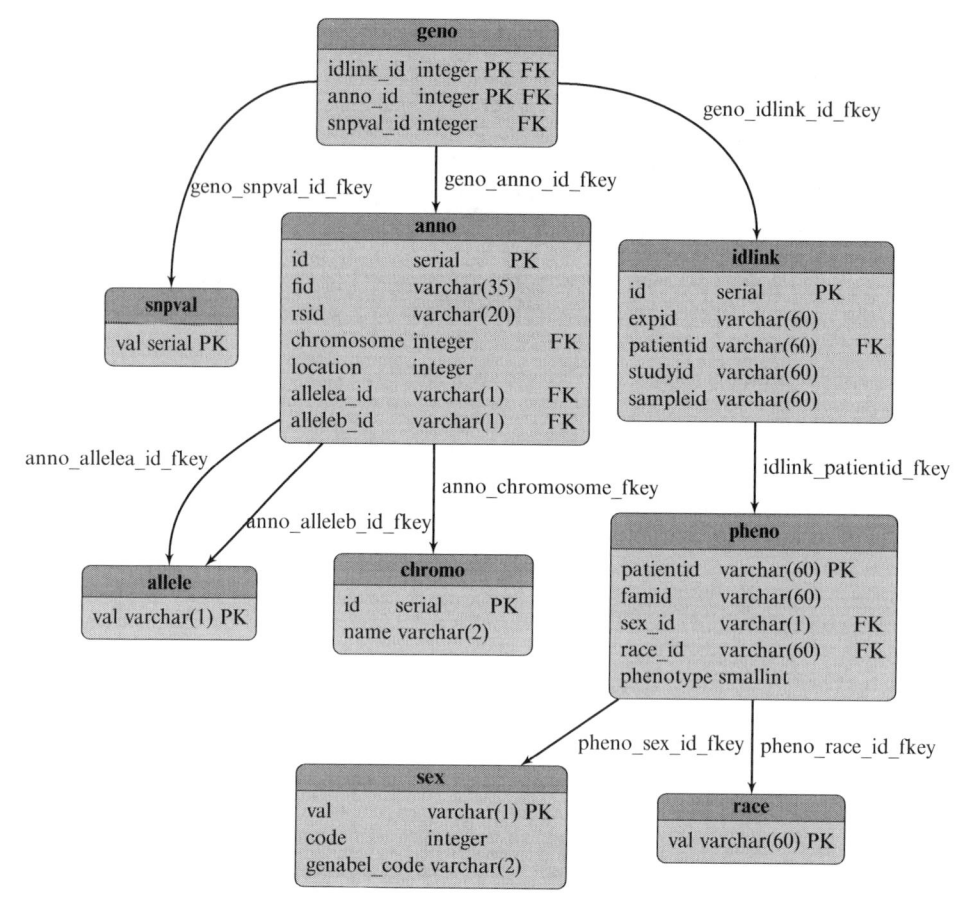

Fig. 2 Data model. Geno single data model for the Affymetrix platform (*see* **Note 1**). In this diagram, the rectangles correspond to database tables, and the rows in each rectangle correspond to table columns. The four columns in a row correspond to, from left to right, database name (column 1), data type (column 2), primary key indicator (column 3), and foreign key indicator (column 4). The arrows correspond to foreign keys. Observe the number of arrows leaving a table is equal to the number of columns that are foreign keys in that table

of Python scripts: (1) *input scripts* for parsing and loading the database tables, and (2) *output scripts* for processing and exporting the data into different downstream formats, using SQL queries. The input scripts are written using object-oriented Python, with classes corresponding to the different platforms. Currently, the system can produce PED/MAP and TPED/TFAM data formats for individual datasets, as well as the merger of multiple datasets. The latter is useful, for example, for doing quality control with HapMap data (*see* Subheading 2). A diagrammatic representation of the overall workflow is shown in Fig. 1.

SNPpy uses two slightly different data models, namely *Geno Single* and a variant called *Geno Shard*. The data model in the *Geno Single* case consists of nine tables.

- The *pheno* table contains the phenotypic data, where the table's primary key (the unique identifier) is the patient id, as chosen by the clinical study.

- The idlink table contains both experimental id and patient id columns. The former identifies samples, and the latter is a foreign key pointing to the *pheno* table and identifies patients. Note that a patient may give more than one sample to a study, and therefore a patient id may correspond to multiple samples. The genotype calls in the *geno* table are obtained from the samples. So this table connects the phenotypic and genotypic information located in the *pheno* and *geno* table respectively.

- The *anno* table contains SNP annotation data, including the chromosome, base-pair position and reference alleles for each SNP. The unique id for this table is the label assigned to the SNPs by the platform manufacturer.

- The *geno* table contains the genotype calls for the experiments and has a composite primary key consisting of two foreign keys, one which points to the *idlink* id and the other to the *anno* id. So each entry in this table is uniquely identified by both an *idlink* and *anno* id.

The data in the *pheno*, *anno*, and *geno* tables naturally correspond to the imported phenotype, annotation, and genotype/calls files. There are a number of auxiliary tables, namely *allele*, *chromo*, *race*, *sex*, and *snpval*, whose sole purpose is to provide data constraints on the main tables (*see* **item 1** in **Note 4**).

The *Geno Shard* data model is similar to the data model above, with the exception that the genotype data is partitioned into multiple tables, such that rows having the same experimental id are placed in the same table. This creates one genotype table per sample. The intention is to optimize both data loading and exporting to the PED file format. The following discussion is restricted to *Geno Shard*. However, using *Geno Single* is similar. *See* **Note 2** for a discussion of pros and cons of these two data models.

SNPpy currently supports two genotyping platforms, Affymetrix and Illumina. It uses both the data models described above for each platform. There are minor differences between the Affy6 and Illumina data models for *Geno Single*, which are restricted to the *anno* table, and similarly for *Geno Shard*. One database is used for each platform. Within each database, each schema (database namespace) corresponds to a dataset (*see* **Note 3**). Figure 3 provides a graphical representation of the data architecture.

SNPpy can load datasets into the database in these data models, using the load_dataset.py script. It can export data as MAP, PED, TPED and TFAM data files for the *Geno Single* layout, and as MAP and PED files for the *Geno Shard* layout.

Additionally, SNPpy provides a script, simdat.py, to generate simulated phenotype and genotype data, and a test script,

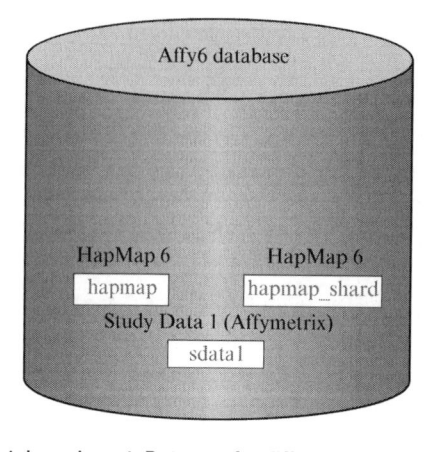

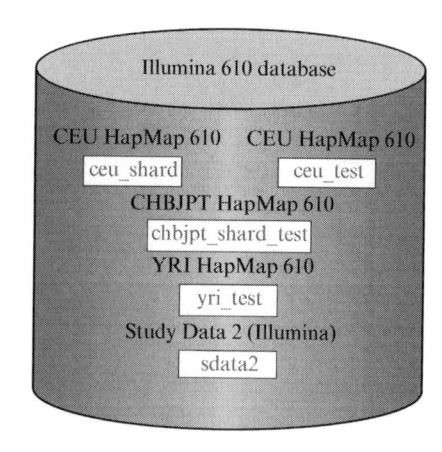

Fig. 3 Database layout. Datasets for different platforms are stored in separate databases, here represented by cylinders. Every dataset is stored in a separate database schema (namespace within a database). The same dataset can be stored in multiple schemas, differing in what options have been selected when loading the dataset. To illustrate this, the figure shows the schema names inside rectangular boxes, and and the dataset names above the boxes. Each of the datasets HapMap 6 and CEU HapMap 610 is stored in two schemas. For further details see the manual

test_simdata.py, which uses the simulated data. The test script currently tests code functionality for both Affymetrix and Illumina platforms using simulated phenotype and genotype data. The simulated data is useful for testing performance for large data sets, and tuning the PostgreSQL configuration for better performance. The test script is useful for quickly checking whether SNPpy runs in a given environment. The test script is used in Subheading 4.3.

It is usual to consider the data in a database to be valuable. However, the system is designed so the database can easily be regenerated from the source files, which are three in number, the phenotype file, the annotation file, and the (upstream) genotype file. Therefore, I advocate a different viewpoint; that the user consider these data files to be valuable, and that necessary changes be made to those files rather than to the database. Archiving flat files and regenerating the database as necessary is much easier than archiving a database. It is also easier if one wants to switch databases, as data in one database is usually not easy to port to another. Generally databases don't correspond to a small set of files as they do here; they are built up from inputs from many sources, which don't necessarily correspond to files at all. So archiving files as an alternative to archiving the database is usually not an option, but should be preferred if possible.

2 Materials

A standard quality control check for GWAS data consists of comparing genetic ancestry based on the study SNP data to self-reported ancestry and to genetic ancestry estimated from the HapMap samples for that platform. This is done by merging the study and HapMap data. This data will also be used in Subheading 3 to illustrate SNPpy usage.

In this section, instructions are provided for downloading and processing HapMap data assayed on the Affymetrix SNP 6.0 and Illumina 610 Quad platforms. In what follows, CEU denotes Utah residents with Northern and Western European ancestry from the Centre d'Etude du Polymorphisme Humain (CEPH) collection. CHB denotes Han Chinese in Beijing, China. JPT denotes Japanese in Tokyo, Japan. YRI denotes Yoruba in Ibadan, Nigeria.

For Affymetrix SNP 6 data, CEU, CHBJPT and YRI CEL files can be preprocessed into a single genotype calls file. For Illumina Human 610-Quad data, separate preprocessed HapMap call files for CEU, CHBJPT and YRI are available.

Check docs/MANUAL in the SNPpy source code repository for updates. These instructions are current as of 1st July 2012.

2.1 Downloading and Processing Affymetrix HapMap Data

NOTE: A free login account is needed to access the files from the Affymetrix page. The following files are required: the genotype calls file, the annotation file, and the phenotype file.

1. Generating the genotype calls file:
   The genotype calls were generated using the birdseed algorithm from the Affymetrix APT suite. This suite is available for download from http://www.affymetrix.com/partners_programs/programs/developer/tools/powertools.affx

   You can use the script apt.py in the SNPpy source to preprocess the files. This script calls the apt-probeset-genotype executable from the APT suite. Before running this script, you will need to change the variables at the top of the file. These are BASEDIR, CELFILEDIR, LIBDIR and RESULTDIR. The latter three directories are subdirectories of BASEDIR. The SNPpy default for BASEDIR is /data/snppy/Hapmap6.0/Geno, for CELFILEDIR is CEL, for LIBDIR is CD_GenomeWideSNP_6_rev3/Full/GenomeWideSNP_6/LibFiles, and for RESULTDIR is result. You will probably not need to change LIBDIR.

   The preprocessing steps and library files used are specific to the birdseed algorithm (version 2). Also see the information provided here:
   http://media.affymetrix.com/support/developer/powertools/changelog/apt-probeset-genotype.html#quickstartbirdseed

   The files required to process the Affymetrix SNP 6.0 HapMap data are:

   (a) The raw intensity data files (CEL files):
       They can be downloaded from http://hapmap.ncbi.nlm.nih.gov/downloads/raw_data/hapmap3_affy6.0

       These should be downloaded into CELFILEDIR. All the files in that directory are required. Warning: The total size of files in this directory is approximately 38 GB. Instead of manually downloading the files individually, you can use (on Unix like systems) wget with the command in Listing 1.

Listing 1. Downloading Affymetrix Hapmap data

```
wget -rce robots=off -np -nd -l1
http://hapmap.ncbi.nlm.nih.gov/
downloads/raw_data/hapmap3_affy6.0/
```

This command should be run in CELFILEDIR. To unzip these files, you can run the command in Listing 2 in the same directory as the files.

Listing 2. Unzipping Affymetrix Hapmap data

```
ls    *.tgz   |   grep   -v   Broad_
hapmap3_r2_Affy6_cels_excluded.tgz | grep
-v GIGAS.tgz | grep -v SCALE.tgz | grep -v
SHELF.tgz | xargs -n1 tar zxvf
```

This unzips all the tarred files except Broad_hapmap3_r2_Affy6_cels_excluded.tgz and GIGAS.tgz, which are not required. Note that these files will be unzipped into the current directory, namely CELFILEDIR.

(b) Library files:
- GenomeWideSNP_6.cdf
- GenomeWideSNP_6.chrXprobes
- GenomeWideSNP_6.chrYprobes
- GenomeWideSNP_6.specialSNPs
- GenomeWideSNP_6.birdseed-v2.models

They can be downloaded from http://www.affymetrix.com/Auth/support/downloads/library_files/genomewidesnp6_libraryfile.zip as a single zip file. This file should be downloaded into BASEDIR. Once unzipped, the library files can be found in the subdirectory CD_GenomeWideSNP_6_rev3/Full/GenomeWideSNP_6/LibFiles.

If you need to use a console Linux browser, w3m worked for me. After logging in, navigate to the link for the file at the bottom, and hit "a." See "How do I download a linked file?" at http://w3m.sourceforge.net/FAQ#www.

After adjusting the variables at the top of apt.py (you may only need to adjust BASEDIR) you can run this script as in Listing 3.

Listing 3. Running apt.py

```
python apt.py
```

The result of this script is the genotype calls file birdseed-v2.calls.txt in the RESULTDIR.

2. The annotation file:
This can be downloaded from http://www.affymetrix.com/support/technical/byproduct.affx?product=genomewidesnp_6 as http://www.affymetrix.com/Auth/analysis/downloads/na32/genotyping/GenomeWideSNP_6.na32.annot.csv.zip (GenomeWideSNP_6 Annotations, CSV format, Release 32 (297 MB, 7/15/11)).

3. The phenotype file:
This is included in the source repository at data/Hapmap6.0/Pheno/hapmappheno.csv.

2.2 Downloading and Processing Illumina HapMap Data

The following files are required: the genotype calls file, the annotation file, and the phenotype file.

1. The preprocessed genotype data can be accessed from the NIH ftp site via FTP:
ftp://ftp.ncbi.nih.gov/pub/geo/DATA/supplementary/series/GSE17205
ftp://ftp.ncbi.nih.gov/pub/geo/DATA/supplementary/series/GSE17206
ftp://ftp.ncbi.nih.gov/pub/geo/DATA/supplementary/series/GSE17207

For example, by using the commands in Listing 4.

Listing 4. Downloading Illumina Hapmap data

```
wget -c
ftp://ftp.ncbi.nih.gov/pub/geo/DATA/
supplementary/series/GSE17205/
GSE17205_Human610-Quadv1_
CEU_Final_Call_Report.csv.gz
ftp://ftp.ncbi.nih.gov/pub/geo/DATA/
supplementary/series/GSE17206/GSE17206_
Human610-Quadv1_CHB+JPT_Final_Call_Report.
csv.gz
ftp://ftp.ncbi.nih.gov/pub/geo/DATA/
supplementary/series/GSE17207/GSE17207_
Human610-Quadv1_YRI_Final_Call_Report.
csv.gz
gunzip*.gz
```

These files are not in the Illumina genotype report format supported by SNPpy. To convert these files into a format which can be used by SNPpy, please use the provided script convert_illumina_hapmap.py as shown in Listing 5.

Listing 5. Converting Illumina Hapmap files (general usage)

```
python        convert_illumina_hapmap.py
originalgenofile convertedgenofile
```

So the call looks as in Listing 6.

Listing 6. Converting Illumina Hapmap files (example)

```
convert_illumina_hapmap.py
GSE17205_Human610-Quadv1_CEU_Final_
Call_Report.csv
GSE17205_Human610-Quadv1_CEU_Final_
Call_Report_converted.csv
```

2. You will find annotation files for the Illumina genotyping platforms at their FTP site ftp://ftp.illumina.com/. The annotation file (Human610-Quadv1_B.csv) used in the chapter is available at ftp://ftp.illumina.com/Whole%20Genome%20Genotyping%20Files/Archived_Human_Products/Human610-Quad_v1_product_files/Human610-Quadv1_B%20files/Human610-Quadv1_B.csv.

You can log into ftp.illumina.com using the username and password in Listing 7.

Listing 7. Login for ftp.illumina.com

```
Username: guest
    Password: illumina
```

You can use this username and password to download the annotation files using wget as in Listing 8.

Listing 8. Downloading Illumina annotation files

```
wget     -c    --ftp-user=guest    --ftp-
password=illumina //ftp.illumina.com/Whole
%20Genome%20Genotyp-ing%20Files/
Archived_Human_Products/Human610-
Quad_v1_product_files/Human610-Quadv1_B%
20files/Human610-Quadv1_B.csv
```

You may also contact the Illumina customer service directly using the contact information provided here http://www.illumina.com/company/contact_us.ilmn.

3. The phenotype files are included in the source repository at data/Hapmap610/Pheno/ceupheno.csv, data/Hapmap610/Pheno/chbjptpheno.csv and data/Hapmap610/Pheno/yripheno.csv, corresponding to the CEU, CHBJPT and YRI datasets respectively.

3 Methods

The material in this section is adapted from the SNPpy user manual (docs/MANUAL in the software repository). To obtain the software repository, *see* **Note 5**. As discussed above, schema denotes a namespace within a database (*see* **Note 3**).

3.1 Creating Databases and Schemas

To create a database, use the script init_db.py as follows.

Listing 9. Creating database (general usage)

```
python init_db.py db dbtype
```

where *db* is the database name and *dbtype* denotes the database type (which is stored as part of the database meta-information, and is either "affy6" or "illumina"). For example, to create the database *affy6_faheem*, use Listing 10.

Listing 10. Creating database for Affymetrix data

```
python init_db.py affy6_faheem affy6
```

In general, creating a database of the form *dbtype_username* is recommended.

This command also adds a schema, *seq*, for storing sequence information. Currently only one sequence is put in the *seq* schema, the *dlink_id_seq* sequence, which is used to make sure the ids (the primary keys) in the *idlink* table are unique across all *idlink* tables in the database, so that datasets are merged correctly. See Subheading 4.1 for further information about creating the database. To create a schema, use the script init_schema.py as follows

Listing 11. Creating a database schema

```
python init_schema.py schema db
```

Table 1
Datasets and schemas

Dataset name	Options	Schema
datasetname	data model = *Geno Single*, dataset = full	datasetname
datasetname	data model = *Geno Shard*, dataset = full	datasetname_shard
datasetname	data model = *Geno Single*, dataset = truncated	datasetname_test
datasetname	data model = *Geno Shard*, dataset = truncated	datasetname_shard_test

where schema is the desired schema name and db is the database in which it is to be created; e.g.,

Listing 12. Creating a schema for Affymetrix data

```
python init_schema.py hapmap affy6_faheem
```

This script will create the schema if it does not exist, and by default, will ask whether to delete and recreate it if it does exist. To omit the request and recreate the schema if it exists, use the option −y. To omit the request and not recreate the schema if it exists, use the option −n. Note: it is not normally necessary to run the init_schema.py separately, as the load_dataset.py script will run it if necessary.

3.2 Relationship Between Datasets and Schemas

As discussed in the Introduction, each platform corresponds to a different database, and within each database, each schema corresponds to a dataset of that platform type. However, the correspondence between datasets and schemas is not one-to-one in general. There may be multiple schemas corresponding to the same dataset. One reason is that one may wish to use different data models with the same dataset, and these different data models need to live in different schemas. As also discussed in the Introduction, two slightly different data models are defined, namely *Geno Single* and *Geno Shard*. Another reason is that one may wish to load only a subset of the dataset for testing purposes. Since these two possibilities may be combined, this gives four different variations on a dataset, as outlined in Table 1. To summarize, datasets can be loaded into the database with two possible options:

- Data model: either *Geno Shard* or *Geno Single*.
- Truncated (test) dataset, or the full dataset.

For information about loading the datasets with these different options, *see* Subheading 3.4. These options are encoded in the name of the schema used to store the dataset, as described in Table 1.

So, for example, if the full dataset *datasetname* is stored in a schema with the *Geno Single* data model, then the schema name is *datasetname*. If the truncated dataset *datasetname* is stored in a schema with the *Geno Shard* data model, then the schema name is *datasetname_shard_test*.

The −s and −t options in most scripts are used to select the appropriate choice of schema from the possibly multiple schemas which contain different versions of the same dataset. Occasionally it is necessary to obtain the original dataset name from the schema name. For example, given the schema name *hapmap_shard_test*, one may need to deduce from that name that the schema corresponds to the *hapmap dataset*. Therefore, avoid using the strings *shard* or *test* in dataset names.

3.3 Config File Layout

SNPpy uses the Python ConfigObj library for configuration. The *config* file should contain one section for every platform. In this case the platforms are *affy6* and *illumina*. Each section contains subsections, one subsection for each dataset of that platform type. Here is the example from *default_conf* for the Affymetrix Hapmap data set. Here the platform used is *affy6* and the dataset is *hapmap*. The full pathnames are given as string values after the configuration options. Note that phenofiles is a list of strings, so this must be a list of paths separated by commas.

Listing 13. Affymetrix config

```
[affy6]
#annotation file
annotfile="/data/snppy/Hapmap6.0/Geno/Geno-
   meWide SNP_6.na29.annot.csv"
[[hapmap]]
phenofiles = "data/Hapmap6.0/Pheno/hapmappheno.
   csv",
genofile = "/data/snppy/Hapmap6.0/Geno/bird-
   seed-v2.calls.txt"
```

Pheno file format. SNPpy expects the pheno file to be of CSV format, with headers as detailed in Table 2.

3.4 Loading Datasets

The script for loading datasets is load_dataset.py. The general usage is

Listing 14. Loading datasets (general usage)

```
python load_dataset.py [-a alleletype] [-j jobs]
   [-n] [-s] [-t] [-y] dataset db
```

Table 2
Pheno file format

Pheno file headers	Notes
expid	Experimental Id
famid	Family Id
patientid	Patient Ids. If patientid is not given, expid is used instead
phenotype	0 or 1, depending on whether the phenotype is present or absent
race	Allowed values correspond to the setting "race_values" in the configuration file
sex	The sex column is restricted to the following values Sex Value Male 1 Female 2 Missing −1
studyid	Description of the study the data belongs to e.g., Hapmap6.0
	If studyid is not given, *unknown* is used

Before running this script, you need to have a subsection in your *config* files corresponding to the dataset to load, containing at least *phenofiles* and *genofile* entries.

Listing 15. Dataset configuration

```
[[dataset]]
phenofiles=...
genofile=...
```

NB: phenofiles is a list of strings, so even if there is only one phenofile, you need to end the string with a comma; e.g.,

Listing 16. Phenofiles entry in dataset configuration

```
phenofiles = "data/Hapmap6.0/Pheno/hapmappheno.
    csv",
```

The *annotfile* entry is also required. This can be placed either in the platform section or the dataset subsection. If it is in both, the entry in the dataset section takes precedence. If it is in the platform section, it looks like

Listing 17. Annotation file entry in dataset configuration

```
[affy6]
annotfile =...
```

You will need to have the files listed available at their path locations. Compile the C++ code used by load_dataset.py with *scons* before running the script. However, the script will attempt to compile the code if necessary before running. NB: the script will create the schema if it does not exist, and by default, will ask whether to delete and recreate it if it does exist. To omit the request and recreate the schema if it exists, use the option −y. To omit the request and not recreate the schema if it exists, use the option −n.

The load_dataset.py has an —*allele-type* (-a short) option for conversion of calls to numerical values, which is for use with the *illumina* platform. For simplicity, only the *Geno Shard* version is discussed here. Below are the commands for the Hapmap examples.

To load the dataset corresponding to the *Geno Shard* data model, use the −s option with load_dataset.py. In this case, the name of the schema is created by appending the string *_shard* to the name of the dataset. With the −j option, *load_dataset* will attempt to run multiple jobs simultaneously for the parts of the process where this is possible. With the −y option, *load_dataset* will delete the schema, including all datasets contained therein, and recreate it. To load the Affymetrix Hapmap data, use Listing 18.

Listing 18. Loading Affymetrix Hapmap data into *Geno Shard* data model

```
python load_dataset.py -syj    4 hapmap affy6_faheem
```

In this case, the schema used will be *hapmap_shard*. The command is similar for Illumina. However, since the Hapmap data is divided into the CEU, CHBJPT and YRI datasets, this command needs to be run 3 times, once for each dataset. For example, with CEU:

Listing 19. Loading Illumina CEU Hapmap data into *Geno Shard* data model

```
python load_dataset.py -syj   4   ceu   illumina_
    faheem -a forward
```

In this case, the schemas used will be *ceu_shard*, *chbjpt_shard*, and *yri_shard*. For a discussion of possible problems with dataset loading, *see* Subheading 4.3.

3.5 Dataset Output: Creating MAP, PED, TPED, and TFAM Files

General usage is of the form:

Listing 20. Writing output file (general usage)

```
python make_output.py fileformat [-s] [-t]
dataset₁... datasetₖ dbname -a alleletype
```

where $dataset_1 \ldots dataset_k$ are k datasets. If multiple datasets are specified, then the output file is calculated based on the merger of those datasets.

The option --*allele-type* specifies the rule for conversion of numerical allele values, and is only needed for Illumina PED and TPED files. It can take the values *forward* and *top*, and has no default value. For more information about stranding, consult doc/NOTES and

1. http://www.openbioinformatics.org/gengen/tutorial_allele_coding.html
2. http://www.illumina.com/documents/products/technotes/technote_topbot.pdf

Output file names are of the form $db.schema_1\text{-}schema_2\text{-}\ldots\text{-}schema_k.fileformat$ or $db.schema_1\text{-}schema_2\text{-}\ldots\text{-}schema_k.alleletype.fileformat$ (for Illumina PED or TPED files), where $schema_1$ is the schema corresponding to $dataset_1$ etc.

NOTE: The R GenABEL package requires the MAP file to be produced with column headers, so there is a -c option for make_output.py to create a MAP with column headers.

In the case of the PED and TPED file creation, it is also possible to run multiple PostgreSQL processes at the same time, which will make the process conclude faster at the cost of higher memory usage. To do this, use the —j option.

Mixing datasets corresponding to different data models (*Geno Single* and *Geno Shard*) will give an error.

To use the *Geno Shard* version of the dataset, use the -s option for make_output.py.

Example usage for Affymetrix Hapmap data is shown in Listings 21 and 22.

Listing 21. Creating MAP file from Affymetrix Hapmap data

```
python make_output.py -s   map hapmap affy6_
faheem
```

In this case, the schema used will be *hapmap_shard*.

Listing 22. Creating PED file from Affymetrix Hapmap data

```
python make_output.py -sj 2 ped hapmap affy6_
faheem
```

As in Listing 21, the schema used will be *hapmap_shard*. Similarly, for Illumina Hapmap data, use the command in Listing 23.

Listing 23. Creating PED file from Illumina Hapmap data

```
python make_output.py -sj 2 ped ceu chbjpt yri
illumina_faheem -a forward
```

In this case, the schemas used will be *ceu_shard*, *chbjpt_shard*, and *yri_shard*.

3.6 More on Dataset Output: Filtering and Merging

The output can be filtered by specifying a filter condition as an SQL expression in terms of the columns of one of the *anno*, *idlink* and *pheno* tables. This SQL expression must be included in the *config* file, within the dataset subsection the filtering should be applied to.

Up to three simultaneous filtering conditions are possible, corresponding to the three tables. These conditions restrict the data export to the selected subset. For instance, the user might export data corresponding to selected chromosomes of Caucasian male patients. Details:

1. There are three filtering options, corresponding to the *anno*, *idlink* and *pheno* tables.

2. The filters should be specified using the SQL standards for expressions.

3. Leave the filter blank if you don't want any filtering to be applied.

4. Remember to put single quotes around strings (VARCHAR, CHAR).

5. The filter expression is a string, so put double quotes around it.

Details of the three filters are

- The *idlink_filter* specifies a filter condition on the *idlink* table. Attributes: *expid*, *patientid*, *studyid*, *sampleid* (all VARCHAR). See examples in Listing 24.

Table 3
Anno table

Attributes: fid	(stored as VARCHAR)
rsid	(stored as VARCHAR)
chromosome	(stored as INTEGER: 1-22, 23=X, 24=Y, 25=XY, 26=MT)
location	(stored as INTEGER)

Listing 24. *idlink_filter* examples

```
idlink_filter="studyid='CEU' OR studyid='YRI'"
idlink_filter="studyid IN ('CEU', 'YRI')"
```
- The *pheno_filter* specifies a filter condition on the *pheno* table.

Attributes: *patientid*, *famid*, *sex_id*, *race_id*, *phenotype* (all VARCHAR except *sex_id* which is CHAR and *phenotype* which is INT). Example:
pheno_filter = "sex_id = 'M' AND race_id = 'white'"

- The *anno_filter* specifies a filter condition on the *anno* table (Table 3). Example:

anno_filter = "rsid IN ('200003', '200005') OR chromosome = 25"

Example of how this looks in a *config* file:

Listing 25. Configuration file section including filters

```
[affy6]
   [...]
   [[hapmap]]
   phenofiles="data/Hapmap6.0/Pheno/hapmap-
       pheno.csv",
   genofile="/data/snppy/Hapmap6.0/Geno/bird-
       seed-v2.calls.txt"
   anno_filter="chromosome=5"
   idlink_filter="studyid='CEU'"
```

Furthermore, it is possible to export data files corresponding to merged datasets. All exporting functionality is performed by the make_output.py script. General usage is of the form:

```
python  make_output.py  [-s][-t]  fileformat
dataset₁ ... datasetₖ dbname -a alleletype
```

where dataset₁ ... datasetₖ are k datasets. If multiple datasets are specified, then the output file is calculated based on the merger of those datasets.

4 Working with Databases

4.1 Creating the Database

All scripts use the same database user, defined as *dbuser* in the configuration, to create the databases and schemas, and write to and read from the databases. This user is initially created by the script init_db.py. This script performs the following steps:

1. It looks for a database user with the same name as the current operating system user. If such a user is available, it will use it. If not, it will attempt to gain root access via *sudo*, and create such a user.

2. Then it will try to connect as that user via unix socket, and create a user with the value of the *dbuser* configuration value, if one does not already exist. The password used is the value of the *password* configuration value.

3. Finally, it will create a database with *dbuser* as owner.

Therefore, the user running init_db.py needs to have *sudo* privileges on the machine, or have a database *superuser* of the same name as the shell account. In addition, the PostgreSQL client authentication configuration may need to be modified. This is located in the file pg_hba.conf. This section has basic instructions about the necessary modifications. For further information about PostgreSQL client authentication, see http://www.postgresql.org/docs/current/static/auth-pg-hba-conf.html. With the Debian and Ubuntu default client authentication configuration located at /etc/postgresql/8.4/main/pg_hba.conf, SNPpy will run without modification. For other Linux distributions, changes may be necessary. For example, the client authentication configuration file is located in Fedora at /var/lib/pgsql/data/pg_hba.conf, and its default values are given in Table 4.

For SNPpy to work, this needs to be changed to read as in Table 5. So for *host type* connections, the method needs to be changed from *ident* to *md5*. Then the PostgreSQL server needs to be restarted.

Listing 26. Restart Fedora PostgreSQL service

```
sudo  service  postgresql  restart
```

Table 4
Default values of Fedora PostgreSQL client authentication *config* file

# TYPE	DATABASE	USER	CIDR-ADDRESS	METHOD
# "local" is for Unix domain socket connections only					
local	all	all		ident
# IPv4 local connections:				
host	all	all	127.0.0.1/32	ident
# IPv6 local connections:				
host	all	all	::1/128	ident

Table 5
Revised values of Fedora PostgreSQL client authentication *config* file

# TYPE	DATABASE	USER	CIDR-ADDRESS	METHOD
# "local" is for Unix domain socket connections only					
local	all	all		ident
# IPv4 local connections				
host	all	all	127.0.0.1/32	md5
# IPv6 local connections				
host	all	all	::1/128	md5

The values in Table 5 are the same as the Debian/Ubuntu default values. The values in Table 4 (*ident*) correspond to the requirement that any database user has the same name as an operating system user, which is unreasonably restrictive.

4.2 Directly Accessing the Database

It is frequently desirable to access the database directly. For PostgreSQL, a good method is to use the *psql* interpreter, which is included as part of the basic PostgreSQL package on Linux, and is available for Windows at http://psql.sourceforge.net/. If PostgreSQL is configured correctly (*see* Subheading 4.1), the commands in Listing 27 should work if run as the same user that ran the other commands. It is necessary to set the search path to *simillum* in the transcript, since the default schema is *public*. Commands such as \dt (list tables) will only print tables associated with the schema in the search path (*see* **Note 3**).

Listing 27. Starting psql

```
psql -d illumina_faheem
psql (8.4.9)
Type "help" for help.
illumina_faheem=# \dn      #list schemas
List of schemas
  Name    |  Owner
----------+----------
[...]
 simillum |  snppy
[...]
# set schema
illumina_faheem=# SET search_path='simillum';
SET
illumina_faheem=# \dt     #list tables
           List of relations
 Schema   | Name  | Type  | Owner
----------+-------+-------+-------
[...]
 simillum | pheno | table | snppy
[...]
(9 rows)
# list values of pheno table
illumina_faheem=# SELECT * from pheno;
 patientid | famid  | sex_id | race_id | phenotype
-----------+--------+--------+---------+----------
 SAMPLE-1  | FAMID1 | M      | Asian   |     0
 SAMPLE-8  | FAMID8 | M      | Asian   |     0
```

4.3 Dataset Loading Errors

When loading datasets, the loading process may exit with an error. A likely reason is that the data violates the restrictions of the database schema. It may be that the error reflects corruption in the data, or that the database restrictions are too rigid. In the latter case, the restrictions can be relaxed if necessary. This section has some examples of data that will produce such errors. When possible, an error is raised before the data reaches the database, since this usually produces more useful errors. When this is not possible, the database raises the error. The built in simulated examples are used, since they run quickly. There are two simulated examples, one for Affymetrix and the other for Illumina. The *config* for the Affymetrix simulated example (*simaffy6*) and Illumina simulated example (*simillum*) are in the *config* file *default_conf*. Here is the *simillum config*,

Listing 28. Illumina *simillum* default *config*

```
[illumina]
[[simillum]]
annotfile = "data/illumina-anno.txt"
phenofiles  =  "data/dummy-n10-K123-illumina-
    pheno.txt",
genofile = "data/dummy-n10-K123-illumina-geno.
    txt"
```

and similarly for the *simaffy6 config*. First, run the basic simulation with Listing 29 to check it works for you. This will create the *geno* and *pheno* data files for the simulated examples in the data directory, and also load the datasets into the database in *Geno Single* mode. For *Geno Shard* use similar commands, except with the −s flag.

Listing 29. Running simulated example

```
python test_simdata.py
```

You can run this command again to reset the *geno* and *pheno* data files to their default values. The data files are obtained by simulation, but this script sets the seeds for the random number generators by default to ensure that the same data files are obtained each time. The default value of the *seed* is 0. To set the value of the *seed* to something else, you can use the *--seed* option to test_simdata.py. The Illumina *simillum config* values are given in *default_conf*. To run either the corresponding Affymetrix or Illumina database loading command in *Geno Single* format, use the commands in Listing 30.

Listing 30. Command to load simulated datasets in Geno Single format

```
python load_dataset.py -y simaffy6 affy6_faheem
python load_dataset.py -y simillum illumina_fa-
    heem -a forward
```

Go to the Illumina Hapmap *pheno* data file referenced above: data/dummy-n10-K123-illumina-pheno.txt, and change the first line from SAMPLE-1, ..., Asian to SAMPLE-1, ..., Martian. So, the race is changed from White to Martian. Now run the second (*simillum*) command in Listing 30. This gives the error:

Listing 31. Error message with illegal race value

```
Error: the value of race in line
SAMPLE-1,FAMID1,
PATID1,0,0,1,1,121.184964607,1.87
89640642,1,Martian
of pheno file data/dummy-n10-K123-illumina-
    pheno.txt is
Martian, which is not listed as a permitted value.
    If you want to allow it, add Martian to the
    race_values list in the config
```

Here, race values are checked against a permitted list in the *config*, namely *race_values* = Oriental, Black, White, Asian, Nat. hawai, Asian/PacificIslander, Nat.amer, American Indian/Alaska, Unknown, Other, CEU, CHD, YRI, CHB, JPT, LWK, ASW, MEX, TSI, GIH, MKK, CHBJPT, before being loaded to the database. Now consider a different change for the same line, from: SAMPLE-1, FAMID1, PATID1, 0, 0, 1, 1, ..., Asian to SAMPLE1, FAMID1, PATID1, 0, 0, 0, 1, ..., Asian; i.e., changing the sixth field, *sex* from 1 to 0. This gives us the error.

Listing 32. Error message with illegal sex value

```
Error: the value of sex in line

SAMPLE-1,   FAMID1,   PATID1,   0,   0,   0,   1,
   121.184964607, 1.8789640642, 1, Asian

of  pheno  file  data/dummy-n10-K123-illumina-
   pheno.txt is 0.

Only sex values of 1, 2 and -1 are allowed.
```

In data/dummy-n10-K123-illumina-pheno.txt change the *expid* in the first line from SAMPLE-1 to SAMPLE-2; i.e., SAMPLE-1, FAMID1, ..., Asian to SAMPLE-2, FAMID1, ..., Asian. Since the sample SAMPLE-2 now occurs twice, and *expid* must be unique, this gives the error in Listing 33. Unlike the other errors, this one comes directly from the database.

Listing 33. Error message with duplicate values of expid

```
sqlalchemy.exc.IntegrityError:
   (IntegrityError)   duplicate   key   value
   violates             unique            constraint
   "ix_simillum_idlink_expid" "INSERT INTO
simillum.idlink(id,expid,patientid,studyid,
   sampleid)
VALUES (nextval(seq.idlink_id_seq), %(expid)s,
   %(patientid)s,
%(studyid)s, %(sampleid)s)" %({expid:SAMPLE-
   1, sampleid:
%DUMMY0,patientid:DUMMY0,%studyid:unknown},
   {expid:
%SAMPLE-2,sampleid:SAMPLE-2,
   %patientid:SAMPLE-2, studyid:
%unknown})... and a total of 11 % bound parameter
   sets
```

Go to the first line (starting with 200003) in data/illumina-anno.txt, and change the chromosome value from 9 to 100; i.e., from 200003-0_B_R_1526882017, 200003, ..., 9, 139026180,

...AAG,18,0,3,0 to 200003_B_R_1526882017, 200003,...,100, 139026180, ... AAG,18,0,3,0. 100 is an illegal chromosome value; allowed values of chromosome are listed in the chromosome table. This gives the database error in Listing 34.

Listing 34. Error message for illegal chromosome value

```
sqlalchemy.exc.IntegrityError:
    (IntegrityError) insert or update on table
    "anno" violates foreign key constraint
    "anno_chromosome_fkey"
DETAIL:Key(chromosome)=(100) is not present in
    table "chromo".
'ALTER TABLE ONLY simillum. anno ADD CONSTRAINT
    anno_chromosome_fkey        FOREIGN      KEY
    (chromosome)   REFERENCES    simillum.chromo
    (id) ON UPDATE CASCADE ON DELETE CASCADE;'{}
```

Go to the first line (starting with 200003) in data/dummy-n10-K123-illumina-geno.txt, and change 200003 GG GG...GG to 200003 TT GG ... GG.

TT is an impossible value, since the only other letter that can occur on that line is C.

This gives the error in Listing 35.

Listing 35. Error message with illegal Illumina genotype call

```
problems detected with call TT in 200003
There were problems converting the genotype
    calls of the following rsids to integers, so
    the calls were discarded: [200003]
```

Removing the first TT entirely gives the error in Listing 36.

Listing 36. Error message with incorrect number of genotype calls

```
Error: The number of genotype calls
    corresponding to rsid 200003 in
    genofiledata/dummy-n10-K123-illumina-
    geno.txt should be the same as the number of
    idlink_ids (10) but is 9
```

Go to the first line (starting with SNP_A-1780419) in data/dummy-n10-K123-affy6-geno.txt, and change SNP_A-1780419 2 2...2 2 to SNP_A-1780419 3 2...2 2. 3 is an impossible value for a genotype call, since the permitted values, as listed in the snpval table, are {-1, 0, 1, 2}. This gives the error message in Listing 37.

Listing 37. Error message with illegal Affymetrix genotype call

```
sqlalchemy.exc.IntegrityError:
    (IntegrityError) insert or update on table
    "geno" violates foreign key constraint
    "geno_snpval_id_fkey"       DETAIL:      Key
    (snpval_id)=(3) is not present in table
    "snpval".'ALTER TABLE ONLY simaffy6. geno
    ADD CONSTRAINT geno_snpval_id_fkey FOREIGN
    KEY (snpval_id) REFERENCES simaffy6.snpval
    (val) ON UPDATE CASCADE ON DELETE CASCADE;'{}
```

5 Notes

1. *Data model.* A data model is used in this chapter to denote the structure of a set of tables in a database, the relationships between them, and the constraints on them. Relationships between tables are typically modeled with foreign keys, which can be thought of as pointers from a table to columns in another table. The main purpose of foreign keys is to ensure values in a column (or a group of columns) match the values appearing in a column (or a group of columns) of another table. See http://www.postgresql.org/docs/current/static/tutorial-fk.html and http://www.postgresql.org/docs/current/static/ddl-constraints.html#DDL-CONSTRAINTS-FK. Foreign keys are important in the context of validation (*see* **Note 4**).

   From a more abstract perspective, a data model can be thought of as a way to organize data. It is often illustrated using a notation called an "Entity-Relationship Diagram." In this chapter, the ERD for the data model is given by Fig. 2. In this diagram, the arrows correspond to foreign keys.

2. *Geno Single* vs. *Geno Shard*. *Geno Single* and *Geno Shard* are two main data models defined in SNPpy. These are very similar; the only difference is that there is one *geno* table in *Geno Single*, whereas *Geno Shard* splits this table up into a collection of small *geno* tables by sample, each of size equal to the number of SNPs in the anno table; i.e., each table in the *Geno Shard* layout contains exactly the genotype data of a single sample.

   Geno Single is the natural way to define this table. The main reason for defining *Geno Shard* is performance reasons, specifically to optimize PED data file export. SNPpy performance issues are discussed in detail in the Subsection Performance and Optimization of the Section Design and Implementation of [3]. The drawbacks of *Geno Shard* are, first, increased complexity of implementation. Second, the performance benefits only accrue when queries or selections from the *geno* tables are made for genotype data from the same sample. When this is the case, as with writing the PED file, the performance benefits over that of *Geno Single* are considerable. If this is not the case, the performance may be no better, or worse, than that of *Geno Single*.

3. *Database schemas. Schema* is a term used by a number of different database projects to denote an object within a database used to create a name space. A PostgreSQL database may contain one or more schemas. These schemas in turn contain tables, as well as other data objects. The same object name can be used in different schemas without conflict; for example, both *schema1* and *myschema* can contain tables named *mytable*. Therefore schemas are used where a set of items can have names that are unique within that name space, but do not have to be globally unique across the database, as in the current application. For further information, see http://www.postgresql.org/docs/current/static/ddl-schemas.html.

By default, tables (and other objects) created without specifying a schema are automatically placed in a schema named *public*. Every database contains this schema. When working with a schema other than *public*, it is necessary to prefix data objects with the schema name and table name separated by a dot, i.e., *schema.table*. This is the qualified name. It is convenient to change the search path to the schema being used, in which case one can just use the table name. This is the unqualified name. Schema is also sometimes used to mean the design of the tables, the relationships between them, and the constraints on them. However, to avoid confusion, the term *data model* will be used for this (*see* **Note 1**). For further information, see http://www.postgresql.org/docs/current/interactive/ddl-schemas.html.

4. *Validation*. Here is a list of database constraint features which are useful for data validation, illustrated with examples of their usage in SNPpy, where applicable. For an explanation of the term data model *see* **Note 1**. SNPpy's data model is illustrated in Fig. 2.

- Column values can be constrained to a fixed set of values defined in auxiliary tables. SNPpy's data model uses the *allele, chromo, race, sex,* and *snpval* tables for providing data constraints on the main tables by restricting certain columns to the contents of the aforementioned tables (as discussed in the 'Design' Subsection of the 'Design and Implementation' Section of [3]. This is typically done using foreign keys (*see* **Note 1**).

- Less restrictively, column values can be confined to a specific type. For instance, a specific column can be configured to only accept integers or character strings of a fixed or minimum or maximum length. SNPpy's data model contains many examples of type constraints on integer and character values. For example, the *chromosome* and *location* columns in the *anno* table are constrained to be integers.

- A more general category of constraint is *check constraints*, which allows the user to specify that the value in a certain column must satisfy a Boolean (truth-value) expression.

Other useful constraints are *unique constraints*, which ensure that the data contained in a column or a group of columns is unique with respect to all the rows in the table. Uniqueness is used in a number of places. For example, all main identifiers, like *fid* (feature id) in the *anno* table, and *name* in the *chromo* table, are constrained to be unique.

As examples of how this works in practice, if a record specifies that a patient's sex is "B" (the only valid sex values are "F" for female and "M" for male), the database will return an error on loading. Similarly, if a string value is specified for a SNP location (only integer values are allowed) the database will return an error. Also, genotype data is converted to and constrained to be stored as integers from the set $\{-1, 0, 1, 2\}$, which makes corruption of this data unlikely. For examples of corrupt data, and actual errors given by the database, *see* Subheading 4.3.

5. Availability and requirements.

    - Project name: SNPpy

    - Project home page: http://bitbucket.org/faheem/snppy. Please submit bugs to the Bitbucket issue tracker at that page. The mailing list is located at http://groups.google.com/group/snppy. A secondary project home page is http://code.google.com/p/snppy-code/. Please use this in case of problems with Bitbucket. However, Bitbucket should still be considered the main home page.

    - Operating system(s): Linux i386 and AMD64. Tested on the following distributions: Debian 5.0 (lenny) and 6.0 (squeeze), Fedora Core 13 (Goddard) and 14 (Laughlin), Ubuntu 10.04.1 LTS (Lucid Lynx), OpenSUSE 11.3.

    - License: GNU General Public License (GPL), version 2 or later.

    - Dependencies. To run this code, you require:

Build Dependencies
- The GNU C++ compiler (http://gcc.gnu.org). Tested with version 4.4 and 4.7. 4.5 and 4.6 should also work.
- The C++ Boost libraries, including Boost Python (http://boost.org). Tested with version 1.42. Later versions should also work.
- SCons (http://scons.org).

Runtime Dependencies
- PostgreSQL version 8.4 or later (http://postgresql.org),
- Python 2.x where x is >=6. The Python development files or header files, usually corresponding to a package called python-dev or python-development (http://python.org).
- SQLAlchemy version 0.7.x (http://sqlalchemy.org). 0.6.x and 0.5.x should also work.
- The ConfigObj Python library, version 4.7.2 or later (http://www.voidspace.org.uk/python/configobj.html).
- Psycopg 2 (http://initd.org/projects/psycopg). PostgreSQL database adapter for Python.

Additionally, the following are needed if you want to run the test_simdata.py script:

- The Python Numpy library (http://numpy.scipy.org/).
- The PLINK whole-genome association analysis toolset (http://pngu.mgh.harvard.edu/~purcell/plink/)

Acknowledgements

This article is adapted from [3]. The author wishes to thank his coauthors on this, the original SNPpy project paper; namely Herodotos Herodotou, Nedyalko Borisov, Chen Jiang, Josh Yoder, and Kouros Owzar. The author also wishes to thank the PostgreSQL community, specifically the denizens of the postgresql-general and postgresql-performance mailing list, and particularly the Freenode IRC channel #postgresql. The people who contributed advice and suggestions are too numerous to list them all, but specific mention goes to Andrew Gierth (RhodiumToad) #postgresql's resident expert on everything PostgreSQL related, Erikjan Rijkers (breinbaas), Jeff Trout (threshar), David Fetter (davidfetter), Jon T Erdman (StuckMojo), David Blewett (BlueAidan), depesz, Casey Allen Shobe (Raptelan), Chua Khee Chin (merlin83), Marko Tiikkaja (johto), Robert Haas and Robert Schnabel. The author also wishes to thank Michael Bayer, the author of SQLAlchemy, who has been extremely helpful in answering numerous queries on the

SQLAlchemy user mailing list. Finally, the author wishes to thank the members of the StackExchange question-answer sites, especially stackoverflow.com, unix.stackexchange.com, and tex.stackexchange.com, for answering many questions in connection with this project. The members of tex.stackexchange.com were particularly helpful with regard to LaTeX issues.

References

1. Purcell S et al (2007) PLINK: a tool set for whole-genome association and population-based linkage analyses. Am J Hum Genet 81(3):559–575
2. Aulchenko YS et al (2007) GenABEL: an R library for genome-wide association analysis. Bioinformatics 23(10):1294–1296
3. Mitha F, Herodotou H, Borisov N, Jiang C, Yoder J, Owzar K (2011) SNPpy - Database Management for SNP Data from Genome Wide Association Studies. PLoS ONE 6(10):e24,982, DOI 10.1371/journal.pone.0024982, URL http://dx.doi.org/10.1371%2Fjournal.pone.0024982

Chapter 5

Quality Control for Genome-Wide Association Studies

Cedric Gondro, Seung Hwan Lee, Hak Kyo Lee, and Laercio R. Porto-Neto

Abstract

This chapter overviews the quality control (QC) issues for SNP-based genotyping methods used in genome-wide association studies. The main metrics for evaluating the quality of the genotypes are discussed followed by a worked out example of QC pipeline starting with raw data and finishing with a fully filtered dataset ready for downstream analysis. The emphasis is on automation of data storage, filtering, and manipulation to ensure data integrity throughput the process and on how to extract a global summary from these high dimensional datasets to allow better-informed downstream analytical decisions. All examples will be run using the R statistical programming language followed by a practical example using a fully automated QC pipeline for the Illumina platform.

Key words Genome-wide association studies, Quality control, Illumina, R statistics

1 Introduction

Data for genome-wide association studies (GWAS) demand a fair amount of preprocessing and quality control (QC), especially SNP genotypes. A couple of good review articles on quality control issues in GWAS are given by Ziegler [1] and Teo [2]. These are high dimensional data, which preclude any meaningful form of manual data evaluation. This means that any QC step will have to be automated with limited opportunity for the researcher to intervene directly in the process. The need for this high level of automation can lead to a suboptimal understanding of the dataset at hand and, while the objective of QC is to remove *bad* data points, there is a risk of adding additional bias through the process. In this chapter we will discuss the most commonly used QC metrics and show some simple code to run these analyses using R. Two underlying themes will run throughout the chapter; the first revolves around the importance of setting up a backbone infrastructure to ensure data integrity/consistency, and the second theme is on automating the QC steps, and also summarizing these results into a *human digestible* format which will allow the data to reveal itself and help guide decisions on the best way forward for downstream analysis. A fully automated pipeline for analysis and reporting of QC results for Illumina SNP data is available at http://www-personal.une.edu.au/~cgondro2/CGhomepage. This pipeline is briefly discussed at the end of the chapter.

2 Platform

In recent years R [3] has become de facto statistical programming language of choice for statisticians and it is also arguably the most widely used generic environment for analysis of high throughput genomic data. We will use R to illustrate the concepts and show how to implement the QC metrics in practice, but it is straightforward to port them to other platforms. Herein we assume the reader is reasonably familiar with R and its syntax. For those who are unfamiliar with it, two excellent texts more focused on the programming aspects of the language are Chambers [4] and Jones et al. [5].

3 Storing and Handling Data

Data management is an important step in any project, but it is particularly critical with large datasets. Unfortunately it is often relegated to second plan. A common workflow in GWAS starts with a phenotypes and samples collection stage, followed by SNP genotyping, data QC, genotypes phasing, imputation, and actual association testing. At each point data gets changed (e.g., SNP and samples are discarded in the QC stage); it is paramount to be able to track these changes and revert back to the original raw data at any stage, if things go wrong—and they rather frequently tend to (*see* **Note 1**).

Databases are by far the best approach to manage large datasets. While for large collaborative projects it is essentially mandatory to have a dedicated database manager to design the database and manage data storage/access, it is still easy to implement robust solutions for smaller projects. To illustrate we will work with a small simulated dataset that can be downloaded from the publisher's website (http://extras.springer.com/). This is a small dataset with a limited number of samples and only 50,000 SNP but still convenient to illustrate the process. This data could quite easily be handled directly in R, but for larger datasets dimensionality can become a problem—the whole dataset would not fit into the memory of common desktop machines. The simplest work around is to store the data in a database and retrieve only the parts of the data that are needed at any given time. R can interface quite easily with databases—SQL queries can be sent straight from R to the database and retrieved records stored as a *data.frame*.

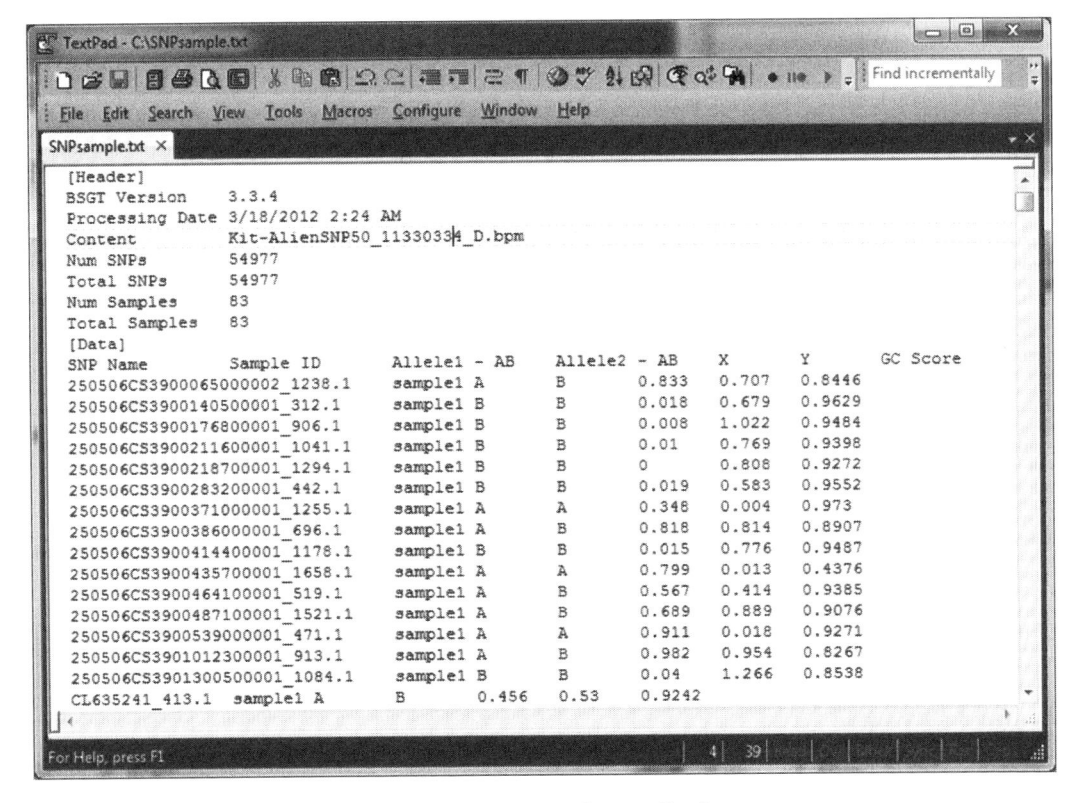

Fig. 1 Example Illumina SNP genotype file exported from GenomeStudio

SNP array data will usually come in two formats: either in a proprietary database structure developed by the chip manufacturer (e.g., the Genome Studio *bim* files) or as a flat file. We will work only with the flat files as it is the most commonly used form. The first step is to build a database from the flat files for further downstream analysis. This can be built straight from R, but of course it does not have to be. There are many options for working with databases. We will use *SQLite*. The key advantages are that it connects well to R—e.g., the annotation packages from Bioconductor (http://www.bioconductor.org) are all built with it; there's no installation involved—a single 500 kb executable is all that's needed; databases can simply be copied across machines without any further installation (*see* **Note 2**). To access an SQLite database from R the *RSQLite* package is needed.

For our example we will use two files: a *genotypes* file and a *map* file. The first contains all genotype calls for all SNP and all samples (Fig. 1); the second holds mapping information of the SNP, e.g., chromosome, physical location, etc.

To build a database directly in SQLite we first need to create a schema (a schema is simply a text file that describes the database structure and is used to create the tables and fields in an empty DB) with the tables and columns we want in it. A schema can be written in any plain text editor. Here the *map* file has only 3 columns: *SNP name, chromosome, and position* but could have additional columns, e.g., actual SNP nucleotides. The *genotypes* file (Fig. 1) has 7 columns: *SNP name, Sample ID, Allele1, Allele2, X, Y,* and *GC Score* (the first 4 columns are intuitive and the last 3 we will discuss later on). Notice that we have usual nuisance information lines (header text), nine lines in this example. A simple schema for these data will consist of two tables, one for each of the files and one column for each source of information. A *snpmap* table with SNP name, chromosome and position, and a *snps* table with the 7 columns from our dataset. The schema could look like

```
CREATE TABLE snpmap(
    name,
    chromosome,
    position
);

CREATE TABLE snps(
    snp,
    animal,
    allele1,
    allele2,
    x,
    y,
    gcscore
);

CREATE INDEX snp_idx ON snps(snp);
CREATE INDEX animal_idx ON snps(animal);
CREATE INDEX chromosome_idx ON snpmap(chromosome);
```

This is simply a plain text file with the structure (description) of the DB that we want to create. The table and columns names are discretionary; here we decided to hold mapping information (3 columns) in *snpmap* and the genotypes in *snps* (7 columns). Note that indices are created at the end to make sure that searches by SNP ID, sample ID, or chromosome are fast to execute (indices make the database slow to build but speed up the queries dramatically). A good source of information for SQLite syntax can be found at http://www.sqlite.org/. Once we've saved our schema in a text file (e.g., *snpDB.sql* or *schema.txt*) we are ready to create the database and populate it with the data. This can be done on the command line (provided of course *sqlite* is installed and in the OS path) with:

```
sqlite SNPsmall < snpDB.sql
```

and this creates a new database with the previous schema (*see* **Note 3**) ready to be populated with the genotype data. It is also simple to create a new database directly in R, and for this case not even the schema is needed. First open R and load the *RSQLite* package

```
library(RSQLite)
```

Then run the following code:

```
con=dbConnect(dbDriver("SQLite"),dbname="SNPsmall")
dbWriteTable(con,"snpmap","SNPmap.txt",header=TRUE,
append=T,sep="\t")
dbWriteTable(con,"snps","SNPsample.txt",append=T,
header=TRUE,skip=9,sep="\t")
```

Let's go over the code. The first line simply creates a new blank database called *SNPsmall* using the *SQLite* database driver (R has DBMS for most common database engines). In the same line a connection (function *dbConnect*) to the new database is created (if the database already existed, R would simply open a connection to it).

Now we have an empty DB similar to what we did using straight *sqlite* but without any tables/fields information. R can simply create tables and fields directly from the flat files themselves. To populate the DB (once connected to it using *dbConnect*) we can upload our flat files of genomic data *SNPmap.txt* and *SNPsample. txt* using the function *dbWriteTable* (once for each file). The function takes quite a few arguments: the database connection *con* that was created in the previous line, the name of the a new table to be created in the DB, the name of the flat file to import, if there is a header in the file it can be used to create the field/column names, if appending or not to the database, the separator used between columns in the data, and how many lines to skip if there are extraneous header files. Note that in this example the top lines from the genotypes file have to be removed (Fig. 1).

We can have a look at the tables and fields in the DB using *dbListTables* and *dbListFields*

```
dbListTables(con)
> "snpmap" "snps"

dbListFields(con,"snpmap")
> "name" "chromosome" "position"

dbListFields(con,"snps")
> "snp" "animal" "allele1" "allele2" "x" "y" "gcscore"
```

The function *dbGetQuery* is used to send an SQL query to the DB and return the data in a single step. A two-step approach is using *dbSendQuery* and fetch, but we will not discuss these here. The syntax for *dbGetQuery* is *dbGetQuery (connection name, "SQLquery")*. For example, to retrieve the number of records in a table:

```
dbGetQuery(con,"select count (*) from snpmap")
> 1   54977
```

```
dbGetQuery(con,"select count (*) from snps")
> 1   4563091
```

That looks about right. There are 54,977 records in *snpmap* and we know the chip has 54,977 SNP so that matched up well. The number of records in *snps* is also fine—it should be the number of samples times the number of SNP. To retrieve sample ids we would do something along the lines of

```
animids=dbGetQuery(con, "select distinct animal from snps")
```

```
dim(animids)
> 83  1
```

```
head(animids)
```

```
    animal
1 sample1
2 sample10
3 sample11
4 sample12
5 sample13
6 sample14
```

Herein we will not discuss details of SQL queries or syntax. Any general SQL book will cover most of the common needs (see for example [6]). All we really need to know is how to use *select* * from *tableName* where *columnName="mysearch"*. For example to retrieve all data associated to the first sample.

```
animids=as.vector(animids$animal)
hold=dbGetQuery(con,paste("select * from snps where animal='",animids[1],"'", sep=""))
```

```
dim(hold)
> 54977   7
```

```
head(hold)
```

```
                        snp animal allele1 allele2
1 250506CS3900065000002_1238.1 sample1       A       B
2 250506CS3900140500001_312.1  sample1       B       B
3 250506CS3900176800001_906.1  sample1       B       B
4 250506CS3900211600001_1041.1 sample1       B       B
5 250506CS3900218700001_1294.1 sample1       B       B
6 250506CS3900283200001_442.1  sample1       B       B
      x   y gcscore
```

```
1 0.833 0.707 0.8446
2 0.018 0.679 0.9629
3 0.008 1.022 0.9484
4 0.010 0.769 0.9398
5 0.000 0.808 0.9272
6 0.019 0.583 0.9552
```

In the first line we just changed the *data.frame* with animal ids to a vector—saves some indexing work. Then we retrieved the data for the first sample from our vector of sample ids. It's quite easy to picture a loop for each sample—read in all genotypes for the sample, run some QC metrics, read in the next sample… Notice the use of *paste* to create a query string and also the rather awkward use of single and double quotes—we need quotes for the R string and we also need to include a single quote for the SQL query in the DB. Just the query string looks like

```
paste("select * from snps where
animal='",animids[1],"'",sep="")
> "select * from snps where animal='sample1'"
```

We already have a vector for the samples. Let's also get a vector of SNP.

```
snpids=as.vector(dbGetQuery(con, "select distinct
name from snpmap")[,1])

length(snpids)
> 54977

head(snpids)
[1] "250506CS3900065000002_1238.1"
[2] "250506CS3900140500001_312.1"
[3] "250506CS3900176800001_906.1"
[4] "250506CS3900211600001_1041.1"
[5] "250506CS3900218700001_1294.1"
[6] "250506CS3900283200001_442.1"
```

And one last thing. When finished with the DB we should close the connection.

```
dbDisconnect(con)
```

4 Quality Control

Now that the data is safely tucked away in a database, we are ready to do some quality control on the genotypes. Various metrics are commonly used and there is still some level of subjectivity in these, particularly when setting thresholds. The statistics are performed either across SNP or across samples. The key concept is to detect

SNP and/or samples that should be removed prior to the association analyses. Let's start with across SNP analyses.

4.1 Genotype Calling and Signal Intensities

SNP alleles are usually coded as A/B in Illumina or the actual nucleotides are used. This changes depending on the platform or laboratory; but preference should be to use a simple reference for alleles and an additional DB table with all pertinent information for the SNP. Let's have a look at the first SNP (*snpids[1]*) in the dataset.

```
con=dbConnect(dbDriver("SQLite"),dbname =
"SNPsmall")

snp=dbGetQuery(con,
    paste("select * from snps where snp='", snpids
    [1],"'",sep=""))

dim(snp)
> 83 7

head(snp)
                                  snp animal allele1 allele2
1 250506CS3900065000002_1238.1 sample1    A       B
2 250506CS3900065000002_1238.1 sample5    A       B
3 250506CS3900065000002_1238.1 sample6    A       B
4 250506CS3900065000002_1238.1 sample7    B       B
5 250506CS3900065000002_1238.1 sample8    B       B
6 250506CS3900065000002_1238.1 sample9    B       B
      x     y gcscore
1 0.833 0.707  0.8446
2 0.829 0.714  0.8446
3 0.816 0.730  0.8446
4 0.031 1.132  0.8446
5 0.036 1.146  0.8446
6 0.037 1.150  0.8446

snp$allele1=factor(snp$allele1)
snp$allele2=factor(snp$allele2)

summary(snp$allele1)
> A B
> 36 47

summary(snp$allele2)
> A B
> 6 77
```

First reopen the DB connection and then send an SQL query to retrieve data for the first SNP. Convert alleles into factors (usually data is returned as *character*) and then summarize the allele information. There are no missing values in the data and as expected, there are only two alleles—A and B. If there were missing values (missing calls, to use the terminology) we would see a third factor

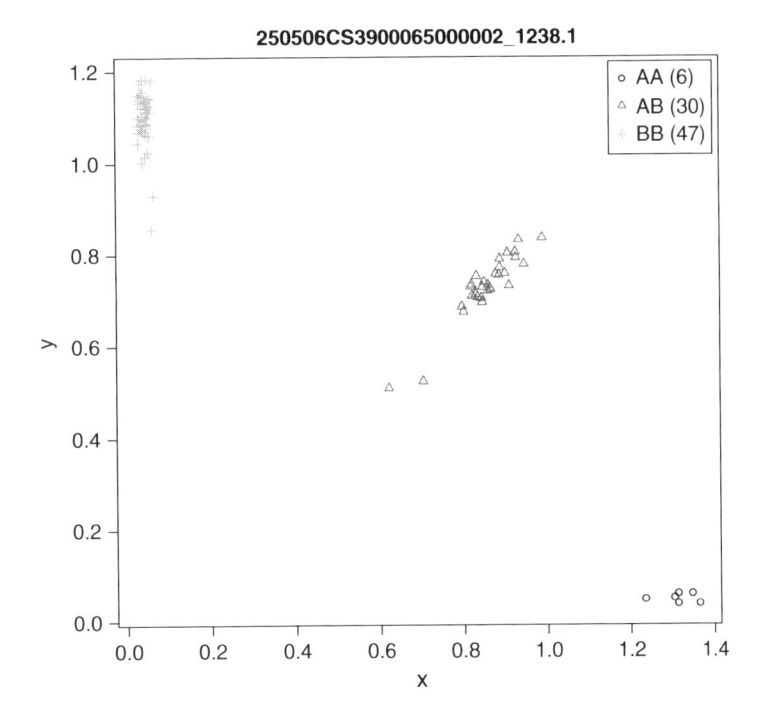

Fig. 2 Clustering of genotype calls based on X/Y coordinates

(e.g., NA or commonly "-"). Our data also has an X and a Y column (this is for the specific case of Illumina data). These are the normalized intensities of the reads for each of the two alleles. Allele calls are assigned based on the signal intensity of the fluorescence read by the scanner. These intensities can be plotted as an XY plot. We would expect that one of the homozygous genotypes would show high X values and low Y values while the other homozygote would be the opposite. Heterozygotes would be somewhere between the two. If the technology was 100 % accurate there would be only three perfect points on the plot and all samples would have the same intensity measures; but since reality gets in the way, what we do observe are three clouds (clusters) of data which will hopefully separate well between each other (Fig. 2). To plot the data:

```
snp=data.frame(snp, genotype=factor(paste(snp$allele1,
snp$allele2,sep=""),levels=c("AA","AB","BB")))
```

```
plot(snp$x,snp$y,col=snp$genotype,pch=as.numeric
(snp$genotype),
xlab="x",ylab="y",main=snp$snp[1],cex.main=0.9)
```

```
legend("bottomleft",paste(levels(snp$geno-
type),"(",summary(snp$genotype),")",sep=""), col=
1:length(levels(snp$genotype)),   pch=   1:length
(levels(snp$genotype)),cex=0.7)
```

The genotypes were color coded and symbols were used to make it easier to distinguish between them. Notice how the genotypes clearly cluster into three discrete groups—an indication of good data. Of course it is not possible to look at each of these plots for every SNP. Common practice is to go back to these plots after the association test and make sure that the significant SNP have clear distinction between clusters. There are some methods to summarize the clusters into an objective measurement e.g., sums of the distances to the nearest centroid of each cluster and the individual calls themselves.

Another metric included in this dataset is the GC score (*see* **Note 4**), without any in-depth details, it is a measure of how reliable the call is (essentially, distance of the call to the centroid as we mentioned above) on a scale from 0 to 1. Some labs will not assign a call (genotype) to GC scores under 0.25. Another common rule of thumb number is to cull reads under 0.6 (and projects working with human data may use even higher thresholds of 0.7–0.8).

```
length(which(snp$gcscore<0.6))
> 0
```

For this particular SNP all GC scores are above 0.6. Figures 3 and 4 exemplify what a good and a bad SNP look like. We might want to cull individual reads based on a threshold GC score value, but we might also remove the whole SNP if, for example, more than 2 or 3 % of its genotyping failed or if the median GC score is below a certain value (say 0.5 or 0.6). Again, the SNP we are analyzing is fine.

```
median(snp$gcscore)
> 0.8446
```

4.2 Minor Allele Frequency and Hardy–Weinberg Equilibrium

Population-based metrics are also employed. A simple one is the minor allele frequency (MAF). Not all SNP will be polymorphic, some will show only one allele across all samples (monomorphic) or one of the alleles will be at a very low frequency. The association between a phenotype and a rare allele might be supported by only very few individuals (no power to detect the association), in this case the results should be interpreted with caution. To avoid this potential problem, SNP filtering based on MAF is often used to exclude low MAF SNP (usual thresholds are between 1 and 5 %), but it is worthwhile to check the sample sizes and estimate an adequate value for your dataset (*see* **Note 5**). Back to the example SNP the allelic frequencies are

```
alleles=factor(c(as.character(snp$allele1),   as.character(snp$allele2)),levels=c("A","B"))

summary(alleles)/sum(summary(alleles))*100
> A       B
25.3012 74.6988
```

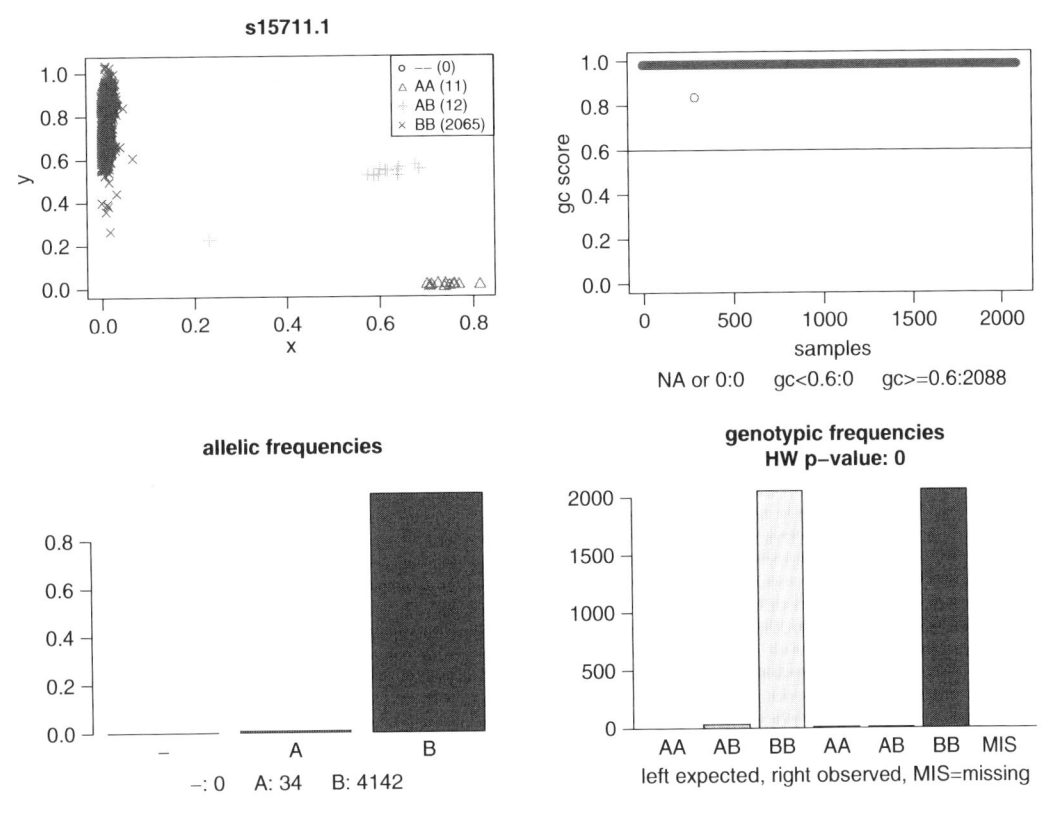

Fig. 3 Example of a good quality SNP. *Top left*: clustering for each genotype (non-calls are shown as *black circles*). *Top right*: GC scores. *Bottom left*: non-calls and allelic frequencies (actual counts are shown under the histogram). *Bottom right*: genotypic counts, on the *left* hand side the expected counts and on the *right* the observed counts; the last block shows number of non-calls

The frequencies are reasonable, around one-quarter A allele and three-quarters B allele. But again the point to consider is the objective of the work and the structure of the actual data that was collected. For example, if QC is being performed on mixed samples with an overrepresentation of one group, it is quite easy to have SNP that are not segregating in the larger population but are segregating in the smaller one—the MAF frequency in this case will essentially be the proportion of the minor allele from the smaller population in the overall sample. And if the objective of the study was to characterize genetic diversity between groups, the interesting SNP will have been excluded during the QC stage.

The next metric is Hardy–Weinberg equilibrium—HW. For a quick refresher, the Hardy–Weinberg principle, independently proposed by G. H. Hardy and W. Weinberg in 1908, describes the relationship between genotypic frequencies and allelic frequencies and how they remain constant across generations (hence also referred to as Hardy–Weinberg equilibrium) in a population of diploid sexually reproducing organisms under the assumptions of

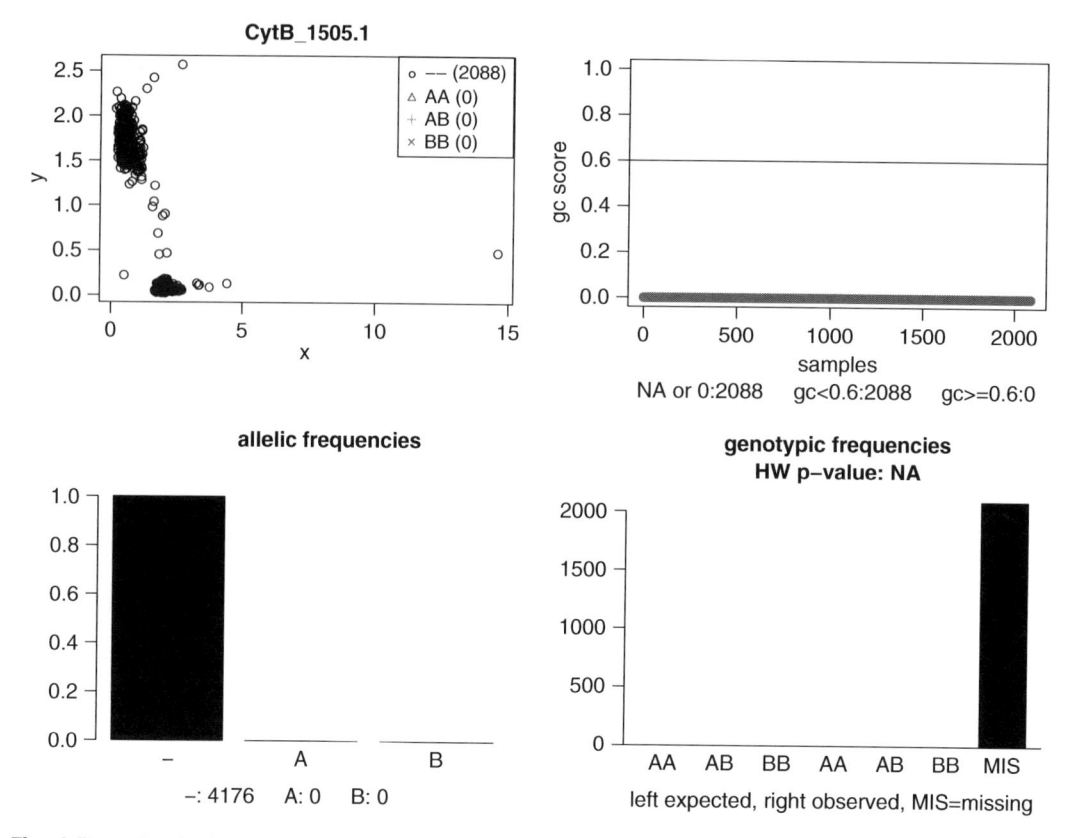

Fig. 4 Example of a bad quality SNP. *Top left*: clustering for each genotype (non-calls are shown as *black circles*—here all samples). *Top right*: GC scores. Bottom left: non-calls and allelic frequencies (actual counts are shown under the histogram). *Bottom right*: genotypic counts, on the *left* hand side the expected counts and on the *right* the observed counts; the last block shows number of non-calls

random mating, an infinitely large population and a few other assumptions.

Consider the bi-allelic SNP with variants A and B at any given locus, there are three possible genotypes: AA, AB and BB. Let's call the frequencies for each genotype D, H, and R. Under random mating (assumption of independence between events) the probability of a cross AA × AA is D^2, the probability for AB × AB is 2DH and the probability for BB × BB is R^2. If p is the frequency of allele A (p = D + H/2) then the frequency of B will be q = 1 − p and consequently the genotypic frequencies D, H, and R will respectively be p^2, 2pq, and q^2. This relationship model in itself is simply a binomial expansion.

Hardy–Weinberg equilibrium can be seen as the null hypothesis of the distribution of genetic variation when no biologically significant event is occurring in the population. Naturally real populations will not strictly adhere to the assumptions for Hardy–Weinberg equilibrium, but the model is however quite robust to deviations. When empirical observations are in a statistical sense significantly different from the model's predictions, there is a strong indication that some biologically relevant factor is acting on this population or there are genotyping errors in the data. This is where HW becomes controversial—it can be hard to distinguish a genotyping error from a real population effect, particularly when dealing with populations from mixed genetic backgrounds. Common p-value thresholds for HW are e.g., 10^{-4} or less (in practice use multiple testing corrected p-values, so much lower cut offs). To calculate HW for a SNP in R:

```
obs=summary(factor(snp$genotype,levels=c("AA","AB",
"BB")))
```

> AA AB BB
   6 30 47

```
hwal=summary(factor(c(as.character(snp$allele1),
as.character(snp$allele2),levels=c("A","B"))))

hwal=hwal/sum(hwal)
```

          A   B
0.2559524 0.7440476

```
exp=c(hwal[1]^2,2*hwal[1]*hwal[2],hwal[2]^2)*sum
(obs)
names(exp)=c("AA","AB","BB")
# chi-square test
# with yates correction

xtot=sum((abs(obs-exp)-c(0.5,1,0.5))^2/exp)

# get p-value
pval=1-pchisq(xtot,1)
```
> 0.8897645

It is a typical χ-square test. The only additional part is the Yates correction used when adding up the χ-square values (*see* **Note 6**). And, in this example, the SNP is in Hardy–Weinberg equilibrium.

4.3 Call Rates and Correlations in Samples

Quality control across samples is similar to what was done with the SNP. If 2 or 3 % of the genotypes are missing (call-rate <0.97) it is probably a good idea to exclude the sample, it is an indication of poor DNA quality. Another criterion that has not been discussed so far is the correlation between samples. If samples show very high correlations they might have to be excluded. Non-related samples would on average show a correlation of 0.5. Of course, again, the structure of the data has to be taken into account—a case–control study with random samples is very different from a half-sib project in livestock. In R correlations are computationally intensive since all data needs to be stored in memory as a matrix to calculate the pairwise correlations (this can be split into pairs but will be painstakingly slow for larger datasets). The R function is simply *cor* (*NameOfGenotypesMatrix*). If you cannot fit the data matrix into

4.4 Heterozygosity

memory consider some workarounds such as estimating correlations from a manageable subset of the SNP at a time.

Another useful metric is heterozygosity, which is simply the proportion of heterozygotes in relation to all genotypes. It is also worthwhile to check heterozygosity on SNP and compare to the expected heterozygosity (or gene diversity), it's just more common to evaluate heterozygosity on the samples. Essentially if a sample's heterozygosity is too high it can be an indication of DNA contamination (and once again: it could also be simply that a small proportion of samples are *truly* very different from the bulk of the data). Removal of samples that depart plus or minus 3 standard deviations (3SD) from the mean is a reasonable approach.

We will need the heterozygosity for all samples before we can look for outliers. We could just go over all samples in the DB, calculate (and store) the heterozygosity for each subject and discard the data from memory. You might have to do that if the dataset is too large, but since the example set is quite small let's build a matrix with the genotype counts (*sumslides*) for all samples and a matrix of SNP × sample (*numgeno*)—this is the entire dataset.

```
sumslides=matrix(NA,83,4)
rownames(sumslides)=animids
colnames(sumslides)=c("-/-","A/A","A/B","B/B")

#hold reshaped (numeric data)
numgeno=matrix(9,54977,83)

for (i in 1:83)
{
   hold=dbGetQuery(con,
   paste("select * from snps where animal='",animids
   [i],"'",sep=""))

   hold=data.frame(hold,
     genotype=factor(paste(hold$allele1,hold
     $allele2,
     sep=""),levels=c("--","AA","AB","BB")))

   hold=hold[order(hold$snp),]

   sumslides[i,]=summary(hold$genotype)

   temp=hold$genotype
   levels(temp)=c(9,0,1,2)

   numgeno[,i]=as.numeric(as.character(temp))

   # change to 9 genotypes under GC score cutoff
   numgeno[which(hold$gcscore<0.6),i]=9
}
```

```
rownames(numgeno)=hold$snp
colnames(numgeno)=animids

head(sumslides)
          -/-    A/A     A/B     B/B
sample1   838    15818   20100   18221
sample10  777    15397   21367   17436
sample11  803    15381   21564   17229
sample12  763    15440   21145   17629
sample13  822    16257   19524   18374
sample14  750    15637   21014   17576

numgeno[1:10,1:3]
                              sample1  sample10  sample11
250506CS3900065000002_1238.1      1        2         1
250506CS3900140500001_312.1       2        1         2
250506CS3900176800001_906.1       2        1         2
250506CS3900211600001_1041.1      2        2         2
250506CS3900218700001_1294.1      2        2         1
250506CS3900283200001_442.1       2        0         1
250506CS3900371000001_1255.1      0        2         2
250506CS3900386000001_696.1       1        1         0
250506CS3900414400001_1178.1      2        2         2
250506CS3900435700001_1658.1      9        9         9
```

We defined a matrix to store genotype counts for each sample (*sumslides*) and gave names to the rows and columns just to make it easier to identify in the output (see above). Notice that we used three genotypes plus -/- for missing genotypes. Another matrix *numgeno* was created to store all genotypic data. Then we made a loop to query the DB and extract data for each animal, sorted the data by SNP (using order) to make sure that the data returned by the DB is always in the same order; summarized the genotypic data in their classes and added the results to *sumslides*. Then we re-levelled the genotypes into numeric format (9—missing, 0–AA, 1–AB, 2–BB), this simply because it is a common format for downstream analysis. And in the last line of the loop we set all genotypes with GC scores under 0.6 as missing. Finally some housekeeping, assign names to the rows and columns so we can identify SNP and samples and check if everything looks alright.

To calculate and plot heterozygosities is quite simple now that we have the data. All that's needed is to divide the number of heterozygotes by the total genotypes; calculate the mean and standard deviation and then calculate the values for 3SD to each side of the mean. Finally plot (Fig. 5) the data and add lines for the mean and 3SD.

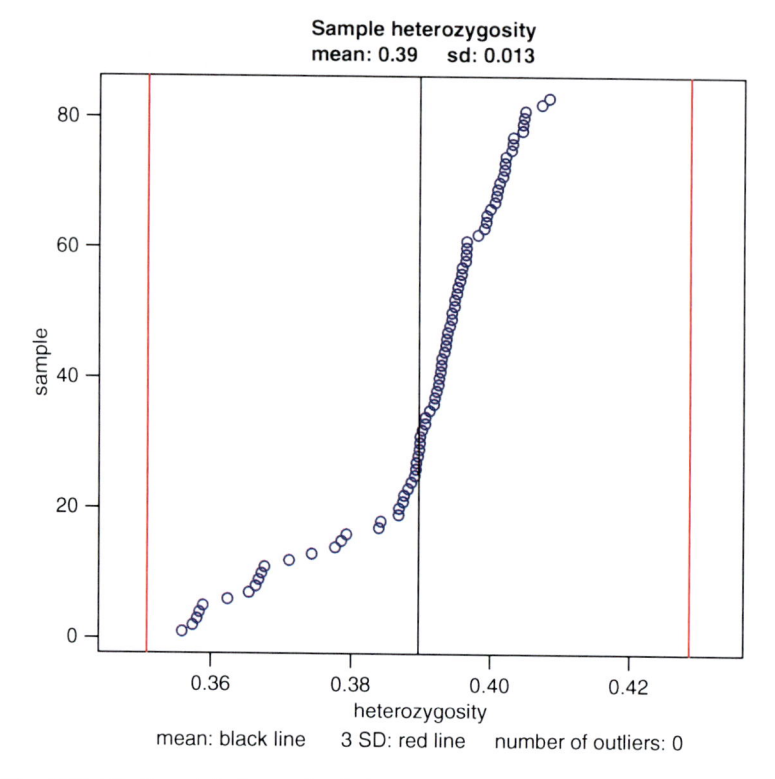

Fig. 5 Distribution plot of samples heterozygosity. All samples are within three standard deviations from the mean

```
samplehetero=sumslides[,3]/(sumslides[,2]+
sumslides[,3]+sumslides[,4])
```

```
# outliers 3 SD
up=mean(samplehetero)+3*sd(samplehetero)
down=mean(samplehetero)-3*sd(samplehetero)
hsout=length(which(samplehetero>up))
```

```
# number of outliers
hsout=hsout+length(which(samplehetero<down))
```

```
plot(sort(samplehetero),1:83,col="blue",cex.
main=0.9,
    cex.axis=0.8,cex.lab=0.8,
    ylab="sample",xlab="heterozygosity",
    main=paste("Sample heterozygosity\nmean:",
    round(mean(samplehetero),3)," sd:",
    round(sd(samplehetero),3)),
    sub=paste("mean: black line    ",3,
    "SD: red line    number of outliers:",hsout),
    cex.sub=0.8)
```

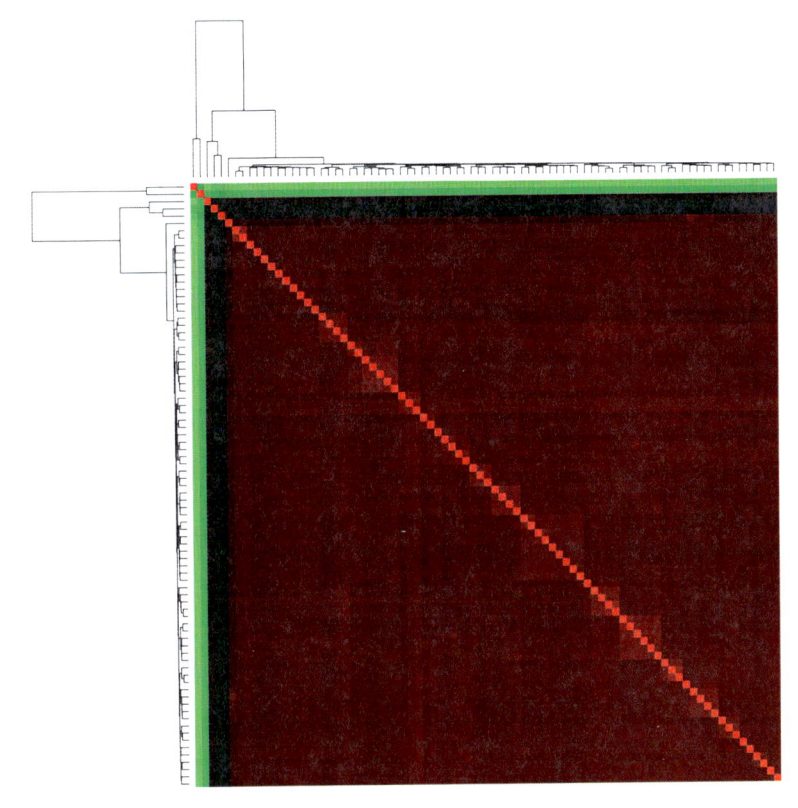

Fig. 6 Heatmap of correlations between samples. Samples on the outer edges are very different from the bulk of the data. A strong indication of bad quality samples

```
abline(v=mean(samplehetero))
abline(v=mean(samplehetero)-3*sd(samplehetero),
col="red")
abline(v=mean(samplehetero)+3*sd(samplehetero),
col="red")
```

We still have not calculated the correlation matrix. The function *cor* will only work with numeric data, hence us changing the genotypes to numeric format (it's also much smaller to store—10 × 208 M in the original file).

```
animcor=cor(numgeno)
```

```
library("gplots")
hmcol=greenred(256)
heatmap(animcor,col=hmcol,symm=T,labRow="          ",
labCol=" ")
```

The first line calculates the correlation matrix, a simple Pearson correlation, and the remaining lines of code are used to plot the results as a heatmap (Fig. 6). Heatmaps are excellent to visualize relationships between data. The library *gplots* also has some nice graphing capabilities. Note that missing data was replaced by 9—this greatly inflates differences between samples and, on the other hand, strongly pulls together samples with a lot of missing data. Keep in mind that this is for QC purposes only; such an approach should not be used to estimate genomic relationships from the data.

A couple of last comments: (1) what was discussed here was across all SNP and/or samples. With case–control studies it is worth considering running these metrics independently on cases and controls and then checking the results for consistency. (2) We have not plotted any results based on mapping information; it is a good idea to plot e.g., HW statistics per chromosome to see if there are any evident patterns such as a block on the chromosome that is consistently out of HW.

5 Fully Automated QC for Illumina SNP Data

On the publisher's website (http://extras.springer.com/), there is a full example of QC report for the dataset used in this chapter. The entire report is automatically generated using an R program and a full dataset of data filtered applying the QC metrics is also output for further analyses and summarized. This way of viewing the data is preferable to simply applying filtering without investigating the actual data structure. Once all metrics are summarized and pulled together into a report it becomes much easier to understand what each of these metrics are doing to the data and it also provides a chance to QC the QC itself.

The program also builds a database with the genotypic data and at the end of the run adds the QC results as additional tables to the database for future reference or fine tuning of filtering parameters based on an evaluation of the QC report. The program and documentation and an example dataset is freely available for download from http://www-personal.une.edu.au/~cgondro2/CGhomepage.

6 Notes

1. To highlight the importance of data integrity and traceable changes, think of a scenario where raw data were QC'ed and all genotypes flagged as bad were excluded from the dataset and later replaced using an imputation algorithm. At a later stage the researcher questions some final results and decides to trace back the genotypes for further evaluation, but finds out the genotypes cannot be distinguished between real and imputed ones anymore.

2. More enterprise database engines might be preferable for larger projects. SQLite can be slower than other engines and concurrency can also become a concern. But the main issue with SQLite databases is on data security due to its easy portability.

3. In this schema the indexes are created at the same time as the database. For large datasets this is inefficient—it is much faster to first populate the tables and then index them.

4. For Illumina data the GC score is arguably the main QC criterion for individual SNP and the most objective one since it relates directly to the genotyping method itself, albeit there is still some influence from the cluster file used to call the genotypes. Most other metrics confound quality problems with population structures and it is not a trivial task to tease these components apart.

5. At a MAF cutoff value of 5 % there would only be 25 homozygous samples in 10,000 (assuming H-W equilibrium). The weightings for the genotypes can be adjusted to minimize these frequency problems but caution should be exercised if it is worthwhile to keep such SNP.

6. Yates correction is appropriate for small sample sizes and can lead to biases in larger datasets. Generally the interpretation of results (which SNP to include/exclude) is the same but actual χ-square values can be quite unexpected.

Acknowledgments

This work was supported by a grant from the Next-Generation BioGreen 21 Program (No. PJ008196), Rural Development Administration, Republic of Korea.

References

1. Ziegler A, Konig IR, Thompson JR (2008) Biostatistical aspects of genome-wide association studies. Biom J 50:8–28
2. Teo YY (2008) Common statistical issues in genome-wide association studies: a review on power, data quality control, genotype calling and population structure. Curr Opin Lipidol 19:133–143
3. R Development Core Team (2012) R: a language and environment for statistical computing. R Foundation for Statistical Computing, Vienna, Austria
4. Chambers JM (2008) Software for data analysis: programming with R. Springer, New York
5. Jones M, Maillardet R, Robinson A (2009) Scientific programming and simulation using R. CRC, Boca Raton, FL
6. Ramalho JA (2000) Learn SQL. Wordware, Plano, TX

Chapter 6

Overview of Statistical Methods for Genome-Wide Association Studies (GWAS)

Ben Hayes

Abstract

This chapter provides an overview of statistical methods for genome-wide association studies (GWAS) in animals, plants, and humans. The simplest form of GWAS, a marker-by-marker analysis, is illustrated with a simple example. The problem of selecting a significance threshold that accounts for the large amount of multiple testing that occurs in GWAS is discussed. Population structure causes false positive associations in GWAS if not accounted for, and methods to deal with this are presented. Methodology for more complex models for GWAS, including haplotype-based approaches, accounting for identical by descent versus identical by state, and fitting all markers simultaneously are described and illustrated with examples.

Key words GWAS, Population structure, Multiple testing

1 Genome-Wide Association Tests Using Single Marker Regression

Genome-wide association studies (GWAS) exploit linkage disequilibrium, which are population level associations between markers and causative mutations (also called quantitative trait loci or QTL). These associations arise because there are small segments of chromosome in the current population which are descended from the same common ancestor. These chromosome segments, which trace back to the same common ancestor without intervening recombination, will carry identical marker alleles or marker haplotypes. If there is a QTL somewhere within the chromosome segment, they will also carry identical QTL alleles. There are a number of statistical methodologies which exploit these associations. The simplest of these is the genome-wide association test using single marker regression.

In a random mating population with no population structure, the association between a marker and a trait can be tested with single marker regression as

$$y = Wb + Xg + e$$

where y is a vector of phenotypes, W is a design matrix assigning records to phenotypes fixed effects, b is a vector of fixed effects (e.g., the mean, population structure effects, and age), X is a design matrix allocating records to the marker effect, g is the effect of the marker, and e is a vector of random deviates $e_{ij} \sim N(0, \sigma_e^2)$, where σ_e^2 is the error variance. In this model the effect of the marker is treated as a fixed effect, and the model is additive, such that two copies of the second allele have twice as much effect as one copy, and no copies have zero effect. The underlying assumption here is that the marker will only affect the trait if it is in linkage disequilibrium with an unobserved QTL.

The null hypothesis is that the marker has no effect on the trait, while the alternative hypothesis is that the marker does affect the trait (because it is in LD with a QTL). The null hypothesis is rejected if $F > F_{\alpha, v1, v2}$, where F is the F statistic calculated from the data and $F_{\alpha, v1, v2}$ is the value from an F distribution at α level of significance and $v1$, $v2$ degrees of freedom.

Consider a small example of ten animals genotyped for a single SNP. The phenotypic and genotypic data is:

Animal	Phenotype	SNP allele 1	SNP allele 2
1	2.03	1	1
2	3.54	1	2
3	3.83	1	2
4	4.87	2	2
5	3.41	1	2
6	2.34	1	1
7	2.65	1	1
8	3.76	1	2
9	3.69	1	2
10	3.69	1	2

We need a design matrix X to allocate both the mean and SNP alleles to phenotypes. In this case we will use an X matrix with number of rows equal to the number of records, and one column for the SNP effect. We will set the effect of the "1" allele to zero, so the SNP effect column in the X matrix is the number of copies of the "2" allele an animal carries (X matrix in bold):

Animal	1_n	X, Number of "2" alleles
1	1	0
2	1	1
3	1	1
4	1	2
5	1	1
6	1	0
7	1	0
8	1	1
9	1	1
10	1	1

In this case the **W** matrix is simply a vector, with each element 1, as each individual gets a dose of the mean. The mean and SNP effect can then be estimated as:

$$\begin{bmatrix} \hat{\mu} \\ \hat{g} \end{bmatrix} = \begin{bmatrix} \mathbf{W}'\mathbf{W} & \mathbf{W}'\mathbf{X} \\ \mathbf{X}'\mathbf{W} & \mathbf{X}'\mathbf{X} \end{bmatrix}^{-1} \begin{bmatrix} \mathbf{W}'\mathbf{y} \\ \mathbf{X}'\mathbf{y} \end{bmatrix}$$

where **y** is the (number of animals) vector of phenotypes.

In the above example the estimate of the mean and SNP effect are

$$\begin{bmatrix} \hat{\mu} \\ \hat{g} \end{bmatrix} = \begin{bmatrix} 2.35 \\ 1.28 \end{bmatrix}$$

This is not far from the real value of these parameters. The data above was "simulated" with a mean of 2, a QTL effect of 1, an r^2 (a standard measure of LD, *see* **Note 1**) between the QTL and the SNP of 1, plus a normally distributed error term.

The *F*-value can be calculated as:

$$F = \frac{(n-1)(\hat{g}\mathbf{X}'\mathbf{y} - 1/n\mathbf{y}'\mathbf{y})}{\mathbf{y}'\mathbf{y} - \hat{g}\mathbf{X}'\mathbf{y} - \hat{u}1_n'\mathbf{y}}$$

Using the above values, the value of *F* is 4.56. This can be compared to the tabulated *F*-value of 5.12 at a 5 % significance value and 1 and 9 (number of records −1) degrees of freedom. So the SNP effect in this case is not significant (not surprisingly with only ten records!). *F*-values can of course be easily transformed into *P* values for comparison with significance thresholds, a topic which is addressed later.

2 Power of Genome-Wide Association Tests Using Single Marker Regression

An important question for GWAS is how big does the study have to be to have any power to detect associations of a given size? The power of the association test to detect a QTL by testing the marker effect depends on:

1. The r^2 between the marker and QTL. Specifically, sample size must be increased by a factor of $1/r^2$ to detect an ungenotyped QTL, compared with the sample size for testing the QTL itself [1].
2. The proportion of total phenotypic variance explained by the QTL, termed h_Q^2.
3. The number of phenotypic records n
4. The allele frequency of the rare allele of the SNP or marker, p, which determines the minimum number of records used to estimate an allele effect. The power becomes particularly sensitive to p when p is small (e.g., <0.1).
5. The significance level α set by the experimenter.

The power is the probability that the experiment will correctly reject the null hypothesis when a QTL of a given size of effect really does exist in the population.

Figure 1 illustrates the power of an association test to detect a QTL with different levels of r^2 between the QTL and the marker and with different numbers of phenotypic records. The power was derived using the formula of [2].

Using both this figure, and the extent of LD in our population, we can make predictions of the number of markers required to detect QTL in a GWAS. For example, an r^2 of at least 0.2 is required to achieve power ≥0.8 to detect a QTL of $h_Q^2 = 0.05$ with 1,000 phenotypic records. To illustrate, in dairy cattle, $r^2 \approx 0.2$ at 100 kb. So assuming a genome length of 3,000 Mb in cattle, we would need at least 15,000 markers in such an experiment to ensure there is a marker 100 kb from every QTL. However this assumes that the markers are evenly spaced, and all have a rare allele frequency above 0.2. In practise, the markers may not be evenly spaced and the rare allele frequency of a reasonable proportion of the markers will be below 0.2. Taking these two factors into account, approximately 30,000 markers would be required. In practise, higher levels of r^2 than 0.2 are desirable; otherwise, it is difficult to distinguish true associations from noise when tens of thousands of markers are tested.

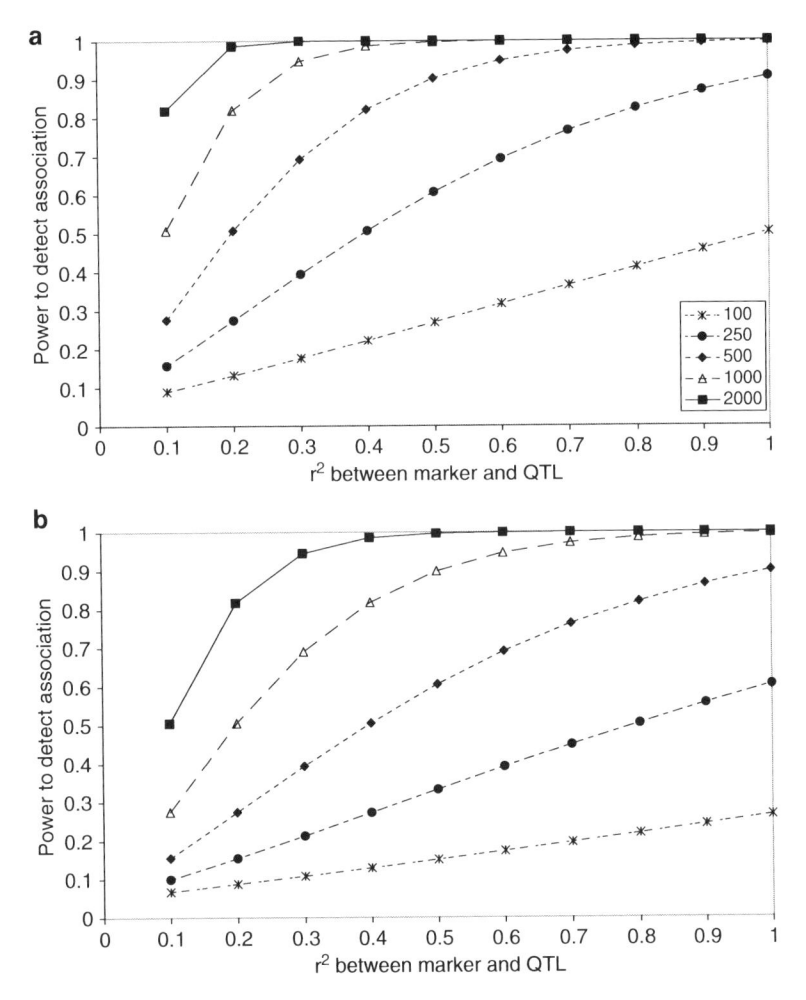

Fig. 1 (**a**) Power to detect a QTL explaining 5 % of the phenotypic variance with a marker. (**b**) Power to detect a QTL explaining 2.5 % of the phenotypic variance with a marker, for different numbers of phenotypic records given in the legend and for different levels of r^2 between the marker and the QTL, with a *P* value of 0.05. Rare allele frequencies at the QTL and marker were both 0.2

3 Choice of Significance Level

With such a large number of markers tested in GWAS, an important question is what value of α to choose. In a GWAS, we will be testing tens of thousands, hundreds of thousands, or with sequence data potentially millions of variants. So a major issue in setting significance thresholds is the multiple testing problem. In most QTL mapping experiments, many positions along the genome or a chromosome are analyzed for the presence of a QTL. As a result, when these multiple tests are performed the "nominal" significance levels of single tests don't correspond to the actual significance levels in the whole experiment, e.g., when considered across a chromosome or across the whole genome. For example, if we set a point-wise significance threshold of 5 %, we expect 5 % of results to be false positives. If we analyze 1,00,000 markers (assuming for the moment these points are independent), we would expect $1,00,000 \times 0.05 = 5,000$ false positive results! Obviously more stringent thresholds need to be set. One option would be to adjust the significance level for the number of markers tested using a Bonferoni correction to obtain an experiment wise P-value of 0.05. However such a correction does not take account of the fact that "tests" on the same chromosome may not be independent, as the markers can be in linkage disequilibrium with each other as well as the QTL. As a result, the Bonferoni correction tends to be very conservative, or requires some decision to be made about how many independent regions of the genome were tested.

Churchill and Doerge [3] proposed the technique of permutation testing to overcome the problem of multiple testing in QTL mapping experiments. Permutation testing is a method to set appropriate significance thresholds with multiple testing (e.g., testing many locations along the genome for the presence of the QTL). Permutation testing is performed by analyzing a large number of simulated data sets that have been generated from the real one, by randomly shuffling the phenotypes across individuals in the mapping population. This removes any existing relationship between genotype and phenotype, and generates a series of data sets corresponding to the null hypothesis. Genome scans can then be performed on these simulated data sets. For each simulated data, the highest value for the test statistic is identified and stored. The values obtained over a large number of such simulated data sets are ranked yielding an empirical distribution of the test statistic under the null hypothesis of no QTL. The position of the test statistic obtained with the real data in this empirical distribution immediately measures the significance of the real data set. For example, if we carry out 1,00,000 analyses of permuted data, the F value for the 5,000th highest value will represent the cut off point for the 5 % level of significance. Significance thresholds can then be set corresponding to 5 % false positives for the entire experiment, 5 % false positives for a single chromosome, and so on. Permutation testing is an excellent method of setting significance thresholds in a random mating population. In populations with some pedigree or other structure, however, randomly shuffling phenotypes across marker genotypes will not preserve any pedigree structure that exists in the data.

In human genetics, permutation testing has been used to determine the number of independent tests, given the SNP on standard SNP panels (typically close to a million, with >10 % MAF), and for widely studied populations. Such studies derive a nominal P value in the order of $<5 \times 10^{-8}$, in order to arrive at an experiment wise P value of 0.05 [4].

An alternative to attempting to avoid false positives is to monitor the number of false positives relative to the number of positive results [5]. The researcher can then set a significance level

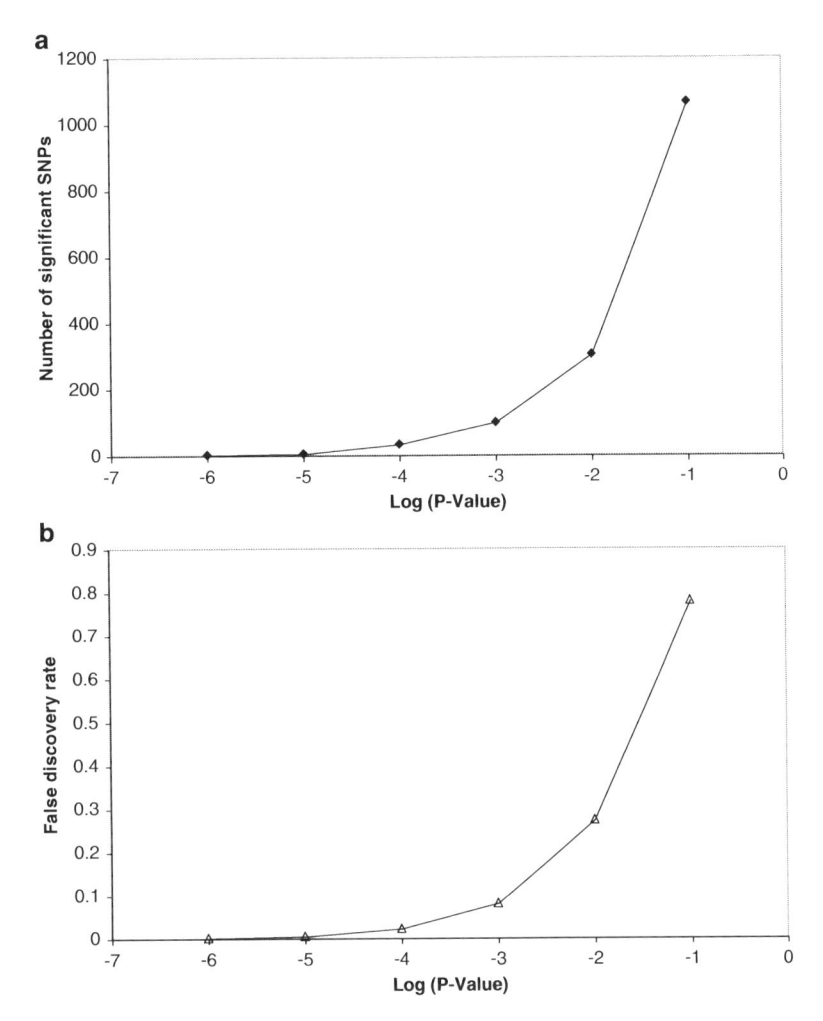

Fig. 2 (**a**) Number of significant markers at different *P* values in a genome-wide association study with 9,918 SNPs, using 384 Angus cattle with phenotypes for feed conversion efficiency. (**b**). False discovery rate at the different *P*-values

with an acceptable proportion of false positives. The false discovery rate (FDR) is the expected proportion of detected QTL that are in fact false positives [6, 7]. FDR can be calculated for a QTL mapping experiment as

$$mP_{max}/n$$

where P_{max} is a chosen P value significance threshold, n is the number of QTL which exceeds the significance threshold, and m is the number of markers tested. Figure 2 shows an example of the FDR in an experiment where 9,918 SNPs were tested for the effect on feed conversion efficiency in 384 Angus cattle. As the significance threshold is relaxed, the number of significant SNPs increases. However, the FDR also increases.

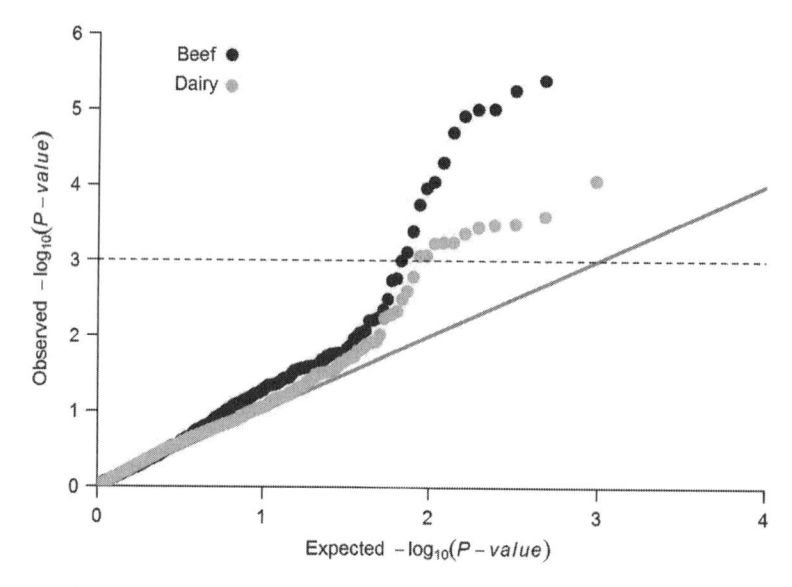

Fig. 3 An example of a quantile-quantile plot of observed against expected by chance *P* values. From Pryce et al. [9], an association test of SNP for effect on stature in cattle, in regions of genes associated with variation in height in other species. The quantile-quantile plot is of *P*-values of 879 SNPs that were 500 kbp either side of 55 orthologous genes found to be associated with height in human populations [38–41]. Using dairy and beef data sets, the phenotype (stature) was regressed on each SNP by using a mixed model that included pedigree (ASReml [17]) and the same approach as Pryce et al. [18]. Observed and expected *P*-values would fall on the gray solid line if there were no association. The *dashed horizontal line* is the threshold selected for significance ($P < 0.001$). Note that a 1-Mbp window was used from which to select SNPs because, in contrast to humans, where LD is expected to persist over only tens of kilobase pairs [42], non-zero levels of LD have been observed up to 1 Mbp in cattle [43]

In this experiment, a *P*-value of 0.001 was chosen as a criteria to select SNPs for further investigation. At this *P*-value, there were 56 significant SNPs. So the FDR was $9{,}918 \times 0.001/56 = 0.18$. This level of false discovery was deemed acceptable by the researchers.

A number of other statistics have been proposed to control the proportion of false positives, including the proportion of false positives—PRP [5], and the positive false discovery rate—pFDR [8].

Quantile-Quantile plots (QQ plots) are widely used to display the proportion of significant results compared to the expected number of significant results at a given *P* value. An example QQ plot [9] is shown in Fig. 3. The figure clearly demonstrated that in their study, at values greater than $P < 0.001$, more significant SNP were observed than expected by chance.

4 Confidence Intervals

Interestingly, there are very few reports in the literature on methods to estimate confidence intervals in GWAS. A method based on cross-validation is briefly described here. To calculate approximate

95 % confidence intervals for the location of QTL underlying the significant SNPs, a GWAS is first conducted as above. The data set is then split into two halves at random (e.g., half the animals in the first data set, the other half in the second data set). The GWAS is then re-run for each half of the data. When each half of the data confirmed a significant SNP in the analysis of the full data (i.e., a significant SNP in almost the same location), then a confidence interval can be calculated in the following way. The position of the most significant SNP from each split data set was designated x_{1i} and x_{2i} respectively, for the ith QTL position (taken as the most significant SNP in a region from the full data set). So for n pairs of such SNPs, the standard error of the underlying QTL is calculated as

$$\text{se}(\bar{x}) = \sqrt{\frac{1}{4n} \sum_{i=1}^{n} x_{1i} - x_{2i}}.$$ The 95 % confidence interval is then the position of the most significant SNP from the full data analysis $\pm 1.96\,\text{se}(\bar{x})$.

Using this approach in a data set with 9,918 SNPs genotyped on 384 Holstein-Friesian cattle, and for the trait protein kg, there were 24 significant SNP clusters (clusters of SNP putatively marking the same QTL, a cluster consists of 1 or more SNPs) in the full data, and the confidence interval for the QTL was calculated as 2 Mb.

5 Avoiding Spurious False Positives Due to Population Structure

Any unaccounted for population structure will result in false positive associations in GWAS [10]. In livestock and plant populations with multiple offspring per parent, selection for specific breeding goals and breeds, strains, or lines within the population all create population structure. A simple example is where the population includes a parent with a large number of progeny in the population. In our example the parent has a significantly higher estimated breeding value than other parents in the population. If a rare allele at a marker anywhere on the genome is homozygous in the parent, the subpopulation made up of it's progeny will have a higher frequency of the allele than the rest of the population. As the parent estimated breeding value is high, his progeny will also have higher than average estimated breeding values. Then in the GWAS, if the number of progeny of the parent is not accounted for, the rare allele will appear to have a (perhaps significant) positive effect.

Spielman et al. [11] proposed the transmission disequilibrium test (TDT) which requires that parents of individuals in the GWAS are genotyped to ensure the association between a marker allele and phenotype is linked to the disease locus, as well as in linkage disequilibrium across the population with it. In this way the TDT test avoids spurious associations due to population structure.

Table 1
Detection of type I errors in data with no simulated QTL [19]

Analysis model	Significance level		
	$p < 0.005$	$p < 0.001$	$p < 0.0005$
Expected type I errors	40	8	4
1. Full pedigree model	39 (SD = 14)	9 (SD = 5)	4 (SD = 3)
2. Sire pedigree model	46* (SD = 21)	11* (SD = 7)	6* (SD = 5.5)
3. No pedigree model	68** (SD = 31)	18** (SD = 11)	10** (SD = 7)
4. Selected 27 %—full pedigree	54** (SD = 18)	12** (SD = 6)	7** (SD = 4)

However the TDT test has a cost in that genotypes of both parents must be collected, and this is often not possible in livestock and plant populations.

An alternative is to remove the effect of population structure using a mixed model:

$$\mathbf{y} = \mathbf{1}_n'\mu + \mathbf{X}g + \mathbf{Z}\mathbf{u} + \mathbf{e}$$

where **u** is a vector of polygenic effect in the model with a covariance structure $u_i \sim N(0, \mathbf{A}\sigma_a^2)$, where **A** is the average relationship matrix built from the pedigree of the population, and σ_a^2 is the polygenic variance. **Z** is a design matrix allocating animals to records. In other words, the pedigree structure of the population is accounted for in the model. Note that this is BLUP, with the marker effect and the mean as fixed effects and the polygenic effects as random effects.

In the study of Macleod et al. [12], they assessed the effect of including or omitting the pedigree on the number of QTL detected in the experiment, in a simulation where no QTL effects were simulated so that all QTL detected were false positives (Table 1). They found a significant increase in the number of false positives, when the polygenic effects were not fully accounted for.

The results indicate that the number of type 1 errors (significant SNPs detected when no QTL exist) is significantly higher when no pedigree is fitted, and even fitting sire does not remove all spurious associations due to population structure.

A problem arises if the pedigree of the population is not recorded, or is recorded with many errors. One solution in this case is to use the markers themselves to infer the genomic relationship matrix **G** [13] or population structure (e.g., [10]). The **G** matrix can then be fitted in the place of **A** in the model above.

Principal components of the genomic are widely used in human GWAS to take account of population structure [14]. In livestock and plant populations, extreme caution is recommended with principal components approaches, as unless they are specifically tested it is unclear what component of variation they are removing [15, 16].

6 Genome-Wide Association Experiments Using Haplotypes

Rather than using single markers, haplotypes of markers could be used in the genome-wide association. The effect of haplotypes in windows across the genome would then be tested for their association with phenotype. The justification for using haplotypes is that marker haplotypes may be in greater linkage disequilibrium with the QTL alleles than single markers. If this is true, then the r^2 between the QTL and the haplotypes is increased, thereby increasing the power of the experiment.

To understand why marker haplotypes can have a higher r^2 with a QTL than an individual marker, consider two chromosome segments containing a QTL drawn at random from the population, which happen to carry identical marker haplotypes for the markers on the chromosome segment. There are two ways in which marker haplotypes can be identical, either they are derived from the same common ancestor so they are identical by descent (IBD), or the same marker haplotypes have been regenerated by chance recombination (identical by state (IBS)). If the "haplotype" consists only of a single SNP, the chance of being IBS is a function of the marker homozygosity. Now as more and more markers are added into the chromosome segment, the chance of regenerating identical marker haplotypes by chance recombination is reduced. So the probability that identical haplotypes carried by different animals are IBD is increased. If the haplotypes are IBD, then the chromosome segments will also carry the same QTL alleles. As the probability of two identical haplotypes being IBD increases, the proportion of QTL variance explained by the haplotypes will increase, as marker haplotypes are more and more likely to be associated with unique QTL alleles. This is particularly true for QTL with rare (low frequency) minor alleles.

Just as for single markers, the proportion of QTL variance explained by the markers can be calculated. Let q_1 be the frequency of the first QTL allele and q_2 be the frequency of the second QTL allele. The surrounding markers are classified into n haplotypes, with p_i the frequency of the ith haplotype. The results can be classified into a contingency table:

	Haplotype			
	1	i	N	
QTL allele 1	$p_1q_1 - D_1$	$p_iq_1 - D_i$	$p_nq_1 - D_n$	Q_1
QTL allele 2	$p_1q_2 + D_1$	$p_iq_2 + D_i$	$p_nq_2 + D_n$	Q_2
	p_1	p_i	p_n	1

For a particular haplotype i represented in the data, we calculated the disequilibrium as $D_i = p_i(q_1) - p_iq_1$, where $p_i(q_1)$ is the proportion of haplotypes i in the data that carry QTL allele 1 (observed from the data), p_i is the proportion of haplotypes i, and q_1 is the frequency of QTL allele 1. The proportion of the QTL variance explained by the haplotypes, and corrected for sampling effects was then calculated as

$$r^2(h,q) = \frac{\sum_{i=1}^{n} \frac{D_i^2}{p_i}}{q_1 q_2}$$

A model for testing haplotypes in an association study could be similar to the model described above:

$$\mathbf{y} = \mathbf{1}_n'\mu + \mathbf{Xg} + \mathbf{Zu} + \mathbf{e}$$

However **g** is now a vector of haplotype effects rather than the effect of a single marker. The haplotypes could be treated as random, as there are likely to be many of them and some haplotypes will occur only a small number of times. The effect of treating the haplotypes as random is to "shrink" the estimates of the haplotypes with only a small number of observations. This is desirable because it reflects the uncertainty of predicting these effects. So $g_i \sim N(0, \mathbf{I}\sigma_h^2)$ where **I** is an identity matrix and σ_h^2 the variance of the haplotype effects. The **g** can be estimated from the equations:

$$\begin{bmatrix} \hat{\mu} \\ \hat{\mathbf{u}} \\ \hat{\mathbf{g}} \end{bmatrix} = \begin{bmatrix} \mathbf{1}_n'\mathbf{1}_n & \mathbf{1}_n'\mathbf{Z} & \mathbf{1}_n'\mathbf{X} \\ \mathbf{Z}'\mathbf{1}_n & \mathbf{Z}'Z + \mathbf{A}^{-1}\lambda_1 & \mathbf{Z}'X \\ \mathbf{X}'\mathbf{1}_n & \mathbf{X}'Z & \mathbf{X}'X + \mathbf{I}\lambda_2 \end{bmatrix}^{-1} \begin{bmatrix} \mathbf{1}_n'\mathbf{y} \\ \mathbf{Z}'y \\ \mathbf{X}'y \end{bmatrix}$$

where $\lambda_1 = \frac{\sigma_e^2}{\sigma_a^2}$, and $\lambda_2 = \frac{\sigma_e^2}{\sigma_h^2}$. Note that this model assumes no covariance between haplotype effects. In practise, the haplotype variance is unlikely to be known, so will need to be estimated. An REML program, such as ASREML [17], can be used to do this. As the haplotypes are fitted as random effects, an F value is no longer appropriate. Rather, the statistic $-2 \times$ (Loglikelihood no haplotype fitted − Loglikelihood haplotype fitted) can be calculated, and compared to an inverse chi square distribution with 1 degree of freedom.

In GWAS in real data, haplotypes may have some advantage. Pryce et al. [18] conducted a GWAS using either 50,000 genome-wide SNP or haplotypes constructed from the alleles of these SNP, in a dairy cattle population. For the trait fertility, significant effects were only detected, and subsequently validated in a different population, when haplotypes were used. There was little difference, in terms of number of effects validated for other traits like milk production.

While the use of haplotypes seems initially attractive, there are a number of factors which potentially limit their value over single markers. These are:

- The requirement that the genotypes must be sorted into haplotypes and this may not be a trivial task.
- The number of effects which must be estimated increases. For a single marker there is one effect to estimate if an additive model is assumed, while for marker haplotypes there are potentially a large number of effects to estimate depending on the number of markers in the haplotype.
- Some simulation results which show benefits of marker haplotypes rely on increasing the density of markers in a given chromosome segment to achieve this. This may not be possible in practise.

7 Mapping QTL with an Identical by Descent Approach

The IBD is quite different from that used in single marker or haplotype regression, as now the effect of a putative QTL is explicitly modelled, rather than assuming the marker is associated with the QTL:

$$y_i = \mu + u_i + vp_i + vm_i + e_i$$

where vp_i and vm_i are the effects of the QTL alleles carried on the ith animal's paternal and maternal chromosome, respectively. In this model, the assumption is that each animal carries two unique QTL alleles, and so there are two QTL effects fitted for each animal.

Then marker haplotype information is used to infer the probability that two individuals carry the same QTL allele at a putative QTL position. As described above, the existence of LD implies there are small segments of chromosome in the current population which are descended from the same common ancestor. These IBD chromosome segments will not only carry identical marker haplotypes; if there is a QTL somewhere within the chromosome segment, the IBD chromosome segments will also carry identical QTL alleles. Therefore if two animals carry chromosomes which are likely to be IBD at a point of the chromosome carrying a QTL, then their phenotypes will be correlated. We can calculate the probability the two chromosomes are IBD at a particular point based on the marker haplotypes and store these probabilities in an

IBD matrix (**G**). Then the v are distributed $v \sim N(0, \mathbf{G}\sigma^2_{\text{QTL}})$, where σ^2_{QTL} is the QTL variance. If the correlation between the animals is proportional to **G,** there is evidence for a QTL at this position.

Consider a chromosome segment which carries ten marker loci and a single central QTL locus. Three chromosome segments were selected from the population at random, and were genotyped at the marker loci to give the marker haplotypes 11212Q11211, 22212Q11111, and 11212Q11211, where Q designates the position of the QTL. The probability of being IBD at the QTL position is higher for the first and third chromosome segments than for the first and second or second and third chromosome segments, as the first and third chromosome segments have identical marker alleles for every marker locus.

This type of information can be used, together with information on recombination rate of the chromosome segment and effective population size, for calculating an IBD matrix, **G**, for a putative QTL position from a sample of marker haplotypes. Element $\mathbf{G}_{ij}$ of this matrix is the probability that haplotype i and haplotype j carry the same QTL allele. The dimensions of this matrix is (2 × the number of animals) × (2 × the number of animals), as each animal has two haplotypes.

Meuwissen and Goddard [19] described a method to calculate the IBD matrix based on deterministic predictions which took into account the number of markers flanking the putative QTL position which are IBS, the extent of LD in the population based on the expectation under finite population size, and the number of generations ago that the mutation occurred.

Now consider a population of effective population size 100, and a chromosome segment of 10 cM with eight markers. Two animals are drawn from this population. Their marker haplotypes are 12222111, 11122111 for the first animal, and 12222111 and 11122211 for the second animal. The putative QTL position is between markers 4 and 5 (i.e., in the middle of the haplotype). The **G** matrix could look something like:

			Animal 1		Animal 2	
			Hap 1	Hap 2	Hap 1	Hap 2
			12222111	11122111	12222111	11122211
Animal 1	Hap 1	12222111	1.00			
	Hap 2	11122111	0.30	1.00		
Animal 2	Hap 1	12222111	0.90	0.30	1.00	
	Hap 2	11122211	0.20	0.40	0.20	1.00

To estimate the additive genetic variance, we could calculate the extent of the correlation between animals with high additive genetic relationships $\mathbf{A}_{ij}$. In practise, we fit a linear model which includes additive genetic value (**u**) with $\mathbf{V(u)} = \mathbf{A}\sigma_a^2$, and then estimate σ_a^2. In a similar way, to estimate the QTL variance at a putative QTL position we fit the following linear model:

$$\mathbf{y} = \mathbf{1}_n\mu + \mathbf{Zu} + \mathbf{Wv} + \mathbf{e},$$

where **W** is a design matrix relating phenotypic records to QTL alleles, **v** is a vector of additive QTL effects, **e** the residual vector, where the random effects v are assumed to be distributed as $v \sim (0, \mathbf{G}\sigma_{QTL}^2)$. An REML program, such as ASREML [17], can be used to estimate the QTL variance and the likelihood of the data given the QTL and polygenic parameters.

QTL mapping then proceeds by proposing a putative QTL position at intervals along the chromosome. At each point, the QTL variance is estimated and the likelihood of the data given the QTL and polygenic parameters is calculated. The most likely position of the QTL is the position where this likelihood is a maximum.

The significance of the QTL at its most likely position can then be tested using a likelihood ratio test by comparing the maximum likelihood of the model with the QTL fitted and without the QTL fitted:

$$\text{LRT} = -2(\text{LogLikelihood}_{\text{no\_QTL\_fitted}} - \text{LogLikelihood}_{\text{QTL\_fitted}})$$

This test statistic has a χ_1^2 distribution. The QTL is significant at the 5 % level if LR > 3.84.

Grapes et al. [20], Grapes et al. [21], and Zhao et al. [22] compared single marker regression, regression on marker haplotypes, and the IBD mapping approach for the power and precision of QTL mapping. Grapes et al. [20] and Grapes et al. [21] did this assuming that a QTL had already been mapped to a chromosome region, Zhao et al. [22] did this in the context of a genome-wide scan for QTL. All three papers compared the approaches using simulated populations. The conclusion from these papers was that single marker regression gives greater power and precision than regression on marker haplotypes, and was comparable to the IBD method. However these results contradict those of Hayes et al. [23], who found that in real data (9,323 SNPs genotyped in Angus cattle) using marker haplotypes would give greater accuracy of predicting QTL alleles than single markers. They also contradict the results of Calus et al. [24], who found that in genomic selection, use of the IBD approach gave greater accuracies of breeding values than using either single marker regression or regression on haplotypes, particularly at low marker densities (discussed further in Subheading 8). The explanation for the contradictory results may be that these authors [20–22] were simulating a situation where single markers had very high r^2 values with the QTL, in which case using marker haplotypes would only add noise to the estimation of the QTL effect.

With current densities of markers in livestock (up to 7,77,000 for cattle) and humans, the high levels of r^2 obtainable would appear to make the IBD approaches redundant. However, this statement does have an implicit assumption that the distribution of allele frequencies of the QTL match that of the markers; otherwise, the LD between QTL and markers will still be limited. For traits where many of the QTL have low minor allele frequencies, using haplotypes or the IBD approach may still have considerable benefits. For example, Browning and Thompson [25] reported rare sequence variants associated with type 1 diabetes that were only detected with an IBD approach.

8 Fitting All Markers Simultaneously

There are two disadvantages of the approaches described above that fit single SNPs, haplotypes, or single genome regions in the analysis. One of these is the multiple testing problem, that is many thousands of tests are run, so the significance level must be very stringent to take this into account. Further, the setting of a significance threshold combined with the testing of so many marker effects means that the markers most likely to exceed the threshold are those with favorable error terms, so that the significant markers have overestimated effects. The second disadvantage, particularly of the single SNP approach, is that a region containing the true mutation can be hard to define, as a large number of SNP can be in LD with the QTL, such that significant SNP span a wide region (e.g., [18]). This is particularly problematic in livestock (and likely some plant species), as low, but non zero, LD extends for Mb. While a partial solution to this second problem is to jointly fit SNP in multiple or conditional regression (e.g., [26]), an even better solution to both these issues is to fit all SNP simultaneously. This involves fitting the same models that have been proposed for genomic prediction (e.g., [27]).

This can be achieved by fitting the SNPs as random effects (e.g., derived from a distribution), with different prior assumptions on the distribution of possible SNP effects (e.g., a Bayesian approach). The model is:

$$\mathbf{y} = \mathbf{1}_n'\mu + \mathbf{Xg} + \mathbf{e}$$

However $\mathbf{g}$ is now a vector of SNP effects. The simplest assumption is that the SNP effect are derived from a normal distribution, in an SNPBLUP (e.g., [27]) or ridge regression. So $g \sim N(0, \mathbf{I}\sigma_g^2)$ where $\mathbf{I}$ is an identity matrix and σ_g^2 due to a single SNP. The $\mathbf{g}$ can be estimated from the equations:

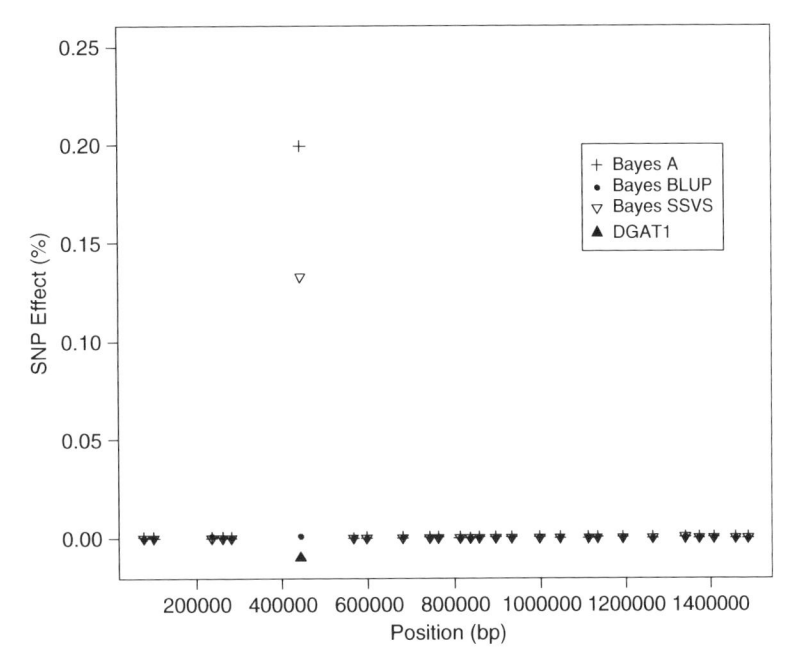

Fig. 4 SNP effects for fat percentage from Bayes A, Bayes BLUP, and Bayes C for the centromeric end of chromosome 14 showing the DGAT1 mutation [28]

$$\begin{bmatrix} \hat{\mu} \\ \hat{\mathbf{g}} \end{bmatrix} = \begin{bmatrix} 1_n'1_n & 1_n'\mathbf{X} \\ \mathbf{X}'1_n & \mathbf{X}'X + \mathbf{I}\lambda \end{bmatrix}^{-1} \begin{bmatrix} 1_n'\mathbf{y} \\ \mathbf{X}'y \end{bmatrix}$$

where $\lambda = \sigma_e^2/\sigma_g^2$, or alternatively λ can be estimated using cross validation for ridge regression.

One disadvantage with the SNPBLUP approach is that the SNP effects will all be shrunk back very hard, as with a large number of SNPs the value of σ_g^2 must be very small.

An alternative is to use a different prior assumption for the SNP effects that results in less shrinkage of SNP with moderate to large effects (e.g., nonlinear regression). One possibility is to assume the prior effects follow a Student's t-distribution, such that there is possibility for SNPs to have a moderate to large effect. That is, the effect of some SNP will not be regressed back as strongly. This method was called BayesA by Meuwissen et al. [27]. If there are QTL of moderate to large effect, the effect sizes estimated by SNPBLUP or BayesA can be dramatically different (Fig. 4).

Even better methods for GWAS are those that give a posterior probability of the SNP being in or out of the model. That is, the SNP has some or no effect on the trait. Such methods assume a prior distribution of SNP effects with a large mass at zero, and either a t-distribution (Bayes SSVS, [28]) or normal distribution for the remaining SNP (BayesCpi, [29]). These methods have been used for GWAS in a number of studies [30–34]. These methods can be extended further with priors of multiple normal distributions, such that SNP can be classified on the basis of their posterior probabilities of being in each distribution as zero effect, small effect, moderate effect, and so on (e.g., [35]).

These methods can be extended to an IBD approach, as described by Meuwissen and Goddard [36]. Computationally this is a very intensive approach, and has not (yet) been attempted for large SNP data sets.

9 The Need for Validation

The only evidence that a significant association detected in a GWAS is "real" (that is truly associated with a QTL affecting the trait) is validation in an independent population. Despite efforts to control for population structure, and use of fairly stringent thresholds, false positives will still occur in GWAS given the enormous number of SNPs tested, which means that the chance that at least one of these is associated with some unaccounted for structure in the data is high. This means that the design of a GWAS experiment includes both discovery and validation. The validation set must be large enough to have sufficient power (e.g., Fig. 1); otherwise, an SNP may fail to validate just because the experiment is underpowered. The relationship between the discovery and validation set should also be carefully considered. For example, if a significant SNP is discovered in a population of dairy bulls, and the SNP is "validated" in their daughters, there is high chance that the same population structure exists in both data sets, leading to apparent validation of what is really a false positive result. In livestock, the most convincing validation is across breeds (as the pedigree structure in the breeds should be independent). However, if SNP fails to validate across breeds, it may be because the underlying QTL is not segregating in both breeds.

10 Notes

1. Consider a chromosome segment which includes two markers A and B, with alleles A1, A2 and B1, B2, respectively. One measure of linkage disequilibrium is D, calculated as:

   $$D = \text{freq}(A1\_B1) * \text{freq}(A2\_B2) - \text{freq}(A1\_B2) * \text{freq}(A2\_B1)$$

   where freq(A1_B1) is the frequency of the A1_B1 haplotype in the population, and likewise for the other haplotypes. The D statistic is very dependent on the frequencies of the individual alleles, and so is not particularly useful for comparing the extent of LD among multiple pairs of loci (e.g., at different points along the genome). Hill and Robertson [37] proposed a statistic, r^2, which was less dependent on allele frequencies.

$$r^2 = \frac{D^2}{\text{freq}(A1) * \text{freq}(A2) * \text{freq}(B1) * \text{freq}(B2)}$$

where freq (A1) is the frequency of the A1 allele in the population, and likewise for the other alleles in the population. Values of r^2 range from 0, for a pair of loci with no linkage disequilibrium between them, to 1 for a pair of loci in complete LD.

As an example, consider a situation where the allele frequencies are

freq(A1) = freq(A2) = freq (B1) = freq (B2) = 0.5

The haplotype frequencies are:

freq(A1_B1) = 0.1
freq(A1_B2) = 0.4
freq(A2_B1) = 0.4
freq(A2_B2) = 0.1
The D = 0.1 × 0.1 − 0.4 × 0.4 = −0.15
And D^2 = 0.0225.
The value of r^2 is then 0.0225/(0.5 × 0.5 × 0.5 × 0.5) = 0.36. This is a moderate level of r^2.

References

1. Pritchard JK, Przeworski M (2001) Linkage disequilibrium in humans: models and data. Am J Hum Genet 69:1–14
2. Luo ZW (1998) Linkage disequilibrium in a two-locus model. Heredity 80:198–208
3. Churchill GA, Doerge RW (1994) Empirical threshold values for quantitative trait mapping. Genetics 138:963–971
4. Dudbridge F, Gusnanto A (2008) Estimation of significance thresholds for genomewide association scans. Genet Epidemiol 32:2227–2234
5. Fernando RL, Nettleton D, Southey BR, Dekkers JCM, Rothschild MF et al (2004) Controlling the proportion of false positives in multiple dependent tests. Genetics 166:611–619
6. Benjamini Y, Hochberg Y (1995) Controlling the false discovery rate: a practical and powerful approach to multiple testing. J R Stat Soc Ser B 57(1):289–300
7. Weller JI, Song JZ, Heyen DW, Lewin HA, Ron M (1998) A new approach to the problem of multiple comparisons in the genetic dissection of complex traits. Genetics 150:1699–1706
8. Storey JD (2002) A direct approach to false discovery rates. J R Stat Soc Ser B 64:479–498
9. Pryce JE, Hayes BJ, Bolormaa S, Goddard ME (2011) Polymorphic regions affecting human height also control stature in cattle. Genetics 187(3):981–984
10. Pritchard JK, Stephens M, Rosenberg NA, Donnelly P (2000) Association mapping in structured populations. Am J Hum Genet 67:170–181
11. Spielman RS, McGinnis RE, Ewens WJ (1993) Transmission test for linkage disequilibrium: the insulin gene region and insulin-dependent diabetes mellitus (IDDM). Am J Hum Genet 52:506–513
12. MacLeod IM, Hayes BJ, Savin KW, Chamberlain AJ, McPartlan HC, Goddard ME (2010) Power of a genome scan to detect and locate quantitative trait loci in cattle using dense single nucleotide polymorphisms. J Anim Breed Genet 127(2):133–142
13. Hayes BJ, Goddard ME (2008) Technical note: prediction of breeding values using marker-derived relationship matrices. J Anim Sci 86(9):2089–2092

14. Patterson N, Price AL, Reich D (2006) Population structure and eigenanalysis. PLoS Genet 2(12):e190
15. McVean G (2009) A genealogical interpretation of principal components analysis. PLoS Genet 5(10):e1000686
16. Daetwyler HD, Kemper KE, van der Werf JH, Hayes BJ (2012) Components of the accuracy of genomic prediction in a multi-breed sheep population. J Anim Sci 2012 May 14 [Epub ahead of print]
17. Gilmour AR, Gogel BJ, Cullis BR, Welham SJ, Thompson R (2006) ASReml user guide release 2.0. VSN International, Hemel Hempstead, UK
18. Pryce JE, Bolormaa S, Chamberlain AJ, Bowman PJ, Savin K, Goddard ME, Hayes BJ (2010) A validated genome-wide association study in 2 dairy cattle breeds for milk production and fertility traits using variable length haplotypes. J Dairy Sci 93(7):3331–3345
19. Meuwissen THE, Goddard ME (2001) Prediction of identity by descent probabilities from marker-haplotypes. Genet Sel Evol 33:605–634
20. Grapes L, Dekkers JC, Rothschild MF, Fernando RL (2004) Genetics 166:1561
21. Grapes L, Firat MZ, Dekkers JC, Rothschild MF, Fernando RL (2006) Genetics 172:1955
22. Zhao HH, Fernando RL, Dekkers JCM (2007) Power and precision of alternate methods for linkage disequilibrium mapping of quantitative trait loci. Genetics 175(1975–1986):27
23. Hayes BJ, Chamberlain AC, McPartlan H, McLeod I, Sethuraman L, Goddard ME (2007) Accuracy of marker assisted selection with single markers and marker haplotypes in cattle. Genet Res 89:215–220
24. Calus MP, Meuwissen TH, de Roos AP, Veerkamp RF (2008) Accuracy of genomic selection using different methods to define haplotypes. Genetics 178(1):553–561
25. Browning SR, Thompson EA (2012) Detecting rare variant associations by identity-by-descent mapping in case-control studies. Genetics 190(4):1521–1531
26. Yang J, Ferreira T, Morris AP, Medland SE, Genetic Investigation of ANthropometric Traits (GIANT) Consortium, DIAbetes Genetics Replication And Meta-analysis (DIAGRAM) Consortium, Madden PA, Heath AC, Martin NG, Montgomery GW, Weedon MN, Loos RJ, Frayling TM, McCarthy MI, Hirschhorn JN, Goddard ME, Visscher PM (2012) Conditional and joint multiple-SNP analysis of GWAS summary statistics identifies additional variants influencing complex traits. Nat Genet 44(4):369–375, S1–3
27. Meuwissen THE, Hayes B, Goddard ME (2001) Prediction of total genetic value using genome-wide dense marker maps. Genetics 157:1819–182933
28. Verbyla KL, Hayes BJ, Bowman PJ, Goddard ME (2009) Accuracy of genomic selection using stochastic search variable selection in Australian Holstein Friesian dairy cattle. Genet Res (Camb) 91(5):307–311
29. Habier D, Fernando RL, Kizilkaya K, Garrick DJ (2011) Extension of the Bayesian alphabet for genomic selection. BMC Bioinformatics 12:186
30. Veerkamp RF, Verbyla KL, Mulder HA, Calus MP (2010) Simultaneous QTL detection and genomic breeding value estimation using high density SNP chips. BMC Proc 4(Suppl 1):S9
31. Peters SO, Kizilkaya K, Garrick DJ, Fernando RL, Reecy JM, Weaber RL, Silver GA, Thomas MG (2012) Bayesian genome wide association analyses of growth and yearling ultrasound measures of carcass traits in Brangus heifers. J Anim Sci 2012 Jun 4. [Epub ahead of print]
32. Zeng J, Pszczola M, Wolc A, Strabel T, Fernando RL, Garrick DJ, Dekkers JC (2012) Genomic breeding value prediction and QTL mapping of QTLMAS2011 data using Bayesian and GBLUP methods. BMC Proc 6(Suppl 2):S7
33. Kizilkaya K, Tait RG, Garrick DJ, Fernando RL, Reecy JM (2011) Whole genome analysis of infectious bovine keratoconjunctivitis in Angus cattle using Bayesian threshold models. BMC Proc 5(Suppl 4):S22
34. Sun X, Habier D, Fernando RL, Garrick DJ, Dekkers JC (2011) Genomic breeding value prediction and QTL mapping of QTLMAS2010 data using Bayesian methods. BMC Proc 5(Suppl 3):S13
35. Erbe M, Hayes BJ, Matukumalli LK, Goswami S, Bowman PJ, Reich CM, Mason BA, Goddard ME (2012) Improving accuracy of genomic predictions within and between dairy cattle breeds with imputed high-density single nucleotide polymorphism panels. J Dairy Sci 95(7):4114–4129
36. Meuwissen TH, Goddard ME (2004) Mapping multiple QTL using linkage disequilibrium and linkage analysis information and multitrait data. Genet Sel Evol 36(3):261–279
37. Hill WG, Robertson A (1968) Linkage disequilibrium in finite populations. Theor Appl Genet 38:226–231

38. Lettre G, Jackson AU, Gieger C, Schumacher FR, Berndt SI et al (2008) Identification of ten loci associated with height highlights new biological pathways in human growth. Nat Genet 40:584–591
39. Gudbjartsson DF, Walters GB, Thorleifsson G, Stefansson H, Halldorsson BV et al (2008) Many sequence variants affecting diversity of adult human height. Nat Genet 40:609–615
40. Weedon MN, Lango H, Lindgren CM, Wallace C, Evans DM et al (2008) Genome wide association study identifies 20 loci that influence human height. Nat Genet 39:1245–1250
41. Kim J-J, Lee H-I, Park T, Kim K, Lee J-E et al (2010) Identification of 15 loci influencing height in a Korean population. J Hum Genet 55:27–31
42. Tenesa A, Navarro P, Hayes BJ, Duffy DL, Clarke GM et al (2007) Recent human effective population size estimated from linkage disequilibrium. Genome Res 17:520–526
43. Bovine Hapmap Consortium (2009) Genome-wide survey of SNP variation uncovers the genetic structure of cattle breeds. Science 24:528–532

Chapter 7

Statistical Analysis of Genomic Data

Roderick D. Ball

Abstract

In this chapter we describe methods for statistical analysis of GWAS data with the goal of quantifying evidence for genomic effects associated with trait variation, while avoiding spurious associations due to evidence not being well quantified or due to population structure.

Single marker analysis and imputation are discussed in Sect. 1, and a Bayesian multi-locus analysis using the `BayesQTLBIC` R package (1, 2) is described in Sect. 2. The multi-locus analysis, applied in a genomic window, enables local inference of the QTL genetic architecture and is an alternative to imputation. Multi-locus analysis with `BayesQTLBIC`, including calculation of posterior probabilities for alternative models, posterior probabilities for number of QTL, marginal probabilities for markers, and Bayes factors for individual chromosomes, is demonstrated for simulated QTL data. Methods for correcting the population structure and the possible effects of population structure on power are discussed in Sect. 3. Section 4 considers analysis combining information from linkage and linkage disequilibrium when sampling from a pedigree. Section 5 considers combining information from two different studies—showing that data from an existing QTL mapping family can be profitably used in combination with an association study—prior odds are higher for candidate genes mapping into a QTL region in the QTL mapping family, and, optionally, the number of markers genotyped in an association study can be reduced. Examples using R and the R packages `BayesQTLBIC`, `ncdf` are given.

Key words Bayesian statistics, Bayes factors, BIC criterion, Posterior probabilities, Power calculations, Bayesian power calculations, Robust detection of effects, Genome-wide association studies (GWAS), Quantitative trait loci (QTL), Linkage, Linkage disequilibrium, Single locus analysis, Multi-locus analysis, Bayesian model selection, Bayesian model averaging, `BayesQTLBIC` R package, Population structure, Confounding, Principal components, Mixed models, `hapgen` software, Large datasets, netcdf databases, `ncdf` R package, Linkage, Linkage disequilibrium, Combined linkage and linkage disequilibrium

1 Single Marker Analysis

Single marker analysis is appropriate when the extent of LD is comparable to marker spacing. A linear model is fitted testing for the effect using a t-test, F-test, or Bayes factor.

When the extent of LD is more than the marker spacing, multiple markers in a region may be associated with the trait due to linkage disequilibrium with a common causal locus. Ideally a multi-locus method would be used, as described in the next section. However, single marker analysis is still commonly used. Haplotypes observed within a small genomic window potentially contain more information about the causal locus. Imputation exploits some of this information by imputing missing values or markers not genotyped in the study population.

Imputation of missing genotypes (1, 3–5) reviewed in (6) can be used to impute the values of missing marker genotypes. Imputation in GWAS uses a model generated in a reference population, e.g., the HapMap population ((6); www.hapmap.org) for predicting each unknown marker genotype from a GWAS study, where a less dense set of markers is genotyped. Then, each imputed marker is tested with the single marker analysis.

Imputation captures some of the information from haplotypes within the extent of LD of the locus being tested.

$$\text{haplotypes} \xrightarrow{\text{imputation model}} \text{imputed locus} \xrightarrow{\text{single-locus model}} \text{trait} \quad (1)$$

One problem with imputation is the implicit assumption that the relationship between haplotypes and causal loci in the study population is the same as the reference population. This may or may not be the case.

When this is not the case or there is no suitable reference population the process can be short circuited. When not using a reference population, there is no information in the imputed genotypes that is not already in the haplotypes. Hence we argue that there is no benefit to imputation while using multi-locus models. Instead of generating an imputed marker value from haplotypes and predicting trait value from the imputed marker value, we can simply predict the trait directly from haplotypes.

$$\text{haplotypes} \xrightarrow{\text{multi-locus model}} \text{trait} \quad (2)$$

2 Multi-Locus Analysis

When marker spacing is substantially less than the extent of LD there is a potential benefit from a multi-locus analysis.

The imputation methods discussed above use a multi-locus analysis— considering multiple linear models used to predict missing marker genotypes. Simply use the missing marker genotype as a trait, and fit the model in the reference dataset, or for partially observed markers the study dataset.

Ball (1) considers all possible models where each model corresponds to a subset of markers representing QTL locations to within the resolution of the marker map. This method is available in the R package `BayesQTLBIC` (2). This is in turn based on the Splus `bicreg` function ((8); R package `BMA`) which in turn uses the R `leaps` package to search through the space of models. Approximate posterior probabilities of models are obtained from the BIC criterion. In our experience, this is reasonably accurate for the sample sizes used in association studies or QTL mapping. The approximation is most accurate when comparing several models with approximately equal posterior probabilities (as when comparing the most likely "top" models) and becoming less accurate when comparing one model with a much lower probability model. This might be the case when making comparisons with the null model, where a strong QTL effect is present. In this case Bayes factors are large so evidence is strong despite the approximation. A Laplace approximation can be used to obtain more accurate estimates for selected models, at the expense of additional computation.

The method was originally developed for QTL mapping to analyze the relationship between markers and traits in families. This utilizes linkage disequilibrium generated within families. The method applies equally well to linkage disequilibrium in an association study, generated by population history. The main difference being that in an association study the extent of LD and hence potential resolution is much less. It is only necessary to apply the method in a genomic window of width comparable to the extent of LD in the populations (4, 5).

For a window of size m bi-allelic markers, there are 2^m possible additive models. Exhaustive evaluation of all models is practical for up to around 30 markers. However in a GWAS, the method needs to be applied repeatedly for hundreds of thousands of markers. Hence it is desirable to reduce the number of combinations considered.

The BayesQTLBIC R package (2) has options to reduce the computing time required per analysis, including restricting the maximum model size to, e.g., 1, 2, 3, or 4. For example, considering only models of size 2 or less requires evaluating approximately $m^2/2$ linear models; considering models of size 3 or less requires evaluating approximately $m^3/6$ models.

We recommend using a maximum model size at least one higher than the number of QTL being contemplated, e.g., when testing for 1 or more QTL consider models of size at least 2. Otherwise markers in LD with a causal locus tend to share the posterior distribution of QTL and the posterior distribution of QTL location is over-dispersed (as in interval mapping compared with composite interval mapping (9); *cf.* (5) Fig. 7).

In a genome-wide application it is possible to scan the genome using models of size 1 or 2 then reexamine promising regions more closely using models of size up to 3 or 4.

Servin and Stephens (7) propose incorporating prior information by considering $\sigma_a^2 = 0.05, 0.1, 0.2, 0.4$ for additive effects on a scale where the phenotypic variance is one, and averaging the resulting Bayes factors. This works when making comparisons with the null model. More appropriate (in general) would be to average posterior probabilities, obtaining results for the mixture prior to these value of σ_a^2 with probability 1/4. The corresponding values of a are 20, 10, 5, 2.5.

More general MCMC-based fine mapping methods (e.g., (10)), can be also applied to selected regions of interest. MCMC models allow more flexibility in model and prior specification at the expense of being more computationally intensive to run, diagnose, and interpret.

R example 6.1: QTL analysis using the `BayesQTLBIC` R package. Simulate a QTL mapping family with 100 progeny. Assume 5 markers on a chromosome at positions 20, 40, 60, 80, and 100cM, and a QTL explaining 15% of the variance located at 55cM. Assume a prior probability of 0.2 per marker. Apply `bicreg.qtl()` to obtain the most probable models and their probabilities, the probabilities for model sizes, and the marginal probabilities per marker. Find the probability for a QTL in the region 30–70cM and a chromosome-wise Bayes factor. Give individual marker effects and effects of allelic substitution if using the marker for selection.

```
> library(BayesQTLBIC)
> options(digits=4)
> set.seed(12345)
> # simulate backcross QTL data
> sim.data <- sim.bc.progeny(n=100, Vp=0.15, map.pos=c
                              (20,40,60,80,
+                              100),qtl.pos=55)
chromosome 1

> cbind(sim.data$x,y=sim.data$y)[1:5,]

      c1m1 c1m2 c1m3 c1m4 c1m5       y
[1,]    1    1    2    1    1   0.8688
[2,]    1    1    1    2    2  -0.3783
[3,]    1    1    1    1    1  -0.7934
[4,]    2    2    1    1    2   0.7186
[5,]    1    1    1    1    1  -0.4956

> additive.contrast <- function(x) ifelse(x==1,-1/2,1/2)
> df1 <- data.frame(y=sim.data$y)
> df1$X <- apply(sim.data$x, 2, additive.contrast)
```

```
> res1 <- bicreg.qtl(x=df1$X, y=df1$y, prior=0.2, nvmax=5,
  nbest=32)
> # 12 models account for 99% of the posterior probability
> summary(res1, nbest=12)
```
R-squared, BIC, and approximate posterior probabilities for individual models:

```
   c1m1 c1m2 c1m3 c1m4 c1m5    R2     BIC postprob cumprob
1     0    0    1    0    0 10.875 -1.90384 0.560620 0.5606
2     0    1    0    0    0  9.199 -0.04079 0.220858 0.7815
3     0    0    0    0    0  0.000  2.23144 0.070910 0.8524
4     0    1    1    0    0 12.668  3.44163 0.038718 0.8911
5     0    0    1    1    0 12.465  3.67381 0.034475 0.9256
6     0    0    1    0    1 11.538  4.72724 0.020359 0.9459
7     1    0    1    0    0 10.909  5.43577 0.014285 0.9602
8     1    1    0    0    0 10.579  5.80549 0.011874 0.9721
9     0    1    0    0    1  9.546  6.95408 0.006686 0.9788
10    0    1    0    1    0  9.218  7.31604 0.005580 0.9844
11    0    0    0    1    0  0.732  8.87450 0.002560 0.9869
12    0    1    1    1    0 14.185  9.06708 0.002325 0.9892
```
marginal probabilities for model sizes
```
         0         1         2         3         4         5
 7.091e-02 7.880e-01 1.322e-01 8.706e-03 2.515e-04 2.365e-06
```
marginal probabilities for individual variables
```
   c1m2    c1m3
 0.2860  0.6708
```
attr,"prior")
[1] 0.2
attr,"intercept")
[1] TRUE

```
> # posterior probability for the region from 30--70 cM
> sum(res1$postprob[apply(res1$which,1,function(u){u[2] ||
u[3]})])
```
[1] 0.9224

```
> # prior probability for 1 or more QTL
> prior.chrom <- 1 - (1-0.2)^5
> prior.chrom
```
[1] 0.6723

```
> # posterior probability for 1 or more QTL
> postprob.chrom <- 1 - res1$postprob.size[1]
> postprob.chrom
```
```
     0
0.9291
```

> # chromosome-wise Bayes factor

```
> priorodds.chrom <- prior.chrom/(1-prior.chrom)
> postodds.chrom <- postprob.chrom/(1-postprob.chrom)
> B.chrom <- postodds.chrom/priorodds.chrom
> B.chrom
       0
   6.386

> # individual marker statistics
> with(res1, rbind(probne0=probne0/100, postmean= postmean
   [-1],
+       postsd=postsd[-1], condpostmean=condpostmean[-1],
+       condpostsd= condpostsd[-1], d.av))
                    c1m1       c1m2     c1m3     c1m4       c1m5
probne0          0.032000   0.2900   0.6770   0.04700    0.033000
postmean        -0.003976   0.1586   0.4224  -0.01100   -0.004473
postsd           0.047503   0.2741   0.3338   0.07423    0.043398
condpostmean    -0.123049   0.5461   0.6243  -0.23174   -0.136890
condpostsd       0.234927   0.2168   0.1965   0.25477    0.198762
d.av             0.175153   0.4036   0.5073   0.27928    0.149570
```

Note:

- 12 models account for approximately 99% of the probability.
- The most probable models are the model with c1m3 with probability 0.56, the model with c1m2 with probability 0.22, followed by the null model with probability 0.07, followed by the model with both c1m2 and c1m3 with probability 0.04.
- This suggests that the most likely QTL configuration is a QTL between c1m2 and c2m3.
- Considering the marker positions and probabilities this suggests a QTL located at $40 + (60 - 40) \times 56/(22 + 56) = 54.4 cM$, close to the simulated QTL.
- When calculating the probability for a QTL to be in a region we consider the marginal probability for a marker m_i to be shared among the points in the vicinity $V(m_i)$ defined as the set of points closer to m_i than any other marker, according to a uniform distribution.
- Hence the probability for a QTL in the region 30–70cM is the probability of a QTL in the vicinity of c1m2 or c1m3.
- In the case of an interval partially overlapping the vicinity of that marker the contributions to the probability are pro-rata according to the proportion overlapped.
- The Bayes factor per chromosome gives a measure of strength of evidence for a QTL independent of the prior probabilities of QTL per marker.
- A Bayes factor of 6 is suggestive but it is not strong evidence.

- The object returned of class `bicreg.qtl` also contains estimates of individual marker effects.
- The posterior mean effects (`postmean`) are model averaged effects for markers.
- The conditional mean effects (`condpostmean`) are effects for markers, conditional on selection.
- The increase in size between posterior mean effects and conditional posterior mean effects gives an indication of selection bias that would occur if not using model averaging.
- The effect of allelic substitution of a marker is `d.av`; this is the posterior mean estimated gain that would be obtained if using the marker for selection.

3 Population Structure and Confounding

Care needs to be taken to avoid spurious or inflated associations due to population structure. Astle and Balding (11) note that the main causes of confounding in GWAS are:

C1. "Population structure" or the existence of major subgroups in the population, where differences in allele frequencies are confounded with subpopulation

C2. "cryptic relatedness," i.e., the existence of small groups (often pairs) of highly related individuals

C3. Environmental differences between subpopulations or geographic locations

C4. Differences in allele call rates between subpopulations.

Note: associations due to LD between a marker and causal locus induced by recent population structure are considered spurious if the marker is not located close to a causal variant. Such "non-causal" correlations have limited benefit in forming predictive models or finding causal variants.

Various methods have been proposed to control for population structure:

S1. Genomic control (12), where χ^2 test statistics are scaled so that the median test statistic equals the expected value (0.455) under H_0.

S2. Structured association (13–15), where a Bayesian population model with individuals either belonging to one of k putative ancestral subpopulations or, in the admixture model, individual genotypes being an admixture of the k ancestral populations is fitted to the marker-allele data. Confounding with population structure is controlled by fitting ancestral population membership as covariates.

S3. Regression on a set of markers (16), where a set of fairly widely spaced markers covering the genome is used as covariates. Over-fitting is avoided by a backwards elimination stepwise regression procedure.

S4. Principal components, where a number (e.g., 10) of principal components (eigenvalues of the matrix XX' for the set of marker genotypes (X)) are used as covariates (17, 18).

S5. Mixed model method, where a set of random effects is fitted for each individual with covariance based either on an estimated kinship matrix, $\hat{K}$ (19), or a known pedigree (20).

For a review see (11). According to Astle and Balding (11), to be effective: genomic control, structured association, and the regression method require $\sim 10^2$ markers; principal components require $\sim 10^4$ SNP markers; and the mixed model with $\hat{K}$ requires 10^4–10^5 markers. In their simulations (20 SNPs with OR=1.18, 1000 cases and 1000 controls, sampled from 3 subpopulations with $n = 6000$ and $F_{ST} = 0.1$), the mixed model or their MCP method performed best when the true K was used. However when K was estimated, principal components (PC10) and the mixed model performed best. Genomic control was conservative and down-rated all test statistics rather than those associated with population confounders.

> "Principal components (PC) adjustment in effect eliminates from test statistics the part of the phenotypic signal that can be predicted from large scale population structure. In particular, if natural selection leads to large allele frequency differences across subpopulations at a particular SNP, and case-control ratios vary across subpopulations, then spurious associations can arise that PC adjustment will control, because the SNP genotypes are strongly correlated with subpopulation, whereas the MM and MCP methods will not. On the other hand, the MM and MCP methods can gain power over PC adjustment because they explicitly model the phenotype-genotype correlations induced by relatedness and genetic drift. For example, they should provide better power than PC adjustment when analyzing data from human population isolates which are homogeneous for environmental exposures." (11).

Astle and Balding (11) note that since there are good methods for allowing for population structure, population association tests are preferable to family-based tests. For example, the TDT test requires three genotypes to give the same information as one case–control pair, so incurs additional cost or lower power, as well as requiring genotypes to be available on parents.

However it is important to note that, with any population association study, the data is observational; no method is guaranteed to adjust for all possible confounding factors, and adjustments are imperfect. For example, there is no guarantee that a given confounder appears in the first 10 principal components. Increasing the number of principal components introduces a trade-off: reducing more possible confounders at the expense of introducing more "noise" from the estimation of the smaller principal components. Family-based designs have an advantage of robustness due to the inherent randomization of independent meioses. Designs with larger families are possible, particularly in plant and animal species, that do not have the same relative efficiency penalty as the trio-based TDT.

The principal components method is implemented in the software `eigenstrat` (18). We illustrate the computations using R below. To efficiently handle the large number of columns in the matrix X of SNP genotypes, we use an incremental approach to calculate the matrix XX', rather than using the R built in function `prcomp`. The matrix X is stored in a `netcdf` database, an array-based database, using the R package `ncdf`.

First, we use `hapgen` (21) to simulate some data. The data is based on CEU and YRI haplotype panels from the HapMap project (7). The `hapgen` program enables generating further samples from a population with similar LD structure. Then, we simulate QTL effects on one set of markers and sample another set of markers representing the observed markers in a study. Then, we calculate the principal components, and do single marker analyses for each marker with 10 principal components as covariates.

Example 6.1. Simulate 2000 individuals with LD structure based on each of the CEU and YRI populations using the program `hapgen`.

```
#
# simulate haplotypes for 2000 individuals
# LD structure from CEU and YRI, chromosome 4
# use a minimal number of cases and rr ~=1 because not
# interested in case-control
#

> hapgen -h genotypes_chr4_YRI_r22_nr.b36_fwd.phase
+  -l genotypes_chr4_YRI_r22_nr.b36_fwd_legend.txt
+  -r genetic_map_chr4.txt  -o YRI_simn2000 -n 1950 50
+  -hap -rr 1.01 1.02 -dl 80123

> hapgen -h genotypes_chr4_CEU_r22_nr.b36_fwd.phase
+  -l genotypes_chr4_CEU_r22_nr.b36_fwd_legend.txt
+  -r genetic_map_chr4.txt  -o CEU_simn2000
+  -n 1950 50 -hap -rr 1.01 1.02 -dl 80123
```

The haplotype data was downloaded from the HapMap web site: http://hapmap.ncbi.nlm.nih.gov/downloads/phasing/2007-08_rel22/phased and the genetic map data was downloaded from: https://mathgen.stats.ox.ac.uk/wtccc-software/recombination_rates/genetic_map_chr4.txt.

After forming a consensus dataset of 142226 markers common to both populations, a random sample of 14000 markers was selected and stored in a netcdf database using the R package ncdf.

R Example 6.2: Simulate 100 additive QTL effects from a Gamma (1,1) distribution and scaled to collectively explain 15% of the variation of a trait.

```
> A.q <-1; B.q <- 1; nQTL <- 100
> ncols <- 142226
> set.seed(792345)
> qtl.effects2 <- rgamma(nQTL, shape=A.q, rate=B.q)
> qtl.signs <- sample(c(-1,1),size=length(qtl.effects2),
+                prob=c(0.5,0.5), replace=TRUE)
> qtl.effects <- sqrt(qtl.effects2)*qtl.signs
> qtl.markers.ind <- sample(1:ncols, size=nQTL)
> qtl.markers.ind <- sort(qtl.markers.ind)
> nc1 <- open.ncdf("ceuyri.sim2000.nc", write=FALSE)
> Xqh <- matrix(0,nrow=8000, ncol=nQTL)
> for(ii in seq(along=qtl.markers.ind)){
+    qi <- qtl.markers.ind[ii]
+    Xqh[,ii] <- get.var.ncdf(nc1, "snp.int", start=c(1,qi),
+                         count=c(nrows,1))
+ }
> Xq <- matrix(aperm(array(Xqh, dim=c(2,4000,nQTL)),c(2,1,3)),
+            ncol=2*nQTL)
> vq <- apply(Xq,2,var)
> # additive model
> bq <- rep(qtl.effects, rep(2,length(qtl.effects)))
> VQ1 <- sum(vq*bq^2)
> qtl.effects <- qtl.effects*sqrt(sigma2.P*h2.Q/VQ1)
> bq <- rep(qtl.effects, rep(2,length(qtl.effects)))
> y.sim <- Xq%*%bq + rnorm(4000)*sqrt((1 - h2.Q)*sigma2.P)

> Xmh <- matrix(0, nrow=nrows, ncol=nmarkers)
> markers.ind <- sample(1:ncols, size=nmarkers)
> for(ii in seq(along=markers.ind)){
+    mi <- markers.ind[ii]
+    Xmh[,ii] <- get.var.ncdf(nc1, "snp.int", start=c(1,mi),
+                         count=c(nrows,1))
+ }
```

Note: haplotypes are stored as a matrix with two rows per genotype or an array 2 x n x m for n genotypes and m markers.

Example 6.2. Calculate the first 10 principal components for the marker data stored in the netcdf database file ceuyri_markers+traits.nc.

```
> library(ncdf)
> source("ncdf_functions.R")
> mt.nc1 <- open.ncdf("ceuyri_markers+traits.nc",
    write=FALSE)
> ev <- prcomp.ncdf(mt.nc1, "Xm", chunk=50, npcs=10)
> PC10 <- ev$vectors[,1:10]
```

The key steps in the function `prcomp.ncdf()` are:

1. To calculate row means

   ```
   > X.rowmeans <- lds.rowmeans.2xRxC(nc1,X.name,
   chunk)
   ```

2. To get `chunk` columns of `Xm` from the database

   ```
   > Xi <- get.var.ncdf(mt.nc1,"Xm",start=c(1,1,
   c1),
   +           count=c(2,ngenotypes,chunk))
   ```

3. To center the columns

   ```
   > Xc <- matrix(aperm(Xi,c(2,1,3)),ncol=2*d3)
   > Xc <- Xc - X.rowmeans
   ```

4. Accumulate the cross-product in `xtx`

   ```
   > xtx <- xtx + crossprod(t(Xc))
   ```

5. When done, calculate the eigenvectors from `xtx`

   ```
   > ev <- eigen(xtx)
   ```

Example 6.3. Single marker analyses for each marker.

```
> verbose <- FALSE
> library(ncdf)
> nmarkers <- 14000
>  mt.nc1  <-  open.ncdf("ceuyri_markers+traits.
   nc",write=FALSE)
> y <- get.var.ncdf(mt.nc1,"y")
> df1 <- data.frame(y=y)
> df1$PC10 <- PC10
> Fstats <- numeric(nmarkers)
> Fstats.raw <- numeric(nmarkers)
> for(jj in 1:nmarkers){
+ cat(".")
+ if(jj%%10==0)cat("X")
+ if(jj%%100==0)cat("C")
+ Xj <- get.var.ncdf(mt.nc1,"Xm",start=c(1,1,jj),
+                 count=c(2,4000,1))
+ df1$x <- apply(Xj,2,sum)
+ fitii.raw <- lm(y ~ x, data=df1)
+ Fstats.raw[jj] <- summary(fitii.raw)$fstatistic
[1]
+ fit0ii <- lm(y ~ PC10, data=df1)
+ fitii <- lm(y ~ x + PC10, data=df1)
+ Fstats[jj] <- anova(fit0ii,fitii)$F[2]
+ }
> Bfs <- SS.oneway.bf(c(2000,2000),Fstats)
> Bfs.raw <- SS.oneway.bf(c(2000,2000),Fstats.raw)
   }
```

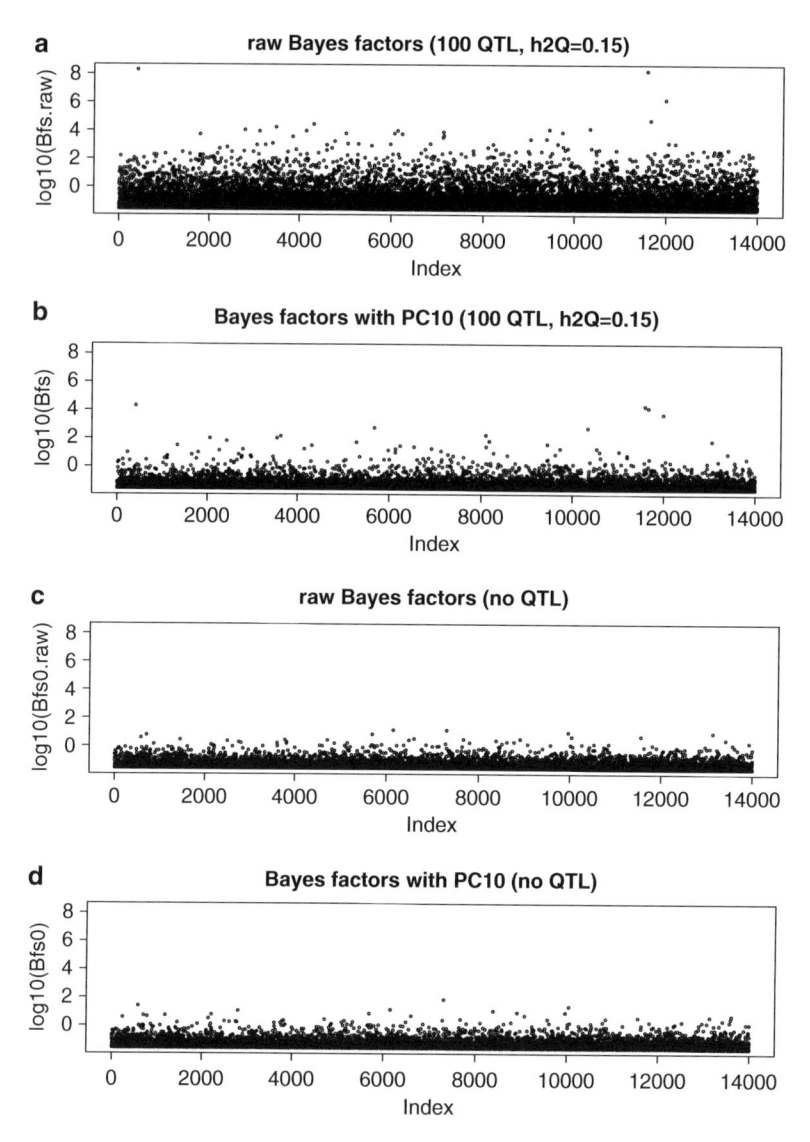

Fig. 1 Manhattan plots for chromosome 4 simulated data Bayes factors. Bayes factors are plotted vs. index (genome order) for each marker. The panels (a), (c) are of "raw" Bayes factors, with no population structure adjustment; panels (b),(d) are Bayes factors obtained when using PC10 to adjust for population structure. Panels (a),(b) are for data simulated with 100 QTL collectively explaining 15% of the phenotypic variation; panels (c), (d) are for the same marker data but with no simulated QTL, and serve as a control

Manhattan plots are shown in Fig. 1. Note the substantial reduction in the largest Bayes factors after correcting for population structure, as previously noted by (11).

3.1 Effect on Power Calculations

The effect of population structure on power calculations is unclear. However it is not unreasonable to suppose that most loci have low correlations with the top 10 principal components, so power to detect these effects would be similar. Effects for loci correlated with the principal components would be partly confounded with the population structure, hence their power would be reduced.

On the other hand, the adjustment for population structure is imperfect— some effects may still be inflated by "residual population structure," or by the errors introduced in the process of estimating the population structure, for example when estimating the kinship matrix. Such effects could increase the background "noise," requiring a stronger signal in order to detect the true effects.

A conservative approach might be to make an adjustment based on the genomic control factor, i.e., the factor by which χ^2 test statistics are reduced when using genomic control in similar studies, if known.

4 Analysis with Pedigree Data: Testing for Linkage and Linkage Disequilibrium

Many plant and animal breeders will have access to populations where pedigree information is available. This material cannot be regarded as an independent population sample because individuals are related. However, these populations may still contain LD useful for association studies.

The methods in this section take into account relatedness between individuals in the analysis of marker-trait associations from a known pedigree. The methodology uses mixed models to allow for correlation between haplotype effects, with correlation structure based on IBD (identical by descent) probabilities, and also to allow for polygenic effects, with covariance structure given by the additive relationship matrix from quantitative genetics. In so doing, the methods combine linkage and linkage disequilibrium information. The linkage or "QTL mapping" information is generated by recombinations within the pedigree, detected by marker genotypes of parents and their offspring. The linkage disequilibrium or association mapping, information, is generated by ancestral recombinations, and detected by population level associations between individuals.

Incorporating polygenic random effects in the model via the additive relationship matrix effectively controls for population structure within the pedigree (10). The pedigree analysis may, however, still be affected by spurious associations from population structure present when the founders were obtained. Relatedness between the founders would probably be unknown and still needs to be checked and/or controlled by methods of Sect. 3. This might happen if a breeding population was obtained from material taken from several native provenances, as is the case for *P. radiata*. If individuals' ancestry cannot be traced back to the provenances, the genomes of currently growing trees may be a mixture of provenances, with unknown mixing probabilities, which can be estimated by the program Structure (13–15).

With a known pedigree, the numerator relationship matrix (A-matrix A) can be used to control for relatedness, with set of a random effects for each genotype. Alternatively an estimated kinship matrix $\hat{K}$ may be used. The kinship matrix has potentially more information than the A-matrix because, due to the stochastic nature of meiosis, individuals may incorporate more or less than the average proportion of DNA from each grandparent, and hence, e.g., some pairs of sibs may have more than 50% of loci IBD (22). On the other hand the error in the estimated kinship matrix $\hat{K}$ can reduce the effectiveness of the adjustment, or possibly introduce spurious associations.

If the founders are not unrelated, or originate from a population with structure, the methods of the previous section, e.g., using principal components, can also be applied in conjunction with the A-matrix or K-matrix. Alternatively, a coalescent-based approach can be used to estimate the genetic covariance between founders (23, 24).

4.1 Frequentist Analysis

Meuwissen et al. (25) use combined linkage disequilibrium and linkage information to fine map a QTL in cattle in a known pedigree. They fit a mixed model:

$$y = \mu + Zh + u + e \qquad (3)$$

$$h \sim N(0, G\sigma_h^2) \qquad (4)$$

$$u \sim N(0, A\sigma_u^2) \qquad (5)$$

$$e \sim N(0, \sigma_e^2) \qquad (6)$$

where μ is the overall mean, h are random haplotype effects, u are random polygenic effects, and e are residual errors. The haplotypes are based on markers close to the QTL locus. An "infinite alleles model" is assumed so that each haplotype potentially has a different effect.

Note that each individual has two haplotypes. The number of haplotypes is greater than the number of individuals, so haplotype effects cannot be estimated individually; however haplotype effects are correlated. Haplotype effects are identical if the corresponding QTL alleles are IBD. It follows that haplotype effects are correlated, with correlation matrix, G, given by the IBD probabilities for the QTL. The correlation between polygenic effects is given by the "additive relationship matrix" A (26).

The IBD probability calculation for founders is based on (24).

- The calculation is different for each putative QTL locus. Haplotypes can be based on up to around 15 closely spaced markers around the putative QTL locus.

- For base haplotypes (1st generation genotyped) Meuwissen and Goddard (23, 24) use a modified coalescent to estimate IBD probabilities. Briefly, assume an effective population size

```
library(nlme)
# QTL analysis
# given: trait y, G, A matrices
# Z-matrices for paternal and maternal haplotypes
# individual: a factor coding individual genotypes
Zhp <- model.matrix(~ individual -1)
Zhm <- model.matrix(~ individual -1)
# Choleski matrix for G
Rg <- chol(G,pivot=FALSE)
Zh <- cbind(Zhp, Zhm) %*% t(Rg)
# Choleski matrix for G
Ra <- chol(A, pivot=FALSE)
Za <- Ra
# model with polygenic effects only
fit0 <- lme(y ~ 1, random=list(all=pdIdent(~Za -1)))
# model with QTL plus polygenic effects
fit1 <- lme(y ~ 1, random=list(all=pdBlocked(list(
         pdIdent(~Zh -1), pdIdent(~Za -1)))))
# compare models, LR test etc.
anova(fit0,fit1)
```

Fig. 2 R code for mixed model QTL analysis (3–6) combining linkage and linkage disequilibrium

of N_e and T generations of random mating. Either simulate the coalescent or use the analytical formulae. In a given simulated coalescent, haplotypes are considered IBD if they have coalesced within the T generations. IBD probabilities for a pair of haplotypes are estimated as the proportion of simulated coalescents where the haplotypes coalesced.

- For subsequent generations estimate IBD, using parental and marker information.

Similar to interval mapping QTL approaches (27), the analysis is repeated for each putative QTL position and likelihood ratios calculated at each position. A *p*-value is obtained by referring the likelihood ratio statistic to its sampling distribution under the null hypothesis of no effect. As demonstrated in previous sections, there are problems with the interpretation of *p*-values.

Note:

1. Meuwissen *et al.* (25) fitted their model using ASREML (28). Analysis using the publicly available `nlme` R package is also possible, by forming the Choleski decomposition of the matrices *G* and *A*, and incorporating the Choleski factor into the *Z*-matrices, effectively transforming the sets of random effects to independent random effects, enabling the model to be fitted using the standard `nlme` covariance matrix classes as in Fig. 2. This technique is used in the `lmeSplines` R package (29).

2. The major computational difficulty in fitting the mixed models is evaluating the inverse of the matrix A, for large pedigrees. Specialized algorithms (not shown) are available for calculating the Choleski decomposition directly.

4.2 Bayesian Analysis

The mixed model of equations 3–6 is almost Bayesian in that random effects have probability distributions. To make a full Bayesian model requires only specifying priors on the variance components $\sigma_b^2, \sigma_a^2, \sigma_e^2$. As in previous sections, Bayes factors and posterior probabilities are used for inference. An MCMC sampler can be generated and Bayes factors and posterior probabilities calculated. ((30); subsection 8.3.4).

Note: There are some similarities between this approach and the "BLADE" method (31). Meuwissen and Goddard (23, 24) simulate or calculate IBD probabilities based on possible ancestral genealogies and use the IBD probabilities in a mixed model analysis, while Liu et al. (31) simulate possible ancestral genealogies from a coalescent process with inference based on analysis of each of the simulated genealogies. The mixed model approach has the advantage of being able to incorporate pedigree information, and control for population structure, but also has the disadvantage of using fixed estimates of IBD probabilities. This means one is effectively conditioning on the IBD probabilities being the true values in the mixed model analysis. This is the price paid for the convenience of using a more standard mixed model, with easier implementation in R or BUGS. The full Bayesian coalescent based model conditions on the population assumptions inherent in the coalescent, as does the Meuwissen and Goddard IBD estimation, but not on possible values of IBD probabilities that might be consistent with these assumptions.

4.3 Summary

A sample from a known pedigree combines QTL and LD mapping information in a single dataset. Mixed model analysis for a pedigree combines haplotype effects (at or around a single locus), with a correlation structure based on IBD probabilities, and polygenic effects. The effectiveness of the pedigree sample for LD mapping depends on the breadth and sample size of individuals from which the pedigree was founded.

Incorporating polygenic effects via the additive relationship matrix controls for population structure generated within the pedigree, but not for population structure when the founders were chosen, since relatedness between the founders is probably unknown. Population structure analysis on the founders is recommended.

For further information on models combining pedigree and LD information see (32–40). Other approaches to calculation of IBD probabilities include a stochastic MCMC method for use in large pedigrees (41, 42), available as a software package Loki;

and deterministic methods (31, 43). The deterministic methods are faster but are approximate, and/or ignore uncertainty in haplotypes.

5 QTL and LD Mapping Combined

In this section we consider combining information from QTL and LD mapping. Unlike Sect. 4, where population and pedigree information are combined in a single dataset, this section considers QTL and LD analysis on distinct datasets, where the results of QTL analysis are used as prior information for the LD analysis.

Brute force genotyping of, e.g., 500,000 SNP markers for large numbers of individuals would be prohibitive. In this section we consider a strategy for reducing the amount of genotyping by combining QTL (linkage) mapping and LD (association) mapping. A QTL mapping family is used to narrow down the range of the genomic region to search for associations, reducing the amount of genotyping required, and in the process increasing the prior odds per marker genotyped.

In the Bayesian paradigm, successive datasets can be analyzed sequentially with the posterior distribution from each analysis being the prior for the next analysis. This make sense logically since the posterior distribution represents our knowledge after the ith analysis which is the same as our knowledge prior to the $i + 1$st analysis. The same posterior distributions are obtained as if the datasets are analyzed jointly in a single model. Utilizing this fact, a natural way to approach combined analysis for separate QTL and LD mapping datasets is to use the posterior distribution from the QTL analysis as the prior distribution for the LD analysis. As before, if QTL mapping data is available, but the LD experiment has not yet been done, we can design the LD experiment with given power to obtain a sufficiently high Bayes factor to obtain a reasonably high posterior probability after the LD analysis. Next, we apply this approach to locating small effect QTL.

Table 1 shows results for sample sizes and amount of genotyping required to detect a QTL explaining 5% of the variation of a trait. Results are given for various sizes ($n_{QTL} = 100, 400, 1000, 3000$) of the QTL mapping family. For each family size the average standard error ($se(\hat{x})$) of the estimate of QTL location was calculated by simulation of an additive QTL. The QTL interval was assumed to be two standard deviations on either side of the estimate, although smaller values could be considered and may be more cost-effective, at the risk of loosing some QTL. The number of SNPs within the QTL region was calculated and average prior odds per SNP were determined from this. Then, the Bayes factor required to obtain the required posterior probability of 0.9 was calculated and the sample sizes (n_{LD}) for this were calculated using the R function `ld.design()` from the `ldDesign` package (44, 45).

Table 1

Sample sizes and amount of genotyping required to locate a QTL when searching the genome using QTL and LD mapping combined. Assume there are 10 QTL each explaining 5% of the variation, $D = 0.1$ or $D = 0.2$, allele frequencies 0.5 for QTL and marker for closest marker to the trait locus, a genome of 3×10^9 bases, extent of LD 6*kb*, 500,000 SNP markers available at a spacing of 6*kb*, giving prior probabilities per marker of 1/50,000. The QTL mapping results assume there are 12 chromosomes and 20 markers per chromosome at a spacing of 10*cM*. Results are given for an overall posterior probability of 0.9 for an association

	Number of QTL progeny genotyped			
	nQTL = 100	*nQTL* = 400	*nQTL* = 1000	*nQTL* = 3000
QTL: PProb(H_0)	0.5	0.2	< 0.001	< 0.001
$se(\hat{x})$	12.2*cM*	6.5*cM*	4.1*cM*	2.4*cM*
Number of SNPs in QTL interval	10,167	5417	3417	2000
Average prior odds per SNP	1/20,333	1/6770	1/3417	1/2000
Bayes factor required from LD	183,000	60,938	30,750	18,000
n_{LD}				
$D = 0.2$	1589	1508	1451	1407
$D = 0.1$	6909	6554	6345	5713
QTL marker genotyping	24,000	96,000	240,000	720,000
LD marker genotyping				
$D = 0.2$	16.2×10^6	8.2×10^6	5.0×10^6	2.8×10^6
$D = 0.1$	70.2×10^6	35.5×10^6	21.7×10^6	11.4×10^6
Total genotyping				
$D = 0.2$	16.2×10^6	8.3×10^6	5.2×10^6	3.5×10^5
$D = 0.1$	70.3×10^6	35.6×10^6	21.9×10^6	12.1×10^6

There are a number of factors which could be varied in searching for an optimal design—we have considered only two special cases here. Nevertheless, the results suggest, with the extent of LD considered, that a significant efficiency gain can be achieved by combining QTL and LD mapping, and that the optimal QTL mapping population size will often be quite large. There are still quite a large number of SNPs to genotype per individual within the QTL region. Hence, the LD genotyping dominated the QTL genotyping except for the largest QTL sample size, and the maximum disequilibrium). Except for QTL population size 100, which had posterior probability of only 0.5, the width of the QTL region decreased gradually in inverse proportion to the square root of the QTL mapping population size. The least amount of total

Table 2

Sample sizes and amount of genotyping required to locate a QTL when searching the genome using QTL and LD mapping combined. Assume there are 10 QTL each explaining 5% of the variation, $D = 0.1$ or $D = 0.2$, allele frequencies 0.5 for QTL and marker for closest marker to the trait locus, a genome of 3×10^9 bases, extent of LD 60kb, 50,000 SNP markers available at a spacing of 60kb, giving prior probabilities per marker of 1/5,000. The QTL mapping results assume there are 12 chromosomes and 20 markers per chromosome at a spacing of 10cM. Results are given for a posterior probability of 0.9

	Number of QTL progeny genotyped			
	nQTL = 100	nQTL = 400	nQTL = 1000	nQTL = 3000
QTL: PProb(H_0)	0.5	0.2	< 0.001	< 0.001
$se(\hat{x})$	12.2cM	6.5cM	4.1cM	2.4cM
Number of SNPs in QTL interval	1017	541	342	200
Average prior odds per SNP	1/2033	1/677	1/342	1/200
Bayes factor required from LD	18,300	6,094	3,075	1800
n_{LD}				
$D = 0.2$	1408	1323	1268	1222
$D = 0.1$	6193	5826	5603	5435
QTL marker genotyping	24,000	96,000	240,000	720,000
LD marker genotyping				
$D = 0.2$	1.4×10^6	7.2×10^5	4.3×10^5	2.4×10^5
$D = 0.1$	6.3×10^6	3.2×10^6	1.9×10^6	1.1×10^6
Total genotyping				
$D = 0.2$	1.5×10^6	8.1×10^5	6.7×10^5	9.6×10^5
$D = 0.1$	6.3×10^6	3.3×10^6	2.1×10^6	1.8×10^6

genotyping was for the largest QTL population of size $n_{QTL} = 3000$, with a fivefold reduction in genotyping compared to $n_{QTL} = 100$. In this case, depending on phenotyping costs, larger QTL mapping populations should be considered before embarking on LD mapping. Values are given for both $D = 0.2$ and $D = 0.1$, with the latter being the minimum disequilibrium expected within the marker interval, by assumption. The total genotyping was still decreasing between $n_{QTL} = 1000$ and $n_{QTL} = 3000$, for both $D = 0.2$ and $D = 0.1$, so the optimum may be even higher.

Similar results are shown in Table 2 where the extent of LD is assumed to be 60kb. In this case the prior odds per SNP have increased tenfold compared to the previous case. The optimal design appears to be approximate when $n_{QTL} \approx 1000$ when $D = 0.2$, and $n_{QTL} \approx 3000$, with up to approximately a threefold reduction in genotyping compared to $n_{QTL} = 100$ when $D = 0.1$.

5.1 Summary   This section shows that QTL mapping and LD mapping analysis and experimental design can be profitably combined.

The posterior distributions from QTL analysis can be used as prior distributions for the LD analysis. The Bayes factor required from the LD mapping population for a given posterior probability is reduced for loci mapping into a QTL region.

Brute force genotyping of all markers in a genome scan for a sufficiently large population is very costly due to the very large amount of total genotyping. One possible strategy is to restrict genotyping of the LD mapping population to QTL regions. We have seen that this can result in reduced overall genotyping compared to a stand-alone LD mapping approach. Considering a single trait, the examples suggest that the optimal strategy is to use even larger QTL mapping populations than those currently used, prior to LD mapping, in order to find small effect genes.

A by-product of this approach is that most spurious associations due to population structure will be eliminated by the QTL mapping study. If the QTL mapping intervals are small, e.g., with a sufficiently large QTL mapping family size, this can be more effective than the TDT.

References

1. Ball, R. D. 2001: Bayesian methods for quantitative trait loci mapping based on model selection: approximate analysis using the Bayesian Information Criterion. Genetics 159: 1351–1364. http://www.genetics.org/cgi/content/abstract/159/3/1351 Accessed 29/5/2012.
2. Ball, R. D. 2009: BayesQTLBIC—Bayesian multi-locus QTL analysis based on the BIC criterion. http://cran.r-project.org/web/packages/BayesQTLBIC/index.html Accessed 29/5/2012.
3. Sen, S. and Churchill, G. A. 2001: A statistical framework for quantitative trait mapping. Genetics 159: 371–387.
4. Marchini, J. Howie, B., Myers, S., McVean, G. and Donnelly, P. 2007: A new multipoint method for genome-wide association studies via imputation of genotypes. Nature Genetics 8: 1750–1761.
5. Servin B. and Stephens, M. 2007: Imputation-based analysis of association studies: Candidate regions and quantitative traits. PLoS Genet 3 (7): 1296–1308. e114. doi: 10.1371/journal.pgen.0030114
6. Stephens, M. and Balding, D. J. 2009: Bayesian statistical methods for association studies. Nat. Rev. Genet 10: 681–690.
7. HapMap project 2012: http://hapmap.ncbi.nlm.nih.gov/ Accessed 31/5/2012.
8. Raftery, A. E. 1995: Bayesian model selection in social research (with Discussion). Sociological Methodology 1995 (Peter V. Marsden, ed.), pp. 111–196, Cambridge, Mass.: Blackwells.
9. Ball, R. D. 2007b: Quantifying evidence for candidate gene polymorphisms: Bayesian analysis combining sequence-specific and quantitative trait loci colocation information. Genetics 177: 2399–2416. http://www.genetics.org/cgi/content/abstract/177/4/2399 Accessed 29/5/2012.
10. Sillanpää, M. J. and Bhattacharjee, M. 2005: Bayesian association-based fine mapping in small chromosomal segments. Genetics 169: 427–439.
11. Astle, W. and Balding, D. 2009: Population structure and cryptic relatedness in genetic association studies. Statistical Science 24: 451–471.
12. Devlin, B. and Roeder, K. 1999: Genomic control for association studies. Biometrics 55: 997–1004.
13. Pritchard, J. K., Stephens, M. and Donnelly, P. 2000a: Inference of population structure using multilocus genotype data, Genetics 155: 945–959.

14. Pritchard, J. K., Stephens, M., Rosenberg, N. A., and Donnelly, P. 2000b: Association mapping in structured populations, Am. J. Hum. Genet. 67: 170–181.
15. Falush, D., Stephens, M. and Pritchard, J. K. 2003: Inference of population structure using multilocus genotype data: linked loci and correlated allele frequencies. Genetics 164: 1567–1587.
16. Setakis, E., Stirnadel, H. and Balding, D. J. 2006: Logistic regression protects against population structure in genetic association studies. Genome Res. 16: 290–296.
17. Zhang, S. Zhu, X. and Zhao, H. 2003: On a semiparametric test to detect associations between quantitative traits and candidate genes using unrelated individuals. Genet. Epidemiol. 24: 44–56.
18. Price, A. L., Patterson, N. J., Plenge, R. M., Weinblatt, M. E., Shadick, N. A. and Reich, D. 2006: Principal components analysis corrects for stratification in genome-wide association studies. Nat. Genet. 38: 904–909.
19. Ritland, K. 1996: Estimators for pairwise relatedness and individual inbreeding coefficients. Genetical Research 67: 175–185.
20. Zhang, Z., Ersoz, E. Lai, C.-Q., . . . and Buckler, E.S. 2011: Mixed linear model approach adapted for genome-wide association studies. Nature Genetics 42: 355–360.
21. Spencer, C. C. A., Su, Z., Donnelly, P., Marchini J. 2009: Designing Genome-Wide Association Studies: Sample Size, Power, Imputation, and the Choice of Genotyping Chip. PLoS Genet 5(5).
22. Weir, B. S., Anderson, A. D., and Hepler, A. B. 2006: Genetic relatedness analysis: modern data and new challenges. Nature Reviews Genetics 7: 771–780.
23. Meuwissen, T. H. E., and Goddard, M. E. 2000: Fine mapping of quantitative trait loci using linkage disequilibria with closely linked marker loci. Genetics 155: 421–430.
24. Meuwissen, T. H. E., and Goddard, M. E. 2001: Prediction of identity-by-descent probabilities from marker haplotypes. Genet. Sel. Evol. 33: 605–634
25. Meuwissen, T. H. E., Karlsen, A., Lien, S., Oldsaker, I., and Goddard, M. 2002: Fine mapping of a quantitative trait locus for twinning rate using combined linkage and linkage disequilibrium mapping. Genetics 161: 373–379.
26. Falconer, D. S., and Mackay, T. F. C. 1996: *Introduction to Quantitative Genetics*. Addison-Wesley Longman, Harlow, England.
27. Lander, E. S. and Botstein, D. 1989: Mapping Mendelian factors underlying quantitative traits using RFLP linkage maps. Genetics 121: 185–199.
28. Gilmour, A. R., Gogel, B. J., Cullis, B. R., and Thompson, R. 2009: *ASReml User Guide Release 3.0* VSN International Ltd, Hemel Hempstead, HP1 1ES, UK. www.vsni.co.uk
29. Ball, R. D. 2003: lmeSplines—an R package for fitting smoothing spline terms in LME models. R News 3/3 p 24–28. http://cran.r-project.org/web/packages/lmeSplines/index.html. Accessed 29/5/2012.
30. Ball, R. D. 2007: Statistical analysis and experimental design Chapter 8, In: Association mapping in plants. N. C. Oraguzie *et al.* editors, Springer Verlag, ISBN 0387358447 (69pp).
31. Liu, J. S., Sabatti, C., Teng, J., Keats, B. J. B. and Risch, N. 2001: Bayesian analysis of haplotypes for linkage disequilibrium mapping. Genome Research 11: 1716–1724.
32. Wu, R. and Zeng, Z.-B. 2001: Joint linkage and linkage disequilibrium mapping in natural populations. Genetics 157: 899–909.
33. Wu, R., Ma, C. X. and Casella, G. 2002: Joint linkage and linkage disequilibrium mapping in natural populations. Genetics 160: 779–792.
34. Farnir, F., Grisart, B., Coppieters, W., Riquet, J., Berzi, P., *et al.* 2002: Simultaneous mining of linkage and linkage disequilibrium to fine map quantitative trait loci in outbred half-sib pedigrees: revisiting the location of a quantitative trait locus with major effect on milk production on bovine chromosome 14. Genetics 161: 275–287.
35. Perez-Enciso, M. 2003: Fine mapping of complex trait genes combining pedigree and linkage disequilibrium information: a Bayesian unified framework. Genetics 163: 1497–1510.
36. Fan, R. and Jung, J. 2002: Association Studies of QTL for multi-allele Markers by mixed models. Hum. Hered. 54: 132–150.
37. Lund, M. S., Sorensen, P., Guldbrandtsen, P., and Sorensen, D. A. 2003: Multitrait fine mapping of quantitative trait loci using combined linkage disequilibria and linkage analysis. Genetics 163: 405–410.
38. Meuwissen, T. H. E., Hayes, B. J., and Goddard, M. E. 2001: Prediction of total genetic value using genome-wide dense marker maps. Genetics 157: 1819–1829.
39. Meuwissen, T. H. E., and Goddard, M. E. 2004: Mapping multiple QTL using linkage disequilibrium and linkage analysis information and multitrait data. Genet. Sel. Evol. 36: 261–279.

40. Lee, S. H. and van der Werf, J. H. J. 2005: The role of pedigree information in combined linkage disequilibrium and linkage mapping of quantitative trait loci in a general complex pedigree. Genetics 169: 455–466.
41. Heath, S. C. 1997: Markov chain Monte Carlo segregation and linkage analysis for oligogenic models. Am. J. Hum. Genet. 61: 748–760.
42. Heath, S. 2003: *Loki 2.4.5—A package for multipoint linkage analysis on large pedigrees using reversible jump Markov chain Monte Carlo.* Centre National de Génotypage, Evry Cedex, France. http://www.stat.washington.edu/thompson/Genepi/Loki.shtml Accessed 31/5/2012.
43. Gao, G. and Hoeschele, I. 2005: Approximating identity-by-descent matrices using multiple haplotype configurations on pedigrees. Genetics 171: 365–376.
44. Ball, R. D. 2004, 2011: ldDesign — design of experiments for genome-wide association studies version 2 incorporating quantitative traits and case-control studies. http://cran.r-project.org/web/packages/ldDesign/index.html Accessed 29/5/2012.
45. Ball, R. D. 2005: Experimental designs for reliable detection of linkage disequilibrium in unstructured random population association studies. Genetics 170: 859–873. http://www.genetics.org/cgi/content/abstract/170/2/859 Accessed 29/5/2012.

Chapter 8

Using PLINK for Genome-Wide Association Studies (GWAS) and Data Analysis

Miguel E. Rentería, Adrian Cortes, and Sarah E. Medland

Abstract

Within this chapter we introduce the basic PLINK functions for reading in data, applying quality control, and running association analyses. Three worked examples are provided to illustrate: data management and assessment of population substructure, association analysis of a quantitative trait, and qualitative or case–control association analyses.

Key words Data management, Population stratification, Genetic association, Binomial trait, Quantitative trait, Multidimensional scaling, PLINK

1 Introduction

PLINK is a freely available, widely used open-source toolset for genetic association that allows for the study of large datasets of genotypes and phenotypes [1]. It was initially developed in 2007, when genome-wide association was a very new concept. Prior to this time the most commonly used method for genome level analysis had been linkage analysis that typically used a set of ~400 markers from which identity by descent information was estimated either directly at the marker or, through inference at a 1–2 cM grid, yielding ~3,500 positions at which analysis would be run. With the development of chip-based genome-wide association studies (GWAS) arrays the volume of data and analyses quickly increased by several orders of magnitude and many researchers were daunted by the sheer volume of data. This historical context has shaped the development of PLINK. The ambitious aim of Purcell et al. [1] was to create a single package that could seamlessly integrate data manipulation, quality control, analysis, and annotation. While there are now many programs that can be used for analyses, there remain only a small number of programs for manipulation and quality control of GWAS data and the use of PLINK has become almost ubiquitous for this type of analysis. The discipline is currently facing an analogous transition point as exome and genome-wide sequence level data become more affordable, and Purcell and colleagues have responded to this new challenge by developing PLINK-seq [1], which we expect will become as popular as PLINK has been.

The omnibus nature of PLINK makes it difficult to provide a comprehensive overview within a single chapter. Our focus here is on four main functional domains: data management, summary statistics, population stratification and estimation of relatedness, and association analysis. We have developed a series of complementary exercises designed to demonstrate these commonly used PLINK functionalities. Readers are referred to the extensive PLINK website for more information on other procedures (http://pngu.mgh.harvard.edu/~purcell/plink/).

1.1 Getting Started

PLINK is a command line program written in C/C++. Binaries and source code can be downloaded from http://pngu.mgh.harvard.edu/~purcell/plink/download.shtml. For information on how to install it on your computer, please refer to the notes on the download page.

As a command line program, all commands involve typing *plink* (*see* **Note 1**) at the command prompt (e.g., DOS window or Unix terminal) followed by the desired options which specify the data files and methods to be used (all prefaced with two minus signs, i.e.,--). In addition to its functions for genetic association, PLINK provides a simple interface for recoding, reordering, merging, flipping DNA-strand, and extracting subsets of data and a number of other versatile functions.

In addition to the command line version, a number of R interfaces have been developed (*see* http://pngu.mgh.harvard.edu/~purcell/plink/rfunc.shtml). There is also a simple Java-based graphic user interface (GUI), gPLINK which provides access to the more commonly used PLINK commands. To learn more about this package, visit: http://pngu.mgh.harvard.edu/~purcell/plink/gplink.shtml

This chapter covers the basics of data management and manipulation in PLINK through three exercises that illustrate how to check for population stratification and basic genetic association analyses. The materials section provides an introduction to the file formats used in PLINK and provides a step-by-step tutorial for data management.

2 Materials

PLINK can accept data in a number of file formats. The most common are generally variations to the basic ASCII-based *linkage file format* PED/MAP and the binary *PLINK file formats* BED/BIM/FAM.

PED files (e.g., genotypes.ped) are white-space (space or tab) delimited text files that contain phenotype and genotype data and are typically structured as follows (*see* **Notes 2** and **3**):

Column 1	Family ID (FID)
Column 2	Individual ID (IID)
Column 3	Paternal ID (PID)
Column 4	Maternal ID (MID)
Column 5	Sex
Column 6	Phenotype
Column 7…n	Genotype(s)

Depending on the nature of the study and other programs that might have been used to produce the data, there are many options that can be included to allow for variations on this format. For example, in many case-control studies data are stored without family or parental IDs, if you wished to use a file without these columns you could do so by adding the --*no-familyID* and --*no-parentalIDs* options to the command line. Similarly, many genotype files do not contain a phenotype, rather than add a dummy phenotype to the file the user can simply add the --*no-pheno* option to the command line.

There are however, some restrictions on the contents of these variables:

- FIDs, IIDs, PIDs, and MIDs must be alphanumeric and unique for each family/individual.
- By default, sex is coded: 1 = male; 2 = female; other=unknown. If an individual's sex is unknown, any character other than 1 or 2 can also be used.
- A PED cannot contain more than one phenotype, and if a phenotype is included it must be placed in the sixth column. A phenotype can be either binary (i.e., affection status; by default, 1 = unaffected, 2 = affected, 0 = missing) or quantitative. PLINK will automatically detect which type (i.e., If a value other than 0, 1, 2, or the missing genotype code is observed, PLINK assumes the phenotype is a quantitative trait).
- For quantitative traits, the missing phenotype value is, by default, −9, but this can be changed to any integer number by using the --*missing-phenotype* option.

By default, PLINK assumes markers are biallelic. Columns 7 and 8 contain the genotype pair at SNP1; Columns 9 and 10 contain the genotype pair at SNP2; and so on. All SNPs (whether haploid or not) must have two alleles specified. In the case of missing data, *both* alleles should be missing (i.e., 0) or neither. For haploid chromosomes, i.e., chromosomes X and Y in males or mitochondrial DNA, genotypes should be entered as homozygotes. The default missing genotype character (0) can be changed with the *--missing-genotype* option. Compound genotypes, that is data in which the two alleles are separated by a slash (i.e., A/C) or concatenated (i.e., AC), can be incorporated by adding the *--compound-genotypes*. PLINK can also read genotypic data that has been coded numerically, either with reference to a given allele (i.e., 1 = major and 2 = minor allele) or in alphabetical order (i.e., 1 = A, 2 = C, 3 = G, 4 = T) using the appropriate options.

PLINK can read other variations on the standard PED file, including transposed files (that are used by BEAGLE and IMPUTE), in which the data for SNPs is contained in rows and each column represents an individual. More information about methods for working with these alternative file-formats can be found on the PLINK website (http://pngu.mgh.harvard.edu/~purcell/plink/data.shtml).

MAP files (e.g., genotypes.map) are white-spaced (space or tab) delimited text files that contain the chromosomal positions of each SNP that has been genotyped, and typically are structured as follows:

Column 1	Chromosome number
Column 2	SNP identifier (rs# or SNP identifier)
Column 3	Genetic distance (in Morgans)
Column 4	Physical base-pair position (bp units)

As with PED files, there are a number of variations available for map files. For example, genetic distance can be specified in centimorgans with the --cm option. Alternatively, you can use a MAP file that does not contain the genetic distance by adding the --map3 option. If working with human data, autosomes are defined by numbers 1–22. Additionally, the following codes are used to specify other chromosome types: X chr. = 23; Y chr. = 24; pseudoautosomal region of X (XY) = 25; and MT chromosome=26. Comments can be added to a PED or MAP file by starting the line with a # character. PLINK also supports several nonhuman species. To this end, the following flags can be added: --dog, --horse, --cow, --sheep, --rice, or --mouse.

Each row in the MAP file corresponds to two (or more, if working with multiploid genomes) columns of the PED file, and these need to be in the same order (e.g., the SNP described in row 1 of the MAP file will be assumed to be the one for which genotypes are contained in Columns 7 and 8 of the PED file, and the SNP described in row 2 of the MAP file will be assumed to correspond to the genotypes contained in Columns 9 and 10 of the PED file, etc.).

PLINK has a number of options for reading data. The most common is the --file function, as in the following example, in which two files (mydata.ped, and mydata.map) will be imported into PLINK:

```
plink --file mydata
```

If the PED and MAP files have different prefixes, they can be specified separately, with the --ped and --map options, as in the following example:

```
plink --ped mydata.ped --map chr1.map
```

To save space and time, it is possible to convert the data to a binary PED file (*.bed). This will store the pedigree/phenotype information in a separate file (*.fam) and create an extended MAP file (*.bim), which contains allelic information that is not stored in the BED file. The prefix of the output files can be specified using the --out option. For example: *plink --file mydata --make-bed --out chr1* would read in the *mydata.ped* and *mydata.map* files and create four files:

chr1.bed	Compressed binary file that contains genotype information
chr1.fam	Contains the first six columns of mydata.ped (FID, IID, PID, MID, sex, and phenotype) if any of these columns are not included in the PED file, they will be set to missing in the FAM file
chr1.bim	Extended MAP file: contains two extra columns = allele names
chr1.log	Contains a running log of the PLINK job which includes basic summary information and a list of the options requested

Often, it is useful to filter out SNPs from datasets based on quality control parameters such as minor allele frequency (MAF) or missingness. This can be achieved using the --*maf* and --*geno* functions, for example:

```
plink --file mydata --make-bed --maf 0.02 --geno 0.1
```

would exclude SNPs with a MAF of 2 % and SNPs with more than 10 % missingness.

To read these file formats (*.bed, *.fam, *.bim) back into PLINK, the --bfile option is used instead of --file, i.e., *plink --bfile mydata*. It is also possible to specify these files separately:

```
plink --bed file1.bed --fam file2.fam --bim file3.bim
```

Data can be converted from *binary* to *linkage* format using the --*recode* function, which will create simple PED and MAP (instead of binary) files. The --*bfile*, --*make-bed*, and --*out* functions are constantly used when subsetting data, performing quality control procedures and analysis. To limit the amount of memory being used, PLINK does not store data for an "analysis session," rather PLINK uses a sequential approach in which each QC or analysis step is performed as a new job. For example, to generate a clean dataset that has been filtered with the --*maf* and --*geno* options so that it can be used for analysis or additional QC the user needs to add the --*make-bed* and --*out* options to their job.

3 Methods

This section comprises two exercises. In the first exercise we will obtain publicly available data for samples from three different populations (CEU, YRI, and JPT + CHB) of the HapMap Project (http://hapmap.ncbi.nlm.nih.gov/) and introduce some data management features before performing multidimensional scaling (MDS) analysis to quantify population structure of the sample. The second exercise focuses on genetic association, with two examples: copy number variation (CNV) data in the Central European cohort (CEU) of the HapMap Project as a quantitative trait, and association in a simulated case–control dataset.

3.1 Exercise 1: Data Management and Population Stratification

Population stratification, also referred to as population substructure, refers to the presence of systematic differences in allele frequencies between subpopulations within a sample, possibly due to different ancestry, which results from nonrandom mating between groups (e.g., this is often explained by physical separation, as in the case of populations of African and European descent) followed by genetic drift of allele frequencies in each group. The presence of population stratification is a problem for association studies, as it increases type 1 error and leads to spurious results. This is particularly true when both the genotypic and phenotypic data differ between populations.

If the structure of a population is known, or a putative structure is found, there are a number of possible ways to control for this in the association studies and thus ameliorate these population biases. Several methods exist, such as genomic control, structured association, or principal component analysis-based methods. In this exercise, publicly available genotype data from three different populations (Central European [CEU], Yoruba in Ibadan, Nigeria [YRI], and the East Asian combined sample of Japanese in Tokyo and Han Chinese in Beijing [JPT + CHB]) collected by the HapMap project will be downloaded, filtered by chromosome, and merged into a single file to estimate pair-wise identity-by-state (IBS) and MDS values. All scripts needed to perform these analyses and to plot the results are provided on the publisher's website (http://extras.springer.com/).

1. Go to the resources section in PLINK's website (data also available from the book's website):

   http://pngu.mgh.harvard.edu/~purcell/plink/res.shtml

   Download the following files:

Population	File
HapMap2 (rel 23) CEU	hapmap_CEU_r23a_filtered.zip
HapMap2 (rel 23) YRI	hapmap_YRI_r23a_filtered.zip
HapMap2 (rel 23) JPT + CHB	hapmap_JPT_CHB_r23a_filtered.zip

   These files contain filtered data for the founders of the CEU, YRI, and JPT + CHB populations in binary PLINK format (*.bed, *.bim, and *.fam), consisting of genome-wide data for 60 CEU, 60 YRI, and 90 JPT + CHB HapMap samples for over 2.5 million SNPs.

2. After downloading and uncompressing the data files, use the --chr and --make-bed functions to subset the data and create a more manageable dataset (composed of only SNPs in chromosome 1) to work with throughout this exercise:

   ```
   plink --bfile hapmap_CEU_r23a_filtered
   --chr 1 --make-bed --out CEU_chr_1
   
   plink --bfile hapmap_YRI_r23a_filtered
   --chr 1 --make-bed --out YRI_chr_1
   
   plink --bfile hapmap_JPT_CHB_r23a_filtered
   --chr 1 --make-bed --out JPT_CHB_chr_1
   ```

   The above commands will create binary PED (.bed) files (and their corresponding .bim and .fam files) for the three populations. Table 1 lists other options PLINK offers to filter datasets by SNPs.

   Likewise, PLINK offers a number of functions to filter a dataset by individuals in the sample. These include the options listed in Table 2 below.

3. To merge two different datasets, PLINK offers the --merge function (and its binary version --bmerge). These functions take the names of the files to be merged in the following order: [.ped, .map] or [.bed, .bim, .fam], respectively.

   ```
   plink --file data1 --merge data2.ped data2.map
   --recode --out data_merged
   
   plink --bfile data1 --bmerge data2.bed data2.bim
   data.fam
   --make-bed --out data_merged
   ```

Table 1
PLINK options to filter by SNPs

Option	Action
--chr VALUE	Only analyze or retain SNPs in chromosome VALUE
--exclude filename	Exclude SNPs in the file, one SNP identifier per line
--include filename	Only analyze or retain SNPs in the file
--exclude filename --range	Exclude regions specified in the file, one region per line. Regions are defined by four fields: chromosome, start, end, and region name. For example, 8 8000000 12000000 R1
--from-kb VALUE/ --to-kb VALUE	Only analyze or retain SNPs between the regions defined. Needs the chromosome options
--thin VALUE	Randomly select SNPs. A value of 0.4 will output only 40 % of the markers

Table 2
PLINK options to filter by individuals

Option	Action
--keep {indlist}	Keep only these individuals
--remove {indlist}	Remove these individuals
--filter-controls	Include only controls
--filter-cases	Include only cases
--filter-males	Include only males
--filter-females	Include only females
--filter-founders	Include only founders
--filter-nonfounders	Include only nonfounders

To merge more than two datasets, use the --merge-list function, which takes a filename as a value. This file must contain the names of the datasets to be merged, one dataset per line. Note that with this function either plain text or binary PLINK files can be used. In our example, the CEU_chr_1 dataset will be merged with the corresponding sets of the YRI and JPT + CHB populations.

The file named **merge_list.txt** contains the following text:

```
JPT_CHB_chr_1.bed             JPT_CHB_chr_1.bim
JPT_CHB_chr_1.fam
YRI_chr_1.bed YRI_chr_1.bim YRI_chr_1.fam
```

Table 3
PLINK options when merging two datasets

1	Only keep genotypes that match
2	Only overwrite calls which are missing in original PED file
3	Only overwrite calls which are not missing in new PED file
4	Never overwrite
5	Always overwrite mode
6[a]	Report all mismatching calls (diff mode—do not merge)
7[a]	Report mismatching non-missing calls (diff mode—do not merge)

[a]These options are particularly useful when computing concordance rates between two datasets with overlapping samples and markers

To create a single merged dataset with data of the three populations, use the following command:

```
plink --bfile CEU_chr_1 --merge-list merge_list.txt
--make-bed --out hapmap_chr_1
```

By default, if an individual is present in two files, SNPs present in one file will be set to missing if not present in one of the others, and any existing genotype data (i.e., in CEU_chr_1.bed) will not be overwritten by data in the second file (JPT_CHB_chr_1.bed). To change this behavior, the --merge-mode {} function provides seven different merging alternatives (see Table 3):

4. The following command will drop any SNPs with missingness ≥5 %. These SNPs with low call rates have mostly resulted from the merging process:

```
plink --bfile hapmap_chr_1 --geno .05
--make-bed --out data
```

5. To estimate population stratification, PLINK offers tools to cluster individuals into homogeneous subsets (which is achieved through complete linkage agglomerative clustering based on pair-wise IBS distance) and to perform classical MDS to visualize substructure and provide quantitative indices of population genetic variation that can be used as covariates in subsequent association analysis to control for stratification, instead of using discrete clusters. Generally, the --*mds-plot* option is used in conjunction with the --*cluster* function.

The following commands will estimate the degree of population stratification within the samples under analysis (see **Note 4**):

```
plink --bfile data_chr1 --genome --out genome_chr1
plink --bfile data_chr1 --read-genome genome_chr1.genome
--cluster --mds-plot 4 --silent --out mds
```

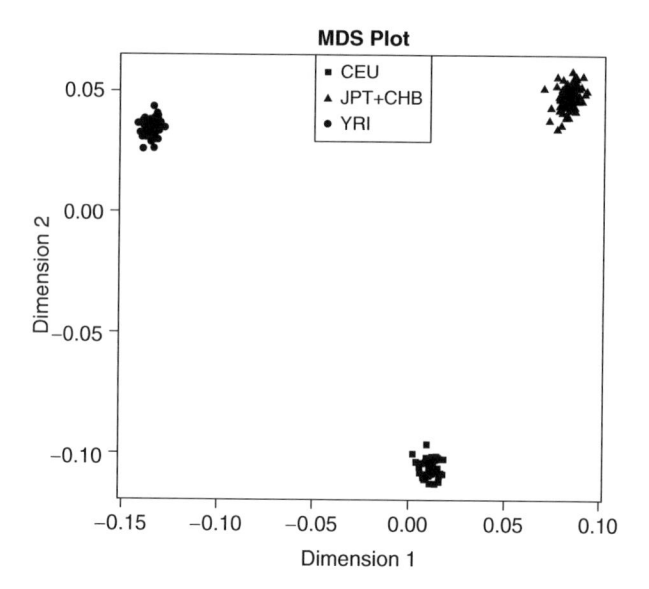

Fig. 1 Multidimensional scaling (MDS) plot of three populations: CEU (Central European), YRI (Yoruba in Ibadan, Nigeria), and JPT + CHB (combined sample of Japanese in Tokyo and Han Chinese in Beijing)

6. Finally, plot the results using R (http://www.r-project.org/) as shown in Fig. 1:

```
d <- read.table("mds.mds",header=T)

cols <- rep("gray",nrow(d))
cols[grepl("^1",d$FID)] <- 'red'
cols[grepl("^NA",d$FID)] <- 'green'
cols[grepl("^Y",d$FID)] <- 'blue'

pchs = rep(19,nrow(d))
pchs[grepl("^1",d$FID)] <- 15
pchs[grepl("^NA",d$FID)] <- 17
pchs[grepl("^Y",d$FID)] <- 19

pdf("Fig1.pdf")

plot(d$C1,d$C2,
   col='black',
   xlab='Dimension 1',
   ylab='Dimension 2',
   main='MDS Plot',
   pch=pchs)

legend('top',
    legend=c("CEU","JPT+CHB","YRI"),
    col='black',
    pch=c(15,17,19))

dev.off()
```

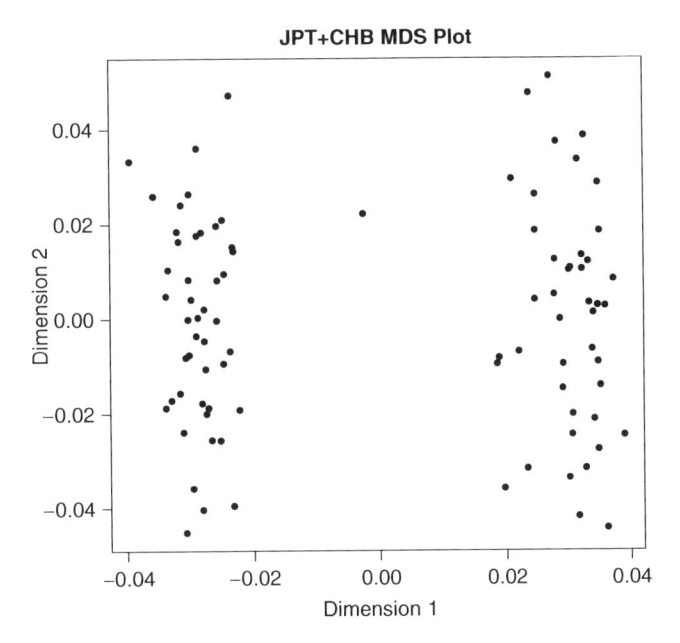

Fig. 2 MDS plot of the combined East Asian (JPT + CHB) populations

As expected, the analyses of these 3 diverse datasets yielded 3 distinct data clusters (on the left). However, in practice, association analyses are generally conducted in rather homogenous populations (e.g., people of European or Asian descent only). For instance, Fig. 2 shows an MDS plot for the JPT + CHB group. Although all individuals were clustered within the same group, the Chinese and Japanese are still distinguishable, and hence MDS coordinates could be used as covariates if controlling for this stratification was needed. The following exercise will provide an example of how to do this.

3.2 Exercise 2: More Data Management Options and Genetic Association

In this section, two genetic association analyses will be conducted. Firstly, we will conduct a quantitative trait GWAS on CNV using publicly available data from the HapMap CEU cohort. This will be followed by a GWAS of a simulated binary affection status (case/control) phenotype.

3.2.1 Quantitative Trait Association

1. CNV data will be used to illustrate how to perform genetic association of a quantitative trait with PLINK. CNV data for the HapMap CEU samples can be found on the HapMap ftp server:

   ftp://ftp.ncbi.nlm.nih.gov/hapmap/cnv_data/hm3_cnv_submission.txt

   The original CNV dataset contains 856 CNV genotypes. For this exercise, we have selected three CNVs (shown in the table below), which are tagged by genotyped SNPs. This data is contained in the **CNV_phenos.txt** file (Table 4).

Table 4
CNV_phenos.txt file example

CNV	Chr	Start pos	End pos
HM3_CNP_35	1	151028547	151035324
HM3_CNP_211	4	34462895	34501120
HM3_CNP_516	9	29084549	29087680

2. Genotype data for this exercise has been prepared and is provided in binary PED format (**CEU_HapMap_GWAS_data.bed**, **CEU_HapMap_GWAS_data.bim** and **CEU_HapMap_GWAS_data.fam**). The original HapMap dataset contains 3,907,239 SNPs for 174 individuals, of which 6 were ethnic outliers and 9 of whom had no CNV data available.

   The script **Exercise_2.sh**, also provided, contains all the commands needed to download and parse the genotype data.

3. The *--pheno* option allows for the specification of alternative (one or more) phenotypes. When using the *--pheno* option, the original PED file must still contain a phenotype Column 6 (even if this is a dummy phenotype, e.g., all missing), unless the *--no-pheno* flag is given. Also, note that the file can contain a header which specifies column name; in this case, the column name can be used with the option *--fillme* to limit the analysis to only this alternative phenotype.

4. To conduct genetic association on all three traits with the *--pheno* option, use the following command:

```
plink --bfile CEU_HapMap_GWAS_data --assoc
--allow-no-sex --pheno CNV_phenos.txt
--all-pheno --missing-phenotype -9 --adjust
--ci 0.95 --out cnv_qtl
```

   Explanations for all PLINK options used in the above command are provided in Table 5:

5. PLINK will generate three pairs of results files (e.g., results for the first CNV [HM3_CNP_35] are written to the files: cnv_qtl.P1.qassoc and cnv_qtl.P1.qassoc.adjusted). The first lines of a *.qassoc output file are shown in Fig. 3 below.

   Table 6 describes the output of each column.

   The *.qassoc.adjusted* files contain adjusted *p*-values with different routines for all SNPs tested and printed in ascending order (most significant SNPs at the top of the file). Figure 4 contains summary results of the six most significant SNPs associated with the HM3_CNP_35 CNV genotype (this corresponds to the first six SNPs in the *adjusted file).

   SNP rs1048535 showed the strongest association with CNV HM3_CNP_35, which is located in chr1: 151028547

Table 5
PLINK commands used for genetic association

Option	Function
--bfile *filename*	Specify *filename*.bed, *filename*.bim, and *filename*.fam files containing binary PED and MAP data
--assoc	Perform association. PLINK will automatically detect whether the phenotype is a binary trait or a continuous trait. If the phenotype contains only 0, 1, or 2 entries, then it is assumed that the trait is binary (0 = missing; 1 = unaffected; 2 = affected). Otherwise, the trait is assumed to be quantitative. In the data provided here, we have added 1 to all entries to prevent PLINK interpreting the trait as binary
--allow-no-sex	Do not include gender as a covariate. If the dataset contains missing gender entries, PLINK will not perform association analysis unless this flag is present
--pheno *filename*	Use the phenotype(s) contained in this file, as opposed to data in the sixth column of the PED/FAM file
--missing-phenotype *value*	The default missing-phenotype value is –9. This option allows for the specification of a different value, which must be an integer
--adjust	Output adjusted *p*-values for multiple testing correction. SNPs are printed in increasing order of significance
--ci *value*	Include 95 % confidence interval of the odds ratio or beta coefficient. This will also force PLINK to print the standard error of the estimate which can then be used for meta-analysis
--out *prefix*	Specify output root filename
--all-pheno	Perform analysis for all phenotypes (all columns) of the file specified with the --pheno option

```
CHR        SNP        BP    NMISS     BETA      SE       R2        T        P
1     rs6650104    554340     158   -0.4968   0.2746   0.02054   -1.809   0.07241
1    rs12565286    711153     154   -0.1962   0.1875   0.00715   -1.046   0.2971
1     rs3094315    742429     157   -0.1063   0.08354  0.08354   -1.273   0.205
1     rs3131972    742584     159   -0.1284   0.08475  0.01441   -1.515   0.1317
```

Fig. 3 Example of a .qassoc output file

–151035324. The *.qassoc.adjusted* file only contains significance values. To extract information about this SNP from the *.qassoc* file (such as chromosome, location, etc.), a search function can be used in a text editor or directly from a command line. For instance (in Linux):

grep rs1048535 cnv_qtl.P1.qassoc

Table 6
Output columns

CHR	Chromosome number
SNP	SNP identifier
BP	Chromosomal position (base-pair)
NMISS	Number of non-missing genotypes
BETA	Regression coefficient
SE	Standard error
R^2	Regression r-squared
T	Wald test (based on t-distribution)
P	Wald test asymptotic p-value

```
 CHR        SNP       UNADJ          GC        BONF       HOLM  SIDAK_SS  SIDAK_SD     FDR_BH     FDR_BY
   1   rs1048535  8.357e-103   1.111e-94   1.092e-96  1.092e-96       INF       INF  1.092e-96  1.601e-95
   1   rs7524281   1.163e-79   5.604e-72   1.519e-73  1.519e-73       INF       INF  7.597e-74  1.114e-72
   1  rs11586156   3.351e-68   6.634e-61   4.379e-62  4.379e-62       INF       INF  6.256e-63  9.172e-62
   1   rs1591077   3.351e-68   6.634e-61   4.379e-62  4.379e-62       INF       INF  6.256e-63  9.172e-62
   1  rs10494277   3.351e-68   6.634e-61   4.379e-62  4.379e-62       INF       INF  6.256e-63  9.172e-62
   1  rs12239774   3.351e-68   6.634e-61   4.379e-62  4.379e-62       INF       INF  6.256e-63  9.172e-62
```

Fig. 4 Summary results for the six most significantly associated SNPs with the HM3_CNP_35 CNV genotype (as reported in the .adjusted output file)

will output the following line:

```
1  rs1048535  151044089  158  −0.957  0.01774  0.9491  −53.95  8.357e-103
```

6. According to the line shown above, the SNP is located in chromosome 1, position 151044089 (or approximately 9 kb from the HM3_CNP_35 CNV). Genotypes at rs1048535 are strongly correlated ($r^2 = 0.9491$) with genotypes at the CNV and it is highly significant ($p = 8.357\text{e-}103$). Also, strong evidence of association was also observed at rs7524281 (second most significant SNP in the .adjusted file. See above), which is also in chromosome 1 and only about 6 kb away from rs1048535. Therefore, it is reasonable to suspect that there is also high correlation between rs7524281 and rs1048535. PLINK features several functions that allow to easily estimate linkage disequilibrium (LD) values between SNPs. For example, the following command computes the LD of rs1048535 to every SNP in a window of size 500 kb around the SNP. The option *--ld-window-r2* 0 ensures all pair-wise LD calculations are printed to the output file.

```
plink --bfile CEU_HapMap_GWAS_data --r2
--ld-snp rs1048535 --ld-window-kb 5000
--ld-window-r2 0 --out rs1048535_ld
```

CHR_A	BP_A	SNP_A	CHR_B	BP_B	SNP_B	R2
1	151044089	rs1048535	1	151046763	rs7411365	0.0340111
1	151044089	rs1048535	1	151046901	rs1930127	0.87627
1	151044089	rs1048535	1	151047146	rs17670505	0.0297426
1	151044089	rs1048535	1	151049879	rs7524281	0.948987
1	151044089	rs1048535	1	15105184	rs7536191	0.0255985
1	151044089	rs1048535	1	151050348	rs12023196	0.0226745
1	151044089	rs1048535	1	151050879	rs11804609	0.0258134
1	151044089	rs1048535	1	151059829	rs7550676	0.0317649
1	151044089	rs1048535	1	151062595	rs6659798	0.0317649
1	151044089	rs1048535	1	151063073	rs7517755	0.0613721
1	151044089	rs1048535	1	151068040	rs11205114	0.145429
1	151044089	rs1048535	1	151092224	rs12565568	0.0317649
1	151044089	rs1048535	1	151114516	rs12022319	0.0486071

Fig. 5 Screenshot of output file rs1048535_ld.ld, which contains all pair-wise LD calculations

The output file rs1048535_ld.ld (Fig. 5) contains all pair-wise LD calculations. While the first three columns show information (chromosome, position, and identifier) on the first SNP, the next three columns show information on the second SNP and the seventh column contains the LD value, r^2. The two SNPs of our interest, rs7524281 and rs1048535, display high linkage-disequilibrium ($r^2 = 0.948987$). Hence, these two SNPs and the CNV are part of the same haplotype block.

7. To illustrate the case when there exists a need to control for population stratification, we can easily repeat the analysis and include the MDS option (see example in the previous section) as covariates. Since PLINK can take covariate values contained in a file (where the first two columns indicate family and individual IDs and subsequent columns contain the covariates), it is possible to use the MDS file generated with PLINK to produce a covariate file with the following command on the linux terminal:

```
awk {print $1,$2,$4,$5} filename.mds > mds_
covar.txt
```

Finally, we will use a command similar to that described in **step 4**, different only for the addition of the covariate option (*--covar*) and the use of *--linear* instead of *--assoc* (*see* **Notes 5** and **6**).

```
plink --bfile CEU_HapMap_GWAS_data --linear
--allow-no-sex --pheno CNV_phenos.txt
--covar mds_covar.tx
--all-pheno --missing-phenotype -9 --adjust
--ci 0.95 --out cnv_qtl
```

PLINK will generate output files, which are similar to those previously described, except that they contain an additional column "TEST." For every SNP, a line containing the results of the additive test is presented, as well as one additional line per analyzed covariate.

8. In summary, this exercise used a CNV as a quantitative trait to exemplify how to run a genome-wide association in PLINK. We described how to navigate the results files generated by PLINK, how to calculate LD between SNPs and how to include covariates and correct for population stratification in our analyses. In the final exercise, you will learn to simulate data and to perform a case vs. control association analysis.

9. In addition to the results files, PLINK also generates a .log file, which contains information about the analysis and the dataset (i.e., analysis start and finish times, number of markers tested, number of individuals, cases/controls, males/females, parameters, options and filters in effect, etc.). We recommend you have a look at the .log files generated by the different analyses you conduct.

3.2.2 Case vs. Control Association Analysis

1. PLINK offers several tests for binary trait association (also known as case–control association) analyses. There exist repositories of real case–control genotype data (e.g., the database of Genotypes and Phenotypes, dbGAP: http://www.ncbi.nlm.nih.gov/gap), but obtaining access to such datasets commonly requires filling out an application form for data release. Given that PLINK also offers a function for simulating genotype and phenotype data, we will use it to generate a case–control dataset. All scripts and data files you need are provided in the publisher's website (http://extras.springer.com/).

2. We will simulate genotype data for 45,030 SNPs, of which 30 are trait-associated and 45,000 are not. This is indicated in the simulate file **gwas.sim** (shown below), which contains the parameters of the desired genotype data.

   Where:

Column 1	Indicates the number of markers to be generated
Column 2	Specifies an SNP identifier prefix
Column 3, 4	MAF window of SNPs to be generated
Columns 5, 6	Odds ratio of the heterozygous genotype and the homozygous minor genotype. Specifying "mult" implies a multiplicative effect for the homozygote genotype, i.e., ORHOM = ORHET * ORHET

20000	nullA	0.00	0.05	1.00	1.00
10000	nullB	0.05	0.10	1.00	1.00
5000	nullC	0.10	0.20	1.00	1.00
10000	nullD	0.20	0.99	1.00	1.00
10	assoc1	0.00	0.05	1.80	Mult
20	assoc2	0.05	0.40	1.20	Mult

3. The following command will generate genotypes for 3,000 cases (--simulate-ncases) and 3,000 controls (--simulate-ncontrols):

```
plink --simulate gwas.sim --make-bed --simulate-ncases 3000 \
--simulate-ncontrols 3000 --simulate-label POP1
--out simulated_gwas
```

4. As mentioned previously, several association tests have been implemented in PLINK. These include:

Test	Option
Allelic test	--assoc
Allelic test using Fisher's exact test	--fisher
Logistic regression	--logistic
Linear regression	--linear
Full-association	--model
Cochran-Armitage trend test	--model --model-trend
Genotypic (2 df) test	--model --model-gen
Dominant gene action (1 df) test	--model --model-dom
Recessive gene action (1 df) test	--model --model-rec

Allelic tests compare frequencies of alleles in groups of cases and controls. If the --model option is specified, PLINK will perform full-model association testing, which includes ALLELIC, TREND, GENO, DOM, and REC for each SNP. The analysis can be restricted to only one of these tests by including an additional option (i.e., --model --model-dom will only compute the dominant gene effect model, and --model --model-rec will only compute the recessive gene effect model). These options can also be combined with the --adjusted option previously described. In GWAS for complex diseases, it is usually assumed that disease alleles have an additive effect and so allelic tests are more widely used. Logistic and linear regression are also commonly used, as they are more flexible than the other tests in that they can account for confounding effects with the use of covariates (e.g., disease onset or gender). Another use of logistic and linear regression is to account for population stratification, as previously shown in the quantitative trait association exercise.

The following command will perform allelic association test on the simulated dataset:

```
plink --bfile simulated_gwas --allow-no-sex --assoc
--adjust --ci 0.95 --out simulated_assoc
```

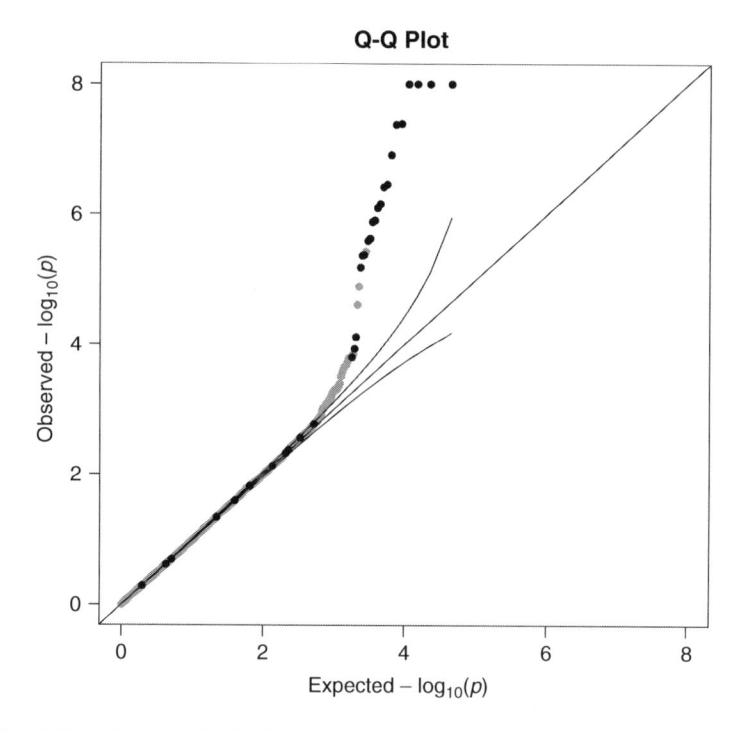

Fig. 6 Quantile-quantile (Q-Q) plot showing expected vs. observed [−log$_{10}$(*P*) values]. Simulated nonassociated SNPs are shown in *gray*, and simulated associated SNPs are shown in *black*

This command is almost identical to that previously used for quantitative trait association. This is because PLINK will automatically assume this is a binary trait when the phenotype column in the PED/FAM file only contain 0, 1, 2 values.

5. The interpretation of whole-genome association studies is deeply facilitated by the use of data visualization tools. For instance, a quantile-quantile plot (Fig. 6), or Q-Q plot, can be used for detecting evidence of systematic bias (be it from unrecognized population structure, genotyping artifacts, etc.). Q-Q plots also show the extent to which the observed distribution of the test statistic follows the expected (null) distribution. The **qq_plot.R** script (provided) will generate a Q-Q plot of the GWAS results obtained from our simulated dataset for chromosome 1:

    In the figure, simulated disease SNPs are colored in red.

6. Similarly, GWAS Manhattan plots display the negative logarithm of the association *p*-value for each SNP (*Y*-axis) against the genomic coordinates (along the *X*-axis). Given that the strongest associations have the smallest *p*-values, the −log$_{10}$ of these *p*-values will have the highest height in the Manhattan plot (Fig. 7). The **manhattan_plot.R** script (also provided) will generate a Manhattan plot of the GWAS results obtained from our simulated dataset for chromosome 1.

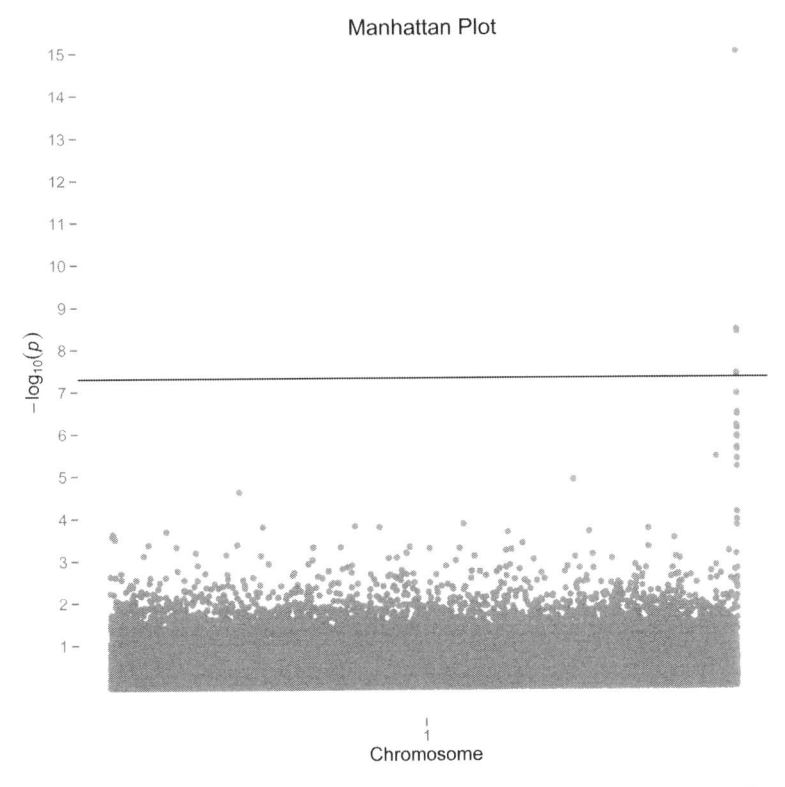

Fig. 7 Manhattan plot of association *p*-values of SNPs in chromosome 1. The *x*-axis shows location and *y*-axis displays the significance of the association ($-\log_{10}(P)$ value)

Note that, in this case, only SNPs from chromosome 1 are plotted. Usually, Manhattan plots also include all other autosomal chromosomes.

On the right hand side of the plot, a column of SNPs with high significance is observed. At least three SNPs display *p*-values above the genome-wide significance threshold. In the case of a true association, we would expect that some of its neighboring SNPs in LD were also associated with the phenotype, since they are expected to be co-inherited in the population.

Although the data used in this exercise are simulated, the same sequence of analyses would be conducted on real data. After finding a significant association, you would typically want to find out more about the genomic context of the SNP. For example, whether there are other SNPs in LD and do they show the expected pattern of *p*-values? Is your SNP, or one in high LD within a transcribed region or a functional element, such as a methylation site? There exist a number of online tools and databases that can help in the annotation and further characterization of GWAS findings.

The Catalog of Published GWAS (http://www.genome.gov/gwastudies/) is a searchable and downloadable database of publications reporting SNP-trait associations. These publications are identified through periodic PubMed searches, NIH-distributed compilations of news and media reports, and occasional comparisons with other GWAS literature databases. The catalogue is searchable by disease/trait, chromosomal region, gene, or SNP. This is a good starting point if one wants to find out whether a gene or genomic region has previously been associated with the same or other traits of interest by a whole-genome association study.

Locus Zoom (http://csg.sph.umich.edu/locuszoom/) is a web-based plotting tool for generating regional plots of association results in their genomic context with publication-ready quality. This enables a quick visual inspection of the strength of association evidence, the position of the associated SNPs relative to genes in the region, and the extent of the association signal and LD.

SNAP (http://www.broadinstitute.org/mpg/snap/) is an online tool that allows for the rapid retrieval of proxy SNPs based on LD, physical distance, and/or membership in commercial genotyping arrays. Given an input of one or more query SNPs, SNAP can report pair-wise LD estimates and generate LD-plots derived from the genomic data from both the International HapMap Project and the 1K Genomes Project.

3.3 Conclusions

In this chapter we have provided a brief practical tutorial on the analysis of GWAS data with PLINK. Given the space limitations, it is not possible to cover all the functionalities of PLINK within a single book chapter. However, the developers of PLINK have provided comprehensive documentation in an online manual, which contains detailed information about all PLINK options and functions:

http://pngu.mgh.harvard.edu/~purcell/plink/pdf.shtml.

4 Notes

It is important to consider the following, while using PLINK:

1. When PLINK starts it will attempt to contact the web, to check whether there is a more up-to-date version available or not. After checking, PLINK writes a file called *.pversion* to the working directory and uses this cached information for the rest of the day. This option can be disabled with the --*noweb* option on the command line. When using PLINK on a machine with no, or a very slow, web connection, it may be desirable to turn this feature off, as no PLINK updates are being made.

2. In family-based analyses, PLINK can only correct for one relationship type. This means that PLINK cannot be used for the analysis of extended families or twin cohorts.

3. In a PED file, quantitative traits with decimal points must be coded with a period/full-stop character and not with a comma, i.e., 2.394 not 2,394.

4. The *--silent* option suppresses output to console window.

5. Covariates can only be used with the linear and logistic commands. However, including the *--covariate* command with other association commands will not yield an error message.

6. When using covariate files, individuals with missing data are ignored by default.

Reference

1. Purcell S et al (2007) PLINK: a tool set for whole-genome association and population-based linkage analyses. Am J Hum Genet 81 (3):559–575

Chapter 9

Genome-Wide Complex Trait Analysis (GCTA): Methods, Data Analyses, and Interpretations

Jian Yang, Sang Hong Lee, Michael E. Goddard, and Peter M. Visscher

Abstract

Estimating genetic variance is traditionally performed using pedigree analysis. Using high-throughput DNA marker data measured across the entire genome it is now possible to estimate and partition genetic variation from population samples. In this chapter, we introduce methods and a software tool called Genome-wide Complex Trait Analysis (GCTA) to estimate genomic relationships between pairs of conventionally unrelated individuals using genome-wide single nucleotide polymorphism (SNP) data, to estimate variance explained by all SNPs simultaneously on genomic or chromosomal segments or over the whole genome, and to perform a joint and conditional multiple SNPs association analysis using summary statistics from a meta-analysis of genome-wide association studies and linkage disequilibrium between SNPs estimated from a reference sample.

Key words GWAS, SNP, Complex trait, Missing heritability, Variance explained, Genomic relationship, REML

1 Introduction

Genome-wide association studies (GWAS) have proven successful in identifying single nucleotide polymorphisms (SNPs) that affect the phenotypic variation in human complex diseases and traits [1]. GWAS was designed to uncover genes and pathways of medical importance to pinpoint the underlying molecular and genetic etiology of diseases but has been criticized for being unable to explain the heritability for most complex traits [2]. We have recently developed a method to estimate the proportion of additive genetic variance that can be captured by considering all the SNPs simultaneously without testing for association of any individual SNP with the trait [3]. We showed by analyses of GWAS data that a large proportion of heritability for quantitative traits such as height [3], body mass index [4], and cognitive ability [5, 6] and for diseases such as schizophrenia [7] can be explained by all the common SNPs. These results suggest that most heritability is hiding rather than missing [8] and that GWAS have not identified the SNPs that explain this proportion of the hidden heritability because the effect sizes of individual SNPs are too small to reach the stringent genome-wide significance level [3]. We further extended the method to partition the genetic variance onto chromosomes and genomic segments. We found that the variance attributed to a chromosome or a DNA segment is proportional to its length, in particular for height [4] and schizophrenia [7], and that SNPs located in genic regions explain more variation than those in intergenic regions. All the results are consistent with a pattern of polygenic inheritance for most complex traits.

In addition, under the assumption of many genes each with a small effect, more genetic variants are expected to be identified with increased sample sizes [9]. There has been a wide range of international collaborations that combine samples from many GWAS cohorts in a meta-analysis in order to obtain efficient statistical power to detect the small effects [10–14]. For a meta-analysis, however, the individual-level genotype and phenotype data are typically unavailable. For a large-scale meta-analysis, it is extremely difficult to organize and perform a conditional analysis and impractical to perform the conditional analysis iteratively. We have developed a method that can perform a conditional and joint analysis of multiple SNPs using the summary statistics from a meta-analysis with the linkage disequilibrium (LD) between SNPs estimated from a reference sample individual-level genotype data [15]. By analyzing the summary data from the GIANT meta-analysis for height we have shown that there are many genetic loci with multiple associated SNPs [15].

We have implemented all the methods mentioned above in a user-friendly software tool called Genome-wide Complex Trait Analysis (GCTA) [16], which is freely available at http://www.complextraitgenomics.com/software. In this chapter, we will introduce the GCTA software in three modules with, in each module, a description of the theories and methods, practical applications of the commands for data analyses, and remarks with respect to caveats for data analysis and interpretation of results.

2 Estimating the Variance Explained by Genome-Wide SNPs

2.1 Equivalent Models

The basic concept behind our method is to fit all the SNPs simultaneously as random effects in a mixed linear model to estimate the variance explained by all the SNPs without testing for associations of any individual SNP with the trait [3]

$$\mathbf{y} = \mathbf{Xb} + \mathbf{Wu} + \mathbf{e} \text{ with } var(\mathbf{y}) = \mathbf{V} = \mathbf{WW'}\sigma_u^2 + \mathbf{I}\sigma_e^2 \quad (1)$$

where **y** is a $n \times 1$ vector of phenotypes with n being the sample size, **b** is a vector of fixed covariates, **u** is a vector of SNP effects with $\mathbf{u} \sim N(0, \mathbf{I}\sigma_u^2)$, **I** is a $n \times n$ identity matrix, and **e** is a vector of residual effects with $\mathbf{e} \sim N(0, \mathbf{I}\sigma_e^2)$. **W** is a standardized SNP genotype matrix with the ijth element $w_{ij} = (x_{ij} - 2p_i)/\sqrt{2p_i(1-p_i)}$, where x_{ij} (coded as 0, 1, or 2) is the number of copies of the reference allele for the ith SNP of the jth individual and p_i is the frequency of the reference allele of SNP i.

If we define $\mathbf{A} = \mathbf{WW}'/N$ and define σ_G^2 as the variance explained by all the SNPs i.e., $\sigma_G^2 = N\sigma_u^2$ with N being the number of SNPs, then Eq. 1 will be equivalent to [17–19]

$$\mathbf{y} = \mathbf{Xb} + \mathbf{g} + \varepsilon \text{ with } \mathbf{V} = \mathbf{A}\sigma_G^2 + \mathbf{I}\sigma_e^2 \qquad (2)$$

where **g** is a $n \times 1$ vector of the total additive genetic effects of the individuals with $\mathbf{g} \sim N(0, \mathbf{A}\sigma_G^2)$, **A** is interpreted as the genetic relationship matrix (GRM) between individuals with its kjth element being $A_{jk} = \frac{1}{N}\sum_{i=1}^{N}\frac{(x_{ij}-2p_i)(x_{ik}-2p_i)}{2p_i(1-p_i)}$, and the other notations are defined as in Eq. 1. The variance components σ_G^2 and σ_e^2 in Eq. 2 can be estimated by the restricted maximum likelihood (REML) approach [20]. Details of the algorithms of the REML approach can be found in Appendix.

We define $h_{SNP}^2 = \sigma_G^2/(\sigma_G^2 + \sigma_e^2)$ as the proportion of phenotypic variance explained by all the SNPs. Since Eq. 2 is mathematically equivalent to Eq. 1, the estimate of h_{SNP}^2 can be interpreted as the phenotypic variance explained by fitting all the genome-wide SNPs simultaneously. Therefore, the estimate of h_{SNP}^2 is directly comparable to the results from genome-wide association (GWA) analyses. For example, for human height, the estimate of h_{SNP}^2 was 0.445 in a sample of 3,925 unrelated individuals, meaning that all the common SNPs explain ~45 % of the phenotypic variance for height, as compared with an estimate of ~10 % of variance explained by 180 genome-wide significant SNPs ($P < 5 \times 10^{-8}$) identified by a meta-analysis of GWAS with ~180,000 individuals [12].

There is an important distinction between h_{SNP}^2 and the narrow-sense heritability (h^2) defined as the proportion of phenotypic variance explained by the additive effects of all causal variants. Strictly speaking, our method is not a method to estimate trait heritability using SNP data because SNPs are not causal variants and common SNPs in the current generation of genotyping arrays are unlikely to be able to perfectly tag all causal variants for a trait. Under the circumstance that the data are from a family study with a substantial proportion of close relatives, the estimate of h_{SNP}^2 would be very similar to the estimate of h^2 from a pedigree analysis, because in this case we basically reconstruct the pedigree using genome-wide SNP markers and the estimate of h_{SNP}^2 is dominated by the close relatives (*see* **Note 1**). Therefore, the estimate h_{SNP}^2 from GCTA in family data cannot be interpreted as the variance explained by all the SNPs because it could be confounded with some possible shared environment effects and the effects of causal variants that are not tagged by the SNPs but captured by pedigree structure. The estimate of h^2_{SNP} from a sample of unrelated individuals is an unbiased estimate of h^2 only if the GRM is generated from all causal variants.

2.2 GCTA Applications

GCTA implements the method in two steps, i.e., generating the GRM between individuals and then estimating the variance explained by all SNPs by a Restricted Maximum Likelihood (REML) analysis of the phenotypes with the GRM.

2.2.1 Estimating the GRM from SNP Data

Suppose that we have a GWAS dataset in PLINK [21] binary PED format (http://pngu.mgh.harvard.edu/~purcell/plink/), e.g., "test.bed", "test.bim", and "test.fam" and the data has been carefully checked through standard quality control procedures. We can use the following command to calculate the genetic relationships between pairwise individuals from all the SNPs on autosomes

```
gcta64 --bfile test --autosome --make-grm --out test
```
[Command 1]

For a large dataset, we can use the following commands

```
gcta64 --bfile test --chr 1 --make-grm --out test_chr1
gcta64 --bfile test --chr 2 --make-grm --out test_chr2
...
gcta64 --bfile test --chr 22 --make-grm --out test_chr22
```
[Command 2]

to calculate the GRM for each individual chromosome and then merge all the GRMs by the following command

```
gcta64 --mgrm grm_chrs.txt --make-grm --out test
```
[Command 3]

Details of the GCTA options and file formats can be found in the online documentation on the GCTA webpage. Each off-diagonal element of the GRM is an estimate of additive genetic relatedness between a pair of individuals. We choose the current population as the base population so that the mean relatedness between one individual and all the individuals in the data is zero. The expected value of self-relatedness of an individual is unity assuming no inbreeding, so that the mean relatedness between an individual and all the other individuals in the data is expected to be $-1/(n-1)$. Therefore, we usually see negative estimates of genetic relatedness, which should be interpreted as a pair of individuals being less related than the mean genetic relationships in the sample (Fig. 1).

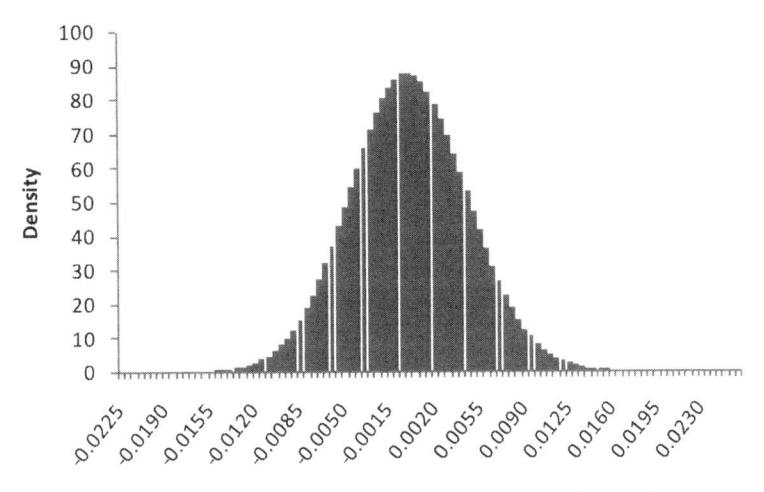

Fig. 1 Distribution of the estimates of genetic relatedness between 3,925 "unrelated" individuals [3]

2.2.2 Removing Cryptic Relatedness

As mentioned above, if there are close relatives remaining in the data, the estimate of h^2_{SNP} would be confounded with shared environment effects and the effects of causal variants that are not tagged by the SNPs. We can use the following command to remove cryptic relatedness from the data:

```
gcta64 --grm test --grm-cutoff 0.025 --make-grm --out test_rm025                                          [Command 4]
```

Here, we choose an arbitrary cutoff value of 0.025 which approximately corresponds to the genetic relationship between third or fourth cousins. In practice, if there are enough data, we could try a range of cutoff values, e.g., 0.025, 0.05, 0.1, and 0.2, to see if there is a change of the estimate of h^2_{SNP} when using less stringent cutoff values. It is notable that this option should only be used for a GRM generated from all the genome-wide SNPs rather than on that from SNPs on a single chromosome (*see* **Note 2**).

2.2.3 REML Analysis

Once we have the GRM, we can implement the REML analysis to estimate h^2_{SNP} with the following command

```
gcta64 --grm test --pheno test.phen --reml --out test                                                     [Command 5]
```

where test.phen is a file containing phenotype data. We can include covariates to control for some fixed effects that might contribute to the phenotypic variation. The covariates can be continuous variables such as eigenvectors calculated from principal component analysis (PCA) or SNPs that are known to have an effect on the trait. The covariates can also be categorical variables such as batches of genotyping and locations from which the samples were collected.

We do not recommend including correlated covariates in the analysis, which may cause a colinearity problem in the **X** matrix and GCTA will report that "the $\mathbf{X'V^{-1}X}$ matrix is invertible." It is recommended to adjust the phenotypes for age and sex effects and standardize them to *z*-scores prior to the REML analysis because for some traits, such as body mass index, the variance in females is, per definition, larger than in males, which cannot be corrected by including the sex effect as a covariate in the REML analysis (*see* **Note 3**). The command to run a REML analysis with both continuous and categorical covariates is given below

```
gcta64 --reml --grm test --pheno test.phen --covar
batch.txt --qcovar test_10PCs.txt --out test
```
[Command 6]

where "batch.txt" is a file to indicate in which batches the samples were genotyped (categorical) and "test_10PCs.txt" is a file containing the first ten eigenvectors from PCA (continuous).

A realized GRM (e.g., **A**) is typically non-positive definite [22]. We therefore solve the mixed linear model based on the **V** matrix because **V** is usually positive definite. However, under some particular circumstances in which **V** is non-positive definite, we use the matrix-bending approach to modify the eigenvalues of the GRM [23, 24] so that the re-formed GRM is positive definite, which will guarantee **V** to be positive definite. In Table 1, we show an example of estimating genetic variance using GCTA when **V** is non-positive definite. In this example, we randomly sampled ten independent SNPs with minor allele frequency (MAF) >1 and <5 % as causal variants from real-world genotype data, and simulated phenotypes of 3,925 unrelated individuals given the effects of the ten SNPs drawn from a normal distribution. The ten SNPs in total explain 50 % of the variance for the simulated trait. We repeated the simulation 10 times by resampling the residual effects but fixing the genetic effects and used both REML and Haseman–Elston Regression [25] to estimate the variance explained by the ten SNPs. We calculated the GRM from all the "causal variants" so that the estimate of variance explained should be an unbiased estimate of the trait heritability, as shown by the results from Haseman–Elston Regression. However, in this special case, the **V** matrix was negative-definite and the estimates from the REML analyses were largely biased. However, when applying the bending procedure as described above, the estimates were unbiased. In practice, GCTA will automatically implement the matrix-bending procedure if **V** is negative-definite.

2.2.4 Adjustment for Imperfect LD

In Yang et al. [3], we proposed an approach to calibrate the error variance in estimating the GRM due to the use of finite number of SNPs, which can also be interpreted as the error variance in predicting the GRM at causal variants by the SNPs. We quantified that

Table 1
Estimating the variance explained by all SNPs when the V matrix is non-positive definite

| | | REML | | | |
| | HE Reg. | No bending | | Bending | |
Rep.	h^2_{SNP}	h^2_{SNP}	SE	h^2_{SNP}	SE
1	0.527	0.053	0.003	0.516	0.112
2	0.511	0.047	0.003	0.501	0.112
3	0.487	0.047	0.003	0.477	0.112
4	0.517	0.046	0.003	0.513	0.113
5	0.503	0.051	0.002	0.491	0.112
6	0.504	0.049	0.003	0.498	0.112
7	0.503	0.044	0.003	0.495	0.113
8	0.498	0.051	0.003	0.488	0.112
9	0.503	0.046	0.003	0.499	0.113
10	0.497	0.053	0.002	0.488	0.112

HE Reg. Haseman–Elston regression; *Rep.* simulation replicate; *SE* standard error

if the causal variants have the same MAF spectrum as the SNPs, the prediction error variance (PEV) is $1/N$, and if the causal variants have lower MAF than the SNPs, e.g., MAF of the causal variants are all $<\theta$, PEV $= 1/N + c$ with c being a constant given a certain value of θ. We provide an option *–grm-adj* in GCTA to adjust the GRM for PEV given a user-specified c value. In practice, however, we do not usually have to use this option unless we are interested in drawing inference about the variance explained by all the causal variants under strong assumptions (*see* **Note 4**). Our method was originally designed to estimate the variance explained by all SNPs in a sample of unrelated individuals rather than the trait heritability. The estimate of variance explained by SNPs after adjusting for PEV is equal to the trait heritability only if the assumption about the allele frequency spectrum of the causal variants is correct.

2.2.5 Significance Test

We assess the significance of h^2_{SNP} by likelihood ratio test (LRT), which is the ratio of likelihood under the alternative hypothesis (H$_1$: $h^2_{SNP} \neq 0$) to that under the null (H$_0$: $h^2_{SNP} = 0$), where twice the log likelihood ratio, LRT $= 2[L(\text{H}_1) - L(\text{H}_0)]$, is distributed as a half probability of 0 and a half probability of chi-squared with 1 degree of freedom (χ^2_1). The LRT and its corresponding *p*-value are reported in the GCTA output file.

2.2.6 Estimating the Variance Explained by a Subset of SNPs

GCTA can also estimate the variance explained a subset of SNPs using the following command

```
gcta64 --bfile test --extract test.snplist --make-grm --out test_subset
gcta64 --grm test_subset --pheno test.phen --reml --out test_subset
```
[Command 7]

There is a caveat that if the SNPs included in the analysis are the top associated SNPs selected from GWA analysis in the same data, the variance explained by these SNPs will be overestimated because the effect sizes of the top associated SNPs are overestimated, known as the "winner's curse" problem (see **Note 5**). To obtain an unbiased estimate of the variance explained by a subset of SNPs, the SNPs should be either ascertained independently of the phenotypes in the data or selected from association analyses in an independent sample.

2.2.7 Case–Control Data

The methodology described above is derived for a quantitative trait, which is also applicable to case–control data, where the estimate of h^2_{SNP} corresponds to variation on the observed 0–1 scale. Under the assumption of a threshold-liability model for a disease, i.e., disease liability on the underlying scale follows standard normal distribution [26], the estimate of h^2_{SNP} on the observed 0–1 scale $\left(\hat{h}^2_{SNP(O)}\right)$ can be transformed to that on the underlying (unobserved) continuous liability scale ($\hat{h}^2_{SNP(L)}$) by a linear transformation [27, 28]. We recently extended the linear transformation to account for ascertainment bias in a case–control study [29], i.e., the proportion of cases in the sample is usually much larger than that in the general population. In GCTA, we analyze the dichotomous case–control status (0 or 1) by the REML approach as if it is a quantitative trait using exactly the same approach described above. We then transform the estimate of $\hat{h}^2_{SNP(O)}$ to $\hat{h}^2_{SNP(L)}$ by the following equation

$$\hat{h}^2_{SNP(L)} = \hat{h}^2_{SNP(O)} \frac{K(1-K)}{z^2} \frac{K(1-K)}{v(1-v)} \quad (3)$$

with the standard error of $\hat{h}^2_{SNP(L)}$ being

$$SE(\hat{h}^2_{SNP(L)}) = SE(\hat{h}^2_{SNP(O)}) \frac{K(1-K)}{z^2} \frac{K(1-K)}{v(1-v)}$$

where K is the disease prevalence in the population and v is the proportion of the cases in the sample and z is the height of the standard normal curve at the truncation point pertaining to a disease prevalence of K. The GCTA command to implement the analysis is given below

```
gcta64 --reml --grm test_cc --pheno test_cc.phen --prevalence 0.1 --out test_cc
```
[Command 8]

where the option --*prevalence* is to specify the prevalence of the disease and to perform the transformation.

There is an important caveat in applying these methods to case–control data. Any batch, plate or other technical artifact, which causes allele frequencies between cases and controls to be, on average, more different than the null hypothesis that cases and controls are sampled from the same population, will contribute to the estimation of spurious "genetic" variation, since cases will appear to be more related to other cases than to controls (*see* **Note 6**). Therefore, stringent quality control is essential when applying GCTA to case–control data, as suggested and demonstrated in Lee et al. [29].

We define $h^2_{SNP(O)} = \sigma^2_{G(O)}/\sigma^2_{P(O)}$, where $\sigma^2_{G(O)}$ and $\sigma^2_{P(O)}$ are the genetic and phenotypic variance on the observed 0–1 scale, respectively. The relationship between genetic and phenotypic variance on the observed scale, however, is not linear and $\sigma^2_{G(O)}$ can be greater than $\sigma^2_{P(O)}$ when $h^2_{SNP(L)}$ is large and K is small, i.e., $h^2_{SNP(O)}$ can be greater than 1. For example, given $K = 0.005$, $v = 0.5$ and $h^2_{SNP(L)} = 0.7$, we have from Eq. 3 that $h^2_{SNP(O)} = h^2_{SNP(L)} \frac{v(1-v)z^2}{K^2(1-K)^2} \approx 1.48$. If this is the case for real data, the estimate of $h^2_{SNP(O)}$ will be constrained at 1 in the REML analysis, which will result in an underestimate of $h^2_{SNP(L)}$. In this case, we would recommend to try the option --*reml-no-constrain* to allow the estimate of $h^2_{SNP(O)}$ to be greater than 1 so as to get an unbiased estimate of $h^2_{SNP(L)}$. Nevertheless, it does not mean that whenever we see an estimate of $h^2_{SNP(O)}$ being constrained at 1, the true parameter of $h^2_{SNP(O)}$ is greater than 1. In fact, if the sample size is small, we will often see an estimate of $h^2_{SNP(O)}$ being constrained, which is usually just because of the large sampling error.

3 Modeling Multiple Components

We have shown above how to fit a mixed linear model with only one genetic component and its implementation in GCTA. We can extend the model to a more general form as

$$\mathbf{y} = \mathbf{Xb} + \sum_{i=1}^{r} \mathbf{g}_i + \mathbf{e} \text{ and } var(\mathbf{y}) = \mathbf{V} = \sum_{i=1}^{r} \mathbf{A}_i \sigma_i^2 + \mathbf{I}\sigma_e^2 \quad (4)$$

where $\mathbf{g}_i$ is a vector of random effects with $\mathbf{g}_i \sim N(\mathbf{0}, \mathbf{A}_i \sigma_i^2)$ and the other notations are defined as in Eq. 1. In practice, we can specify different models by including different lists of file names for the GRMs in the input file with the --*mgrm* option.

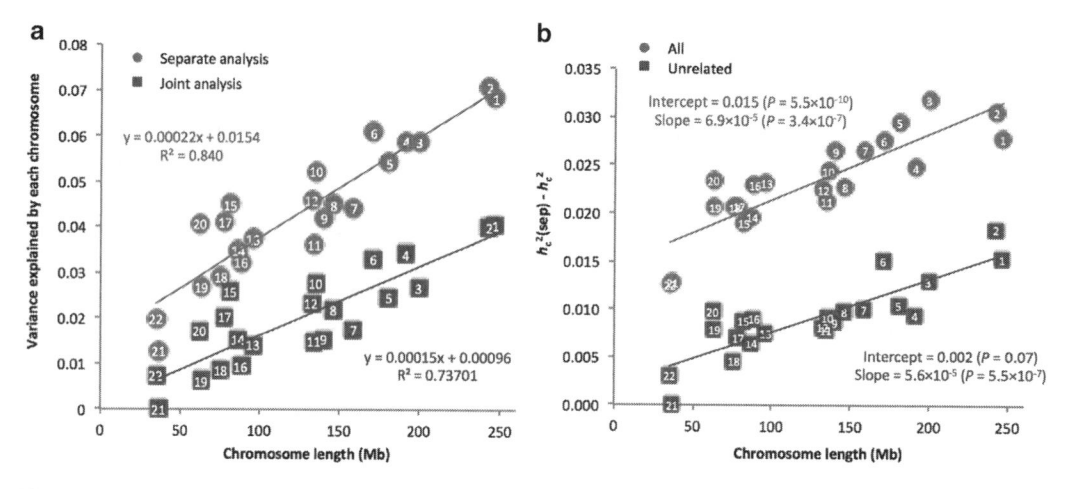

Fig. 2 Estimate of the variance explained by each chromosome for height in a sample of 14,347 European Americans genotyped on 565,040 SNPs [4]. The numbers in the *circles/squares* are the chromosome numbers. In (**a**), the estimates from separate analyses (*circles*) and joint analysis (*squares*) are plotted against chromosome length. In (**b**), the difference between the estimates from separate analysis and from joint analysis are plotted against chromosome length using all the individuals (*squares*) and a set of 11,586 unrelated individuals (*squares*) after excluding one of each pair of individuals with an estimated genomic relatedness >0.025

3.1 Partitioning the Genetic Variance onto Individual Chromosomes

One example is to partition the genetic variance onto each of the 22 autosomes, in which case, $r = 22$ and $\mathbf{A}_i$ is a GRM calculated from the SNPs on the ith chromosome. The corresponding GCTA command is

```
gcta64 --reml --mgrm grm_chrs.txt --pheno test.phen
--out test_chrs                                     [Command 9]
```

The proportion of variance explained by the SNPs on the ith chromosome is defined as $h^2_{C(i)} = \sigma^2_i / (\sum_i \sigma^2_i + \sigma^2_e)$. It is notable that h^2_C is typically small so that a large sample size (e.g., >10,000) is usually required to get an estimate of h^2_C with useful precision. In this analysis, all the chromosomes are fitted in one model, which is a joint analysis. Alternatively, we can also fit one chromosome at a time by using Eq. 2 and Command 6, which is called a separate analysis. Shown in Fig. 2a is an example where the estimates of h^2_C from joint and separate analyses are plotted against the chromosome length (L_C). The estimates of h^2_C from separate analyses, denoted as $h^2_C(\text{sep})$, will be larger than those from a joint analysis if SNPs on different chromosomes are correlated more than expected by chance [4]. The inter-chromosomal SNP correlations occur for two reasons: (1) cryptic relatedness (e.g., unexpected cousins) because closely related individuals will share SNPs identical by descent on more than one chromosome; (2) systematic difference in allele frequencies between subpopulations (population stratification). The joint analysis has the advantage of protecting against such inter-chromosomal correlations because the estimate of each h_C^2 is conditional on the other chromosomes in the model so that the estimates of variance explained by different chromosomes are independent of each other. We therefore can model the variance attributable to population structure as $h_C^2(\text{sep}) - h_C^2 = b_0 + b_1 L_C + \varepsilon$, where the slope b_1 allows the possibility that longer chromosomes track population structure better than smaller chromosomes and the intercept b_0 appears to be due to cryptic relatedness because when we eliminate relatives with a relationship >0.025, b_0 declines to zero (Fig. 2b). The variance attributed to cryptic relatedness is irrespective of chromosome length because it does not require many SNPs per chromosome to detect close relatives. The regression slope b_1 appears to be due to population stratification because longer chromosomes are likely to have more ancestry informative markers (AIMs), assuming that the AIMs are randomly distributed across the genome. The difference between $h_C^2(\text{sep})$ and h_C^2 represents the overall effect of all the other 21 chromosomes on one chromosome. Therefore, the proportion of variance attributed to population structure (cryptic relatedness and population stratification) across the whole genome is approximately equal to $b_0 22/21 + b_1 \sum_{C=1}^{22} L_C/21$.

3.2 Estimating the Variance Explained by Genic and Intergenic SNPs

Another example is to partition the variance explained by all the SNPs into the genic and intergenic regions of the whole genome, in which case, $r = 2$ and there are two GRMs with $\mathbf{A}_1$ and $\mathbf{A}_2$ being the GRMs calculated from all the genic and intergenic SNPs, respectively. The GCTA command is the same as Command 9 expect that there are only two file names for GRMs specified in the input file with the --*mgrm* option.

We can further partition the genetic variance into genic and intergenic SNPs on each chromosome, where $r = 44$ and the GCTA command is the same as Command 9 with 44 GRMs being specified by the option --*mgrm*.

3.3 Estimating the Variance of Genotype–Environment Interaction

To estimate the variance of genotype–environment interaction effects (σ_{gE}^2), the model can be specified as

$$\mathbf{y} = \mathbf{X}\boldsymbol{\beta} + \mathbf{g} + \mathbf{gE} + \mathbf{e} \text{ with } \mathbf{V} = \mathbf{A}_G \sigma_G^2 + \mathbf{A}_{GE} \sigma_{GE}^2 + \mathbf{I}\sigma_e^2,$$

where $\mathbf{gE}$ is a vector of genotype–environment interaction effects for all the individuals with $\mathbf{A}_{GE} = \mathbf{A}_g$ for the pairs of individuals in the same environment and $\mathbf{A}_{GE} = \mathbf{0}$ for the pairs of individuals in different environments. The environment factor, for example, could be sex or medical treatment. σ_{GE}^2 is interpreted as the environmental-specific additive genetic variance captured by all SNPs with the phenotypic covariance between individuals being $\text{cov}(y_i, y_k) = A_{jk}(\sigma_G^2 + \sigma_{GE}^2)$ in the same environment and being $\text{cov}(y_i, y_k) = A_{jk}\sigma_G^2$ in different environments. The GCTA command to implement the analysis is as follows

```
gcta --reml --grm test --pheno test.phen --gxe test.
gxe --reml-lrt 2 --out test                [Command 10]
```

where the option "*--reml-lrt 2*" is to test the significance of the second variance component in the model (σ^2_{GE}).

3.4 User-Specified Covariance Matrix

In Eq. 4, there is no design matrix for any of the random effects because our method was originally developed to model the total additive effects of all the individuals (**g**) for which the design matrix is an identity matrix. In fact, we are able to include in the model a random effect for which the design matrix is not an identity matrix, i.e.,

$$\mathbf{y} = \mathbf{Xb} + \mathbf{g} + \mathbf{Zd} + \mathbf{e} \text{ and } \mathbf{V} = \mathbf{A}\sigma^2_G + \mathbf{ZRZ}'\sigma^2_d + \mathbf{I}\sigma^2_e$$

where **d** is a vector of random effects with **Z** being its design matrix, $\mathbf{d} \sim N(\mathbf{0}, \mathbf{R}\sigma^2_d)$. In this case, we need to create the **ZRZ'** matrix manually in, for example, *R*, save it in the required format for GCTA (*.grm.gz and *.grm.id files) and include it in the analysis with the *--mgrm* option.

3.5 Estimating the Variance Explained by SNPs in Family Data

We have mentioned above that the estimate of h^2_{SNP} in family data is similar to the estimate of h^2 from a pedigree analysis and cannot be interpreted as the variance explained by all SNPs because genetic variants (including rare variants) that are not well tagged by the SNPs will be captured by the pedigree structure and because the estimate would possibly be confounded by some nonadditive genetic and/or nongenetic effects shared between close relatives in the same families. For a GWAS sample collected from family studies, we can estimate h^2_{SNP} while controlling for the pedigree structure using the following model

$$\mathbf{y} = \mathbf{Xb} + \mathbf{g} + \mathbf{f} + \mathbf{e} \text{ and } \mathbf{V} = \mathbf{A}\sigma^2_G + \mathbf{F}\sigma^2_f + \mathbf{I}\sigma^2_e$$

where **f** is a vector of family effects with $\mathbf{f} \sim N(\mathbf{0}, \mathbf{F}\sigma^2_f)$ and **F** is the correlation matrix for **f** with its elements being 1 for pairs of individuals in the same families and 0 for those in different families. The covariance between a pair of individuals j and k is $\text{cov}(y_j, y_k) = A_{jk}\sigma^2_G + \sigma^2_f$ in the same family and $\text{cov}(y_j, y_k) = A_{jk}\sigma^2_G$ in different families, where A_{jk} is the jkth element of **A**. From this analysis, the estimate of $\sigma^2_G/(\sigma^2_G + \sigma^2_f + \sigma^2_e)$ can therefore be interpreted as the variance explained by all SNPs, which should be similar to the estimate of h^2_{SNP} based on Eq. 2 in a sample of unrelated individuals.

4 Conditional and Joint Analysis of GWAS Summary Statistics

For a large-scale meta-analysis of GWAS, the individual-level genotype data are typically unavailable and it is therefore extremely difficult to perform a conditional analysis. In this section, we will introduce an approximate conditional and joint analysis approach that only uses summary data from the meta-analysis with LD structure between SNPs estimated from a reference sample with individual-level genotype data. This section largely follows the methods section in Yang et al. [15].

4.1 Multi-SNP Association Analysis

In an association analysis of multiple SNPs, we usually fit the SNPs in a multiple regression model as

$$\mathbf{y} = \mathbf{Xb} + \mathbf{e} \quad (5)$$

where $\mathbf{y}$ a vector of phenotypes, $\mathbf{X} = \{x_{ij}\}$ is the SNP genotype matrix with $x_{ij} = -2p_j$, $1 - 2p_j$, or $2 - 2p_j$ for the jth SNP of the ith individual, p_j is the allele frequency of a SNP j, $\mathbf{b}$ is a vector of SNP effects (fixed effects) and $\mathbf{e}$ is a vector of residual (random effects) with $\mathbf{e} \sim N(\mathbf{0}, \mathbf{I}\sigma_J^2)$. We can subtract the mean of the phenotype from each element of $\mathbf{y}$ such that we do not need to fit the intercept term in the model. The least-squares estimate of $\mathbf{b}$ is

$$\hat{\mathbf{b}} = (\mathbf{X'X})^{-1}\mathbf{X'y} \text{ and } var(\hat{\mathbf{b}}) = \sigma_J^2(\mathbf{X'X})^{-1} \quad (6)$$

If we ignore all the correlations between SNPs, the multi-SNP analysis will be equivalent to single SNP-based separate analyses

$$\hat{\mathbf{b}}_M = \mathbf{D}^{-1}\mathbf{X'y} \text{ and } var(\hat{\mathbf{b}}_M) = \sigma_M^2 \mathbf{D}^{-1} \quad (7)$$

where $\boldsymbol{\beta}$ is a vector of marginal SNP effects, $\mathbf{D}$ the diagonal matrix of $\mathbf{X'X}$ with $D_j = \sum_i^n x_{ij}^2$ and σ_M^2 is the residual variance in the single SNP analyses. With the summary statistics from single SNP analyses and individual-level genotype data of the discovery sample, we can convert the marginal effects to joint effects without using the phenotype data. We know from Eq. 7 that $\mathbf{X'y} = \mathbf{D}\hat{\mathbf{b}}_M$, we therefore can rewrite Eq. 6 as

$$\hat{\mathbf{b}} = (\mathbf{X'X})^{-1}\mathbf{D}\hat{\mathbf{b}}_M \text{ and } var(\hat{\mathbf{b}}) = \sigma_J^2(\mathbf{X'X})^{-1} \quad (8)$$

The $\mathbf{X'X}$ is a variance–covariance matrix of SNP genotypes, which can be estimated from the allele frequencies in the meta-analysis sample along with LD correlations between SNPs from a reference sample with reasonably large sample size (e.g., >4,000). Let $\mathbf{W} = \{w_{ij}\}$ denote the genotype matrix of the reference sample with sample size of m, where $w_{ij} = -2f_j$, $1 - 2f_j$, or $2 - 2f_j$ for the three genotypes with f_j being the allele frequency of a SNP j in the reference sample, and let $\mathbf{D}_W$ denote the diagonal matrix of $\mathbf{W'W}$ with $D_{W(j)} = \sum_i^m w_{ij}^2$. Assuming the reference sample and the whole meta-analysis sample are from the same population, the LD correlation between a pair of SNPs j and k estimated from these two samples should be similar, i.e., $\frac{\sum_i^n x_{ij}x_{ik}}{\sqrt{\sum_i^n x_{ij}^2 \sum_i^n x_{ik}^2}} \approx \frac{\sum_i^m w_{ij}w_{ik}}{\sqrt{\sum_i^m w_{ij}^2 \sum_i^m w_{ik}^2}}$, so that $\mathbf{X'X}$ is approximately equal to $\mathbf{B}$ with $B_{jk} \approx \sqrt{\frac{D_j D_k}{D_{W(j)} D_{W(k)}}} \sum_i^m w_{ij}w_{ik}$. We have defined above that $D_j = \sum_i^n x_{ij}^2$

whilst x_{ij} is unavailable, we thus take $D_j = 2p_j(1-p_j)n$ assuming Hardy–Weinberg equilibrium. In matrix form, $\mathbf{B} = \mathbf{D}^{1/2}\mathbf{D}_W^{-1/2}\mathbf{W'WD}_W^{-1/2}\mathbf{D}^{1/2}$. We therefore can approximate a multi-SNP analysis as

$$\tilde{\mathbf{b}} = \mathbf{B}^{-1}\mathbf{D}\hat{\mathbf{b}}_M \text{ and } var(\tilde{\mathbf{b}}) = \sigma_J^2\mathbf{B}^{-1} \qquad (9)$$

where $\mathbf{b}$ is a vector of approximate estimates of joint SNP effects. The residual variance can be estimated as

$$\hat{\sigma}_J^2 = \frac{(1-R_J^2)\mathbf{y'y}}{n-N} = \frac{\mathbf{y'y} - \hat{\mathbf{b}}'\mathbf{D}\mathbf{b}_M}{n-N} \qquad (10)$$

where R_J^2 is coefficient of determination of the multiple regression model. In single SNP association analysis, $\hat{\sigma}_{M(j)}^2 = \frac{\mathbf{y'y} - D_j\hat{b}_{M(j)}^2}{n-1}$ and the squared standard error of the estimate of the effect size is $S_j^2 = \hat{\sigma}_{M(j)}^2/D_j$ so that $\mathbf{y'y} = D_jS_j^2(n-1) + D_j\hat{b}_{M(j)}^2$. Despite phenotypes usually being standardized to z-scores, we take the median of $D_jS_j^2(n-1) + D_j\hat{b}_{M(j)}^2$ across all the SNPs to calculate $\mathbf{y'y}$ instead of relying on the variance being known.

The sample size (n) varies for different SNPs due to imputation failures for different SNPs in different participating studies. Therefore, n is no longer constant across different SNPs and we need to scale the elements of $\mathbf{B}$ and $\mathbf{D}$ specifically according to the different effective sample sizes of different SNPs. For any SNP j, $\hat{n}_j = \mathbf{y'y}/D_jS_j^2 - \hat{b}_{M(j)}^2/S_j^2 + 1$ where we take the variance explained by a single SNP into account considering that the effect sizes of some particular SNPs are large for some traits. We use the estimated effective sample size rather than the reported sample size because the effective sample size will be smaller than the reported sample size if there is some degree of relatedness in the data. We then adjust the jkth element of $\mathbf{B}$ for the sample size variability of the SNPs as $B_{jk} = \min(\hat{n}_j, \hat{n}_k)2\sqrt{\frac{p_j(1-p_j)p_k(1-p_k)}{\sum_i^m w_{ij}^2 \sum_i^m w_{ik}^2}}\sum_i^m w_{ij}w_{ik}$ and adjust the jth diagonal element of $\mathbf{D}$ as $D_j = 2p_j(1-p_j)\hat{n}_j$.

4.2 Conditional Association Analysis

For a multi-SNP association analysis, the least-squares estimate of a SNP effect conditional on a group of selected SNPs ($b_2|\mathbf{b}_1$) is

$$\hat{b}_2|\hat{\mathbf{b}}_1 = \hat{b}_{M(2)} - \mathbf{X'}_2\mathbf{X}_1(\mathbf{X'}_1\mathbf{X}_1)^{-1}\mathbf{D}_1\hat{\mathbf{b}}_{M(1)}/D_2$$

$$var(\hat{b}_2|\hat{\mathbf{b}}_1) = \sigma_C^2[D_2 - \mathbf{X'}_2\mathbf{X}_1(\mathbf{X'}_1\mathbf{X}_1)^{-1}\mathbf{X'}_1\mathbf{X}_2]/D_2^2$$

where $\hat{b}_{M(2)}$ is the estimate of marginal effect of the target SNP and $D_2 = \sum_i^n x_{2i}^2$ for the target SNP, σ_C^2 is the residual variance in the conditional analysis and $\mathbf{X}$ is the SNP genotype matrix defined as in Eq. 5 with the subscript "2" indicating the target SNP and "1" indicating the group of selected SNPs.

Similarly, we can estimate the LD correlations from the reference sample and approximate a conditional analysis as

$$\tilde{b}_2|\tilde{\mathbf{b}}_1 = \hat{b}_{M(2)} - \mathbf{C}\mathbf{B}_1^{-1}\mathbf{D}_1\hat{\mathbf{b}}_{M(1)}/D_2 \text{ and } var(\tilde{b}_2|\tilde{\mathbf{b}}_1)$$
$$= \sigma_C^2[D_2 - \mathbf{C}\mathbf{B}_1^{-1}\mathbf{C}']/D_2^2 \quad (11)$$

where $\mathbf{B}_1$ is defined similarly as in Eq. 9 and $\mathbf{C} \approx \mathbf{X}'_2\mathbf{X}_1$ with its kth element being $C_k = \min(\hat{n}_2, \hat{n}_{1k})2\sqrt{\frac{p_2(1-p_2)p_{1k}(1-p_{1k})}{\sum_i^m w_{2i}^2 \sum_i^m w_{1ik}^2}} \sum_i^m w_{2i}w_{1ik}$. The residual variance in the conditional analysis can be estimated as $\hat{\sigma}_C^2 = \frac{\mathbf{y}'\mathbf{y} - \tilde{\mathbf{b}}'_1\mathbf{D}_1\hat{\beta}_1 - (\tilde{b}_2|\tilde{\mathbf{b}}_1)'D_2\hat{\beta}_2}{(n-N_1-1)}$ with N_1 being the number of selected SNPs.

4.3 Selecting the Top Associated SNPs Iteratively by Conditional Analyses

We use a stepwise model selection strategy to select the top associated SNPs iteratively based on the p-values calculated from the approximate conditional analysis. We start with a model with the most significant SNP in the meta-analysis with p-value less than a cutoff p-value, e.g., 5×10^{-8}. At the tth step, we calculate the conditional p-values of all the remaining SNPs conditional on the SNPs that have been selected in the model based on Eq. 11. To avoid problems due to colinearity, if the squared multiple correlation between the target SNP and the selected SNPs is larger than a cutoff value, e.g., 0.9, the conditional p-value for the target SNP will be set to 1. We then select the SNP with minimum conditional p-value that is less than the cutoff p-value. However, if adding the new SNP causes new colinearity problems between any of the selected SNPs and the others, we drop the new SNP. We repeat this process until no SNPs can be selected into or dropped out of the model. We finally fit all the selected SNPs in the model to estimate the joint effects of these SNPs based on Eq. 9.

A multiple regression analysis with model selection might suffer from over-fitting of effects because the residual variance decreases as more and more SNPs are included in the model, so that the false positive rate for the inclusion of new SNPs in the model would be inflated. In GCTA, we keep the residual variance constant being the phenotypic variance, even if we add significant SNPs into the model. This may be too conservative and therefore we lose power to detect additional associated variants, but it has the benefit of keeping the false positive rate in the conditional and joint analysis at the same level as that of the meta-analysis.

The GCTA command to implement the model selection for the top associated SNPs by the joint and conditional analysis is

```
gcta --bfile test --massoc-file test.ma --massoc-slct --out test                    [Command 11]
```

This analysis will report all the SNPs that are jointly associated at $P <$ threshold, where the default threshold p-value is 5e−8 and can be changed using the option --*massoc-p*. If there are not many SNPs that pass the significance level, usually there will be few secondary SNPs detected by this approach.

We also provide two additional options in GCTA to perform conditional or joint analysis without model selection. The command for performing a conditional analysis without model selection is

```
gcta --bfile test --massoc-file test.ma --massoc-cond
cond.snplist --out test                       [Command 12]
```

In this analysis, GCTA will calculate the approximate conditional *p*-value of each of the SNPs across the whole genome conditioning on the SNPs given in the "cond.snplist" file. Those target SNPs that are in extremely high LD with the given SNPs and therefore cause colinearity problems for the model are ignored in the analysis. The command for performing a joint analysis of multi-SNPs without model selection is

```
gcta --bfile test --extract test.snplist --massoc-
file test.ma --massoc-joint --out test    [Command 13]
```

where the list of SNPs to be fitted in the model is given in the file "test.snplist". The analysis will fail if some of the given SNPs are in high LD which causes a colinearity problem for the model.

In a genetically homogenous population of large effective size, the expected value of LD correlation between two SNPs on different chromosomes or a large distance apart is approximately zero and the observed LD correlations between such pairs of SNPs in a sample are just due to random sampling. Therefore, it is inappropriate to represent a sampling correlation observed in the meta-analysis sample by the sampling correlation observed in the reference sample. We know from empirical data that the observed LD correlation between SNPs more than 10 Mb distant is consistent with what we would expect by chance (Fig. 3). By default in GCTA, we use the expected values (zeros) in the matrix **B** for pairs of SNPs that are more than 10 Mb apart. We also provide an option in GCTA (--*massoc-wind*) to allow users to specify the distance.

4.4 GWAS Data with Individual-Level Genotype Data

The method was originally developed but not limited to analyze summary data with LD estimated from a reference sample. In fact, it can also be applied to data with individual-level genotype information, i.e., the reference sample is the sample from which the summary statistics for genotype–phenotype association are estimated. In this case, the method is equivalent to a multi-SNP association analysis (i.e., stepwise linear multiple regression), where we take into account the correlations between all SNPs and the reduction of residual variance for the inclusion of SNPs in the model. The GCTA commands to implement model selection to select the top associated SNPs, conditional GWA without model selection, and joint association analysis of multiple SNPs are the

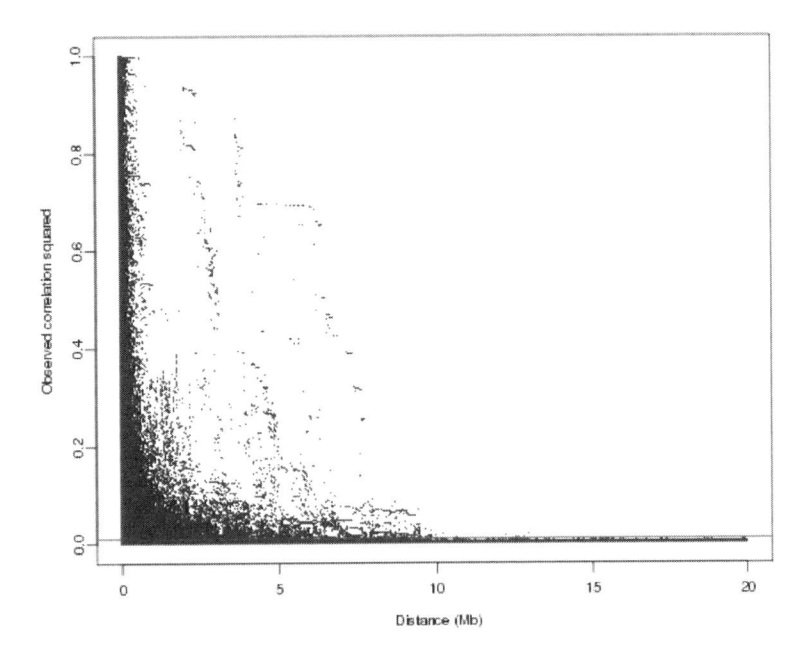

Fig. 3 Plot of the observed r^2 (squared LD correlation) values between pairs of SNPs against their physical distances in a sample of 3,925 unrelated individuals [15]. We randomly sampled 1,000 SNPs across the whole genome as target SNPs. For each target SNP, we calculated its r^2 value with all the other SNPs in 20 Mb distance in either direction. We then plotted the observed r^2 values against the distances between the target SNPs and the other SNPs. The figure shows that in general LD degrades exponentially with the increase of distance except for a number of outliers. The major outliers, e.g., $r^2 > 0.4$ when distance >2 Mb and $r^2 > 0.2$ when distance >5 Mb, are located at a small number of long-range LD regions [30]. When the distance is >10 Mb, the observed r^2 values are consistent with what we would expect by chance (the line, 95 % confidence interval adjusted for the number of calculations)

same as in Commands 11, 12, and 13 respectively, but adding the option --massoc-actual-geno.

4.5 Case–Control GWAS Data

The methods of the approximate conditional and joint association analysis for a quantitative trait can be directly applied to case–control data for a disease. This is justified as follows. In an association analysis of case–control data, the effect size of a SNP on a disease is estimated on the observed scale (affected or unaffected), i.e., the odds ratio (OR). Assuming a continuous underlying liability to a disease, we can transform the liability to the probability of being affected or unaffected for an individual by a logistic regression analysis. In logistic regression, the probability of an individual i being affected or unaffected given SNP data is

$$f_i = \exp(l_i)/[1 + \exp(l_i)] \text{ with } l_i = \mathbf{x}_i \mathbf{b} + e_i$$

where l_i is the liability of the ith individual, $\mathbf{x}_i$ is the ith row of the SNP genotype matrix $\mathbf{X} = \{x_{ij}\}$ as defined in Eq. 5 and $\mathbf{b} = \{b_j\}$ is a vector of SNP effects and e_i is the residual. Importantly, according to the definition of a logistic regression model, b_j is interpreted as the log(OR). We know from the derivations in Subheading 4.1 that the scale of measurement of a quantitative trait is not important as it can be dropped from the equations. We therefore can assume that the disease liabilities of all the individuals are known, and model the effects of multiple SNPs on log(OR) scale as

$$\mathbf{L} = \mathbf{X}\mathbf{b} + \mathbf{e}$$

where $\mathbf{L} = \{l_i\}$. This model is in exactly the same form as in Eq. 5 so that all the methods for a quantitative trait described above can be directly applied to case–control data, as long as the effect size of a SNP and its standard error are expressed on the log(OR) scale. Although the residuals follow a logistic distribution, the least-squares estimates of effect sizes are unbiased because the least-squares approach does not rely on a normality assumption.

5 Future Developments

The generality and flexibility of the mixed model methodology facilitates new applications, for example multivariate analysis [6], random regression and combined between and within family analysis.

6 Notes

1. If the data are from a family study, the estimate of variance explained by all SNPs from GCTA will be very similar to the estimate of narrow-sense heritability from a pedigree analysis, because the GCTA basically reconstructs the pedigree structure using genome-wide SNP data. In this case, the estimate from GCTA cannot be interpreted as the variance explained by all SNPs.

2. The *--grm-cutoff* is only applicable to the GRM estimated from genome-wide SNPs rather than that estimated from the SNPs on a single chromosome. The variance in genetic relationships estimated from the SNPs on a single chromosome depends on the length of the chromosome and is much larger than the GRM estimated from genome-wide SNPs.

3. We recommend to adjust the phenotype for covariates such as age and date of birth and standardize it to z-score prior to GCTA rather than including all the covariates in the REML analysis by the *--covar* (or *--qcovar*) option. This can be done by regressing the phenotypes on the covariates and then converting the residuals into z-scores by standard normal

transformation. It is also recommended to stratify the adjustment on gender and case–control status because for trait such as BMI the variance in males is smaller than in females and such difference in variance is unable to be adjusted by the regression approach.

4. In practice, we do not usually use the option --*grm-adj* because it relies on strong assumptions of the allele frequency distribution of the causal variants. This option is to adjust for the sampling error in predicting the genetic relationship at the causal variants by the SNPs. The adjustment (the input parameter c), however, depends on the assumption about the difference in allele frequency spectrum between the causal variants and the SNPs. For example, $c = 0$ if we assume the causal variants have the same allele frequency spectrum as the SNP and $c \neq 0$ and is genotype data dependent if we assume their allele frequency spectrum are different.

5. If we select a subset of SNPs by association *p*-values and use GCTA to estimate the variance explained by these selected SNPs in the same data, the variance explained will be overestimated. In association analysis, the estimate of a SNP effect ($\hat{b}$) can be written as $\hat{b} = b + e$ where b is the true effect of the SNP and e is the estimation error. If the SNPs are selected based on *p*-values, it will create a positive correlation between b and e so that the effects of the selected SNPs will tend to be overestimated. On the other hand, GCTA is equivalent to an analysis of modeling multiple SNPs, therefore the variance explained by the selected SNPs will be overestimated with GCTA. There will be no such problem if the GCTA is performed in an additional independent data.

6. When estimating the variance explained by all SNPs in case–control data, the estimate will be more subject to technical artifact as compared with that for a quantitative trait. Any artifacts, which cause spurious difference in allele frequencies between cases and controls, will contribute to the estimation of "genetic" variation. We have suggested a more stringent quality control procedure in Lee et al. [29].

Appendix A

In Eq. 4, we have

$$\mathbf{y} = \mathbf{Xb} + \sum_{i=1}^{r} \mathbf{g}_i + \mathbf{e} \text{ and } var(\mathbf{y}) = \mathbf{V} = \sum_{i=1}^{r} \mathbf{A}_i \sigma_i^2 + \mathbf{I}\sigma_e^2,$$

of which Eq. 2 is a special case with $r = 1$. By default in GCTA, we use the average information (AI) REML algorithm [31] to obtain the estimates the variance components σ_i^2 and σ_e^2 through iteration. In the tth iteration, $\mathbf{q}^{(t)} = \mathbf{q}^{(t-1)} + (\mathbf{H}^{(t-1)})^{-1} \frac{\partial L}{\partial \mathbf{q}} | \mathbf{q}^{(t-1)}$, where $\mathbf{q}$ is a vector of the estimates of variance components $(\hat{\sigma}_1^2, \ldots, \hat{\sigma}_r^2$ and $\hat{\sigma}_e^2)$; L is the log likelihood function of the mixed linear model (ignoring the constant), $L = -1/2(\log|\hat{\mathbf{V}}| + \log|\mathbf{X}'\hat{\mathbf{V}}^{-1}\mathbf{X}| + \mathbf{y}'\mathbf{P}\mathbf{y})$ with $\hat{\mathbf{V}} = \sum_{i=1}^{r} \mathbf{A}_i \hat{\sigma}_i^{2(t-1)} + \mathbf{I} \hat{\sigma}_e^{2(t-1)}$ and $\mathbf{P} = \hat{\mathbf{V}}^{-1} - \hat{\mathbf{V}}^{-1}\mathbf{X}(\mathbf{X}'\hat{\mathbf{V}}^{-1}\mathbf{X})^{-1}\mathbf{X}'\hat{\mathbf{V}}^{-1}$; $\mathbf{H}$ is the average of the observed and expected information matrices [22],

$$\mathbf{H} = \frac{1}{2} \begin{bmatrix} \mathbf{y}'\mathbf{P}\mathbf{A}_1\mathbf{P}\mathbf{A}_1\mathbf{P}\mathbf{y} & \cdots & \mathbf{y}'\mathbf{P}\mathbf{A}_1\mathbf{P}\mathbf{A}_r\mathbf{P}\mathbf{y} & \mathbf{y}'\mathbf{P}\mathbf{A}_1\mathbf{P}\mathbf{P}\mathbf{y} \\ \vdots & \vdots & \vdots & \vdots \\ \mathbf{y}'\mathbf{P}\mathbf{A}_r\mathbf{P}\mathbf{A}_1\mathbf{P}\mathbf{y} & \cdots & \mathbf{y}'\mathbf{P}\mathbf{A}_r\mathbf{P}\mathbf{A}_r\mathbf{P}\mathbf{y} & \mathbf{y}'\mathbf{P}\mathbf{A}_r\mathbf{P}\mathbf{P}\mathbf{y} \\ \mathbf{y}'\mathbf{P}\mathbf{P}\mathbf{A}_1\mathbf{P}\mathbf{y} & \cdots & \mathbf{y}'\mathbf{P}\mathbf{P}\mathbf{A}_r\mathbf{P}\mathbf{y} & \mathbf{y}'\mathbf{P}\mathbf{P}\mathbf{P}\mathbf{y} \end{bmatrix};$$

and $\frac{\partial L}{\partial \mathbf{q}}$ is a vector of first derivatives of the log likelihood function with respect to each variance component,

$$\frac{\partial L}{\partial \mathbf{q}} = -\frac{1}{2} \begin{bmatrix} tr(\mathbf{P}\mathbf{A}_1) - \mathbf{y}'\mathbf{P}\mathbf{A}_1\mathbf{P}\mathbf{y} \\ \vdots \\ tr(\mathbf{P}\mathbf{A}_r) - \mathbf{y}'\mathbf{P}\mathbf{A}_r\mathbf{P}\mathbf{y} \\ tr(\mathbf{P}) - \mathbf{y}'\mathbf{P}\mathbf{P}\mathbf{y} \end{bmatrix}$$

We also provide in GCTA two optional algorithms to estimate the variance components, which we call the direct REML and EM-REML. For the direct REML algorithm, the variance components in the tth iteration are estimated as

$$\mathbf{q}^{(t)} = \begin{bmatrix} tr(\mathbf{P}\mathbf{A}_1\mathbf{P}\mathbf{A}_1) & \cdots & tr(\mathbf{P}\mathbf{A}_1\mathbf{P}\mathbf{A}_r) & tr(\mathbf{P}\mathbf{A}_1\mathbf{P}) \\ \vdots & \vdots & \vdots & \vdots \\ tr(\mathbf{P}\mathbf{A}_r\mathbf{P}\mathbf{A}_1) & \cdots & tr(\mathbf{P}\mathbf{A}_r\mathbf{P}\mathbf{A}_r) & tr(\mathbf{P}\mathbf{A}_r\mathbf{P}) \\ tr(\mathbf{P}\mathbf{P}\mathbf{A}_1) & \cdots & tr(\mathbf{P}\mathbf{P}\mathbf{A}_r) & tr(\mathbf{P}\mathbf{P}) \end{bmatrix}^{-1} \begin{bmatrix} \mathbf{y}'\mathbf{P}\mathbf{A}_1\mathbf{P}\mathbf{y} \\ \vdots \\ \mathbf{y}'\mathbf{P}\mathbf{A}_r\mathbf{P}\mathbf{y} \\ \mathbf{y}'\mathbf{P}\mathbf{P}\mathbf{y} \end{bmatrix}$$

The direct REML algorithm is generally more robust but computationally less efficient than AI-REML. For the EM-REML algorithm, each variance component is estimated as

$$\sigma_i^{2(t)} = [\hat{\sigma}_i^{4(t-1)} \mathbf{y}'\mathbf{P}\mathbf{A}_i\mathbf{P}\mathbf{y} + \mathrm{tr}(\sigma_i^{2(t-1)}\mathbf{I} - \sigma_i^{4(t-1)}\mathbf{P}\mathbf{A}_i)]/n$$

The EM-REML is robust, which guarantees increased likelihood after each iteration, but is extremely slow to converge. We therefore do not recommend choosing EM-REML in GCTA unless we know that the starting values are very close to the estimates. The GCTA option for choosing different REML algorithm is *--reml-alg* with the input value 0 for AI-REML (default), 1 for the direct REML algorithm and 2 for EM-REML. At the beginning of the iteration process, all the variance components are initialized by an arbitrary value, i.e., $\sigma_i^{2(0)} = \sigma_P^2/(r+1)$, which is subsequently updated by the EM-REML algorithm $\sigma_i^{2(1)} = [\sigma_i^{4(0)}\mathbf{y}'\mathbf{PA}_i\mathbf{Py} + \text{tr}(\sigma_i^{2(0)}\mathbf{I} - \sigma_i^{4(0)}\mathbf{PA}_i)]/n$. The EM-REML algorithm is used as an initial step to determine the direction of the iteration updates because it is robust to poor starting values. We also provide options (*--reml-priors* and *--reml-priors-var*) in GCTA for users to specify starting values. After one EM-REML iteration, GCTA switches to the AI-REML algorithm (or the other two algorithms) for the remaining iterations until the iteration converges with the criteria of $L^{(t)} - L^{(t-1)} < 10^{-4}$ where $L^{(t)}$ is the log likelihood of the *t*th iteration. By default, any variance component that escapes from the parameter space (i.e., its estimate is negative) will be set to $10^{-6} \times \sigma_P^2$. If a component keeps escaping from the parameter space, it will be constrained at $10^{-6} \times \sigma_P^2$. There is an option in GCTA (*--reml-no-constrain*) that allows the estimates of variance components to be negative. This is justified because if a parameter is zero, an unbiased estimate of this parameter will have half chance being negative. In practice, however, a negative variance component is usually difficult to interpret. We also provide an option (*--reml-maxit*) for users to specify the maximum number of iterations at which the iteration process will stop without convergence.

References

1. Hindorff LA, Sethupathy P, Junkins HA et al (2009) Potential etiologic and functional implications of genome-wide association loci for human diseases and traits. Proc Natl Acad Sci U S A 106(23):9362–9367
2. Maher B (2008) Personal genomes: the case of the missing heritability. Nature 456(7218):18–21
3. Yang J, Benyamin B, McEvoy BP et al (2010) Common SNPs explain a large proportion of the heritability for human height. Nat Genet 42(7):565–569
4. Yang J, Manolio TA, Pasquale LR et al (2011) Genome partitioning of genetic variation for complex traits using common SNPs. Nat Genet 43(6):519–525
5. Davies G, Tenesa A, Payton A et al (2011) Genome-wide association studies establish that human intelligence is highly heritable and polygenic. Mol Psychiatry 16(10):996–1005
6. Deary IJ, Yang J, Davies G et al (2012) Genetic contributions to stability and change in intelligence from childhood to old age. Nature 482(7384):212–215
7. Lee SH, Decandia TR, Ripke S et al (2012) Estimating the proportion of variation in susceptibility to schizophrenia captured by common SNPs. Nat Genet 44(3):247–250
8. Gibson G (2010) Hints of hidden heritability in GWAS. Nat Genet 42(7):558–560
9. Visscher PM, Brown MA, McCarthy MI, Yang J (2012) Five years of GWAS discovery. Am J Hum Genet 90(1):7–24
10. Teslovich TM, Musunuru K, Smith AV et al (2010) Biological, clinical and population relevance of 95 loci for blood lipids. Nature 466(7307):707–713
11. Heid IM, Jackson AU, Randall JC et al (2010) Meta-analysis identifies 13 new loci associated with waist-hip ratio and reveals sexual dimorphism in the genetic basis of fat distribution. Nat Genet 42(11):949–960
12. Lango Allen H, Estrada K, Lettre G et al (2010) Hundreds of variants clustered in genomic loci and biological pathways affect human height. Nature 467(7317):832–838
13. Speliotes EK, Willer CJ, Berndt SI et al (2010) Association analyses of 249,796 individuals reveal 18 new loci associated with body mass index. Nat Genet 42(11):937–948
14. Ripke S, Sanders AR, Kendler KS et al (2011) Genome-wide association study identifies five

new schizophrenia loci. Nat Genet 43 (10):969–976
15. Yang J, Ferreira T, Morris AP et al (2012) Conditional and joint multiple-SNP analysis of GWAS summary statistics identifies additional variants influencing complex traits. Nat Genet 44(4):369–375
16. Yang J, Lee SH, Goddard ME, Visscher PM (2011) GCTA: a tool for genome-wide complex trait analysis. Am J Hum Genet 88 (1):76–82
17. Hayes BJ, Visscher PM, Goddard ME (2009) Increased accuracy of artificial selection by using the realized relationship matrix. Genet Res 91(1):47–60
18. Strandén I, Garrick DJ (2009) Technical note: derivation of equivalent computing algorithms for genomic predictions and reliabilities of animal merit. J Dairy Sci 92(6):2971–2975
19. VanRaden PM (2008) Efficient methods to compute genomic predictions. J Dairy Sci 91 (11):4414–4423
20. Patterson HD, Thompson R (1971) Recovery of inter-block information when block sizes are unequal. Biometrika 58(3):545–554
21. Purcell S, Neale B, Todd-Brown K et al (2007) PLINK: a tool set for whole-genome association and population-based linkage analyses. Am J Hum Genet 81(3):559–575
22. Lee SH, van der Werf JH (2006) An efficient variance component approach implementing an average information REML suitable for combined LD and linkage mapping with a general complex pedigree. Genet Sel Evol 38 (1):25–43
23. Jorjani H, Klei L, Emanuelson U (2003) A simple method for weighted bending of genetic (co)variance matrices. J Dairy Sci 86 (2):677–679
24. Hill WG, Thompson R (1978) Probabilities of non-positive definite between-group or genetic covariance matrices. Biometrics 34:429–439
25. Haseman JK, Elston RC (1972) The investigation of linkage between a quantitative trait and a marker locus. Behav Genet 2:2–19
26. Lynch M, Walsh B (1998) Genetics and analysis of quantitative traits. Sinauer Associates, Sunderland, MA
27. Falconer DS (1965) The inheritance of liability to certain diseases, estimated from the incidence among relatives. Ann Hum Genet 29:51–71
28. Dempster ER, Lerner IM (1950) Heritability of threshold characters. Genetics 35(2):212–236
29. Lee SH, Wray NR, Goddard ME, Visscher PM (2011) Estimating missing heritability for disease from genome-wide association studies. Am J Hum Genet 88(3):294–305
30. Price AL, Weale ME, Patterson N et al (2008) Long-range LD can confound genome scans in admixed populations. Am J Hum Genet 83 (1):132–135
31. Gilmour AR, Thompson R, Cullis BR (1995) Average information REML: an efficient algorithm for variance parameters estimation in linear mixed models. Biometrics 51:1440–1450

Chapter 10

Bayesian Methods Applied to GWAS

Rohan L. Fernando and Dorian Garrick

Abstract

Bayesian multiple-regression methods are being successfully used for genomic prediction and selection. These regression models simultaneously fit many more markers than the number of observations available for the analysis. Thus, the Bayes theorem is used to combine prior beliefs of marker effects, which are expressed in terms of prior distributions, with information from data for inference. Often, the analyses are too complex for closed-form solutions and Markov chain Monte Carlo (MCMC) sampling is used to draw inferences from posterior distributions. This chapter describes how these Bayesian multiple-regression analyses can be used for GWAS. In most GWAS, false positives are controlled by limiting the genome-wise error rate, which is the probability of one or more false-positive results, to a small value. As the number of test in GWAS is very large, this results in very low power. Here we show how in Bayesian GWAS false positives can be controlled by limiting the *proportion of false-positive* results among all positives to some small value. The advantage of this approach is that the power of detecting associations is not inversely related to the number of markers.

Key words GWAS, Bayesian multiple-regression, Genomic prediction, MCMC sampling, R-scripts

1 Introduction

The most widely used approach for genome-wide association analyses (GWAS) tests each marker for association with the trait of interest. This approach has been useful to detect many associations, but significant associations explain only a small fraction of the genetic variance of quantitative traits [1–3]. In contrast, methods that simultaneously fit all markers as random effects are able to account for most of the genetic variance [4–6].

In most GWAS, the number k of markers available for analysis greatly exceeds the number n of observations. Thus, multiple regression analysis based on ordinary least-squares cannot be used to simultaneously estimate the effects of all the k markers. Bayesian regression methods, however, which combine prior information on the k marker effects with the n observed trait phenotypes, can be used to estimate all marker effects [7]. Although these methods were originally proposed for whole-genome prediction [7], they can also be used for GWAS [6, 8].

In most GWAS, false positives from multiple tests are controlled by managing the genome-wise error rate (GWER), which is the probability of one or more false-positive results among all tests across the whole genome. As the number of tests in a GWAS can be very large, controlling the GWER results in very low power. An alternative approach to account for multiple tests is to control the proportion of false positives (PFP) among all positive results from the study [9, 10], which is closely related to controlling the false discovery rate (FDR; *see* **Note 1**) [11]. An advantage of this approach is that the power of detecting associations is not inversely related to the number of markers tested [10, 12]. In this chapter we will describe how Bayesian regression methods that are used for whole-genome prediction can also be used for GWAS, accounting for multiple tests by controlling PFP.

2 Methods

Although Bayesian methods can be applied for prediction [13] or GWAS [14] under non-additive models, here we consider GWAS only under additive gene action as in the methods proposed by Meuwissen et al. [7]. The starting point for these methods is a mixed linear model of the form:

$$y = X\beta + Z\alpha + e, \qquad (1)$$

where y is an $n \times 1$ vector of trait phenotypic values, X is an $n \times p$ incidence matrix relating the vector β of non-genetic fixed effects to y, Z is an $n \times k$ matrix of genotype covariates (coded as 0, 1, or 2) for k SNP markers, α is a $k \times 1$ vector of random partial regression coefficients of the k SNPs (which are more commonly referred to as the marker effects), and e is a vector of residuals.

To proceed with Bayesian regression, prior distributions must be specified for β, α, and e. In all the models considered here a flat prior [15] is used for β, and conditional on the residual variance, σ_e^2, a normal distribution with null mean and covariance matrix $R\sigma_e^2$ is used for the vector of residuals, where R is a diagonal matrix. Further, σ_e^2 is treated as an unknown with a scaled inverse chi-square prior. The alternative methods discussed here differ only in the prior used for α. The priors that lead to what is called ridge-regression BLUP (RR-BLUP), BayesA and BayesB, which were all proposed by Meuwissen et al. [7] for genomic selection, and BayesC and BayesCπ [16], which are variations of BayesB, are discussed next.

2.1 Priors

2.1.1 RR-BLUP

Best linear unbiased prediction (BLUP) requires only specifying the first and second moments of the distribution of the random effects. In their BLUP model, Meuwissen et al. [7] assumed α has null mean and covariance matrix $I\sigma_\alpha^2$ for a known value of σ_α^2 (*see* **Note 2**). If the ratio $\lambda = \frac{\sigma_\alpha^2}{\sigma_e^2}$ is known, it can be shown that the best linear unbiased predictor (BLUP) of α can be obtained by solving Henderson's mixed model equations [17] that correspond to the model equation (Eq. 1).

In addition to this, for GWAS based on Bayesian methods, we specify that, conditional on the value of σ_α^2, α has a multivariate-normal distribution with null mean and covariance matrix $I\sigma_\alpha^2$. Then, it can be shown that BLUP is the posterior mean of α [15, 18, 19].

Further, σ_α^2 can be treated as an unknown quantity that has a scaled inverse chi-square prior with scale parameter S_α^2 and ν_α degrees of freedom. Then it can be shown (*see* **Note 3**) that the marginal prior for α has a multivariate-t distribution with null mean, scale matrix $S_\alpha^2 I$ and ν_α degrees of freedom [15]. Two of the consequences of this are:

1. The variance of α becomes

$$Var(\alpha) = \frac{\nu_\alpha}{\nu_\alpha - 2} S_\alpha^2 I, \qquad (2)$$

for $\nu_\alpha > 2$, and

2. The posterior mean of α is no longer a linear function of y (*see* **Note 4**).

As shown later, Markov Chain Monte Carlo (MCMC) sampling [15] will be used to make inferences from the posterior distribution of α.

In studies where linkage disequilibrium is simulated, BayesA and BayesB have had higher accuracy of prediction than RR-BLUP [7, 20]. In real applications, however, accuracy of RR-BLUP is very close and sometimes even higher than those for BayesA, BayesB, or BayesC [16]. The reason for this could be that when LD between markers is not strong, most of the accuracy is due to relationships, and RR-BLUP, which shrinks all markers equally, better captures these relationships [20, 21]. RR-BLUP is also expected to perform better when the trait is controlled by a very large number of genes evenly dispersed across the genome. On the other hand, the other approaches give better results in GWAS for detecting the largest associations [22].

2.1.2 BayesA

In BayesA, the prior assumption is that marker effects have identical and independent univariate-t distributions each with a null mean, scale parameter S_α^2 and ν degrees of freedom. As shown in [23], this is equivalent to assuming that the marker effect at locus i has a univariate normal with null mean and unknown, locus-specific variance σ_i^2, which in turn is assigned a scaled inverse chi-square prior with scale parameter S_α^2 and ν_α degrees of freedom. This is the form of the prior given in the original paper by Meuwissen et al. [7], which is also the basis for constructing the MCMC sampler.

The accuracy of prediction and power of detecting associations for BayesA, which also assumes non-null effects for all loci, falls somewhere between RR-BLUP and BayesB.

2.1.3 BayesB

In BayesB, the prior assumption is that marker effects have identical and independent mixture distributions, where each has a point mass at zero with probability π and a univariate-t distribution with probability $1 - \pi$ having a null mean, scale parameter S_α^2, and ν degrees of freedom. Thus, BayesA is a special case of BayesB with $\pi = 0$. Further, as in BayesA, the t-distribution in BayesB is equivalent to a univariate normal with null mean and unknown, locus-specific variance, which in turn is assigned a scaled inverse chi-square prior with scale parameter S_α^2 and ν_α degrees of freedom.

Here, we introduce a third form of the BayesB prior that is equivalent to the two given above in that all three produce the same posterior for locus effects. To do so, we introduce a Bernoulli variable δ_i for locus i that is 1 with probability $1 - \pi$ and zero with probability π. Then, the effect of locus i is written as

$$\alpha_i = \zeta_i \delta_i, \qquad (3)$$

where ζ_i has a normal distribution with null mean and locus-specific variance σ_i^2, which in turn has a scaled inverse chi-square prior with scale parameter S_α^2 and ν_α degrees of freedom.

When some loci have a large effect on the trait, BayesB will outperform RR-BLUP and BayesA, especially in small data sets. The reason is that with π close to 1.0 in BayesB, the number of nonzero marker effects that are estimated can be made small.

2.1.4 BayesC and BayesCπ

In BayesC, the prior assumption is that marker effects have identical and independent mixture distributions, where each has a point mass at zero with probability π and a univariate-normal distribution with probability $1 - \pi$ having a null mean and variance σ_α^2, which in turn has a scaled inverse chi-square prior with scale parameter S_α^2 and ν_α degrees of freedom. Then with $\pi = 0$, the marginal distribution of locus effects becomes a multivariate-t distribution with null mean, scale matrix $S^2 \mathbf{I}$ and ν_α degrees of freedom. Thus, BayesC with $\pi = 0$ is identical to RR-BLUP when σ_α^2 is treated as unknown with a scaled inverse chi-square prior.

In addition to the above assumptions, in BayesCπ, π is treated as unknown with a uniform prior.

The performance of BayesC and BayesCπ relative to the other methods will depend on the actual distribution of the marker effects. In most real analyses, when appropriate values of π and S_α^2

are used, BayesB performs better than BayesC or BayesCπ. As explained in Subheading 3.1 and **Note 2**, we use an estimate of the total genetic variance, σ_g^2 to obtain a value for S_α^2. When a good estimate of σ_g^2 is not available, BayesCπ can be used to estimate π and σ_g^2. Then, these can be used in BayesB.

2.2 Bayesian Inference

In Bayesian analyses, inferences of unknowns are based on their posterior distributions. Thus, inferences of marker associations are based on the posterior distribution of marker effects: $f(\alpha \mid y)$. Under the priors given above, however, closed-form expressions are not available for making inferences from $f(\alpha \mid y)$. Thus, as described later, samples drawn from $f(\alpha \mid y)$ are used to make inferences.

2.3 MCMC Sampling

In most Bayesian analyses, for computational reasons, samples are not directly drawn from the posterior of interest. Rather, samples are obtained using MCMC techniques, where the resulting Markov chain (*see* **Note 5**) is such that as chain length increases statistics computed from the chain converge to those from the posterior of interest [15, 24, 25].

The most widely used method to construct such a Markov chain is the Gibbs sampler. Let θ denote all the unknowns in the model. In BayesA, for example, this will include the non-genetic fixed effects, the random marker effects, the locus-specific variances of the marker effects, and the residual variance. Then, in the single-site Gibbs sampler, samples are drawn for each element i of the vector θ from its full-conditional posterior: $f(\theta_i \mid \theta_{-i}, y)$, which is the conditional distribution of θ_i given the current values of the other unknowns (θ_{-i}) and the trait phenotypes.

This full-conditional posterior $f(\theta_i \mid \theta_{-i}, y)$ can be written as

$$f(\theta_i \mid \theta_{-i}, y) = \frac{f(\theta_i, \theta_{-i}, y)}{f(\theta_{-i}, y)} \quad (4)$$
$$\propto f(\theta_i, \theta_{-i}, y),$$

because the denominator of the first line in Eq. 4 is free of θ_i [15]. To identify the form of the full-conditional distribution, the joint density in Eq. 4 is written as

$$f(\theta_i, \theta_{-i}, y) = f(y \mid \theta) f(\theta_i) f(\theta_{-i}), \quad (5)$$

where $f(y \mid \theta)$ is the density function of the conditional distribution of *y* given the values of the unknowns specified by θ, and $f(\theta_i)$ and $f(\theta_{-i})$ are the densities of the prior distributions of θ_i and θ_{-i}. Dropping factors that are constant with respect to θ_i gives the kernel of the full-conditional posterior. If the full-conditional belongs to a standard family of distributions, efficient algorithms are available to draw samples from the full-conditional. As an example, the full-conditional posterior for α_i is derived next under the priors of BayesA.

Under all the priors considered here, the conditional distribution of y given all the unknowns is multivariate normal with mean $X\beta + Z\alpha$ and covariance matrix $R\sigma_e^2$. The density function for this multivariate normal is:

$$f(y|\theta) = |2\pi R\sigma_e^2|^{-1/2}$$
$$\times \exp\left[-\frac{(y - X\beta - Z\alpha)'R^{-1}(y - X\beta - Z\alpha)}{2\sigma_e^2}\right] \quad (6)$$

The prior for β is flat, and thus the density function is a constant and will not contribute to the full-conditional posterior of α_i. The prior distribution for marker effects conditional on the locus-specific variances of marker effects are independent normals with density function:

$$f(\alpha|\xi) = \prod_{j=1}^{k} (2\pi\sigma_j^2)^{-\frac{1}{2}} \exp\left[-\frac{\alpha_j^2}{2\sigma_j^2}\right], \quad (7)$$

where ξ is the vector of locus-specific variances. The prior for ξ are identical and independent scaled inverse chi-square distributions with density function:

$$f(\xi|\nu_\alpha S_\alpha^2) = \prod_{j=1}^{k} \frac{(\nu_\alpha S_\alpha^2/2)^{\nu_\alpha/2}}{\Gamma(\nu_\alpha/2)} (\sigma_j^2)^{-(\nu_\alpha+2)/2} \exp\left[-\frac{\nu_\alpha S_\alpha^2}{2\sigma_j^2}\right], \quad (8)$$

and the density function for the scaled inverse chi-square prior of the residual variance is

$$f(\sigma_e^2|\nu_e, S_e^2) = \frac{(\nu_e S_e^2/2)^{\nu_e/2}}{\Gamma(\nu_e/2)} (\sigma_e^2)^{-(\nu_e+2)/2} \exp\left[-\frac{\nu_e S_e^2}{2\sigma_e^2}\right]. \quad (9)$$

Now, from Eqs. 4 and 5, the full-conditional posterior for α_j is proportional to the product of Eqs. 6–9. Next, to recognize the kernel of this full-conditional, factors that do not depend on α_i are dropped. First, note that the priors for the variance components given by Eqs. 8 and 9 are free of α_i. Also in Eq. 7, the normal densities of α_j for $j \ne i$ are free of α_i. Thus, the full-conditional for α_i is proportional to the product of Eq. 6 and the normal density of α_i, where α_i appears only in the exp() functions of these two densities. Thus, the full-conditional can be written as

$$f(\alpha_i|\text{ELSE}) \propto \exp\left[-\frac{(y - X\beta - Z\alpha)'R^{-1}(y - X\beta - Z\alpha)}{2\sigma_e^2}\right] \exp\left[-\frac{\alpha_i^2}{2\sigma_i^2}\right]$$

$$\propto \exp\left[-\frac{1}{2\sigma_e^2}\left\{(w - z_i\alpha_i)'R^{-1}(w - z_i\alpha_i) + \alpha_i^2\frac{\sigma_e^2}{\sigma_i^2}\right\}\right]$$

$$\propto \exp\left[-\frac{1}{2\sigma_e^2}\left\{w'R^{-1}w - 2z_i'R^{-1}w\alpha_i + z_i'R^{-1}z_i\alpha_i^2 + \frac{\sigma_e^2}{\sigma_i^2}\alpha_i^2\right\}\right]$$

where ELSE denotes conditioning on y and all unknowns other than α_i,

$$w = y - X\beta - \sum_{j \neq i} z_j \alpha_j \tag{10}$$

is the vector of trait phenotypes corrected for all effects other than that being sampled, and z_i is column i of the matrix z. Completing the square with respect to α_i in the exponent of the previous expression for $f(\alpha_i \mid \text{ELSE})$ gives

$$f(\alpha_i | \text{ELSE}) \propto \exp\left[-\frac{1}{2\sigma_e^2}\left\{w'R^{-1}w - c_i\hat{\alpha}_i^2 + c_i(\alpha_i - \hat{\alpha}_i)^2\right\}\right], \tag{11}$$

where $c_i = (z_i'R^{-1}z_i + \frac{\sigma_e^2}{\sigma_i^2})$ is the left-hand side of the mixed model equation for α_i, $\hat{\alpha}_i = \frac{z_i'R^{-1}w}{c_i}$, and $z_i'R^{-1}w$ is the right-hand side. Dropping terms from the exponent of Eq. 11 that are free of α_i, the full-conditional becomes

$$f(\alpha_i|\text{ELSE}) \propto \exp\left[-\frac{1}{2}\frac{(\alpha_i - \hat{\alpha}_i)^2}{\frac{\sigma_e^2}{c_i}}\right]. \tag{12}$$

Now, Eq. 12 can be recognized as the kernel of a normal distribution with mean $\hat{\alpha}_i$ and variance $\frac{\sigma_e^2}{c_i}$. Efficient algorithms that are available for sampling from a univariate normal distribution can be used to draw samples for α_i from its full-conditional posterior.

When the kernel of the posterior for some variable does not have a standard form, the Metropolis–Hastings algorithm [15, 26] can be used to draw samples. Here, a candidate sample is drawn from some proposal distribution that is easy to sample from. Then, the candidate is either accepted or rejected according to a rule that is constructed using the full-conditional posterior and proposal distributions. In the paper that introduced BayesB, the Metropolis–Hastings algorithm was used to sample the locus-specific variance of marker effects. Here, we will use a Gibbs sampler to draw locus-specific variances by using the third form of the BayesB prior introduced previously.

In the following sections, the full-conditional posteriors are given for the unknown variables under the alternative priors considered here. Sampling these variables iteratively yields a Markov chain that can be used to make inferences from the marginal posteriors of these unknowns.

2.3.1 RR-BLUP

The non-genetic fixed effects, β, appear only in the exp() function of Eq. 6. Thus, dropping all terms that do not involve β_i from this exponent and rearranging, it can be shown that the full-conditional posterior for β_i is univariate normal with mean $\hat{\beta}_i$ and variance $\frac{\sigma_e^2}{c_i}$, where $\hat{\beta}_i = \frac{x'_i R^{-1} w}{c_i}$, $c_i = x'_i R^{-1} x_i$ is the left-hand side of the least-squares equations for β_i, $x'_i R^{-1} w$ is the right-hand side, and

$$w = y - \sum_{j \neq i} x_j \beta_j - Z\alpha$$

is the vector of trait phenotypes corrected for all effects other than that being sampled.

In RR-BLUP, the prior distribution for marker effects conditional on the common variance, σ_a^2, are independent normals with density function:

$$f(\alpha|\sigma_a^2) = \prod_{j=1}^{k} (2\pi\sigma_a^2)^{-\frac{1}{2}} \exp\left[-\frac{\alpha_j^2}{2\sigma_a^2}\right], \quad (13)$$

for $\sigma_a^2 > 0$. Combining this prior with Eq. 6 and deriving the full-conditional for α_i are almost identical to that given previously under the BayesA prior. The only difference is that here $c_i = (z'_i R^{-1} z_i + \frac{\sigma_e^2}{\sigma_a^2})$, because in RR-BLUP all loci are assumed to have a common variance, σ_a^2, for marker effects.

The scaled inverse chi-square prior for σ_a^2 has density function

$$f(\sigma_a^2|\nu_a S_a^2) = \frac{(\nu_a S_a^2/2)^{\nu_a/2}}{\Gamma(\nu_a/2)} (\sigma_a^2)^{-(\nu_a+2)/2} \exp\left[-\frac{\nu_a S_a^2}{2\sigma_a^2}\right], \quad (14)$$

for $\nu_a > 0$ and $S_a^2 > 0$. The unknown variance component, σ_a^2, appears only in its prior given above and in Eq. 13. To obtain the full-conditional for σ_a^2, the exponents of σ_a^2 and the exp() expressions in Eqs. 13 and 14 are combined, and constant factors are dropped, giving

$$f(\sigma_a^2|\text{ELSE}) \propto (\sigma_a^2)^{-(k+\nu_a+2)/2} \exp\left[-\frac{\sum_j \alpha_j^2 + \nu_a S_a^2}{2\sigma_a^2}\right], \quad (15)$$

which can be recognized as the kernel of a scaled inverse chi-square distribution with $\tilde{\nu}_a = k + \nu_a$ degrees of freedom and scale parameter $\tilde{S}_a^2 = \frac{\sum_j \alpha_j^2 + \nu_a S_a^2}{\tilde{\nu}_a}$. A sample from this full-conditional posterior is obtained as $\frac{\tilde{\nu}_a \tilde{S}_a^2}{\chi^2_{\tilde{\nu}_a}}$, where $\chi^2_{\tilde{\nu}_a}$ is a chi-square random variable with $\tilde{\nu}_a$ degrees of freedom.

Finally, to derive the full-conditional of σ_e^2, note that σ_e^2 appears only in its prior given by Eq. 9 and in Eq. 6. Dropping constant factors from the product of these two densities and rearranging, gives

$$f(\sigma_e^2|\text{ELSE}) \propto (\sigma_e^2)^{-(n+\nu_e+2)/2} \exp\left[-\frac{SSE + \nu_e S_e^2}{2\sigma_e^2}\right], \quad (16)$$

which can be recognized as the kernel of a scaled inverse chi-square distribution with $\tilde{\nu}_e = n + \nu_e$ degrees of freedom and scale parameter $\tilde{S}_e^2 = \frac{SSE + \nu_e S_e^2}{\tilde{\nu}_e}$, where

$$SSE = (y - X\beta - Z\alpha)' R^{-1} (y - X\beta - Z\alpha).$$

A sample from this full-conditional posterior is obtained as $\frac{\tilde{\nu}_e \tilde{S}_e^2}{\chi^2_{\tilde{\nu}_e}}$, where $\chi^2_{\tilde{\nu}_e}$ is a chi-square random variable with $\tilde{\nu}_e$ degrees of freedom.

2.3.2 BayesA

A flat prior is used for β in all the models. Thus, the full-conditional for β_i is identical to that given for RR-BLUP. The full-conditional for α_i under the BayesA prior was derived in detail previously.

The locus-specific variance, σ_i^2, appears only in its own prior density in Eq. 8 and in the conditional density of α_i in Eq. 7. Dropping constant factors from the product of these two densities and rearranging, gives

$$f(\sigma_i^2 | \text{ELSE}) \propto (\sigma_i^2)^{-(1+\nu_a+2)/2} \exp\left[-\frac{\alpha_i^2 + \nu_a S_a^2}{2\sigma_i^2}\right], \quad (17)$$

which can be recognized as the kernel of a scaled inverse chi-square distribution with $\tilde{\nu}_a = 1 + \nu_a$ degrees of freedom and scale parameter $\tilde{S}_a^2 = \frac{\alpha_i^2 + \nu_a S_a^2}{\tilde{\nu}_a}$. A sample from this full-conditional posterior is obtained as $\frac{\tilde{\nu}_a \tilde{S}_a^2}{\chi^2_{\tilde{\nu}_a}}$, where $\chi^2_{\tilde{\nu}_a}$ is a chi-square random variable with $\tilde{\nu}_a$ degrees of freedom.

Again, as the prior for σ_e^2 is the same in all models, the full-conditional posterior for σ_e^2 is identical to that derived for RR-BLUP.

2.3.3 BayesB

The full-conditionals for β and for σ_e^2 are as in RR-BLUP. The third form of the BayesB prior, which has three unknowns, ζ_i, δ_i, and σ_i^2, per locus, will be used here. The Gibbs sampler will be used to draw samples for these unknowns (*see* **Note 6**). Our interest is in the effects of the markers, α_i, but this can be written as $\alpha_i = \zeta_i \delta_i$ in terms of the sampled values of ζ_i and δ_i.

When $\delta_i = 1$, $\zeta_i = \alpha_i$ and the full-conditional for ζ_i is identical to that derived for α_i under BayesA. When $\delta_i = 0$, ζ_i appears only in its own prior density, and thus, the prior becomes the full-conditional posterior.

The Bernoulli variable δ_i appears only in Eq. 6 and in its own prior, which is

$$\Pr(\delta_i) = (1 - \pi)^{\delta_i} + \pi^{(1-\delta_i)}. \quad (18)$$

Thus, the full-conditional posterior for δ_i is

$$\Pr(\delta_i = 1 | \text{ELSE}) = \frac{h_1(1 - \pi)}{h_0 \pi + h_1(1 - \pi)}, \quad (19)$$

where h_1 is Eq. 6 evaluated with $\delta_i = 1$, and h_0 is Eq. 6 evaluated with $\delta_i = 0$.

The variance, σ_i^2, of ζ_i appears only in the normal density of ζ_i and in its own prior. Thus, dropping constant factors from the product of these two densities and rearranging gives

$$f(\sigma_i^2|\text{ELSE}) \propto (\sigma_i^2)^{-(1+\nu_a+2)/2} \exp\left[-\frac{\zeta_i^2 + \nu_a S_a^2}{2\sigma_i^2}\right], \quad (20)$$

which can be recognized as the kernel of a scaled inverse chi-square distribution with $\tilde{\nu}_a = 1 + \nu_a$ degrees of freedom and scale parameter $\tilde{S}_a^2 = \frac{\zeta_i^2 + \nu_a S_a^2}{\tilde{\nu}_a}$.

2.3.4 BayesC and BayesCπ

The full-conditionals for β and for σ_e^2 are as in RR-BLUP. The Bernoulli variable δ_i will be used to indicate if marker α_i has a normal distribution ($\delta_i = 1$) or if it is null ($\delta_i = 0$).

Here, we will deviate from using the single-site Gibbs sampler and sample δ_i, and α_i jointly, conditional on the remaining variables. The strategy employed is to first sample δ_i from

$$\Pr(\delta_i|\text{ELSE}) = \int f(\alpha_i, \delta_i|\text{ELSE}) d\alpha_i, \quad (21)$$

and then sample α_i from $f(\alpha_i \mid \delta_i, \text{ELSE})$, where ELSE denotes conditioning on the phenotypes and all unknowns other than α_i and δ_i. As these two variables appear only in Eq. 6 and their own priors, Eq. 21 can be written as

$$\Pr(\delta_i = 1|\text{ELSE}) \propto \int f(w|\alpha_i, \sigma_e^2) f(\alpha_i|\sigma_a^2) \Pr(\delta_i = 1) d\alpha_i, \quad (22)$$

where $f(w \mid \alpha_i, \sigma_e^2)$ is the density function for conditional distribution of w given α_i, which is multivariate-normal distribution with

$$E(w|\alpha_i, \sigma_e^2) = z_i \alpha_i, \quad Var(w|\alpha_i, \sigma_e^2) = R\sigma_e^2,$$

and $f(\alpha_i \mid \sigma_a^2)$ is the density function for the conditional distribution of α_i given σ_a^2, which is univariate-normal with $E(\alpha_i) = 0$ and $Var(\alpha_i) = \sigma_a^2$. After integration, Eq. 22 becomes

$$\Pr(\delta_i = 1|\text{ELSE}) \propto f_1(w|\sigma_a^2, \sigma_e^2) \Pr(\delta_i = 1), \quad (23)$$

where $f_1(w \mid \sigma_a^2, \sigma_e^2)$ is the density function for the marginal distribution of w when $\delta_i = 1$, which is still multivariate-normal with

$$E(w|\sigma_a^2, \sigma_e^2) = 0, \quad Var(w|\sigma_a^2, \sigma_e^2) = z_i z_i' \sigma_a^2 + R\sigma_e^2.$$

Similarly,

$$\Pr(\delta_i = 0|\text{ELSE}) \propto f_0(w|\sigma_e^2) \Pr(\delta_i = 0), \quad (24)$$

where $f_0(w \mid \sigma_e^2)$ is the density function for the distribution of w when $\delta_i = 0$, which is multivariate-normal with

$$E(w|\sigma_e^2) = 0, \quad Var(w|\sigma_e^2) = R\sigma_e^2.$$

So, samples can be drawn using

$$\Pr(\delta_i = 1|\text{ELSE})$$
$$= \frac{f_1(w|\sigma_a^2, \sigma_e^2)\Pr(\delta_i = 1)}{f_0(w|\sigma_e^2)\Pr(\delta_i = 0) + f_1(w|\sigma_a^2, \sigma_e^2)\Pr(\delta_i = 1)}. \quad (25)$$

Evaluating the multivariate-normal density $f_1(w \mid \sigma_a^2, \sigma_e^2)$, however, can be computationally intense as it involves computing the determinant and the inverse of the non-diagonal covariance matrix: $Var(w|\sigma_a^2, \sigma_e^2) = z_i z_i' \sigma_a^2 + R\sigma_e^2$. A computationally efficient formula for Eq. 25 can be obtained by observing that the contribution from w to the full-conditional posterior of α_i given by Eq. 12 is through $r_i = z_i' R^{-1} w$. Thus, r_i contains all the information from w about α_i, and so Eq. 25 can be computed efficiently as

$$\Pr(\delta_i = 1|\text{ELSE})$$
$$= \frac{f_1(r_i|\sigma_a^2, \sigma_e^2)\Pr(\delta_i = 1)}{f_0(r_i|\sigma_e^2)\Pr(\delta_i = 0) + f_1(r_i|\sigma_a^2, \sigma_e^2)\Pr(\delta_i = 1)}, \quad (26)$$

where $f_1(r_i \mid \sigma_a^2, \sigma_e^2)$ is the density function of the distribution of r_i when $\delta_i = 1$, which is univariate-normal with

$$E(r_i|\sigma_a^2, \sigma_e^2) = 0, \quad Var(r_i|\sigma_a^2, \sigma_e^2) = (z_i' z_i)^2 \sigma_a^2 + (z_i' z_i)\sigma_e^2,$$

and $f_0(r_i \mid \sigma_e^2)$ is the density function of the distribution of r_i when $\delta_i = 0$, which is univariate-normal with

$$E(r_i|\sigma_e^2) = 0, \quad Var(r_i|\sigma_e^2) = (z_i' z_i)\sigma_e^2.$$

When $\delta_i = 1$, conditional on σ_a^2, α_i has a normal prior distribution. So, its full-conditional distribution is identical to that under RR-BLUP. When $\delta_i = 0$, the prior for $\alpha_i = 0$ is a point mass at 0. Thus in this case, the sampled value is set to 0.

The common marker effect variance, σ_a^2, appears only in the normal density functions of α_j for which $\delta_j = 1$ and its own prior. Dropping constant factors from the product of these densities and rearranging gives

$$f(\sigma_a^2|\text{ELSE}) \propto (\sigma_a^2)^{-(m+\nu_a+2)} \exp\left[-\frac{\sum_{j,\delta_j=1} \alpha_j^2 + \nu_a S_a^2}{2\sigma_a^2}\right], \quad (27)$$

where $m = \sum_j \delta_j$ is the number of markers included in the model. This expression 27 can be recognized as the kernel of a scaled inverse chi-square distribution with $\tilde{\nu}_\alpha = m + \nu_\alpha$ degrees of freedom and scale parameter $\tilde{S}_\alpha^2 = \frac{\sum_{j,\delta_j=1} \alpha_j^2 + \nu_\alpha S_\alpha^2}{\tilde{\nu}_\alpha}$.

The probability π that a marker has a null effect appears only in

$$\Pr(\delta|\pi) = \pi^{(k-m)}(1-\pi)^m, \tag{28}$$

which is the kernel of a beta distribution with shape parameters $a = k - m + 1$ and $b = m + 1$. Thus, in BayesCπ, where π is treated as an unknown with a uniform prior, it can be sampled from a beta distribution with shape parameters $a = k - m + 1$ and $b = m + 1$.

2.4 Inference of Association

In the Bayesian multiple-regression models that are used for genomic selection, hundreds of thousands of SNP markers are fitted simultaneously in the model. Further, it is expected that markers that are close to each other are highly correlated. Thus, any single marker may not show a strong association with the trait. The markers in a genomic window, however, are expected to jointly capture most of the variability at a nearby trait locus. Therefore, in these multiple-regression models, inference of associations is based on genomic windows rather than on single markers [4–6, 27].

Here, we will show how to construct the posterior distribution for the proportion of the genetic variance explained by the markers in a genomic window relative to the total genetic variance. To do so, the component of the genotypic value that is attributed to a genomic window w is defined as

$$g_w = z_w \alpha_w, \tag{29}$$

where z_w is a matrix with the genotypic covariates of markers in window w, and α_w is the vector of effects for these markers. Then, g_w is the vector of the genotypic values that correspond to the genomic window w. The variance among the elements of g_w computed as

$$\tilde{\sigma}_{g_w}^2 = \left(\sum_{j=1}^n g_{w_j}^2\right)/n - \left[\left(\sum_{j=1}^n g_{w_j}\right)/n\right]^2 \tag{30}$$

is defined as the genetic variance explained by markers in the genomic window w. Similarly, the total genetic variance is computed as

$$\tilde{\sigma}_g^2 = \left(\sum_{j=1}^n g_j^2\right)/n - \left[\left(\sum_{j=1}^n g_j\right)/n\right]^2, \tag{31}$$

where the vector of genotypic values is defined as

$$g = Z\alpha. \tag{32}$$

Now, the proportion of the genetic variance that is explained by markers in window w is defined as

$$q_w = \frac{\tilde{\sigma}^2_{g_w}}{\tilde{\sigma}^2_g}.$$

The posterior distribution for q_w is estimated by saving q_w computed from every Mth MCMC sample of α.

Suppose we are interested in detecting associations for genomic windows that explains more than 1 % of the total genetic variance. To compute the posterior probability that window w explains more than 1 % of the total genetic variance, one needs to count the number C of samples where $q_w > 0.01$. Then, the required posterior probability is given by the ratio $\frac{C}{N}$, where N is the total number of samples saved. This posterior probability of association of the genomic window w (WPPA) has the following interpretation. In the results that accumulate over E independent experiments, among all windows that have WPPA $= 0.95$, for example, the proportion of windows that have no association will converge to 0.05 as E goes to infinity (*see* **Note 7**). Thus, using a threshold of WPPA $= 0.95$ to declare an association will result in controlling PFP ≤ 0.05. Figure 1 shows WPPA computed from 200 MCMC samples using the R script given in Subheading 5.2.

3 Implementation of GWAS in R

Scripts are given in Subheading 5 (also available for download from the publishers web site together with example genotype and phenotype files) to implement GWAS using the R language [28], which is one of the most widely used languages for statistical computing and graphics. These scripts are given for illustrative purposes only, and they are not suitable for analyses with large numbers of markers and observations. Software suitable for large analyses is available at "http://bigs.ansci.iastate.edu," which is a web-based interface to the GenSel package [29] for analysis of genomic data.

To simplify the R scripts, we have assumed that the residual covariance matrix $R\sigma^2_e = I\sigma^2 e$, and that the vector of non-genetic fixed effects β contains only one element, μ, the overall mean. The Bayesian methods that are used for GWAS here were originally developed for genomic prediction of breeding values [7], where first, marker genotypes and trait phenotypes are used to estimate the marker effects (training), and then the estimated effects are used together with marker genotypes of candidates to predict their breeding values. These predictions are called genomic estimated breeding values (GEBV). If trait phenotypes or breeding values estimated from phenotypic data are available, the GEBV can be correlated with these to evaluate the accuracy of prediction (testing).

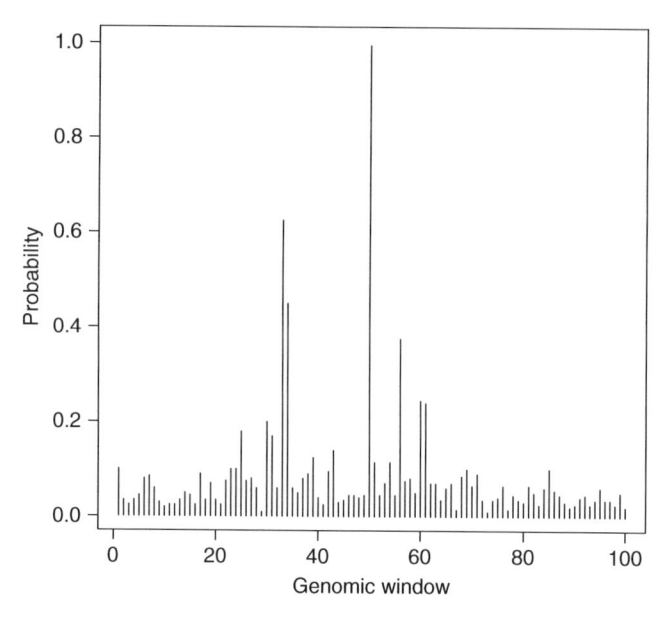

Fig. 1 Posterior probability that genomic window explains more than 1 % of the total genetic variance computed using BayesB with $\pi = 0.98$ for windows of ten consecutive markers on a simulated chromosome of one Morgan in length. The data set consisted of 100 observations, each with a trait phenotype simulated from 100 QTL with normally distributed effects and a normally distributed residual in addition to 1,000 marker genotypes. Random mating was simulated for 1,000 generations in a population of size 100 to generate linkage disequilibrium between loci

In GWAS, a single data set with marker genotypes and trait phenotypes is used. In order to demonstrate the prediction capabilities of these analyses in addition to GWAS, in the scripts given here, two data sets are created—one for training and a second for prediction and testing—from a single genotype file and two phenotype files. The formats of these files are described in the chapter by Garrick and Fernando in this volume.

3.1 RR-BLUP

The input parameters for the analysis are specified in lines 11–18 in script file RR-BLUP.R. These include the length of the MCMC chain, the degrees of freedom for the scaled inverse chi-square priors for marker effect and residual variances, and plausible values for the genetic and residual variances.

Under some assumptions, it can be shown (*see* **Note 2**) that the genetic variance is related to the marker effect variance as

$$\sigma_\alpha^2 = \frac{\sigma_g^2}{\sum_j p_j(1-p_j)}. \tag{33}$$

This relationship is used to get a plausible value for σ_α^2 (line 58 of script) from the input value for σ_g^2. Further, the expected value of a random variable U that has a scaled inverse chi-square distribution with scale parameter S^2 and degrees of freedom ν is

$$E(U) = \frac{\nu S^2}{\nu - 2}, \qquad (34)$$

for $\nu > 2$. So if $\nu > 2$, the scale parameter can be written as

$$S^2 = \frac{E(U)(\nu - 2)}{\nu}. \qquad (35)$$

Thus the scale parameters for the scaled inverse chi-square priors are obtained as

$$S_\alpha^2 = \frac{\sigma_\alpha^2(\nu_\alpha - 2)}{\nu_\alpha}, \qquad (36)$$

and

$$S_e^2 = \frac{\sigma_e^2(\nu_e - 2)}{\nu_e}, \qquad (37)$$

in lines 60 and 61 of the script.

Line 17 of the script specifies the number of consecutive markers in a genomic window, and line 18 specifies that the accuracy of prediction, genetic variance, and window variance should be computed (lines 109–127) every $M = 10$ samples. The names of the genotype and the phenotype files are specified on lines 20–22.

The genotypes are read into a table in line 26. This table contains genotypes for both training and testing animals. The phenotypes for training animals are read into a table in line 27, and the data set for training is obtained by merging these tables by animal ID (line 29). Similarly, the data set for testing is obtained by merging the genotype table with the phenotype table for testing (line 30).

Due to the simplifying assumption that the overall mean is the only non-genetic fixed effect, the x matrix reduces to a column vector of ones, and thus computations involving x will be undertaken without explicitly constructing it. The vector of trait phenotypes and the matrix of genotype covariates for training are extracted in lines 37 and 38. The genotype covariates are transformed to 0, 1, and 2 from $-10, 0, 10$ on line 39. These are repeated for the testing in lines 44–46. In lines 53–54, the genotype covariates are centered. This improves mixing by making the columns of z orthogonal to the column vector of ones in x. In lines 67–75, the scalar, vector, and matrix variables that are used in the calculations are initialized.

At this point, the overall mean, μ, has been set to the mean of the trait phenotypes (line 70) and the vector of marker effects, α, has been set to the null vector (line 67). Thus in line 80, the variable ycorr is the vector of trait phenotypes corrected or adjusted for all location effects in the model.

The loop for MCMC sampling begins at line 83 and ends at line 132, where the residual variance is sampled first on line 86 as described at the end of Subheading 2.3.1. This is followed by sampling of μ and related calculations (lines 89–95). Recall that to sample any fixed effect, y has to be adjusted for all location effects other than that being sampled (beginning of Subheading 2.3.1). At this point, however, y has been adjusted for all effects including μ. Thus, before sampling μ, y has to be "unadjusted" for μ (line 89). The right-hand side for μ can be computed as sum(ycorr) (line 90) and the inverse of the left-hand side as 1.0/nrecords (line 91) because x is a vector of ones and $R = I$. The actual sampling of μ from the full-conditional distribution is on line 93. Following this, y is adjusted for the sampled value of μ in preparation for sampling the next location effect (line 94).

The loop for sampling marker effects and related calculations starts at line 98 and ends at line 108. On line 99, y is unadjusted for the effect of the current locus, leaving y adjusted for all location effects other than that of the current locus. The right-hand side of the mixed model equation corresponding to α_i is computed on line 100 and the left-hand side on lines 101–102. The solution to the mixed model equation is obtained on lines 103–104, and α_i sampled from its full-conditional distribution on line 105. Following this, y is adjusted for the sampled value of α_i on line 106.

Every $M($ outputFrequency$)$ samples: (1) genotypic values are sampled for all animals and the variance among these genotypic values is computed on line 118; (2) the proportion of the variance explained by each genomic window is computed on line 125 as described in Subheading 2.4. Finally, σ_α^2 is sampled from its full-conditional distribution on line 130.

After all samples are drawn, plots are produced for the posterior mean of the variance explained by the genomic windows (lines 151–155) and the posterior probability that proportion of window variance is larger than 1/numberWindows (lines 159–165).

3.2 BayesA and BayesB

The script for BayesA and BayesB is given in the file BayesBL.R. The input parameters for BayesB (lines 14–23) has two extra lines from those for RR-BLUP. The probability π (probFixed) that a marker effect is null is specified on line 17. If bayesType on line 14 is specified as BayesA, π is reset to zero on line 25, resulting in a BayesA analysis.

Data input, sampling of μ, and of σ_e^2 are as in RR-BLUP. The locus-specific variance components of marker effects, for both BayesA and BayesB, are sampled from their full-conditionals on line 109 as described in Subheading 2.3.2. In this script, the vector variable alpha is used to store the sampled values of ζ.

The loop for sampling δ_i and ζ_i for all loci starts at line 114 and ends at line 142. If $\delta_i = 0$, then from Eqs. 3 and 10, y was not adjusted for the sampled value of ζ_i. So, y is unadjusted for the sampled value of $\alpha_i = \zeta_i$ only when $\delta_i = 1$ (lines 115 and 117). Following the `if` statement that starts at line 115 and ends at line 120, the vector `w1` has *y* adjusted for all location effects including ζ_i. On line 123, the value of $\log(h_0\pi)$ is stored in the variable `logDelta0`, and on line 124, the value of $\log(h_1(1-\pi))$ is stored in the variable `logDelta1`. Thus, the expression

```
probDelta1 = 1.0/(1.0 + exp(logDelta0-logDelta1))
```

on line 125 is equivalent in value to Eq. 19 but is numerically more stable. A uniform random variable sampled on line 126 is used to set $\delta_i = 1$ with probability `probDelta` and $\delta_i = 0$ with probability $1 -$ `probDelta` (lines 127, 136 and 140). When $\delta_i = 1$, $\alpha_i = \zeta_i$ is sampled from its full-conditional distribution (lines 129–132) as described in Subheading 2.3 for BayesA. In this case, *y* is adjusted for the sampled value of $\alpha_i = \zeta_i$. On the other hand, when $\delta_i = 0$, ζ_i is sampled from its prior on line 139. In this case, $\alpha_i = 0$ and thus, *y* is not adjusted. Other calculations are as in RR-BLUP.

3.3 BayesC and BayesCπ

The script for BayesC and BayesCπ is given in file BayesCPi.R. The input parameters for the analysis are on lines 12–21. Setting the input parameter `estimatePi` on line 15 to "yes" will result in π being sampled in a BayesCπ analysis. Otherwise, the value specified on line 14 for π (`probFixed`) will be used in a BayesC analysis.

Sampling of μ and of σ_e^2 are as in RR-BLUP. The loop for sampling the Bernoulli variable δ_i and the marker effect α_i starts at line 108 and ends at line 133. The variance of r_i, the right-hand side for α_i, is computed on line 112 with $\delta_i = 0$ and on line 113 with $\delta_i = 1$. These variances are used to compute $\log[f_0(r_i \mid \sigma_e^2) \Pr(\delta_i = 0)]$ on line 114 and $\log[f_1(r_i \mid \sigma_\alpha^2, \sigma_e^2)\Pr(\delta_i = 1)]$ on line 115. These quantities are used on line 116 to compute the $\Pr(\delta_i = 1 \mid \text{ELSE})$, but using a more numerically stable expression than given by Eq. 26 as was also done in Subheading 3.2 for BayesB. This probability is used to sample δ_i from its full-conditional distribution. If the sampled value of $\delta_i = 1$, α_i is sampled as in RR-BLUP (lines 120–123). If the sampled value of $\delta_i = 0$, α_i is set to zero (line 130). The common variance σ_α^2 of marker effects is sampled from its full-conditional distribution on line 136. Next, if `estimatePi="yes"`, π is sampled from its full-conditional posterior distribution on lines 140–142. Other calculations are as in RR-BLUP.

4 Notes

1. PFP, which is an extension of the posterior type I error rate (PER; *see* **Note 8**) [30] to a multiple test setting [10], is closely related to FDR, but they are not identical. Let *V* be the number

of false-positive results in an experiment and let R be the total number of positive results. Then, PFP is defined as

$$\text{PFP} = \frac{E(V)}{E(R)},$$

and FDR is defined as

$$\text{FDR} = E\left(\frac{V}{R}\bigg| R > 0\right)\Pr(R > 0).$$

Storey [31] defined the "positive false discovery rate" (pFDR) as

$$\text{pFDR} = E\left(\frac{V}{R}\bigg| R > 0\right),$$

because it is closer to the type of error control that is desired. Fernando et al. [10] showed that control of PFP across multiple experiments will lead to control of the proportion of false-positive results among all positive results in the long run. They also gave a hypothetical example to show that the other error measures (FDR, pFDR, and GWER) do not necessarily share this property.

2. The variance σ_α^2 of the marker effects is distinct from the genetic variance at a locus [20, 23], but as shown below, under some assumptions it can be related to the total genetic variance.

The genotypic value of a randomly sampled individual i is given by

$$g_i = z_i'\alpha, \qquad (38)$$

where z_i' are its marker covariates. Note that in Eq. 38, α is common to all randomly sampled individuals, and thus, the variance $\sigma_g^2 = \text{Var}(g_i)$ stems entirely from variability in the genotype covariates z_i'. On the other hand, suppose a locus j is randomly sampled from among the k loci. Then, the effect α_j of this locus is a random variable with variance σ_α^2.

To relate σ_α^2 to σ_g^2, we assume the loci are in linkage equilibrium. Then, the additive genetic variance is

$$\sigma_g^2 = \sum_j^k 2p_j(1-p_j)\alpha_j^2, \qquad (39)$$

where p_j is the gene frequency at locus j. Letting $U_j = 2p_j(1-p_j)$ and $V_j = \alpha_j^2$, this variance can be expressed as

$$\sigma_g^2 = \sum_j^k U_j V_j. \qquad (40)$$

Assuming loci are equally likely to be sampled, we use this expression to write the covariance between U_j and V_j as

$$Cov(U_j, V_j) = \frac{\sum_j U_j V_j}{k} - \left(\frac{\sum_j U_j}{k}\right)\left(\frac{\sum_j V_j}{k}\right)$$

$$= \frac{\sigma_g^2}{k} - \left(\frac{\sum_j 2p_j(1-p_j)}{k}\right)\left(\frac{\sum_j \alpha_j^2}{k}\right). \quad (41)$$

If α_j have a mean of zero,

$$\sigma_\alpha^2 = \frac{\sum_j^k \alpha_j^2}{k},$$

and rearranging Eq. 41 gives

$$\sigma_\alpha^2 = \frac{\sigma_g^2 - kCov(U_j, V_j)}{\sum_j 2p_j(1-p_j)}. \quad (42)$$

If the marker effects are independent of the gene frequencies, Eq. 42 becomes

$$\sigma_\alpha^2 = \frac{\sigma_g^2}{\sum_j 2p_j(1-p_j)}, \quad (43)$$

which is the relationship given in Fernando et al. [32]. A similar expression with fewer assumptions is given in [23] but with σ_g^2 in the numerator of Eq. 43 replaced by $E(\sigma_g^2)$, where the expectation is taken over random vectors of α.

3. On pages 60–61 of [15], it is shown that the multivariate-t distribution with scale matrix Σ and ν degrees of freedom is an infinite mixture of multivariate normal distributions with covariance matrices Σw^{-1}, where mixing is through w that has a gamma distribution with a shape parameter of $\nu/2$ and rate parameter $\nu/2$. To see how this result applies to the prior for α used in RR-BLUP, note that: (1) the marginal prior for α in RR-BLUP is an infinite mixture of normals with covariance matrices $I\sigma_\alpha^2$, where mixing is through σ_α^2 that has a scaled inverse chi-square distribution with scale parameter S_α^2 and degrees of freedom ν_α; (2) if w has a gamma distribution with shape parameter a and a rate parameter b, $\frac{1}{w}$ has an inverse gamma distribution with shape parameter a and scale parameter b; and (3) a scaled inverse chi-square distribution with

degrees of freedom ν_α and scale parameter S_α^2 is an inverse gamma with shape parameter $a = \nu_\alpha/2$ and scale parameter $b = \nu_\alpha S_\alpha^2/2$. So, the marginal prior for α in RR-BLUP can be described as an infinite mixture of normals with covariance matrices $\Sigma \left(\frac{1}{\sigma_\alpha^2}\right)^{-1}$, where $\Sigma = I$ and $\frac{1}{\sigma_\alpha^2}$ has a gamma distribution with a shape parameter of $a = \nu_\alpha/2$ and a rate parameter of $b = \nu_\alpha S_\alpha^2/2$. Now, following the calculations on pages 60–61 in [15], this infinite mixture can be shown to be a multivariate-t distribution with scale matrix $S_\alpha^2 I$ and degrees of freedom ν_α.

4. In RR-BLUP, when the variance of marker effects is treated as unknown with a scaled inverse chi-square prior, the posterior mean of the marker effects is not linear in *y*. So, although BLUP stands for the best *linear* unbiased predictor, in this case RR-BLUP is not a linear predictor.

5. A Markov chain is a sequence of random variables, where the distribution at time t conditional on all the previous values depends only on that at time $t - 1$. A Markov chain obtained using the Gibbs sampler as described in Subheading 2.3 has the property that the posterior distribution of the unknown variables in the model is its stationary distribution. This means that if the sample at time t is from the posterior distribution the sample at time $t + 1$ will also be from the same distribution. If in addition to this property the chain is also irreducible (*see* **Note 9**), samples from the chain can be used to make inferences from the posterior distribution [15, 25, 33]. For example, posterior probabilities, means, variances estimated from the chain will converge to their true values as the length of the chain goes to infinity [25, 33]. In practice, however, chains of finite length have to be used for inference, and determining the length of the chain that is sufficient for useful inference is important. One approach is based on examining several chains with widely dispersed starting points [34]. A family of tests based on between and within chain variability is used to test for convergence. It is common to discard a number of samples at the beginning of the chain as "burn-in," but if the chain is sufficiently long this is not necessary.

6. The Gibbs sampler used here for BayesB can be obtained by applying the composite model space described by Godsill [35], which is very similar to the model proposed by Carlin and Chib [36], to the variable selection problem (see Section 2.6 of [35]). In their approach, the prior for parameters that are not in the model is allowed to be different from the prior for parameters included in the model. It can be shown that the prior used for the parameters that are not in the model, called the "pseudo prior," does not affect the posterior. Thus, it can

be chosen to improve mixing without any effect on the posterior. In the BayesB algorithm given here, for simplicity, we have not used a pseudo prior. An alternative approach to improve mixing is to use a blocking Gibbs sampler that samples the Bernoulli variable δ_i and the effect ψ_i jointly as we did for BayesC. This blocking Gibbs sampler is used in GenSel for BayesB.

7. The posterior probabilities computed in a Bayesian analysis will have a frequentist interpretation when priors used in the analysis for unknown variables in the model are identical to those used to generate these variables. The frequency interpretation of probability for some event is the limiting value in E trials of the relative frequency of the event as E goes to infinity. Thus in results from E independent whole genome analyses, among genomic intervals with WPPA $= 0.95$, the proportion of true associations will converge to 0.95 as E tends to infinity.

    In most real applications, however, the priors distributions used in the analysis will not exactly match the distributions used in generating the data. Fortunately, as data size increases, the posterior distributions from an analysis will become increasingly independent of the priors used. Thus, with sufficiently large data sets we can expect that using WPPA $= 0.95$, for example, will result in a PFP ≤ 0.05.

8. In the current context, PER [30] is the conditional probability of a false declaration of association of a genomic locus with the trait given that a declaration of association at that locus has been made. This is different from the usual type I error rate, which is the conditional probability of declaring association at a locus given that the true status of association at that locus is false.

9. A Markov chain of discrete variables is said to be irreducible if it can move from any point i to any other point j in the stationary distribution in a finite number of steps. Although conceptually similar, the definition of irreducibility for the continuous case is more technical, but chains of continuous variables obtained from most Gibbs samplers are irreducible and can be used for inference [33]. On the other hand, irreducibility is a bigger problem for chains of discrete variables such as multi-locus marker genotypes [37], and constructing irreducible chains for such variables can be more challenging [38, 39].

5 The R Scripts

5.1 RR-BLUP

```
# This code is for illustrative purposes and not efficient for large problems
# Real life data analysis (using the same file formats) is available at
# bigs.ansci.iastate.edu/login.html based on GenSel cpp software implementation
#
#         Rohan Fernando    (rohan@iastate.edu)
#         Dorian Garrick    (dorian@iastate.edu)
#         copyright August 2012

setwd("/Users/rohan/latex/papers/GWASBookChapter/R")  # change this to point to your working directory
# Parameters
seed              = 10    # set the seed for the random number generator
chainLength       = 200   # number of iterations
dfEffectVar       = 4     # hyper parameter (degrees of freedom) for locus effect variance
nuRes             = 4     # hyper parameter (degrees of freedom) for residual variance
varGenotypic      = 1     # used to derive hyper parameter (scale) for locus effect variance
varResidual       = 1     # used to derive hyper parameter (scale) for residual variance
windowSize        = 10    # number of consecutive markers in a genomic window
outputFrequency   = 10    # frequency for reporting performance and for computing genetic variances

markerFileName         = "genotypes.dat"
trainPhenotypeFileName = "trainPhenotypes.dat"
testPhenotypeFileName  = "testPhenotypes.dat"

set.seed(seed)

genotypeFile       = read.table(markerFileName, header=TRUE)              # this is not efficient for large files!
trainPhenotypeFile = read.table(trainPhenotypeFileName, skip=1)[,1:2]     # skip the header as R dislikes # character
testPhenotypeFile  = read.table(testPhenotypeFileName,  skip=1)[,1:2]     # skip the header as R dislikes # character
commonTrainingData = merge(trainPhenotypeFile, genotypeFile, by.x=1, by.y=1)  # Only use animals with genotype and phenotype
commonTestData     = merge(testPhenotypeFile,  genotypeFile, by.x=1, by.y=1)  # Only use animals with genotype and phenotype

remove(genotypeFile)                # Free up space
remove(trainPhenotypeFile)          # Free up space
remove(testPhenotypeFile)           # Free up space
animalID = unname(as.matrix(commonTrainingData[,1]))     # First field is animal identifier
y        = commonTrainingData[, 2]                        # Second field is trait values
Z        = commonTrainingData[, 3: ncol(commonTrainingData)]   # Remaining fields are GenSel-coded genotypes
Z        = unname(as.matrix((Z + 10)/10));                # Recode genotypes to 0, 1, 2 (number of B alleles)
```

```
markerID = colnames(commonTrainingData)[3:ncol(commonTrainingData)]  # Remember the marker locus identifiers
remove(commonTrainingData)

testID  = unname(as.matrix(commonTestData[, 1]))                     # First field is animal identifier
yTest   = commonTestData[, 2]                                        # Second field is trait values
ZTest   = commonTestData[, 3: ncol(commonTestData)]                  # Remaining fields are GenSel-coded genotypes
ZTest   = unname(as.matrix((ZTest + 10)/10));                        # Recode genotypes to 0, 1, 2 (number of B alleles)
remove(commonTestData)

nmarkers = ncol(Z)                                                   # number of markers
nrecords = nrow(Z)                                                   # number of animals

# center the genotype matrix to accelerate mixing
markerMeans = colMeans(Z)                                            # compute the mean for each marker
Z = t(t(Z) - markerMeans)                                            # deviate covariate from its mean
p = markerMeans/2.0                                                  # compute frequency B allele for each marker
mean2pq = mean(2*p*(1-p))                                            # compute mean genotype variance

varEffects = varGenotypic/(nmarkers*mean2pq)                         # variance of locus effects is computed from genetic variance
                                                                     #(e.g. Fernando et al., Acta Agriculturae Scand Section A, 2007;57:192-195)
scaleVar  = varEffects*(dfEffectVar-2)/dfEffectVar;                  # scale factor for locus effects
scaleRes  = varResidual*(nuRes-2)/nuRes                              # scale factor for residual variance

numberWindows = nmarkers/windowSize                                  # number of genomic windows
numberSamples = chainLength/outputFrequency                          # number of samples of genetic variances

alpha         = array(0.0, nmarkers)        # reserve a vector to store sampled locus effects
meanAlpha     = array(0.0, nmarkers)        # reserve a vector to accumulate the posterior mean of locus effects
modelFreq     = array(0.0, nmarkers)        # reserve a vector to store model frequency
mu            = mean(y)                     # starting value for the location parameter
meanMu        = 0                           # reserve a scalar to accumulate the posterior mean
geneticVar    = array(0,numberSamples)      # reserve a vector to store sampled genetic variances
                                            # reserve a matrix to store sampled proportion of variance due to window
windowVarProp = matrix(0,nrow=numberSamples,ncol=numberWindows)
sampleCount   = 0                           # initialize counter for number of samples of genetic variances
```

```
# adjust y for the fixed effect (ie location parameter)
ycorr = y - mu
# mcmc sampling
for (iter in 1:chainLength){

# sample residual variance
    vare = ( t(ycorr)%*%ycorr + nuRes*scaleRes )/rchisq(1,nrecords + nuRes)

# sample intercept
    ycorr  = ycorr + mu                         # Unadjust y for the previous sample of mu
    rhs    = sum(ycorr)                         # Form X'y
    invLhs = 1.0/(nrecords)                     # Form (X'X)-1
    mean   = rhs*invLhs                         # Solve (X'X) mu = X'y
    mu     = rnorm(1,mean,sqrt(invLhs*vare))    # Sample new location parameter
    ycorr  = ycorr - mu                         # Adjust y for the new sample of mu
    meanMu = meanMu + mu                        # Accumulate the sum to compute posterior mean

# sample effect for each locus
    for (locus in 1:nmarkers){
        ycorr = ycorr + Z[,locus]*alpha[locus]
        rhs   = t(Z[,locus])%*%ycorr                   # phenotypes are adjusted for all but this locus
        zpz   = t(Z[,locus])%*%Z[,locus]               # rhs of MME adjusted for all but this locus
        mmeLhs = zpz + vare/varEffects                 # OLS component of MME for this locus
        invLhs = 1.0/mmeLhs                            # Form the coefficient matrix of MME
        mean   = invLhs*rhs                            # Invert the coefficient matrix
        alpha[locus]= rnorm(1,mean,sqrt(invLhs*vare))  # Solve the MME for locus effect
        ycorr = ycorr - Z[,locus]*alpha[locus];        # Sample the locus effect from data
        meanAlpha[locus] = meanAlpha[locus] + alpha[locus];  # Adjust the data for this locus effect
    }                                                  # Accumulate the sum for posterior mean

    if(iter %% outputFrequency == 0){

        cat (sprintf("iteration = %5d ",iter))
        aHatTest = ZTest %*% meanAlpha/iter            # compute genomic breeding values in test data
        corr = cor(aHatTest,yTest)
        regr = corr*sqrt(var(yTest)/var(aHatTest))     # regress yTest on aHatTest
        RSquared = corr*corr
        cat(sprintf(" Corr = %8.5f Regr = %8.5f R2 = %8.5f \n", corr, regr, RSquared,"\n"))

        sampleCount = sampleCount + 1
```

```
        geneticVar[sampleCount] = var(Z%*%alpha)                              # variance of genotypic values
        wEnd = 0
        for (window in 1:numberWindows){
            wStart = wEnd + 1                                                 # start of current window
            wEnd   = wEnd + windowSize                                        # end of current window
            if (wEnd > nmarkers) wEnd = nmarkers                              # last window may be smaller than windowSize
            windowVarProp[sampleCount,window] = var(Z[,wStart:wEnd]%*%alpha[wStart:wEnd])/geneticVar[sampleCount]
                                                                              # proportion of variance due to this window
        }
    }
    # sample the common locus effect variance
    varEffects = ( scaleVar*dfEffectVar + sum(alpha^2) )/rchisq(1,dfEffectVar+nmarkers)
}

meanMu      = meanMu/chainLength
meanAlpha   = meanAlpha/chainLength;
modelFreq   = modelFreq/chainLength;

cat(" Genomic Prediction in Validation (test) Animals \n")
nTestAnimal=nrow(ZTest)
cat(" Animal            GBV \n")
for (animals in 1:nTestAnimal){
    cat(sprintf("%15s %15.7f \n",testID[animals],aHatTest[animals]))
}

cat(" GWAS locus effect results\n")
cat(" Marker      Effect         modelFreq\n")
for (locus in 1:nmarkers){
    cat(sprintf("%15s %15.7f %10.7f \n", markerID[locus],meanAlpha[locus],modelFreq[locus]))
}

cat(" GWAS genomic window results")
plot(colMeans(windowVarProp),
     ylab="Proportion",
     xlab="Genomic window",
     main ="Proportion of variance explained by genomic window"
)
cutoff = 1/numberWindows
```

```
quartz()
# posterior probability that window explains more than 1/numberWindows proportion of genetic variance
posteriorProb = colSums(windowVarProp>cutoff)/numberSamples
plot(posteriorProb,
     xlab="Genomic window",
     type="h",
     ylab="Probability",
     main=sprintf("Posterior probability that proportion of window variance > %5.3f", cutoff)
)
```

5.2 BayesA and BayesB

```
# Gibbs sampler for BayesB (Long Qu and Kadir Kizilkaya)
# Godsill (2001) Journal of Computational and Graphical Statistics, Volume 10, Number 2, Pages 1-19
#
# This code is for illustrative purposes and not efficient for large problems
# Real life data analysis (using the same file formats) is available at
# bigs.ansci.iastate.edu/login.html based on GenSel cpp software implementation
#
#                    Rohan Fernando    (rohan@iastate.edu)
#                    Dorian Garrick    (dorian@iastate.edu)
#                    copyright August 2012

setwd("/Users/rohan/latex/papers/GWASBookChapter/R") # change this to point to your working directory
# Parameters
bayesType       = "BayesB"# "BayesA" or "BayesB"
seed            = 10      # set the seed for the random number generator
chainLength     = 2000    # number of iterations
probFixed       = 0.98    # parameter "pi" the probability SNP effect is zero
dfEffectVar     = 4       # hyper parameter (degrees of freedom) for locus effect variance
nuRes           = 4       # hyper parameter (degrees of freedom) for residual variance
varGenotypic    = 1       # used to derive hyper parameter (scale) for locus effect variance
varResidual     = 1       # used to derive hyper parameter (scale) for residual variance
windowSize      = 10      # number of consecutive markers in a genomic window
outputFrequency = 10      # frequency for reporting performance and for computing genetic variances

if(bayesType != "BayesB") probFixed = 0 # BayesA is BayesB with probFixed=0
markerFileName    = "genotypes.dat"
```

```r
trainPhenotypeFileName = "trainPhenotypes.dat"
testPhenotypeFileName  = "testPhenotypes.dat"

set.seed(seed)

genotypeFile        = read.table(markerFileName, header=TRUE)              # this is not efficient for large files!
trainPhenotypeFile  = read.table(trainPhenotypeFileName, skip=1)[,1:2]     # skip the header as R dislikes # character
testPhenotypeFile   = read.table(testPhenotypeFileName, skip=1)[,1:2]      # skip the header as R dislikes # character
commonTrainingData  = merge(trainPhenotypeFile, genotypeFile, by.x=1, by.y=1) # Only use animals with genotype and phenotype
commonTestData      = merge(testPhenotypeFile,  genotypeFile, by.x=1, by.y=1) # Only use animals with genotype and phenotype

remove(genotypeFile)                                                       # Free up space
remove(trainPhenotypeFile)                                                 # Free up space
remove(testPhenotypeFile)                                                  # Free up space
animalID = unname(as.matrix(commonTrainingData[,1]))                       # First field is animal identifier
y        = commonTrainingData[, 2]                                         # Second field is trait values
Z        = commonTrainingData[, 3: ncol(commonTrainingData)]               # Remaining fields are GenSel-coded genotypes
Z        = unname(as.matrix((Z + 10)/10));                                 # Recode genotypes to 0, 1, 2 (number of B alleles)
markerID = colnames(commonTrainingData)[3:ncol(commonTrainingData)]        # Remember the marker locus identifiers
remove(commonTrainingData)

testID = unname(as.matrix(commonTestData[,1]))                             # First field is animal identifier
yTest  = commonTestData[, 2]                                               # Second field is trait values
ZTest  = commonTestData[, 3: ncol(commonTestData)]                         # Remaining fields are GenSel-coded genotypes
ZTest  = unname(as.matrix((ZTest + 10)/10));                               # Recode genotypes to 0, 1, 2 (number of B alleles)
remove(commonTestData)

nmarkers = ncol(Z)                                                         # number of markers
nrecords = nrow(Z)                                                         # number of animals

# center the genotype matrix to accelerate mixing
markerMeans = colMeans(Z)                                                  # compute the mean for each marker
Z = t(t(Z) - markerMeans)                                                  # deviate covariate from its mean
p = markerMeans/2.0                                                        # compute frequency B allele for each marker
mean2pq = mean(2*p*(1-p))                                                  # compute mean genotype variance
# variance of locus effects is computed from genetic variance
#(e.g. Fernando et al., Acta Agriculturae Scand Section A,2007;57:192-195)
varEffects = varGenotypic/(nmarkers*(1-probFixed)*mean2pq)
```

```
scaleVar      = varEffects*(dfEffectVar-2)/dfEffectVar;     # scale factor for locus effects
scaleRes      = varResidual*(nuRes-2)/nuRes                 # scale factor for residual variance
logPi         = log(probFixed)                              # compute these once since probFixed does not change
logPiComp     = log(1-probFixed)

numberWindows = nmarkers/windowSize                         # number of genomic windows
numberSamples = chainLength/outputFrequency                 # number of samples of genetic variances

alpha         = array(0.0, nmarkers)        # reserve a vector to store sampled locus effects
meanAlpha     = array(0.0, nmarkers)        # reserve a vector to accumulate the posterior mean of locus effects
modelFreq     = array(0.0, nmarkers)        # reserve a vector to store model frequency
mu            = mean(y)                     # starting value for the location parameter
meanMu        = 0                           # reserve a scalar to accumulate the posterior mean
geneticVar    = array(0,numberSamples)      # reserve a vector to store sampled genetic variances
windowVarProp = matrix(0,nrow=numberSamples,ncol=numberWindows)
                                            # reserve a matrix to store sampled proportion proportion of variance due to window
sampleCount   = 0                           # initialize counter for number of samples of genetic variances

locusEffectVar = array(varEffects,nmarkers) # vector containing the starting locus effect variances
delta          = array(1,    nmarkers);     # vector to indicate locus effect in model or not

# adjust y for the fixed effect (ie location parameter)
ycorr = y - mu

# mcmc sampling
for (iter in 1:chainLength){

# sample residual variance
    vare = ( t(ycorr)%*%ycorr + nuRes*scaleRes )/rchisq(1,nrecords + nuRes)

# sample intercept
    ycorr  = ycorr + mu                           # Unadjust y for the previous sample of mu
    rhs    = sum(ycorr)                           # Form X'y
    invLhs = 1.0/(nrecords)                       # Form (X'X)-1
    mean   = rhs*invLhs                           # Solve (X'X) mu = X'y
    mu     = rnorm(1,mean,sqrt(invLhs*vare))      # Sample new location parameter
    ycorr  = ycorr - mu                           # Adjust y for the new sample of mu
    meanMu = meanMu + mu                          # Accumulate the sum to compute posterior mean
```

```
# sample locus effect variance for each locus
    locusEffectVar = (scaleVar*dfEffectVar+alpha*alpha)/rchisq(nmarkers,dfEffectVar+1)

# sample delta and then the effect for each locus
    nLoci = 0;                                              # Counter for number of loci fitted this iteration
    for (locus in 1:nmarkers){
        if(delta[locus]){
            w1 = ycorr                                      # at this point, phenotypes are adjusted for this and all fitted loci
            ycorr = ycorr + Z[,locus]*alpha[locus]          # now phenotypes are adjusted for all but this locus
        } else {
            w1 = ycorr - Z[,locus]*alpha[locus]             # phenotypes are adjusted for last sample if not accepted
        }
        rhs  = t(Z[,locus])%*%ycorr                         # rhs of MME adjusted for all but this locus
        zpz  = t(Z[,locus])%*%Z[,locus]                     # OLS component of MME for this locus
        logDelta0 = -0.5*sum(ycorr*ycorr)/vare     + logPi  # This locus not fitted
        logDelta1 = -0.5*sum(w1*w1)/vare           + logPiComp # This locus fitted
        probDelta1 = 1.0/(1.0 + exp(logDelta0-logDelta1))   # near 0 if locus poor, 1 if locus very good
        u = runif(1)                                        # Sample uniform
        if(u < probDelta1) {                                # set delta = 1 with probability probDelta1
            nLoci = nLoci + 1                               # Increment a counter for loci fitted this iteration
            mmeLhs = zpz + vare/locusEffectVar[locus]       # Form the coefficient matrix of MME
            invLhs = 1.0/mmeLhs                             # Invert the coefficient matrix
            mean = invLhs*rhs                               # Solve the MME for locus effect
            alpha[locus]= rnorm(1,mean,sqrt(invLhs*vare))   # Sample the locus effect from data
            ycorr = ycorr - Z[,locus]*alpha[locus]          # Adjust the data for this locus effect
            meanAlpha[locus] = meanAlpha[locus] + alpha[locus] # Accumulate the sum for posterior mean
            modelFreq[locus] = modelFreq[locus] + 1         # Accumulate counter for acceptance of locus
            delta[locus] = 1                                # record that this locus was fitted
        }
        else {
            alpha[locus] = rnorm(1,0.0,sqrt(locusEffectVar[locus])); # Sample the locus effect from prior
            delta[locus] = 0                                # record that this locus not fitted
        }
    }
    if(iter %% outputFrequency == 0){
        cat (sprintf("iteration = %5d number of loci in model = %5d ",iter, nLoci))
        aHatTest = ZTest %*% meanAlpha/iter                 # compute genomic breeding values in test data
        corr = cor(aHatTest,yTest)
        regr = corr*sqrt(var(yTest)/var(aHatTest))          # regress yTest on aHatTest
```

```
             RSquared = corr*corr
             cat(sprintf(" Corr = %8.5f Regr = %8.5f R2 = %8.5f \n", corr, regr, RSquared,"\n"))
             sampleCount = sampleCount + 1
             geneticVar[sampleCount] = var(Z%*%(alpha*delta))                # variance of genotypic values
             wEnd = 0
             for (window in 1:numberWindows){
                 wStart = wEnd + 1                                            # start of current window
                 wEnd   = wEnd + windowSize                                   # end of current window
                 if (wEnd > nmarkers) wEnd = nmarkers                         # last window may be smaller than windowSize
                 windowVarProp[sampleCount,window] =                          # proportion of variance due to this window
                     var(Z[,wStart:wEnd]%*%(alpha[wStart:wEnd]*delta[wStart:wEnd]))/geneticVar[sampleCount]
             }
         }

meanMu      = meanMu/chainLength
meanAlpha   = meanAlpha/chainLength;
modelFreq   = modelFreq/chainLength;

cat(" Genomic Prediction in Validation (test) Animals \n")
nTestAnimal=nrow(ZTest)
cat(" Animal           GBV \n")
for (animals in 1:nTestAnimal){
    cat(sprintf("%15s %15.7f \n",testID[animals],aHatTest[animals]))
}

cat(" GWAS locus effect results\n")
cat(" Marker       Effect        modelFreq\n")
for (locus in 1:nmarkers){
    cat(sprintf("%15s %15.7f %10.7f \n", markerID[locus],meanAlpha[locus],modelFreq[locus]))
}

cat(" GWAS genomic window results")
plot(colMeans(windowVarProp),
     ylab="Proportion",
     xlab="Genomic window",
     main ="Proportion of variance explained by genomic window"
    )
cutoff = 1/numberWindows
```

```
quartz()
# posterior probability that window explains more than 1/numberWindows proportion of genetic variance
posteriorProb = colSums(windowVarProp>cutoff)/numberSamples
plot(posteriorProb,
     xlab="Genomic window",
     type="h",
     ylab="Probability",
     main=sprintf("Posterior probability that proportion of window variance > %5.3f", cutoff))
```

5.3 BayesC and BayesCπ

```
#   This code is for illustrative purposes and not efficient for large problems
#   Real life data analysis (using the same file formats) is available at
#   bigs.ansci.iastate.edu/login.html based on GenSel cpp software implementation
#
#                    Rohan Fernando    (rohan@iastate.edu)
#                    Dorian Garrick    (dorian@iastate.edu)
#                    copyright August 2012

setwd("/Users/rohan/latex/papers/GWASBookChapter/R")  # change this to point to your working directory
# Parameters
seed             = 10       # set the seed for the random number generator
chainLength      = 2000     # number of iterations
probFixed        = 0.95     # parameter "pi" the probability SNP effect is zero
estimatePi       = "yes"    # "yes" or "no"
dfEffectVar      = 4        # hyper parameter (degrees of freedom) for locus effect variance
nuRes            = 4        # hyper parameter (degrees of freedom) for residual variance
varGenotypic     = 1        # used to derive hyper parameter (scale) for locus effect variance
varResidual      = 1        # used to derive hyper parameter (scale) for residual variance
windowSize       = 10       # number of consecutive markers in a genomic window
outputFrequency  = 10       # frequency for reporting performance and for computing genetic variances

markerFileName        = "genotypes.dat"
trainPhenotypeFileName = "trainPhenotypes.dat"
testPhenotypeFileName  = "testPhenotypes.dat"

set.seed(seed)
```

```
genotypeFile       = read.table(markerFileName, header=TRUE)           # this is not efficient for large files!
trainPhenotypeFile = read.table(trainPhenotypeFileName, skip=1)[,1:2]  # skip the header as R dislikes # character
testPhenotypeFile  = read.table(testPhenotypeFileName,  skip=1)[,1:2]  # skip the header as R dislikes # character
commonTrainingData = merge(trainPhenotypeFile, genotypeFile, by.x=1, by.y=1) # Only use animals with genotype and phenotype
commonTestData     = merge(testPhenotypeFile,  genotypeFile, by.x=1, by.y=1) # Only use animals with genotype and phenotype

remove(genotypeFile)                                                   # Free up space
remove(trainPhenotypeFile)                                             # Free up space
remove(testPhenotypeFile)                                              # Free up space
animalID = unname(as.matrix(commonTrainingData[,1]))                   # First field is animal identifier
y        = commonTrainingData[, 2]                                     # Second field is trait values
Z        = commonTrainingData[, 3: ncol(commonTrainingData)]           # Remaining fields are GenSel-coded genotypes
Z        = unname(as.matrix((Z + 10)/10));                             # Recode genotypes to 0, 1, 2 (number of B alleles)
markerID = colnames(commonTrainingData)[3:ncol(commonTrainingData)]    # Remember the marker locus identifiers
remove(commonTrainingData)

testID = unname(as.matrix(commonTestData[,1]))                         # First field is animal identifier
yTest  = commonTestData[, 2]                                           # Second field is trait values
ZTest  = commonTestData[, 3: ncol(commonTestData)]                     # Remaining fields are GenSel-coded genotypes
ZTest  = unname(as.matrix((ZTest + 10)/10));                           # Recode genotypes to 0, 1, 2 (number of B alleles)
remove(commonTestData)

nmarkers = ncol(Z)                                                     # number of markers
nrecords = nrow(Z)                                                     # number of animals

# center the genotype matrix to accelerate mixing

markerMeans = colMeans(Z)                                              # compute the mean for each marker
Z = t(t(Z) - markerMeans)                                              # deviate covariate from its mean
p = markerMeans/2.0                                                    # compute frequency B allele for each marker
mean2pq = mean(2*p*(1-p))                                              # compute mean genotype variance

varEffects = varGenotypic/(nmarkers*(1-probFixed)*mean2pq)             # variance of locus effects is computed from genetic variance
                                                                       # (e.g. Fernando et al., Acta Agriculturae Scand
                                                                       # Section A, 2007; 57: 192-195)
scaleVar = varEffects*(dfEffectVar-2)/dfEffectVar;                     # scale factor for locus effects
scaleRes = varResidual*(nuRes-2)/nuRes                                 # scale factor for residual variance
logPi    = log(probFixed)                                              # compute these once since probFixed does not change
logPiComp = log(1-probFixed)
```

```
numberWindows  = nmarkers/windowSize                              # number of genomic windows
numberSamples  = chainLength/outputFrequency                      # number of samples of genetic variances

alpha          = array(0.0, nmarkers)                             # reserve a vector to store sampled locus effects
meanAlpha      = array(0.0, nmarkers)                             # reserve a vector to accumulate the posterior mean of locus effects
modelFreq      = array(0.0, nmarkers)                             # reserve a vector to store model frequency
mu             = mean(y)                                          # starting value for the location parameter
meanMu         = 0                                                # reserve a scalar to accumulate the posterior mean
geneticVar     = array(0,numberSamples)                           # reserve a vector to store sampled genetic variances
windowVarProp  = matrix(0,nrow=numberSamples,ncol=numberWindows)  # reserve a matrix to store sampled proportion of variance due to window

sampleCount    = 0                                                # initialize counter for number of samples of genetic variances
piMean         = 0                                                # initialize scalar to accumulate the sum of Pi

delta          = array(1,  nmarkers);     # vector to indicate locus effect in model or not

# adjust y for the fixed effect (ie location parameter)
ycorr = y - mu

# mcmc sampling

for (iter in 1:chainLength){
# sample residual variance
        vare = ( t(ycorr)%*%ycorr + nuRes*scaleRes )/rchisq(1,nrecords + nuRes)

# sample intercept
        ycorr   = ycorr + mu                                      # Unadjust y for the previous sample of mu
        rhs     = sum(ycorr)                                      # Form X'y
        invLhs  = 1.0/(nrecords)                                  # Form (X'X)-1
        mean    = rhs*invLhs                                      # Solve (X'X) mu = X'y
        mu      = rnorm(1,mean,sqrt(invLhs*vare))                 # Sample new location parameter
        ycorr   = ycorr - mu                                      # Adjust y for the new sample of mu
        meanMu  = meanMu + mu                                     # Accumulate the sum to compute posterior mean

# sample delta and then the effect for each locus
        nLoci = 0                                                 # Counter for number of loci fitted this iteration
```

```
for (locus in 1:nmarkers){
    ycorr = ycorr + Z[,locus]*alpha[locus]    # phenotypes are adjusted for all but this locus
    rhs   = t(Z[,locus])%*%ycorr               # rhs of MME adjusted for all but this locus
    zpz   = t(Z[,locus])%*%Z[,locus]           # OLS component of MME for this locus
    v0    = zpz*vare                           # Var(rhs|delta=0)
    v1    = (zpz^2*varEffects + zpz*vare)      # Var(rhs|delta=1)
    logDelta0 = -0.5*(log(v0) + rhs^2/v0) + logPi       # this locus not fitted
    logDelta1 = -0.5*(log(v1) + rhs^2/v1) + logPiComp   # this locus fitted
    probDelta1 = 1.0/(1.0 + exp(logDelta0-logDelta1))   # near 0 if locus poor, 1 if locus very good
    u = runif(1)                               # Sample uniform
    if(u < probDelta1) {                       # Accept the sample with Pr(delta=1|ELSE)
        nLoci = nLoci + 1                      # Increment a counter for loci fitted this iteration
        mmeLhs = zpz + vare/varEffects         # Form the coefficient matrix of MME
        invLhs = 1.0/mmeLhs                    # Invert the coefficient matrix
        mean = invLhs*rhs                      # Solve the MME for locus effect
        alpha[locus]= rnorm(1,mean,sqrt(invLhs*vare))  # Sample the locus effect from data
        ycorr = ycorr - Z[,locus]*alpha[locus];        # Adjust the data for this locus effect
        meanAlpha[locus] = meanAlpha[locus] + alpha[locus]  # Accumulate the sum for posterior mean
        modelFreq[locus] = modelFreq[locus] + 1        # Accumulate counter for acceptance of locus
        delta[locus] = 1                       # record that this locus was fitted
    }
    else {
        alpha[locus] = 0                       # Sample the locus effect from prior
        delta[locus] = 0                       # record that this locus not fitted
    }
}
# sample common locus effect variance
varEffects = (scaleVar*dfEffectVar + sum(alpha^2))/rchisq(1,dfEffectVar+nLoci)

if (estimatePi=="yes"){
    # sample Pi

    aa = nmarkers-nLoci + 1
    bb = nLoci + 1
    pi = rbeta(1, aa, bb)
    piMean = piMean + pi                       # sample pi from full-conditional
    logPi     = log(pi)                        # Accumulate sum for posterior mean of pi
    logPiComp = log(1-pi)
}
```

Bayesian GWAS 271

```
    if(iter %% outputFrequency == 0){
        cat (sprintf("iteration = %5d number of loci in model = %5d ",iter, nLoci))
        aHatTest = ZTest %*% meanAlpha/iter                          # compute genomic breeding values in test data
        corr = cor(aHatTest,yTest)
        regr = corr*sqrt(var(yTest)/var(aHatTest))                   # regress yTest on aHatTest
        RSquared = corr*corr
        cat(sprintf(" Corr = %8.5f Regr = %8.5f R2 = %8.5f \n", corr, regr, RSquared,"\n"))

        sampleCount = sampleCount + 1
        geneticVar[sampleCount] = var(Z%*%alpha)                     # variance of genotypic values
        wEnd = 0
        for (window in 1:numberWindows){
            wStart = wEnd + 1                                        # start of current window
            wEnd   = wEnd + windowSize                               # end of current window
            if (wEnd > nmarkers) wEnd = nmarkers                     # last window may be smaller than windowSize
            windowVarProp[sampleCount,window] =                      # proportion of variance due to this window
                          var(Z[,wStart:wEnd]%*%alpha[wStart:wEnd])/geneticVar[sampleCount]
        }
    }
}

meanMu       = meanMu/chainLength
meanAlpha    = meanAlpha/chainLength;
modelFreq    = modelFreq/chainLength;

cat(" Genomic Prediction in Validation (test) Animals \n")
nTestAnimal=nrow(ZTest)
cat(" Animal       GBV \n")
for (animals in 1:nTestAnimal){
    cat(sprintf("%15s %15.7f \n",testID[animals],aHatTest[animals]))
}

cat(" GWAS locus effect results\n")
cat(" Marker       Effect        modelFreq\n")
for (locus in 1:nmarkers){
    cat(sprintf("%15s %15.7f %10.7f \n", markerID[locus],meanAlpha[locus],modelFreq[locus]))
}
```

```
cat(" GWAS genomic window results")
plot(colMeans(windowVarProp),
     ylab="Proportion",
     xlab="Genomic window",
     main ="Proportion of variance explained by genomic window"
)
cutoff = 1/numberWindows
quartz()
# posterior probability that window explains more than 1/numberWindows proportion of genetic variance
posteriorProb = colSums(windowVarProp>cutoff)/numberSamples
plot(posteriorProb,
     xlab="Genomic window",
     type="h",
     ylab="Probability",
     main=sprintf("Posterior probability that proportion of window variance > %5.3f", cutoff)
)

cat ("Posterior mean of Pi: ", piMean/chainLength)
```

References

1. Maher B (2008) The case of the missing heritability. Nature 456:18–21
2. Manolio TA, Collins FS, Cox NJ, Goldstein DB, Hindorff LA, Hunter DJ, McCarthy MI, Ramos EM, Cardon LR, Chakravarti A, Cho JH, Guttmacher AE, Kong A, Kruglyak L, Mardis E, Rotimi CN, Slatkin M, Valle D, Whittemore AS, Boehnke M, Clark AG, Eichler EE, Gibson G, Haines JL, Mackay TF, McCarroll SA, Visscher PM (2009) Finding the missing heritability of complex diseases. Nature 461(7265):747–753. doi:10.1038/nature08494. http://www.hubmed.org/display.cgi?uids=19812666
3. Visscher PM, Yang J, Goddard ME (2010) A commentary on 'common SNPs explain a large proportion of the heritability for human height' by Yang et al. (2010). Twin Res Hum Genet 13(6):517–524. doi:10.1375/twin.13.6.517. http://www.hubmed.org/display.cgi?uids=21142928
4. Onteru SK, Fan B, Nikkilä MT, Garrick DJ, Stalder KJ, Rothschild MF (2010) Whole-genome association analyses for lifetime reproductive traits in the pig. J Anim Sci. doi: 10.2527/jas.2010-3236. http://jas.fass.org/content/early/2010/12/23/jas.2010-3236.abstract
5. Hayes BJ, Pryce J, Chamberlain AJ, Bowman PJ, Goddard ME (2010) Genetic architecture of complex traits and accuracy of genomic prediction: coat colour, milk-fat percentage, and type in Holstein cattle as contrasting model traits. PLoS Genet 6(9):e1001139. doi: 10.1371/journal.pgen.1001139. http://dx.doi.org/10.1371\%2Fjournal.pgen.1001139
6. Fan B, Onteru SK, Du Z-Q, Garrick DJ, Stalder KJ, Rothschild MF (2011) Genome-wide association study identifies loci for body composition and structural soundness traits in pigs. PLoS One 6(2):e14726. doi:10.1371/journal.pone.0014726. http://dx.doi.org/10.1371\%2Fjournal.pone.0014726
7. Meuwissen THE, Hayes BJ, Goddard ME (2001) Prediction of total genetic value using genome-wide dense marker maps. Genetics 157:1819–1829
8. Sun X, Habier D, Fernando RL, Garrick D, Garrick DJ, Dekkers JCM (2011) Genomic breeding value prediction and QTL mapping of QTLMAS-2010 data using Bayesian methods. BMC Proc 5(Suppl 3):S13
9. Southey BR, Fernando RL (1998) Controlling the proportion of false positives among significant results in QTL detection. In: Proceedings of the 6th world congress on genetics applied to livestock production, vol 26, Armidale, pp 221–224
10. Fernando RL, Nettleton D, Southey B, Dekkers J, Rothschild M, Soller M (2004) Controlling the proportion of false positives in multiple dependent tests. Genetics 166:611–619
11. Benjamini Y, Hochberg Y (1995) Controlling the false discovery rate: a practical and powerful approach to multiple testing. J R Stat Soc B 57:289–300
12. Stephens M, Balding DJ (2009) Bayesian statistical methods for genetic association studies. Nat Rev Genet 10(10):681–690. doi:10.1038/nrg2615. http://www.hubmed.org/display.cgi?uids=19763151
13. Gianola D, Fernando RL, Stella A (2006) Genomic assisted prediction of genetic value with semi-parametric procedures. Genetics 173:1761–1776
14. Yi N, Xu S, Allison DB (2003) Bayesian model choice and search strategies for mapping interacting quantitative trait loci. Genetics 165:867–883
15. Sorensen DA, Gianola D (2002) Likelihood, Bayesian, and MCMC methods in quantitative genetics. Springer, New York
16. Habier D, Fernando RL, Kizilkaya K, Garrick D (2011) Extension of the Bayesian alphabet for genomic selection. BMC Bioinform 12:186
17. Henderson CR (1984) Applications of linear models in animal breeding. Univ. Guelph, Guelph
18. Gianola D, Fernando RL (1986) Bayesian methods in animal breeding. J Anim Sci 63:217–244
19. Fernando RL, Gianola D (1986) Optimal properties of the conditional mean as a selection criterion. Theor Appl Genet 72:822–825
20. Habier D, Fernando RL, Dekkers JCM (2007) The impact of genetic relationship information on genome-assisted breeding values. Genetics 177(4):2389–2397. doi:10.1534/genetics.107.081190. http://www.genetics.org/cgi/content/abstract/177/4/2389
21. Habier D, Tetens J, Seefried F-R, Lichtner P, Thaller G (2010) The impact of genetic relationship information on genomic breeding values in German Holstein cattle. Genet Sel Evol 42(1):5. ISSN 1297-9686. doi:10.1186/1297-9686-42-5. http://www.gsejournal.org/content/42/1/5
22. Zeng J, Pszczola M, Wolc A, Strabel T, Fernando R, Garrick D, Dekkers J (2012) Genomic breeding value prediction and qtl

mapping of qtlmas2011 data using Bayesian and gblup methods. BMC Proc 6(Suppl 2): S7. ISSN 1753-6561. doi:10.1186/1753-6561-6-S2-S7. http://www.biomedcentral.com/1753-6561/6/S2/S7

23. Gianola D, de los Campos G, Hill WG, Manfredi E, Fernando R (2009) Additive genetic variability and the Bayesian alphabet. Genetics 183(1):347–363. doi:10.1534/genetics.109.103952. http://www.hubmed.org/display.cgi?uids=19620397

24. Gilks WR, Roberts GO (1996) Strategies for improving MCMC. In: Gilks WR, Richardson S, Spielgelhalter DJ (eds) Markov chain Monte Carlo in practice, 1st edn. Chapman and Hall, London, pp 1–19

25. Norris JR (1997) Markov chains. Cambridge series on statistical and probabilistic mathematics. Cambridge University Press, New York

26. Hastings WK (1970) Monte Carlo sampling using Markov chains and their applications. Biometrika 57:97–109

27. Sahana G, Guldbrandtsen B, Janss L, Lund MS (2010) Comparison of association mapping methods in a complex pedigreed population. Genet Epidemiol 34:455–462

28. R Development Core Team (2008) R: a language and environment for statistical computing. R Foundation for Statistical Computing, Vienna. ISBN 3-900051-07-0. http://www.R-project.org

29. Fernando RL, Garrick DJ (2008) GenSel—user manual for a portfolio of genomic selection related analyses. Animal Breeding and Genetics, Iowa State University, Ames

30. Morton N (1955) Sequential tests for the detection of linkage. Am J Hum Genet 7:277–318

31. Storey JD (2002) A direct approach to false discovery rates. J R Stat Soc Ser B 64:479–498

32. Fernando RL, Habier D, Stricker C, Dekkers JCM, Totir LR (2007) Genomic selection. Acta Agric Scand Sect A Anim Sci 57(4):192–195. http://www.informaworld.com/10.1080/09064700801959395

33. Tierney L (1996) Introduction to general state-space Markov chain theory. In: Gilks WR, Richardson S, Spielgelhalter DJ (eds) Markov chain Monte Carlo in practice. Chapman and Hall, London

34. Brooks SP, Gelman A (1998) General methods for monitoring convergence of iterative simulations. Comput Graph Stat 7:434–455

35. Godsill SJ (2001) On the relationship between Markov chain Monte Carlo methods for model uncertainty. J Comput Graph Stat 10(2):230–248

36. Carlin BP, Chib S (1995) Bayesian model choice via Markov-chain Monte-Carlo methods. J R Stat Soc Ser B Methodol 57(3):473–484. ISSN 0035-9246

37. Cannings C, Sheehan N (2002) On a misconception about irreducibility of the single-site Gibbs sampler in a pedigree application. Genetics 162:993–996

38. Fernández SA, Fernando RL, Gulbrandtsen B, Stricker C, Schelling M, Carriquiry AL (2002) Irreducibility and efficiency of ESIP to sample marker genotypes in large pedigrees with loops. Genet Sel Evol 34:537–555

39. Abraham KJ, Totir L, Fernando R (2007) Improved techniques for sampling complex pedigrees with the Gibbs sampler. Genet Sel Evol 39(1):27–38. ISSN 1297-9686

Chapter 11

Implementing a QTL Detection Study (GWAS) Using Genomic Prediction Methodology

Dorian J. Garrick and Rohan L. Fernando

Abstract

Genomic prediction exploits historical genotypic and phenotypic data to predict performance on selection candidates based only on their genotypes. It achieves this by a process known as training that derives the values of all the chromosome fragments that can be characterized by regressing the historical phenotypes on some or all of the genotyped loci. A genome-wide association study (GWAS) involves a genome-wide search for chromosome fragments with significant association with phenotype. One Bayesian approach to GWAS makes inferences using samples from the posterior distribution of genotypic effects obtained in the training phase of genomic prediction. Here we describe how to do this from commonly used Bayesian methods for genomic prediction, and we comment on how to interpret the results.

Key words Genome-wide association studies, Genomic prediction, Quantitative trait locus, R programming

1 Introduction

Conventional approaches to GWAS have tended to use models that estimate the effect of one genotype at a time, in order to find any genotype that has an experiment-wise statistically significant effect on the phenotypic trait [see catalog of findings in ref. [1]]. Such analyses can easily be biased by stratification in the population [2], for example due to admixture of different founder groups such as races or breeds, or due to family structure. In these analyses, population structure needs to be taken into account and one way of achieving this is to fit a polygenic effect along with the fixed effect of the genotype being tested for significance [3].

Genomic prediction [4] differs from conventional GWAS analyses in that all genotypes, or certain subsets of genotypes, are considered in the model together, and are treated as random rather than fixed effects. Among other considerations this simultaneous fitting of genotypes at different loci has the advantage that some of the genotypes will simultaneously account for known or unknown population structure even in admixed populations [5]. The effects estimated by simultaneously fitting genotypes will be partial effects of the genotype, adjusted for all the other genotypes in the model [6].

Development of methods and efficient computing algorithms for practical implementation of genomic prediction and GWAS has been a major research initiative at Iowa State University by the authors of this chapter, along with their colleagues, postdoctoral fellows and graduate students. A companion chapter [7] provides the Bayesian background and introduces some of the computing algorithms that form the basis of GenSel software (http://bigs.ansci.iastate.edu/), whereas this chapter details the practical aspects of implementing GWAS using methods in that chapter.

2 Materials and Methods

A Bayesian GWAS analysis requires genotypic information, phenotypic information, a model that describes the factors that influence phenotype, and specification of prior distributions. This is the case because Bayesian techniques combine prior information about parameters (known before the analysis), along with information contained in the data, in order to construct posterior distributions (known after the analysis) that can be used for inference [8]. Preferably the posterior distributions to be used for inference will not be unduly influenced by the prior information, provided the sample size (*see* **Note 1**) is sufficiently large.

We will develop these concepts within the context of implementing GWAS analysis using GenSel software. The basic computing algorithms are included as R scripts from Fernando and Garrick [7] available for download from the publisher's website (http://extras.springer.com/), but practical analysis is more efficient using compiled software (e.g., web-based analysis accessible at http://bigs.ansci.iastate.edu/) that optimizes the computing algorithms for better "mixing" to reduce the number of iterations required, the computational time per iteration, and to reduce rounding errors in order to improve numerical stability.

2.1 Genotypes

Most current GWAS analyses in humans and livestock involve SNP genotypes from bi-allelic loci. Depending upon the genotyping platform there are a number of options for which the alleles will be called and reported from the system. In livestock, the Illumina platform is used more widely than competing options and the preferred genotypes to use are the Illumina A/B calls [9] as these are more likely than the other genotypes to be consistently called across genotyping providers and panel versions. Called genotypes

Table 1
Example Illumina genotype file

[Header]	
GSGT Version	1.9.4
Processing Date	7/16/2012 12:20 PM
Content	BovineHD_B.bpm
Num SNPs	777,962
Total SNPs	777,962
Num Samples	254
Total Samples	3,048
[Data]	

SNP Name	Sample ID	Allele1-Forward	Allele 1-Top	Allele 1-AB	Allele 2-AB	GC Score	X	Y
ARS-BFGL-BAC-10172	408272	G	G	B	B	0.9274	0.015	1.001
ARS-BFGL-BAC-1020	408272	G	G	B	B	0.9360	0.025	0.598
ARS-BFGL-BAC-10245	408272	T	A	A	B	0.7861	1.263	0.925
ARS-BFGL-BAC-10345	408272	A	A	A	B	0.9485	0.750	0.710
ARS-BFGL-BAC-10365	408272	G	C	B	B	0.5292	0.009	0.935
ARS-BFGL-BAC-10591	408272	A	A	A	A	0.9355	0.832	0.029
ARS-BFGL-BAC-10793	408272	G	C	A	B	0.9589	0.567	0.565
ARS-BFGL-BAC-10867	408272	G	C	A	A	0.9558	0.775	0.004
ARS-BFGL-BAC-10951	408272	T	A	A	A	0.4966	0.431	0.021
ARS-BFGL-BAC-10952	408272	A	A	A	A	0.8084	1.309	0.062

will then be represented as AA, AB, or BB at any particular locus. Some loci will not be called, but this is typically a very small percentage (<1 %) and differs from animal to animal. An example of the first 20 lines of the final report from an Illumina genotyping run is shown in Table 1.

Along with the genotypes, there should be a file with the map locations of the SNP markers. An example of the head of that file (SNP_map.txt) is shown in Table 2.

Genotypes should be filtered for quality control purposes (e.g., see ref. 10). It is common to remove loci with low call rates, animals with low call rates, and loci below a threshold minor allele frequency (MAF). Some researchers test for Hardy–Weinberg equilibrium (HWE) and filter out loci that depart significantly

Table 2
Example Illumina SNP_map.txt file showing SNP marker information

Index	Name	Chromosome	Position	GenTrain Score	SNP	ILMN Strand	Customer Strand	NormID
1	ARS-BFGL-BAC-10172	14	6371334	0.8904	[A/G]	TOP	TOP	2
2	ARS-BFGL-BAC-1020	14	7928189	0.8998	[T/C]	BOT	TOP	1
3	ARS-BFGL-BAC-10245	14	31819743	0.7814	[T/C]	BOT	BOT	2
4	ARS-BFGL-BAC-10345	14	6133529	0.9149	[A/C]	TOP	TOP	2
5	ARS-BFGL-BAC-10365	14	27005721	0.9173	[A/C]	TOP	BOT	2
6	ARS-BFGL-BAC-10591	14	17544926	0.8993	[A/G]	TOP	TOP	1
7	ARS-BFGL-BAC-10793	14	29259114	0.9288	[C/G]	TOP	BOT	101
8	ARS-BFGL-BAC-10867	14	34639444	0.9245	[G/C]	BOT	BOT	102
9	ARS-BFGL-BAC-10951	10	17911906	0.8921	[T/G]	BOT	BOT	2

from HWE. However, some datasets will have been subject to selection (only having samples from widely used animals) or migration (representing first-cross or admixed populations), in which case some loci would not be expected to follow HWE. Such filters will likely remove different numbers of loci in different datasets.

2.2 Missing Genotypes

Missing genotypes can be taken into account in Bayesian analyses by conditioning on one or more of the pedigree, the observed genotypes or the phenotypes. Conditioning the missing genotypes on the phenotypes adds an additional level of complexity to the analysis, and in most cases the proportion of missing genotypes is <1 %. In these circumstances it is more common to replace missing genotypes with the most likely genotype, or its expected value [11], and ignore the uncertainty of the estimated marker genotypes. Missing genotypes can be imputed based on haplotypes (see review of methods in ref. 12], or the allele frequency at that locus which can be estimated in the animals with observed genotypes in that population.

An awk-based script for Mac or Unix processing of a file of Illumina genotypes is available from the publisher's website (http://extras.springer.com/). That program requires the Sample_Map.txt file that contains the marker names and positions. It creates a map file ordered by genomic location that includes columns of the observed means and number of counts for each marker. Another awk-based script available from the publisher's website uses this map file along with the file of Illumina genotypes to create a matrix of genotypes that comprises a row for each animal and a column for each marker locus. That file is in the format required for GenSel software and for the R

Table 3
An example genotype file in GenSel format

SNP_Name	ARS-BFGL-BAC-16973	ARS-BFGL-BAC-17968	ARS-BFGL-BAC-19439	ARS-BFGL-BAC-19440
900002301	0	−10	10	5
900002305	10	−10	10	−10
900002309	0	−10	10	−10
900002316	0	−10	10	−10
900002364	10	−10	10	−10
900002401	10	−10	10	0

scripts that implement Bayesian analyses provided with this book [7]. The space-delimited file containing the matrix of genotypes has a first column of animal identifiers, and first row containing the alphanumeric names of the markers, sorted by genomic order to facilitate QTL detection using marker windows as discussed later in this chapter. The genotype covariate values themselves are stored as −10, 0, and 10 for AA, AB, and BB genotypes with missing genotypes replaced by the means for that locus.

An example of a Gensel format genotype file is in Table 3, showing a header row followed by an animal in each data row and a SNP locus in each column, with missing genotypes replaced with their locus mean (e.g., 5 for animal 900002301).

In using GenSel software, the keyword *markerFileName* is used to denote the name of the genotype file. The name of the map file for genomic locations is denoted by the keyword *mapOrderFileName*. These and other keywords that specify the nature of analysis are collected in an input file used in running GenSel.

2.3 Phenotypes

Raw phenotypic observations of genotyped individuals may often be available for GWAS analyses. These could first be adjusted for nongenetic effects, and the adjusted data used in the training. This approach has the disadvantage that subsequent analyses of the preadjusted data cannot account for any uncertainty associated with estimates of the nongenetic effects. Alternatively, it is relatively straightforward to include various fixed effects including covariates or class effects for concurrent estimation with the random SNP effects. Additional random effects may also be fitted in the analyses in some software, for example to fit a residual polygenic effect with a correlation structure based on pedigree relationships, or to fit one or more uncorrelated random effects such as might account for litter, pen, or cage. The keyword *trainPhenotypeFileName* is used to denote to GenSel the name of the phenotype file (*see* **Note 2**).

2.4 Estimated Breeding Values as Phenotypes

Raw phenotypic observations of genotyped individuals may not be available for GWAS analyses for many reasons. In plant and animal breeding circumstances, it is not uncommon to have databases of perhaps millions of phenotypic records that are used to estimate breeding values (EBV) using conventional pedigree-based analyses for the purposes of ranking selection candidates. In that situation, EBV of genotyped individuals obtained from those prior analyses (such as progeny testing) could be used as observations for GWAS. Typically, these EBV for widely used parents contain much more information than a raw phenotypic observation. As a rule of thumb for a trait with moderate heritability, it seems to take about four genotyped animals with individual records to provide the same information content as a single genotyped individual with progeny test information, confirming theoretical arguments [13]. However, when these EBV include progeny test records they will be dominated by the additive effects of each locus and will therefore preclude estimation of dominance, which in theory could be estimated using raw phenotype observations.

Mixed model procedures used to obtain EBV shrink the observed data according to the amount of information on the individual and its relatives and the corresponding genetic and residual variances. If different individuals have differing accuracies for their EBV, the data will have been shrunk to varying degrees. Accordingly it makes sense to deregress the EBV to standardize their genetic variance [14], and it has been shown to provide more accurate prediction than training on EBV [15]. If the EBV or deregressed EBV vary in accuracy, the residual variance will be heterogeneous in the deregressed data. This can be taken into account by specifying the inverse of the residual variance of each observation so this can be used in a weighted analysis of the deregressed observations. Calculation of these weights are in [14] and are used in the web-based version of GenSel software, but are not supported by the simpler R scripts supplied with this book. The keyword *rinverse* when used in the header row of the phenotype file is a reserved word for giving weights in a weighted analysis.

An example of a phenotype file containing deregressed data and corresponding weights are in Table 4. The example shows the mean fitted as a fixed effect, *rinverse* weights that are near unity for individuals with their own record, and larger numbers for sires with progeny. The sire identifier (sire of the bull) is included in the file as a remark (trailing #) and will not be used in the analysis.

In some cases raw data may be available that includes observations on relatives of animals without genotypes, has additional measurements on correlated traits of genotyped individuals, includes traits that exhibit heterogeneity of variance, or the observations may be of categorical nature. In all these circumstances, it may be appealing to use mixed model methodology (e.g., ASREML, ref. 16] with a pedigree-based relationship matrix to

Table 4
Example GenSel phenotype file for weighted deregressed data

Bull_ID	DYD_trait	mu	rinverse	Sire#
19204198	12.9269	1	2.99223	19019899
41147604	8.96998	1	3.08021	19408550
42281040	6.57562	1	3.15373	41040609
23751598	14.5179	1	3.26115	23495300
42270385	8.11082	1	3.05222	41011011
42370242	3.89984	1	3.02282	41114417
42815321	7.01446	1	0.87442	42573557
42717067	3.67418	1	2.95946	42410409
42620083	4.71934	1	2.89676	42409445
42892264	−3.98766	1	1.93729	42460503

obtain EBV for the trait of interest accounting for all these circumstances, and then deregress the resulting EBV for a weighted analysis, rather than trying to account for all this complexity directly in the Bayesian GWAS analysis. This approach can be used for GWAS of maternal effects, when the maternal effects are estimated in a mixed model and subsequently deregressed and weighted in GWAS analysis as described here. Among other factors, the deregression and weighting approach accounts for information lost in estimating fixed effects, which can result in rinverse weights less than 1 as for Bull_ID 42815321 in Table 4.

2.5 Model Equation

The model equation describes the factors to be included in the analysis. All the factors that might be required in the model equation should be included in the phenotype file. In most cases, the model equation should at the very least include an effect for the mean. In GenSel software this requires coding a column of 1's along with any corresponding title (e.g., mu) as the header in the phenotype file (see above example, Table 4).

The model equation should include effects to remove population structure, provided the population structure is not caused by the genetic effects that are being estimated.

GenSel fits a mixed model that includes uncorrelated random regression coefficients for each locus included in the genotype file (*see* **Note 3**). All other effects except for the residual are assumed to be fixed and are specified through the header line in the phenotype file (*see* **Note 4**). The phenotype file must contain animal identifiers in the first column, and the trait phenotypes in the second column.

Table 5
Example GenSel phenotype file for observations with fixed effects

HealthID	PUFA	ContempGrp$	Birthdate
20080111C30255	5.29	C2008011102	103
20080721D10037	2.41	D2008072102	120
20080728D20050	5.67	D2008072802	90
20080825D40052	5.01	D2008082502	110
20080825D40111	3.96	D2008082502	91
20030520I32497	4.17	I2003052001	110
20030520I32499	6.77	I2003052001	100

Remaining columns can contain class or covariate information to specify other fixed effects to be fitted in the model. Every column must contain a corresponding alphanumeric descriptor in the header line. If that descriptor includes a trailing $ (e.g., herd-year $), it signifies that column corresponds to a class variable. The column may then contain alphanumeric values which will be used by GenSel to construct the incidence matrix for that class variable. Columns whose descriptor ends in a trailing # signify the column is to be ignored in the analysis. A column headed *rinverse* denotes values that will be used as the diagonal elements of the weight matrix for analyses that involve heterogeneous residual variance. All other columns are fitted as covariates based on the numeric values in that column. There may be as many class and covariates as the user specifies (*see* **Notes 5** and **6**). The fixed effects are not required to be full rank.

An example phenotype file is shown in Table 5, with alphanumeric subject identification, followed by the trait and two fixed effects, representing an alphanumeric class variable and a covariate.

2.6 Categorical Data

One approach to analyzing categorical data is to assume the observed categories are represented by an underlying continuous liability, such that a particular score results from the underlying liability falling between particular thresholds. Ordered binomial or multinomial categorical data are analyzed in GenSel by sampling the liability and using the sampled liability as data.

In GenSel, categorical phenotypes must be denoted by integer values starting at 1, and keyword value pair of *isCategorical yes* specified in the input parameter file that contains the keywords and corresponding names for the genotype file, phenotype file, map order file, and other analytical options.

Categorical data creates some special challenges in estimation of variance parameters (*see* **Note 7**).

2.7 Some Alternative Bayesian Models

Two major factors discriminate some commonly used Bayesian models. First, the model might fit all random factors simultaneously (e.g., BayesA), or assume a mixture model that classifies random factors as belonging to one of two distributions (e.g., BayesB). In the mixture model, one distribution has zero effect, and only the markers that are sampled as having nonzero effects are fitted simultaneously. The parameter that specifies the proportion of loci with zero effect is denoted as π [4]. Second, the variance parameter assumed for the random factors may be specific to each fitted locus (e.g., BayesA, BayesB), or common for every locus (e.g., RR-BLUP).

In the original formulation [4], the variance parameters for each locus were estimated using Bayesian methods for BayesA and BayesB, but the common locus variance and residual variance were assumed known in the method referred to as BLUP, more commonly described today as RR-BLUP. In GenSel we refer to RR-BLUP as BayesC0 where all random factors are fitted simultaneously, and BayesC when a mixture model is applied to RR-BLUP. In contrast to RR-BLUP of [4], we use Bayesian methods to estimate the variance parameters in all our methods.

In GenSel we specify the value of π using the keyword value pair, for example, *probFixed 0.95* (*see* **Notes 8–10**). The mixture parameter π may be estimated using Bayesian methods in an extension of BayesC we refer to as BayesCπ.

In GenSel we specify these methods with the keyword value pairs *analysisType Bayes* and *bayesType BayesA* (or BayesB, BayesC, BayesCπ). Other Bayesian methods are sometimes used for GWAS, e.g., Bayesian Lasso, but those methods are not implemented in GenSel. Bayesian Lasso is available as the BLR R package [17].

2.8 Specification of Hyperparameters of Prior Distributions

In mixed linear models, estimation of fixed effects and prediction of random effects is straightforward where the variance parameters are known [18]. In Bayesian methods, variance parameters can be treated as unknown, but with assumed prior distribution. The prior is characterized by its hyperparameters which for the inverse chi-squared prior used by [4] involves the scale factor and an associated degrees of freedom.

2.8.1 Prior for the Residual Variance

The expected value of a scaled inverse chi-squared random variable is $(v/v-2)S^2$, where S^2 is the scale parameter and $v > 2$ is the degrees of freedom (df). With a given df, published estimates of the residual variance can be used to derive a value for the scale parameter based on this relationship. Estimates of residual variance can alternatively be obtained by multiplying the phenotypic variance by $1-h^2$. The posterior distribution of the residual variance is obtained from pooling information contained in the prior with that from the data. Each phenotypic observation contributes 1 df to the posterior so the posterior distribution is barely influenced by the prior when $n \gg v$ (*see* **Note 11**).

In GenSel we specify the residual variance to be 120, for example, using the keyword value pair *resVariance* 120 and its associated df to be 10 using the keyword value pair *nuRes* 10. This would result in the scale parameter being 96.

2.8.2 Prior for Locus Effect Variance(s)

The same expected value used to derive the scale parameter for the residual variance can be used to derive a scale parameter for the locus effect variance. However, there are some subtle distinctions.

The scale parameter is related to the expected variance of one fitted locus effect, rather than the overall genetic variance. Accordingly, the genetic variance needs to be decomposed into its expected contribution for a particular locus, before solving for the scale parameter. If there are many loci contributing to the genetic variance (e.g., polygenic genetic model), then the average scale factor should be a small fraction of the genetic variance. If the trait is monogenic or oligogenic, the scale factor will be a large fraction of the genetic variance. The size of the scale factor also depends upon the allele frequency of the QTL, as a given locus effect will contribute more or less to the population genetic variance according to its allele frequency [7, 19].

The genetic variance of a typical locus is therefore estimated by dividing the population genetic variance by the number of markers multiplied by $2pq(1-\pi)$. The scale parameter is then obtained by multiplying that result by $(v - 2/v)$ for $v > 2$. In GenSel, example keyword value pairs would be *varGenotypic 80* and *degreesFreedomEffectVar 4* for the genetic variance and degrees of freedom, respectively.

In BayesC, a common locus variance is assumed and is estimated by combining information from the prior and the data. Each fitted locus contributes to estimation of the common locus variance from the data. If df = 4, the information from the data will overwhelm the prior if many loci are fitted.

In contrast to BayesC, in BayesA and BayesB, the locus variance is specific for each locus. Accordingly, the data itself can only contribute 1 df information to any particular locus, regardless of the number of observations. This means that for $v > 2$ the prior will provide at least twofold the information available from the data (*see* **Note 11**). The posterior for any particular locus variance does not follow Bayesian learning, and has lead to some criticism of these methods [20]. Some researchers [21] have specified $0 < v < 2$ but this makes it problematic to determine the prior for the scale factor. Other researchers [22] have even used $v < 0$ but that results in an improper prior that does not have finite integral. Improper priors can result in improper posterior distributions, and should generally be avoided. However, in GWAS we are not interested in inference about the distribution of variance of locus effects as much as we are interested in inference of locus effects (*see* **Notes 12 and 13**).

One solution to the prior for the locus effect variance outweighing the information from the data for estimating the variance parameters in BayesA and BayesB is to treat the scale parameter for the locus effect variance as unknown [23]. This approach is implemented in GenSel and is invoked using the keyword value pair *FindScale yes* (*see* **Note 14**). An alternative is to use BayesC where the prior for the scale factor will typically be overwhelmed by the information from the data (*see* **Note 15**).

2.9 Markov Chain Monte Carlo

Some simple Bayesian models have closed form solutions and the posterior can be directly obtained. More commonly, inferences about the posterior distribution are made from samples (*see* **Notes 16 and 17**). These samples comprise a Monte Carlo Markov Chain (MCMC) that can be used for inference provided certain circumstances exist [8]. The MCMC samples can be obtained using a number of different strategies. Two commonly used strategies from [4] are Metropolis–Hastings (BayesB) and Gibbs sampling (BayesA). Gibbs sampling is often faster but requires being able to draw samples from the full-conditional posterior distributions. Metropolis–Hastings has the advantage that it does not involve sampling directly from the full conditional posterior, but instead involves sampling from some candidate distribution and accepting or rejecting those samples according to the Metropolis–Hastings acceptance rule. GenSel implements Gibbs sampling approaches to all its Bayesian methods [7].

Some researchers discard the initial MCMC samples to allow for so-called burn in. This is useful if you start the sampler from unlikely coordinates, but is not required if the starting values are from the target distribution. However, even if the sampler is started from unusual coordinates, burn in is not required if the chain is sufficiently long. The keyword value pair to specify the number of samples to discard is for example *burnIn 1000*. We have not seen any evidence of improved performance by using large values of *burnIn* relative to the length of the chain (*see* **Notes 18 and 19**).

The computing effort is directly related to the length of the chain. In GenSel, we specify the total length to be 41,000, for example, including burn in, using the keyword value pair *chainLength 41000*.

3 Inference About Individual Effects

One of the major interests from a GWAS analysis is the identification of SNP or chromosomal locations that have the greatest impact on observed variation in the trait. There are a number of approaches for ranking the covariates to identify the loci with the largest effects. When covariates are labeled for example as having 0, 1, or 2 copies of the Illumina B allele, it is arbitrary in relation to whether the more favorable allele is labeled with an A or a B. Accordingly, the absolute value of the effect is a more appropriate measure for comparison than the estimated effect. In models that fit many SNP simultaneously, it is the partial effects that are estimated, and these tend to be quite small if the model includes other markers in high Linkage Disequilibrium (LD). This will be the case if the genotypes are dense so that most markers have at least some high LD counterparts and the markers are fitted simultaneously.

In analyses using high-density markers, there are typically too many loci and samples of each locus to store them all. However, it is straightforward to accumulate the posterior mean and variance for locus effects, and locus effect variances. In applications other than those assuming an infinitesimal model for genetic effects, it is reasonable to fit a mixture model. Commonly-used marker densities in livestock include 50,000 SNPs and a mixture fraction $\pi = 0.99$ would typically result in about 500 loci being sampled at any particular position in the chain. Across the entire chain, virtually every locus is likely to include at least one nonzero sample. One obvious statistic to represent the information content for a locus is the proportion of samples that the locus had a nonzero effect. We refer to this statistic as model frequency, and its a priori expected value is $1-\pi$. A randomly chosen locus for $\pi = 0.95$ would a priori be expected to have 0.95 samples of zero, and 0.05 nonzero samples. Posterior to the analysis, a particularly good marker will have much higher model frequency, and the researcher may be interested in making inference about the markers with highest model frequency.

Among markers with model frequency of 0.95 or higher there will be a proportion of false positives (PFP) of 0.05 or less [7]. In some analyses, the best markers may only have model frequencies of 0.60 or less. In those cases the best markers have a PFP of 0.40 or less. If PFP at 0.05 is used as the level of significance, such a locus will not be significant, and the researcher may find that none of their loci are significant. As the density of marker loci increase, say from 50 to 700 k, it will become even more difficult to detect significant effects at individual loci. This is particularly evident when nearby markers exhibit high levels of LD, as in that case any one of a family of high LD markers might fit similarly well. Across the samples in the chain, different members of that family will be in different samples, depressing their individual model frequencies. The worst-case scenario in a mixture model exists if two or more markers have identical genotypes, as in that case they will fit equally well and MCMC samples with good mixing will fit each of those markers in some of the models. A particular marker might have less than extreme model frequency because it is not well associated with the trait, or because there are many other markers with which it is in very high LD. Inference about individual effects from a model that fits markers simultaneously is therefore difficult.

An example of estimated marker effects from a GenSel output is in Table 6 for a BayesC analysis with $\pi = 0.95$. The second column

Table 6
Posterior means of estimated marker effects and other statistics

SNP_Name	Effect	EffectVar	ModelFreq	WindowFreq	GeneFreq	GenVar	EffectDelta1	SDDelta1	t-like	Shrink
ARS-BFGL-NGS-16466	1.561e-03	1.009854e-03	0.0440	0.2883	0.505	1.217711e-06	3.54846e-02	1.32922e-01	0.267	0.607
ARS-BFGL-NGS-19289	−4.563e-05	1.146293e-03	0.0501	0.3208	0.000	0.000000e+00	−9.09933e-04	1.51014e-01	0.006	0.000
ARS-BFGL-NGS-105096	−5.872e-06	1.060389e-03	0.0461	0.3522	0.650	1.568249e-11	−1.27425e-04	1.46597e-01	0.001	0.217
BTA-07251-no-rs	5.898e-05	1.116235e-03	0.0485	0.4116	0.104	6.468007e-10	1.21545e-03	1.50593e-01	0.008	0.072
ARS-BFGL-NGS-98142	−4.704e-03	1.244645e-03	0.0543	0.4088	0.869	5.031188e-06	−8.65494e-02	1.60261e-01	0.540	0.328
ARS-BFGL-NGS-114208	−1.353e-03	1.069951e-03	0.0466	0.4089	0.859	4.430744e-07	−2.90181e-02	1.42649e-01	0.203	0.321
ARS-BFGL-NGS-66449	4.413e-04	9.774085e-04	0.0427	0.4117	0.253	7.354440e-08	1.03343e-02	1.33730e-01	0.077	0.490
ARS-BFGL-NGS-51647	1.622e-03	1.155577e-03	0.0504	0.4156	0.650	1.197015e-06	3.21546e-02	1.48786e-01	0.216	0.222
ARS-BFGL-BAC-32770	−4.845e-04	9.838698e-04	0.0432	0.4154	0.693	9.990410e-08	−1.12276e-02	1.31500e-01	0.085	0.552
ARS-BFGL-NGS-65067	−1.496e-03	1.124425e-03	0.0492	0.4148	0.076	3.148541e-07	−3.04206e-02	1.47881e-01	0.206	0.224
ARS-BFGL-BAC-31497	1.630e-03	1.018125e-03	0.0444	0.4165	0.269	1.045444e-06	3.67499e-02	1.37724e-01	0.267	0.516
ARS-BFGL-BAC-32722	−2.887e-03	1.142505e-03	0.0499	0.4178	0.099	1.481089e-06	−5.78836e-02	1.50306e-01	0.385	0.269
ARS-BFGL-BAC-34682	3.021e-03	1.126382e-03	0.0493	0.4212	0.902	1.620247e-06	6.13487e-02	1.57219e-01	0.390	0.269
ARS-BFGL-NGS-3964	1.990e-03	1.098510e-03	0.0481	0.4210	0.923	5.614394e-07	4.13962e-02	1.47674e-01	0.280	0.226
ARS-BFGL-NGS-98203	−4.117e-03	1.212605e-03	0.0529	0.4238	0.156	4.473430e-06	−7.77650e-02	1.65757e-01	0.469	0.440
ARS-BFGL-BAC-36895	−2.364e-04	1.106941e-03	0.0482	0.4251	0.001	1.065684e-10	−4.90486e-03	1.46130e-01	0.034	0.003

denotes the posterior mean of the substitution effects (B allele minus A allele). The effect variance represents the posterior mean of the locus effects variance (*see* **Note 20**). Model frequency is the proportion of fitted models that included this particular marker. Window frequency denotes the proportion of models that included any locus in the window centered on the current locus extending five loci in either direction, gene frequency is the frequency of the B allele. GenVar is the marginal contribution of this locus to genetic variance, computed from the allele frequency and posterior mean of the substitution effect. EffectDelta1 is the posterior mean of the locus effects from just those models where it was fitted. That value, multiplied by the model frequency gives the effect in column 2. SDDelta1 is the standard deviation of sampled effect for those iterations where this locus was fitted. The t-like statistic is the posterior mean of the sampled effect divided by its standard deviation when fitted in the model. It does not follow a t-distribution, but can be a useful indicator of loci with consistently large sampled effects. The last column, shrink, denotes the shrinkage of the mixed model estimate for this locus relative to its least squares estimate (*see* **Note 21**).

4 Inference About Major Genes

A useful property about Bayesian inference from MC samples is that those samples can be used to reconstruct samples for other parameters of interest. For example, given samples of residual variances and genetic variances, we can construct samples of phenotypic variance or genomic heritability to make inference using those constructed posterior distributions.

Rather than using samples of locus effects to make inference about a particular locus, one could use the samples to make inference about a particular genomic location that encompasses a number of contiguous loci. The underlying concept of a genomic breeding value (GBV) is that it results from the sum of the additive QTL effects across all the QTL that influence the trait, but we don't know the QTL and instead treat the marker loci effects like QTL, and sum up their effects to get the GBV. This same concept can be applied to subunits of the genome, for example to a particular chromosome, or a smaller particular window of the genome.

The alleles at a QTL locus are likely to be in LD with a number of marker loci, to varying extents. Ideal marker data would exhibit high LD between the alleles at the QTL and alleles at nearby marker loci, and that LD would rapidly diminish with genomic distance from the marker locus. Even in that ideal situation, the QTL effect is likely to be distributed across a number of the marker loci. However, these individual locus effects are collectively capturing the same QTL effect. A GBV for a window that encompasses many of these marker loci could recover more of the QTL effect than any particular marker locus.

In GenSel, genomic windows are constructed based on the chromosome and base-pair positions denoted in the marker map file (*see* **Note 22**). Each window consists of those markers on the same chromosome falling within the same 1 Mb (*see* **Note 23**). The keyword value pair *windowBV yes* is used to instruct GenSel to form these windows and to compute sample values for the variance of the GBV for each window. The frequency with which these samples are computed is determined by the *outputFreq* keyword.

The individual samples of window variance, expressed as a proportion of the genomic variance, are stored in an output file for inference purposes. A separate column is used for each 1 Mb window. Summary statistics are also produced, in sorted order according to the descending window variance, that express the proportion of samples with nonzero variance for that window, and the proportion of samples with variance greater than would be expected if the genomic architecture was infinitesimal, and every window accounted for an average amount of variation.

An example of the summary statistics for the window variance output is in Table 7 for an analysis using BayesB with $\pi = 0.95$ using numerically coded marker loci. Windows are numbered sequentially along the genome, with the output sorted by the variance contributed by each 1 Mb window (*see* **Note 24**). In this example, the largest window accounts for 7.99 % genetic variance, and consists of 28 marker loci. The column $p > 0$ reflects the proportion of models where this window was included, and therefore accounted for more than 0 % genetic variance. The column $p >$ Average reflects the proportion of models where this window accounted for more than the amount of variance that would be explained if every window had the same effect. The largest three windows are therefore all significant at PFP < 0.05 in terms of both null hypotheses, namely that (1) this window accounts for genetic variation, and (2) this window accounts for more genetic variation than expected under an infinitesimal model. None of the windows accounting for lesser amounts of variation than these three windows are significant in this dataset. Only four windows individually accounted for more than 1 % genetic variance, but collectively the three largest windows explained 20 % genetic variance.

Individual samples stored for the largest ten windows will be automatically used via the keyword value pair *plotPosteriors yes* to generate plots of the posterior distributions of percentage variance contributed by that window. An example plot for the largest window shown in Table 7 is depicted below in Fig. 1.

A smoother plot can be generated by reducing *outputFreq* or by increasing *chainLength*. Both options would increase the analysis run time.

Table 7
Estimates of the percentage of genetic variance for the 1 Mb windows with largest effects

Window	Start	End	#SNPs	%Var	Cum%Var	p > 0	p > Average	map_pos0	map_posn	chr_Mb
1974	42826	42852	28	7.99	7.99	1.000000	1.000000	20_4043933	20_4989460	20_4
701	15129	15153	22	7.07	15.06	0.986000	0.975000	6_38042011	6_38939012	6_38
876	19064	19073	11	5.85	20.90	1.000000	1.000000	7_93007435	7_93886136	7_93
702	15154	15181	27	1.41	22.31	0.795000	0.556000	6_39034201	6_39837065	6_39
2083	45295	45316	20	0.76	23.07	0.598000	0.357000	21_41134760	21_41995360	21_41
1500	32469	32495	26	0.56	24.24	0.597000	0.274000	13_78023758	13_78982992	13_78
1201	26072	26098	27	0.52	24.76	0.574000	0.264000	10_86003549	10_86994261	10_86
2077	45161	45185	24	0.47	25.24	0.554000	0.228000	21_35003844	21_35999673	21_35

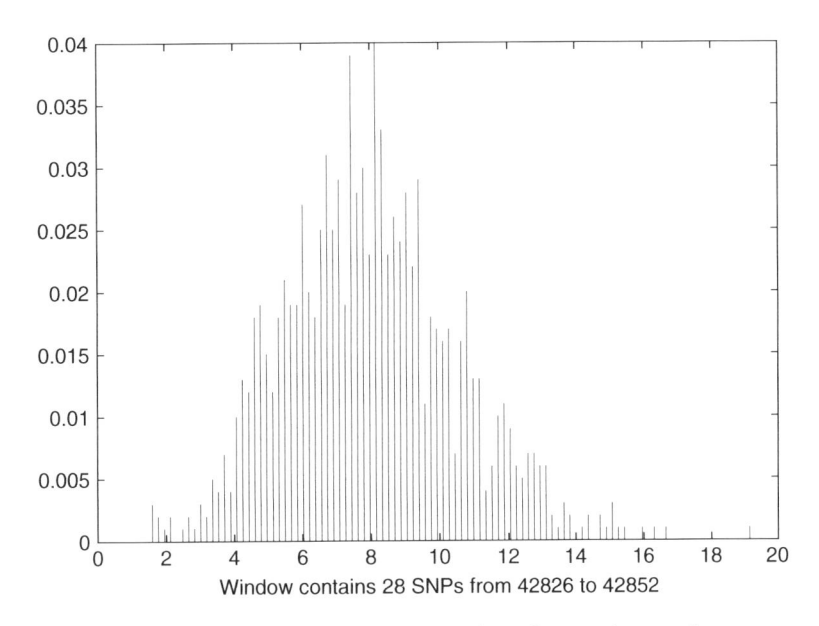

[Figure: histogram plot with x-axis from 0 to 20, y-axis from 0 to 0.04]

Window contains 28 SNPs from 42826 to 42852

Fig. 1 Example plot of the posterior distribution of percentage variance contributed by a window of 28 SNPs

5 Notes

1. Generally, the aim should be to have at least 1,000 genotyped animals in any analyses. Large QTL effects can sometimes be found with several hundred animals, but such datasets can be problematic for analyses, particularly for binomial score data.

2. The phenotype file can contain individual identifiers that are not present in the genotype file and vice versa. The identifiers in each row of the genotype and phenotype files can be in any order in each file, and only those alphanumeric identifiers that match in both files will be used for analysis. This means that a single large genotype file can be created with all samples from several studies, and the phenotype files which are more easily manipulated can be edited for different analyses.

3. The model equation used in GenSel can be readily modified to filter the loci to consider in the model without modifying the genotype file. Specific loci to be included or excluded can be listed by marker name in a file whose name is specified using the keywords *includeFileName* or *excludeFileName*. In addition to specifying specific loci to include or exclude, filtering can also be achieved using chromosome names and corresponding genomic locations as defined in the map file. The keywords *markerWindowIncFileName* (or *markerWindowExcFileName*) denote the names of files that contain in each row information about a region to include (or exclude). If the row contains only

an alphanumeric chromosome name, all markers on that chromosome would be candidates for inclusion (or exclusion). Alternatively, following the chromosome name, the row may contain a starting and ending base-pair position in which case all markers within that range will be included (or excluded). The use of these options to include or exclude markers by name or location is not mutually exclusive. Any combination of these options can be used and include options are processed before exclude options.

4. The model equation used in GenSel can be readily modified in terms of fixed effects. Fixed effects that are coded in the phenotype file can be omitted by adding a trailing # in their name on the header row of the phenotype file. Random effects potentially include all those columns in the genotype file, as named in the header row. However, a file can list one locus name per row to denote specific markers to be included (or excluded) in an analysis. The relevant keywords are *includeFileName* and *excludeFileName* to denote the names of those files.

5. Nongenetic effects that are not included in the analysis, but are important, will cause bias. First, they may lead to an inflated residual variance, which will lead to an inflated variance ratio, causing locus effects to be shrunk towards zero more than they would otherwise. This is particularly obvious when no mean (or intercept) is fitted so that the sum of squares for the mean contributes directly to the sum of squares for the residual. Second, locus effects will tend to pick up some nongenetic effects not included in the model, and this may lead to certain locus effects having large effects or high model frequency that might be incorrectly interpreted as representing genetic associations.

6. Effects that account for genetic differences should not normally be fitted as fixed effects. For example sire, family, or breed differences will include locus effects that are typically the object of the GWAS. It does not make sense to fit breed as a fixed effect in an admixed analysis, provided many locus effects are being fitted together. This contrasts with single locus analyses often undertaken in human GWAS studies where it is critical to account for population structure to avoid spurious conclusions. Fitting many or all loci concurrently has been shown to achieve this same result in genomic prediction without the structure needing to be explicitly taken into account [5].

7. Analysis of case–control and data with categorical scores can work well with large datasets (>1,000 observations) but sometimes results in inflated estimates of heritability, particularly <500 observations. The high or even near perfect heritability is a result of the extreme category problem that also makes it difficult to estimate variance components from threshold analyses fitted an animal model where the same

data gives sensible estimates of variance components when applied to a sire model [24]. Categorical analyses that result in much higher heritabilities than expected from pedigree-based estimates of variance components should be treated with caution. BayesC may work better in these circumstances. Part of the problem is that with 50 k or more loci and <1,000 observations, there can be a high probability that one or a few covariates are almost perfectly correlated with the categorical phenotype by chance alone.

8. The assumed mixture parameter, π, can in some cases have considerable influence on results. Convergence tends to be faster with higher values of π, whereas more locus effects are fitted in every model with lower values of π. Values of π that are too high will tend to fit too few markers and can reduce the posterior mean of the genetic variance. Low values of π are seldom problematic for genomic prediction of continuous traits as evidenced by similarity of BayesA and BayesB results, unless the trait is largely controlled by a few genes with large effect in which case high values of π result in better predictive ability [25].

9. Parameter π can be estimated using BayesCπ. This reflects the finding that when π is too small, the number of markers fitted is $1-\pi$ in product with the number of markers available, but when π is too large, more markers will be fitted than expected, demonstrating the data contains information on the mixture frequency. We have been unable to estimate π in BayesB.

10. One approach to GWAS is to estimate π using BayesCπ [23] and then use that value in subsequent BayesB or BayesC analyses. Cross validation shows that lower values of π than those estimated using BayesCπ can sometimes provide higher correlations in genomic prediction applied to close relatives, presumably because the additional markers are better at picking up relationships. Conversely, higher values of π can sometimes be more discriminating in the identification of the largest QTL.

11. Assumed hyperparameters for the genetic and residual variance and the associated degrees of freedom can have considerable impact on results in BayesA, BayesB, and to a much lesser extent in BayesC.

12. In admixed data where population structure such as breed is not removed as fixed effects, the posterior genetic variance includes breed differences. Accordingly, if the breeds differ widely in their means, the estimated genetic variance can be much larger than would normally be expected in a within breed analysis. This will reduce the shrinkage of locus effects unless the prior reflects the within breed genetic variance and the degrees of freedom for the prior are inflated. The extent of inflation required will depend upon the Bayesian model assumed and nature of the data.

13. Increasing the degrees of freedom associated with the hyperparameter will reduce the impact of the data in the estimation of genetic and residual variances. Very high values for the locus effect degrees of freedom in BayesC are equivalent to prior specification of the variance ratio. High values for the locus effect degrees of freedom in BayesA and BayesB make those models more like BayesC.

14. Jointly estimating the scale parameter does not typically seem to influence results compared to using a scale parameter determined from a reasonable estimate of the genetic variance. It is worth noting that the derivation of the scale parameter in a mixture model depends upon π, and this results in the scale parameter not being independent of π in BayesCπ, as is the case in BayesB and BayesC where π is a constant. The option *FindScale thruPi* takes this dependency into account and is recommended in BayesCπ.

15. In the absence of good prior information, for example in GWAS of a novel trait, BayesC is recommended as a good starting point, with π chosen to fit no more markers in any one iteration than there are observations. BayesC typically outperforms BayesA and BayesB when those methods are used with poorly chosen priors. BayesCπ may also be worth fitting, to examine the posterior distribution for π. The shape of that posterior depends upon the amount of data and the genomic architecture, so BayesCπ is not always the best option for GWAS. If the priors are appropriate, for example based on prior analyses of the same trait in other datasets, BayesB may be a better choice than BayesC as it will shrink small effects a little more than BayesC and not shrink the largest effects as much. BayesB can be used with the prior for the genetic variance set from the posterior obtained in BayesC.

16. GenSel estimates the genetic variance using the samples of breeding values that are implied by the genotypes of each training individual and the samples of its locus effects. The posterior means for genetic and residual variances usually seem to converge within 10,000 or fewer iterations. The proportion of variation explained by the markers is the ratio of the genetic variance divided by the sum of the genetic and residual variance. That proportion, the "genomic heritability" exploits information due to LD, relationships, and co-segregation, and is often similar to the heritability obtained in pedigree analyses provided the data includes related individuals. It provides an upper limit for the correlation between genomic prediction and true genetic merit that would be obtained in a validation analysis.

17. Posterior means of the genetic and residual variances should be carefully scrutinized after the analysis. If the proportion of

variance explained by markers is very high or very low it likely indicates problems with the model being fitted, including the assumed hyperparameters, or the analysis. Values that are close to the heritability of the trait are typical in properly formulated models with sufficient numbers of observations.

18. The starting samples in the Markov chain relevant to a particular dataset are determined by the seed used to start the random number generator. The keyword value pair seed 1234 will set the starting value. Changing the seed will result in a new set of MCMC samples whose distribution should converge provided the number of samples defined by *chainLength* is sufficiently large. Analysts can check for convergence by changing the seed and comparing the inferences that would be made from one chain compared to the other.

19. Successive elements of the MCMC samples are correlated. Some researchers therefore only save samples every so often. In GenSel we use every sample to compute the posterior mean of locus effects, locus variances, and residual variances. Genetic variance, the proportion of variance explained by markers, and window variance are expensive to compute using the entire chain of locus effects. For example, the keyword value pair *outputFreq 100* is used to specify these more expensive functions of effects are only computed from every 100th sample. It might make sense to choose *outputFreq* so that 1,000 samples are available for inference.

20. BayesC uses the same sampled locus effect variance for all the loci fitted in the model in a particular iteration. However, the posterior mean of the locus effect variance will only be the same for all loci in BayesC0. The posterior mean of the locus effect variance will be different for each locus in BayesC for $\pi < 1$ because a different locus effect variance is sampled each iteration, and the combination of loci fitted in each iteration typically varies.

21. Bayesian methods commonly used for genomic prediction and GWAS can be characterized as shrinking marker effects in two different manners. First, marker effects are fitted as random rather than fixed effects. Practically, this involves adding a variance ratio to the diagonal element of the coefficient matrix corresponding to that locus. The ratio is the residual variance divided by the variance for that locus. The residual variance is influenced by how well the training data is described by the markers and other fixed effects. The locus variance depends upon the method (it is the same for all markers in BayesC and different for every marker in BayesA and BayesB), and the number of markers being fitted (depends upon π and the number of available markers). The impact of adding the variance ratio to the diagonal of the coefficient matrix depends

upon the size of the variance ratio in relation to the sums of squares of the covariate values for that locus which is the basis of the coefficient matrix in a least squares analysis that fits markers as fixed effects. In BayesA, the shrinkage from fitting markers as random can be substantial, but is much less so in BayesB, BayesC, and BayesCπ, particularly if π is large.

The second manner in which marker effects can be shrunk towards zero is by assuming a mixture model. In BayesA, which is not a mixture model, the posterior distribution of a marker effect comprises all the samples of that marker effect. The best estimate of the marker effect might be the posterior mean, or average of the samples of that marker effect. The posterior mean would therefore be a typical value observed in the Markov Chain. In contrast, in a mixture model with $\pi > 0$, the posterior distribution of a marker effect would typically reflect some density of zero effects, corresponding to those iterations that the marker was assumed to belong to the distribution with zero effect, and the remaining density would represent the iterations where the marker was included in the model and had a sampled effect. The posterior mean would be a weighted estimate of the two distributions, with the effect that the posterior mean of the sampled effects when included in the model would be shrunk towards zero, the amount of shrinkage depending upon the frequency that marker was sampled as belonging to the distribution with zero effect. In this context, shrinkage towards zero is achieved by assigning the marker effect to the distribution with zero effect, as well as by shrinking the marker effect by fitting it as a random rather than fixed effect.

If there are enough observations, the data will eventually overwhelm the shrinkage caused by the variance ratio so that estimates of locus effects, or accuracy of genomic predictions, will not be strongly dependent on the prior. However, the size of commonly used datasets for GWAS may not be sufficiently large and in practice many inferences may be influenced by prior assumptions.

22. The map file that specifies chromosomal and base-pair locations may contain such assignments from more than one genome build. The relevant columns headings must contain trailing strings of _chr and _pos (e.g., UMD_pos and UMD_chr) to be recognized by GenSel. The keyword value pair *linkageMap UMD* for example would specify that GenSel uses the UMD map positions in forming 1 Mb windows.

23. Users can customize the width of windows by modifying the scale of the base-pair positions in the map file. Simulation studies using observed beef cattle 50 k genotypes show that the simulated QTL location can be up to 2 Mb on either side of the 1 Mb windows that show the largest variance. Researchers

should not assume a causal mutation would always be within the significant window itself.

24. The ranking of windows that account for the most variance are relatively robust to assumed Bayesian model, hyperparameters and π. However, the actual reported proportion of variance explained by each window is very sensitive to these assumptions. Very high values of π will fit few locus effects, favoring those with biggest effects. Accordingly, they will account for a larger proportion of variation than will be the case if lower values of π are used. BayesB shrinks smaller locus effects towards zero by using a larger variance ratio for those loci, compared to BayesC which uses an average variance ratio for all loci. Conversely, BayesB shrinks very large effects less than BayesC. Accordingly, the use of BayesB will result in higher estimates of the proportion of variation explained by windows with large effects.

25. An example of an input file for a GenSel run that combines the required parameters for GWAS is below. This file is required as a command line option to run GenSel. The keywords are case specific but can be entered in any order.

```
markerFileName          Trials_1_to_10.newbin
excludeFileName         Exclude_low_MAF_SNP
phenotypeFileName                     BMI.txt
mapOrderFileName                Map_build10_3
linkageMap                            build10
analysisType                            Bayes
bayesType                              BayesB
chainLength                             41010
burnin                                   1000
outputFreq                                 40
probFixed                                0.97
varGenotypic                             0.27
degreesFreedomEffectVar                     4
varResidual                              0.70
nuRes                                      10
seed                                     1234
plotPosteriors                            yes
FindScale                                  no
windowBV                                  yes
```

References

1. Hindorff LA, MacArthur J (European Bioinformatics Institute), Wise A, Junkins HA, Hall PN, Klemm AK, Manolio TA (2012). A catalog of published genome-wide association studies. www.genome.gov/gwastudies. Accessed 16 May 2012

2. Hao K, Chudin E, Greenawalt D, Schadt EE (2010) Magnitude of stratification in human populations and impacts on genome wide association studies. PLoS One 5(1):e8695

3. Snelling WM, Allan MF, Keele JW, Kuehn LA, McDaneld T, Smith TPL, Sonstegard TS,

Thallman RM, Bennett GL (2010) Genome-wide association study of growth in crossbred beef cattle. J Anim Sci 88:837–848

4. Meuwissen THE, Hayes BJ, Goddard ME (2001) Prediction of total genetic value using genome-wide dense marker maps. Genetics 157:1819–1829

5. Toosi A, Fernando RL, Dekkers JCM (2010) Genomic selection in admixed and crossbred populations. J Anim Sci 88:32–46

6. Allen DM, Cady FB (1982) Analyzing experimental data by regression. Lifetime Learning, Belmont, CA, p 64

7. Fernando RL, Garrick DJ. Bayesian methods applied to GWAS. In: Gondro C (ed) Genome-wide association studies and genomic prediction. Methods Mol Biol 1019:235–272

8. Sorenson D, Gianola D (2002) Likelihood, Bayesian and MCMC methods in quantitative genetics. Springer, New York, p 740. ISBN 0-387-954406

9. Illumina (2006) Technical Note: "TOP/BOT" strand and "A/B" allele. http://www.illumina.com/documents/products/technotes/technote_topbot.pdf. Accessed 16 May 2012

10. Pongpanich M, Sullivan PF, Tzeng JY (2010) A quality control algorithm for filtering SNPs in genome-wide association studies. Bioinformatics 26:1731–1737

11. Haley CS, Knott SA (1992) A simple regression method for mapping quantitative trait loci in line crosses using flanking markers. Heredity 69:315–324

12. Marchini J, Howie B (2010) Genotype imputation for genome-wide association studies. Nat Rev Genet 11:499–511

13. Weller JI, Kashi Y, Soller M (1990) Power of daughter and granddaughter designs for determining linkage between marker loci and quantitative trait loci in dairy cattle. J Dairy Sci 1990 (73):2525–2537

14. Garrick DJ, Taylor JF, Fernando RL (2009) Deregressing estimated breeding values and weighting information for genomic regression analyses. Genet Sel Evol 41:55

15. Ostersen T, Christensen OF, Henryon M, Su G, Madsen P (2011) Deregressed EBV as the response variable yield more reliable genomic predictions than traditional EBV in pure-bred pigs. Genet Sel Evol 43:38

16. Gilmour AR, Gogel BJ, Cullis BR, Thompson R (2008) ASReml user guide release 3.0. VSN, Hemel Hempstead

17. de los Campos G, Naya H, Gianola D, Crossa J, Legarra A, Manfredi E, Weigel K, Cotes J (2009) Predicting quantitative traits with regression models for dense molecular markers and pedigree. Genetics 182:375–385

18. Henderson CR (1984) Applications of linear models in animal breeding. University of Guelph, Guelph, ON

19. Fernando RL, Habier D, Stricker C, Dekkers JCM, Totir LR (2007) Genomic selection. Acta Agric Scand A Anim Sci 57:192–195

20. Gianola D, de los Campos G, Hill WG, Manfredi E, Fernando R (2009) Additive genetic variability and the Bayesian alphabet. Genetics 183:347–363

21. ter Braak CJF, Boer MP, Bink MCAM (2005) Extending Xu's Bayesian model for estimating polygenic effects using markers of the entire genome. Genetics 170:1435–1438

22. Xu S (2007) An empirical Bayes method for estimating epistatic effects of quantitative trait loci. Biometrics 63:513–521

23. Habier D, Fernando RL, Kizilkaya K, Garrick DJ (2011) Extension of the Bayesian alphabet for genomic selection. BMC Bioinformatics 12:186

24. Hoeschele I, Tier B (1995) Estimation of variance components of threshold characters by marginal posterior modes and means via Gibbs sampling. Genet Sel Evol 27: 519–540

25. Hayes BJ, Pryce J, Chamberlain AJ, Bowman PJ, Goddard ME (2010) Genetic architecture of complex traits and accuracy of genomic prediction: coat colour, milk-fat percentage, and type in Holstein cattle as contrasting model traits. PLoS Genet 6(9):e1001139

Chapter 12

Genome-Enabled Prediction Using the BLR (Bayesian Linear Regression) R-Package

Gustavo de los Campos, Paulino Pérez, Ana I. Vazquez, and José Crossa

Abstract

The BLR (Bayesian linear regression) package of R implements several Bayesian regression models for continuous traits. The package was originally developed for implementing the Bayesian LASSO (BL) of Park and Casella (J Am Stat Assoc 103(482):681–686, 2008), extended to accommodate fixed effects and regressions on pedigree using methods described by de los Campos et al. (Genetics 182(1):375–385, 2009). In 2010 we further developed the code into an R-package, reprogrammed some internal aspects of the algorithm in the C language to increase computational speed, and further documented the package (Plant Genome J 3(2):106–116, 2010). The first version of BLR was launched in 2010 and since then the package has been used for multiple publications and is being routinely used for genomic evaluations in some animal and plant breeding programs. In this article we review the models implemented by BLR and illustrate the use of the package with examples.

Key words Genomic selection, Whole-genome prediction, Bayesian, Shrinkage and genomic prediction

1 Models, Algorithms, and Data

In whole-genome regression (WGR, e.g., [1–4]) models, phenotypes $\mathbf{y} = \{y_i\}_{i=1}^{n}$ are regressed on large number of markers concurrently [1–3]. The number of markers (p) can vastly exceed the number of individuals with records (n) and implementing these "*large-p-with-small-n*" regressions requires using estimation methods that perform variable selection, shrinkage of estimates, or a combination of both. Bayesian procedures are commonly used for variable selection and for deriving shrunken estimates of effects in regressions models (*see* ref. 5 for a review of methods commonly used in WGR). The BLR package of R ([6], http://www.r-project.org/) implements several types of Bayesian regressions. In this section we describe the models implemented and the algorithms used by BLR.

Bayesian inference is based on the posterior density of the parameters (θ) given the data ($\mathbf{y}$). Following Bayes' theorem, this is proportional to the product of the conditional distribution of the data given the unknowns (or Bayesian likelihood) times the prior density assigned to θ, these objects are described next.

Bayesian Likelihood. In BLR the conditional density of the phenotypes is a multivariate normal density. The data equation is:

$$y_i = \mu + \sum_{j=1}^{p_F} x_{Fij}\beta_{Fj} + \sum_{j=1}^{p_R} x_{Rij}\beta_{Rj} + \sum_{j=1}^{p_L} x_{Lij}\beta_{Lj} + u_i + \varepsilon_i$$
$$= \eta_i + \varepsilon_i \quad (i = 1, \ldots, n) \qquad (1)$$

where

- y_i is a phenotypic record.
- μ is an intercept.
- x_{Fij}, x_{Rij}, and x_{Lij} are covariates whose effects are assigned flat priors (β_{Fj}), normal identical-independent priors (β_{Rj}), and identical-independent double exponential priors (β_{Lj}), respectively.
- u_i is a Gaussian random effect with mean zero and covariance matrix given by the user (e.g., a numerator relationship matrix).
- ε_i are model residuals, each following independent normal densities with mean zero and variance equal to $w_i^2 \sigma_\varepsilon^2$, where w_i is a user-defined weight used to accommodate heterogeneous residual variances, and σ_ε^2 is a variance parameter (if weights are not provided, by default, the w_i's are all set equal to one).
- $\eta_i = \mu + \sum_{j=1}^{p_F} x_{Fij}\beta_{Fj} + \sum_{j=1}^{p_R} x_{Rij}\beta_{Rj} + \sum_{j=1}^{p_L} x_{Lij}\beta_{Lj} + u_i$ is the linear predictor used to model the conditional expectation function. Any of the variables in the linear predictor can be omitted, except for the intercept, which is included by default.

From the above assumptions, it follows that the conditional density of the data given the unknowns is:

$$p(\mathbf{y}|\mu, \beta_F, \beta_R, \beta_L, \mathbf{u}, \sigma_\varepsilon^2) = \prod_{i=1}^{n} N(y_i|\eta_i, w_i^2 \sigma_\varepsilon^2) \qquad (2)$$

where $\beta_F = \{\beta_{Fj}\}$, $\beta_R = \{\beta_{Rj}\}$, $\beta_L = \{\beta_{Lj}\}$, and $\mathbf{u} = \{u_i\}$ are vectors of effects.

Prior. In BLR, μ and β_F are assigned flat priors, $p(\mu, \beta_F) \propto 1$; therefore, these unknowns are estimated using information from the likelihood solely. The remaining effects, β_R, β_L, and $\mathbf{u}$, are assigned informative priors, yielding shrinkage of estimates of these effects. The effects $\{\beta_{Rj}\}$ are assigned identical-independent normal priors centered at zero and with variance σ_R^2, i.e., $p(\beta_R|\sigma_R^2) = \prod_{j=1}^{p_R} N(\beta_{Rj}|0, \sigma_R^2)$. This prior, combined with a Gaussian likelihood, yields estimates of marker effects similar to those of a

Ridge Regression [7]; therefore, we denote this model as BRR (Bayesian Ridge Regression). The priors assigned to effects $\{\beta_{Lj}\}$ are identical-independent double-exponential densities, represented, as in the Bayesian LASSO (BL) of Park and Casella [1], as infinite mixtures of scaled normal densities, $p(\beta_L|\lambda, \sigma_\varepsilon^2) = \prod_{j=1}^{P_L} \int N(\beta_{lj}|0, \sigma_\varepsilon^2 \tau_j^2) \text{Exp}(\tau_j^2|\lambda) \partial \tau_j^2$, where τ_j^2 is a marker-specific scale parameter, $\text{Exp}(\tau_j^2|\lambda)$ is an exponential density assigned to τ_j^2, with rate parameter $\lambda^2/2$. This parameter controls the extent of shrinkage of estimates of effects in the same way that the ratio $\sigma_\varepsilon^2 \sigma_R^{-2}$ does in the BRR. The vector of effects **u** are assigned a multivariate normal density with mean zero and a covariance matrix proportional to **A**, i.e., $p(\mathbf{u}|\sigma_u^2) = N(0, \mathbf{A}\sigma_u^2)$. For pedigree-regressions, **A** could be a numerator relationship matrix computed from a pedigree (*see* **Note 1** for an example of how to use the package pedigreemm, [8, 9] for creating **A** from a pedigree), but in principle **A** can be any symmetric positive definite matrix.

The unknown variance parameters $\{\sigma_\varepsilon^2, \sigma_R^2, \sigma_u^2\}$ are assigned Scaled-inverse Chi-square priors. In the parameterization used in BLR, the prior expectation is $E(\sigma^2|df, S) = S/(df - 2)$, *see* **Note 2** for further details about this parameterization. The regularization parameter of the BL (λ) can be either fixed at a user-specified value or treated as random. When λ is treated as unknown one can either assign a gamma density to λ^2 with rate and shape parameters r and s, respectively, $p(\lambda^2|r, s) = G(\lambda^2|r, s)$ or, alternatively, the ratio of λ to a prespecified upper-bound of λ (max) could be assigned a Beta prior (*see* ref. 2, for further discussion about this prior).

Posterior density. Collecting the aforementioned assumptions, we obtain the joint posterior density of the unknowns, given the data:

$$p(\mu, \beta_F, \beta_R, \beta_L, \mathbf{u}, \sigma_\varepsilon^2, \sigma_R^2, \sigma_u^2, \lambda | \mathbf{y}) \propto \prod_{i=1}^{n} N(y_i|\eta_i, w_i^2 \sigma_\varepsilon^2)$$

$$\times \left\{ \prod_{j=1}^{P_R} N(\beta_{Rj}|0, \sigma_R^2) \right\} \left\{ \prod_{j=1}^{P_L} N(\beta_{lj}|0, \sigma_\varepsilon^2 \tau_j^2) \text{Exp}(\tau_j^2|\lambda) \right\}$$

$$N(\mathbf{u}|0, \mathbf{A}\sigma_u^2) \times \chi^{-2}(\sigma_\varepsilon^2|df_\varepsilon, S_\varepsilon) \chi^{-2}(\sigma_R^2|df_R, S_R) \chi^{-2}(\sigma_u^2|df_u, S_u) p(\lambda)$$

(3)

Algorithms The posterior density of Eq. 3 does not have closed form; however, all the fully conditional densities (except that of λ when λ/max is assigned a Beta prior) have closed form and are easy to sample from. Therefore, in BLR samples are drawn using a Gibbs sampler (*see* **Note 3**) with scalar updates. At each iteration of the Gibbs sampler, the algorithm loops over predictors. In *large-p-regressions*, this can be computationally intensive. Therefore, we have developed a C-function, which produces samples of effects. The sampling of variance parameters can be vectorized; therefore, these updates can be done efficiently using existing R functions such as `rchisq` of the base package or `rinvGauss` of the Supp-Dists R-package (http://cran.r-project.org/web/packages/Supp-Dists/index.html).

Data. A genomic wheat dataset is made available with the package. This dataset is from CIMMYT's (http://www.cimmyt.org) early spring trials and comprises phenotypic, genotypic and pedigree information from 599 pure lines of wheat. The following code illustrates how this data can be loaded into the R session (Box 1).

Box 1: Loading the wheat dataset
1 `library(BLR)`
2 `data(wheat)`
3 `objects()`

Once the dataset is loaded, the following objects become available in the environment:

- **Y** (599 × 4) a matrix containing an evaluation of the grain yields of 599 wheat lines in four different environments. Phenotypes represent average performance of two years and were centered and standardized to a unit sample variance.

- **A** (599 × 599) a numerator relationship matrix derived from pedigree data (the matrix was computed using the methods described in [10]).

- **X** (599 × 1,279) a matrix containing the genotypes of the 599 wheat lines at 1,279 Diversity Array Technology (DART) markers generated by Triticarte Pty. Ltd. (Canberra, Australia; http://www.triticarte.com.au). Since the wheat data consists of pure lines, the two homozygous were coded as 0 and 1.

- **sets** (599 × 1) a vector assigning the wheat lines into tenfolds, this could be used for cross-validation.

2 Using BLR

In this section we introduce several examples that illustrate the use of the BLR package for regressions using molecular markers and pedigree. The main function of the BLR-package is `BLR()`.

Inputs to BLR include **phenotypes** (denoted as y, numeric vector), **design matrices** for fixed (XL, numeric matrix) and random effects (XR for BRR, XL for BL, both numeric matrices), **prior hyper-parameters** (prior, a list; further description given below), MCMC parameters, including number of iterations (nIter, integer), number of iterations to be discarded as burn-in (burnIn, integer), thinning interval (thin, integer), and a string (saveAt, character) that can be used to provide a path and append a prefix to the files containing the samples generated by the algorithm. A complete description of the inputs is given in R-help (accessible typing help(BLR) in the R-console or at http://cran.r-project.org/web/packages/BLR/index.html). All arguments, except the vector of phenotypes, have default values. By default, BLR runs an intercept model with improper priors; we discourage the use of improper priors for variance parameters and for λ, because when improper priors are used for these unknowns the posterior density is not guaranteed to be a proper density. Further details about inputs and how to choose appropriate values for inputs are given in the various examples included in this section.

Processes. As BLR runs, two processes occur: (a) the function **prints to the screen information on the iterations of the sampler** (iteration number, time per iteration, residual variance, etc.) and (b) **samples** of fixed effects, of variances and of λ **are saved to the hard drive** (*see* **Note 4** for further information about the files that are created by BLR).

Return. As the Gibbs sampler is running, BLR computes posterior means and other statistics. Once the program finishes, BLR returns a list containing estimates of posterior means and posterior deviations, measures of goodness of fit and information on the model that was fitted. Further details are given in the following examples.

2.1 Fitting an Intercept Model

The simplest model that can be fitted using BLR is the intercept model; an example is given in Box 2. The first line of the code removes any object which may have been saved from a previous R-session, the second line loads BLR, the third line sets the seed for the random number generator, and data is simulated in line 4 by drawing samples from a normal density with mean 25 and variance equal to 4. The fifth line fits the intercept model using BLR. The argument y in BLR is used to provide phenotypes; it must be a numeric vector and missing values (NA) are allowed. In the example, in addition to phenotypes, we indicate the number of iterations of the Gibbs sampler (1,200) and the number that we want to discard as burn-in (200 in the example). Note that in the example in Box 2 we did not provide any predictor variable; the intercept is included by default.

Box 2. Fitting an intercept model (improper prior)

```
1   rm(list=ls())
2   library(BLR)
3   set.seed(12345)
4   y<-rnorm(1000,mean=25,sd=2)
5   fm<-BLR(y=y,nIter=1200,burnIn=200,
6           saveAt="box2_")
7   str(fm)
8   fm$mu
9   fm$varE
10  mean(y)
11  var(y)
```

You may have noticed that immediately after the welcome message and before the Gibbs sampler started, the program printed the following warning:

===================================
No prior was provided; BLR is running with improper priors.
===================================

This occurs because we did not provide prior hyper-parameters; by default BLR runs improper priors.

2.2 Fitting Fixed-Effects Models

Box 3 provides an example of a model that includes an intercept and 5 fixed effects; in this example, we arbitrarily chose to fit the "fixed" effects of markers whose genotypes are given in columns 100, 200, 300, 400, and 500 of the matrix containing the wheat genotypes (**X**). BLR assigns flat priors (i.e., proportional to a constant) for fixed effects and for the intercept. For the residual variance, the prior is a scaled inverse chi square density. This density has two hyper-parameters: the degree of freedom (df) and the scale (S). If we do not provide them, BLR sets these to zero, which yields an improper prior (*see* example in Box 2).

Choosing the prior for the residual variance. We recommend setting df equal to a small value to reduce the influence of the prior on inferences (e.g., $df = 5$; any value greater than zero is valid, but values less than 4 will yield infinite prior expected value of σ_ε^2), and suggest choosing a scale parameter to match our prior expectations about the residual variance. The prior expectation for this parameter in BLR is: $E(\sigma_\varepsilon^2 | df_\varepsilon, S_\varepsilon) = S_\varepsilon / (df_\varepsilon - 2)$ (*see* **Note 2** for details about the parameterizations of the scaled-inverse chi square density used in BLR). Therefore, if V_y represents an estimate of the phenotypic variance and x is the proportion of that variance that we expect to account for with the model, we can set the scale parameter to be $S_\varepsilon = V_y (1 - x)(df_\varepsilon - 2)$. In Box 3, we apply this approach using $x = 0.5$. After the prior is defined (lines 11–13 of Box 3) the model is fitted (lines 16 and 17 of Box 3). In lines 19–23 of Box 3 we show how to extract parameter estimates (estimated posterior means) and in lines 25–30 we show how to read into R and plot samples collected by the Gibbs sampler (*see* **Note 3**).

Box 3. Fitting a fixed effects model

```
1   rm(list=ls())
2   library(BLR) ; library(coda)
3   set.seed(12345)
4   data(wheat)
5
6   # Extract phenotype and markers
7    y<-Y[,1]
8   Z<-X[,c(100,200,300,400,500)]
9
10  # Defines the prior
11  DF<-5
12  S<-0.5*var(y)*(DF-2)
13  prior=list( varE=list(df=DF,S=S) )
14
15  # Fits the model
16  fm<-BLR(y=y,XF=Z,nIter=5500,burnIn=500,
17      prior=prior,saveAt="box3_")
18
19  str(fm)
20  fm$mu
21  fm$varE
22  fm$bF
23  fm$SD.bF
24
25  # Trace plot of  the effect of the (1st marker)
26  B<-read.table("box3_bF.dat")
27  plot(B[,1],type="o",col="red")
28  HPDinterval(as.mcmc(B),prob=.95)
```

For further information about how BLR handles and stores the samples *see* **Note 4**, and for additional information about how to create incidence matrices for fixed or random effects *see* **Note 5**.

2.3 Fitting Marker Effects as Random

In WGR models [4], phenotypes are regressed on genome-wide markers. With current genotyping technologies, the number of markers can vastly exceed the number of phenotypes. When effects are fitted as fixed, the sampling variance of estimates, and consequently the mean-square error (MSE), increases fast as the number of markers does. To overcome this problem, WGR are usually implemented using shrinkage estimation procedures such as penalized or Bayesian regressions. These procedures introduce bias but reduce the variance of estimates and, when the number of predictors is large relative to the number of data-points, the use of shrinkage yields smaller MSE than that of standard estimation procedures such as ordinary least squares, or maximum likelihood. For regressions on markers, BLR implements two Bayesian shrinkage procedures: one uses Gaussian priors (BRR) and the other one uses double-exponential prior densities (BL). Each of these methods is described next.

2.3.1 Bayesian Ridge Regression

Box 4 provides an example where phenotypes from the wheat dataset are regressed on all available markers using a BRR. In this model, we need to provide hyper parameters for the prior of residual variance and for that of the variance of marker effects, σ_R^2. The scale and *df* parameters for the residual variance are defined using the same principles discussed in the example in Box 3.

We use a similar principle for the variance of marker effects. Noting that the variance of the regression on markers is $\mathrm{Var}(\sum_{j=1}^{p_R} x_{Rij}\beta_{Rj}) = \sigma_R^2 \sum_{j=1}^{p_R} x_{Rij}^2$ and summing over individuals and dividing by n, we get $n^{-1}\sum_{i=1}^{n} \mathrm{Var}\left(\sum_{j=1}^{p_R} x_{Rij}\beta_{Rj}\right) = \sigma_R^2 \times MS_x$, where $MS_x = n^{-1}\sum_{i=1}^{n}\sum_{j=1}^{p_R} x_{Rij}^2$ is the average sum of squares of the genotypes. We can then equate $\sigma_R^2 \times MS_x$ to the product of our prior expectation about trait heritability times (an estimate of) the phenotypic variance, to get $\sigma_R^2 = h^2 V_y / MS_x$. Equating this to the prior expectation of the variance parameter and solving for the scale yields

$$S_R = \frac{h^2 V_y (df_R - 2)}{MS_x} \quad (4)$$

In Box 4, we use this formula and $df_R = 5$ to define the prior for σ_R^2. The model includes an intercept and usually we measure variance using squared-deviations from the center of the sample; therefore, we can compute MS_x as

$$MS_x = n^{-1} \sum_{i=1}^{n} \sum_{j=1}^{p_R} \left(x_{Rij} - \bar{x}_{Rj}\right)^2, \quad (5)$$

where $\bar{x}_{Rj}$ is mean of the jth marker in the sample.

After fitting the model, we provide plots for squared estimates of marker effects (line 22) and predicted genetic values (lines 24–26). The term `fm$yHat` is the estimated posterior mean of the linear predictor η_i, which in the case of the example in Box 4 is $\eta_i = \mu + \sum_{j=1}^{p_R} x_{Rij}\beta_{Rj}$. Note that relative to observed phenotypes, predictions are shrunk towards zero; in this example, the regression of predictions on phenotypes is roughly 0.45 (see line 27, Box 4).

2.3.2 Bayesian LASSO

The code needed to fit the Bayesian LASSO (BL) is very similar to that in Box 4. The most important modifications are (a) in line 19 of Box 4 we replace XR with XL (this indicates to the program that BL instead of BRR needs to be fitted), and, (b) we need to provide hyper-parameters for the BL.

Setting hyper-parameters in the BL. The hyper-parameters of the BL include the scale and *df* parameters assigned to the residual variance and the regularization parameter λ. Hyper-parameters corresponding to the residual variance can be set using the same principles

```
Box 4. Fitting a Bayesian Ridge Regression
1  rm(list=ls())
2  library(BLR)
3  set.seed(12345)
4  data(wheat)
5
6  # Extract phenotype
7   y<-Y[,1]
8
9  # Defines the prior
10 DF<-5
11 Vy<-var(y)
12 h2<-0.5
13 Se<-Vy*(1-h2)*(DF-2)
14 MSx<-sum(apply(FUN=var,MARGIN=2,X=X))
15 Sr<-Vy*h2*(DF-2)/MSx
16 prior=list( varE=list(df=DF,S=Se) , varBR=list(df=DF,S=Sr ) )
17
18 # Fits the model
19 fm<-BLR(y=y,XR=X,
20 nIter=12000,burnIn=2000,prior=prior,saveAt="box4_")
21
22 # Plot of squared-estimated effects
23 plot(fm$bR^2,col=2,ylab=expression(paste(beta[j]^2)),main="BRR")
24 # Plot of estimated genetic values versus phenotypes
25 plot( fm$yHat~y,col=2,ylab="Genomic Value",
26       xlab="Phenotype", ylim=c(-3,3),xlim=c(-3,3))
27 abline(b=1,a=0,lty=2) ; abline(h=0,lty=2,v=0)
28 reg<-lm(fm$yHat~y)$coef ; abline(a=reg[1],b=reg[2],col=4);
29 print(reg)
```

outlined in the previous example (*see* Box 4). For λ BLR has three options: (a) assign a **gamma prior** to λ^2 (*see* ref. 1 and information provider below for further details), (b) assign a **Beta prior** to the ratio of $\tilde{\lambda} = \lambda/\max$ where max is a user prespecified upper-bound to λ (*see* ref. 2 for further details about this approach), or (c) fit the model with λ **fixed** at a value specified by the user (see below).

The list that specifies the prior for λ has the following arguments: $value (numeric), which is the initial value that will be used by the sampler, and $type (character), which informs the algorithm whether λ should remain fixed at the initial value ($type="fixed") or updated from its fully conditional density ($type="random"), $rate and $shape, in case λ^2 is assigned a gamma prior, or $max, $shape1 and $shape2, in case a beta prior is assigned to $\tilde{\lambda} = \lambda/\max$. Here we focus on the case where λ^2 is assigned a gamma prior. In this case, we need to specify the rate (r) and shape (s) parameters of the gamma density assigned to λ^2. To do this, one possibility is to first find a "focal point" by computing a "guess" about λ^2 as a function of trait heritability and marker information. For instance, using $h^2 V_y = \text{Var}(\beta_{jL}|\lambda^2) \times MS_x$ and noting that in the BL $\text{Var}(\beta_{jL}|\lambda^2) = 2(\sigma_\varepsilon^2/\lambda^2)$, we obtain $h^2 V_y = 2(\sigma_\varepsilon^2/\lambda^2)MS_x$, and solving for λ^2, we get:

$$\lambda^2 = 2\frac{(1-h^2)}{h^2} MS_x \qquad (6)$$

```
                    Box 5. Displaying the Density of λ² in the BL
 1    rm(list=ls())
 2    library(BLR)
 3    data(wheat)
 4
 5    # computes MSx and noise-signal ratios
 6    MSx<-0
 7    for(i in 1:ncol(X)){ MSx<-MSx+mean((X[,i]-mean(X[,i]))^2) }
 8    h2<-c(0.3,0.5,0.7)
 9    noiseToSignal<-(1-h2)/h2
10    lambda<-1:50
11
12    # hyper-parameters
13    shape<-1.01
14    rate<-(shape-1)/(2*noiseToSignal[2]*MSx)
15
16    # Density Plot
17    plot( y=dgamma(lambda^2,rate=rate,shape=shape),
18          x=(lambda^2),col=2,type="l",
19          ylab="Density",xlab=expression(paste(lambda^2)))
20    abline(v= (2*(1-h2)/h2*MSx),lty=2)
```

We can then use Eq. 6 to compute a prior "guess" about λ^2. Subsequently we choose the rate (r) and shape (s) parameters of the gamma density so that this density has a mode and is relatively uninformative in the neighborhood of the value computed using Eq. 6. If $\lambda^2 \sim G(\lambda^2|r,s)$, the prior mode is $r^{-1}(s-1)$ (for $s > 1$) and the prior coefficient of variation is $s^{-1/2}$. Our recommendation is to choose s so it is slightly greater than 1 (e.g., $s = 1.01$); this guarantees the existence of the prior mode and gives a large coefficient of variation (i.e., a relatively uninformative prior). Then, one can solve for the rate parameter so that the prior mode matches the value computed using Eq. 6, that is $r = (s-1)(h^2/2(1-h^2)MS_x)$. An example of this is given in Box 5.

We are now ready to fit the BL. The code is given in Box 6.

2.4 Fitting Models for Molecular Markers and Pedigree

We now extend the marker-based regression in Box 6 by adding a regression on the pedigree. BLR has an argument GF (for "grouping factor") that allows fitting a random effect (see u_i in Eq. 1). This argument is specified using a two-components list. The first component of the list, GF$ID, is an integer vector of dimension $n \times 1$, and each entry gives the corresponding level (if there are q-levels, GF$ID must take values from 1 to q) of the random effect. The second entry of the list, GF$A, is a numeric (symmetric, positive definite) matrix of dimensions $q \times q$ providing a (co)variance structure between levels of the random effect (see **Note 6** for examples of how to specify GF with and without repeated measures). An example of a model including markers and pedigree is given in Box 7.

2.4.1 Evaluation of Goodness of Fit, Model Complexity, and Deviance Information Criterion

The posterior means of the residual variances of the models fitted in Boxes 6 and 7 are obtained by typing `fmM$varE` and `fmMP$varE`; these are 0.55 and 0.46, respectively, indicating that adding the pedigree to the model increased the goodness-of-fit of the training dataset. Adding the pedigree can also affect the estimates of marker effects; in this case, when we add the pedigree, the posterior mean of λ increases only slightly (from 19.7, `fmM$lambda` to 20.7, `fmMP$lambda`), suggesting a slight increase in shrinkage of estimates of marker effects. Moreover, the correlations between estimates of marker effects and predictions derived from the model with and without the pedigree, `cor(fmM$bL,fmMP$bL)` and `cor(fmM$yHat,fmMP$yHat)`, were both very high, on the order of 0.97.

Further information on model goodness-of-fit and complexity, Deviance Information Criterion (DIC) and effective number of parameters (pD) [11], is given in the `$fit` entry of the fitted models: adding the pedigree to the model (`fmMP` vs. `fmM`) increased the estimated number of effective parameters by about 55 units (`fmMP$fit$pD-fmM$fit$pD`) and decreased DIC by roughly 50 units. Therefore, DIC ("smaller is better") favors the model that includes marker and pedigree information relative to the one using markers only.

2.5 Evaluation of Predictive Performance

Evaluating models based on predictive performance has become a common practice in genomic selection. Next we describe alternative ways of estimating measures of prediction accuracy.

2.5.1 Training–Testing Evaluation

The training–testing (TRN–TST) validation scheme is the simplest approach we can use; here we partition the data into two disjoint sets (TRN and TST), fit the model to the TRN dataset and evaluate predictive performance in the TST dataset. Therefore, when fitting the model, we need to "mask" all the data in the TST dataset to BLR. We can do this in two different ways: one possibility is to subset the data and give BLR only the data in the TRN dataset. The second possibility is to replace the observed phenotypes in the TST dataset with missing values. Both ways are illustrated in the code of Box 8 for a marker-regression model using the BL.

2.5.2 Replicated TRN–TST Evaluations

The code in Box 9 illustrates how we can replicate the TRN–TST evaluation in Box 8 by randomly generating multiple partitions. The code is implemented using a loop; however, as we discuss later on in this chapter, these partitions could be run in parallel. The number of replicates is defined in line 11 of Box 9. Using `nReplicates<-100` we obtained an average correlation of 0.5 with a 90 % CI given by [0.43, 0.58]. The variability in measures of prediction accuracy observed across replicates reflects uncertainty about estimates of prediction accuracy due to sampling of TRN and TST datasets.

Box 6. Fitting the Bayesian LASSO to genome-wide markers

```
1   rm(list=ls()); library(BLR); set.seed(12345); data(wheat)
2   # Extract phenotype
3   y<-Y[,1]
4
5   # Defines the prior
6   DF<-5
7   Vy<-var(y)
8   h2<-0.5
9   Se<-Vy*(1-h2)*(DF-2)
10  MSx<-sum(apply(FUN=var,MARGIN=2,X=X))
11  lambda<-sqrt(2*(1-h2)/h2*MSx)
12  shape<-1.01
13  rate<-(shape-1)/(lambda^2)
14
15  prior=list(  varE=list(df=DF,S=Se) ,
16               lambda=list( type="random",value=lambda,
17                            shape=shape,rate=rate
18                          )
19             )
20
21  # Fits the model
22  fmM<-BLR(y=y,XL=X,
23      nIter=12000,burnIn=2000,prior=prior,saveAt="box6_")
24  str(fmM)
25  # (continues in Box 7)
```

Box 7. Fitting the Bayesian LASSO to genome-wide markers and pedigree

```
1   #(continued from Box 6)
2
3   # Defines the prior
4   DF<-5
5   Vy<-var(y)
6   h2<-0.5
7   Se<-Vy*(1-h2)*(DF-2)
8   Su<-Vy*h2/3
9   MSx<-sum(apply(FUN=var,MARGIN=2,X=X))
10  lambda<-sqrt(2*(1-(h2*2/3))/(h2*2/3)*MSx)
11  shape<-1.01
12  rate<-(shape-1)/(lambda^2)
13
14  prior=list(  varE=list(df=DF,S=Se) ,
15               lambda=list( type="random",value=lambda,
16                            shape=shape,rate=rate
17                          ),
18               varU=list(df=DF,S=Su)
19             )
20
21  # Fits the model
22  fmMP<-BLR(y=y,XL=X,GF=list(ID=1:599,A=A),
23          nIter=12000,burnIn=2000,prior=prior,saveAt="box7_")
24  str(fmMP)
```

Genomic Prediction Using BLR   311

Box 8. Evaluating predictive performance in training-testing partitions

```
rm(list=ls()); library(BLR); set.seed(12345); data(wheat)

# Extracts phenotypes
y<-Y[,1]

#  Prior   (see Boxes 5 and 6 for details)

prior=list(  varE=list(df=5,S=1.5) ,
             lambda=list(type="random",value=20,
                         rate=2e-4,shape=2)
           )

# Random assignment of data into training and testing
tst<-sample(x=1:599,size=150,replace=FALSE)

## Method 1: Introduce NAs in TST
yNA<-y ; yNA[tst]<-NA
fm1<-BLR(y=yNA,XL=X,
  nIter=12000,burnIn=2000,prior=prior,saveAt="box8_1_")


## Method 2: Subsets the data
XTrn<-X[-tst,] ; yTrn<-y[-tst]
XTst<-X[tst,] ; yTst<-y[tst]
fm2<-BLR( y=yTrn,XL=XTrn,nIter=12000,
          burnIn=2000,prior=prior,saveAt="box8_2_")
yHatTst<-fm2$mu+XTst%*%fm2$bL
plot(yHatTst~fm1$yHat[tst],col=2,
     main="Predictions in TST set by method",
     xlab="Introducting Missing Values",
     ylab="Subseting the data")
```

Box 9. Replicated training-testing partitions

```
rm(list=ls()); library(BLR); set.seed(12345);data(wheat)
# Extracts phenotypes
y<-Y[,1]

#  Prior (see Boxes 5 and 6 for details).

prior=list(  varE=list(df=5,S=1.5) ,
             lambda=list(type="random",value=20,
                         rate=2e-4,shape=2)
           )

trnTstCor<-numeric()
nReplicates<-10

for(i in 1:nReplicates){
    tst<-sample(x=1:599,size=150,replace=FALSE)
    yNA<-y ; yNA[tst]<-NA
    fm<-BLR(y=yNA,XL=X,nIter=6000,burnIn=1000,prior=prior,
           saveAt= paste("box10_rep_",i,"_",sep=""))
    trnTstCor[i]<-cor(y[tst],fm$yHat[tst])
}
summary(trnTstCor)
```

```
Box 10. Example of a 5-fold cross-validation
1  rm(list=ls()); library(BLR); set.seed(12345); data(wheat)
2
3  # Extracts phenotypes
4  y<-Y[,1]
5
6  # Prior    (see Boxes 5 and 6 for details)
7
8  prior=list(   varE=list(df=5,S=1.5) ,
9                lambda=list(type="random",value=20,
10                           rate=2e-4,shape=2)
11             )
12
13 cvCor<-numeric()
14 folds<-rep(1:5,120)[order(runif(599))]
15
16 for(i in 1:5){
17     tst<-which(folds==i)
18     yNA<-y ; yNA[tst]<-NA
19     fm<-BLR(y=yNA,XL=X,nIter=6000,burnIn=1000,
20            prior=prior,saveAt= paste("box10_",i,"_",sep=""))
21     cvCor[i]<-cor(y[tst],fm$yHat[tst])
22 }
23 cvCor
```

2.5.3 Cross-Validation

Cross-validation (CV) is another commonly used method for estimating prediction accuracy. In a CV we assign data points into disjoint sets (e.g., 5 for a fivefold cross-validation, CV). Subsequently, for each of these folds we conduct a TRN–TST evaluation where all individuals assigned to the fold are used for TST and the rest are used for TRN. An example of how to implement this sequentially is given in Box 10.

2.6 Some Useful Commands for Executing R in a Linux Environment

Some of the tasks we have discussed in the previous section can be performed in parallel. For instance, if we have access to several processors or nodes, we could compute each of the TRN–TST evaluations in Box 9 or the CV in Box 10 in parallel. In high-performance clusters, we may do this by creating arrays of jobs which are simultaneously sent to several nodes. In performing these tasks, it is usually convenient to provide some arguments to R using the command line. To do this, we first need to include the script we want to execute (e.g., the code in Box 10) in a text file (let's assume we have named this file `exBox10.R`); then we can execute these instructions from the command line by typing:

```
R CMD BATCH exBox10.R  out &
```

In the above command:

1. `R CMD BATCH` tells the system that we want to run an R command in batch mode.

2. `exBox10.R` is the name of the file with the R-script printed in Box 10 included.
3. `out` is a log file where the system will print all the outputs that are printed to the R-console when the script is executed.
4. `&` sends the job to the background and frees the terminal. In this way, you can continue to work in the command line while the job is executed.

For parallel computing, it is sometimes useful to set values to R objects from the command line. For instance, in a tenfold CV, it would be useful to execute in parallel the analyses required for each fold. This could be done if one of the variables in the R environment (e.g., fold number) is defined at the command line. In order to be able to set variables using arguments provided from the command line, first, we need to include, at the beginning of the R-script, the following instructions (see `help(commandArgs)` in the R-console for further information about the `commandArgs` function)

```
args=(commandArgs(TRUE))
for(i in 1:length(args)){
        eval(parse(text=args[[i]]))
}
```

These instructions allow reading arguments from the command line and set those arguments as variables in the R-session. Subsequently, we execute the script and pass arguments to R using the following command:

```
R CMD BATCH "--args fold=1" exBox11.R out &
```

When we execute the above command, the variable `fold` is set equal to 1 within the R-session. Box 11 gives a modified version of the R-code of Box 10, which could be run in parallel by executing R from the command line. Lines 2–3 read arguments from the command line. Lines 4–14 are exactly as in Box 10. In the remaining lines, instead of running a loop over folds, we just fit a model for onefold (the one specified in the command line). The code in line 16 creates the TST set for that particular fold. Since we are running this in BATCH mode, before quitting R we save the model. In line

```
Box 11. R-script for running 1 fold of a CV (fold specified in the command line)
1   rm(list=ls())
2   args=(commandArgs(TRUE))
3   for(i in 1:length(args)){  eval(parse(text=args[[i]])) }
4   library(BLR); set.seed(12345);data(wheat)
5   y<-Y[,1]
6
7   if(is.null(fold)){ stop("Fold was not provided") }
8
9   # Prior  (see Boxes 5 and 6 for details)
10
11  prior=list(   varE=list(df=5,S=1.5) ,
12                lambda=list(type="random",value=20,
13                            rate=2e-4,shape=2)
14             )
15
16  cvCor<-numeric()
17  folds<-rep(1:5,120)[order(runif(599))]
18
19  tst<-which(folds==fold)
20  yNA<-y ; yNA[tst]<-NA
21  fm<-BLR(y=yNA,XL=X,nIter=6000,burnIn=1000,
22          prior=prior,saveAt=paste("fold_",fold,"_",sep=""))
23          )
24  save(fm,file= paste("fm_fold_",fold, ".rda",sep="") )
```

24 we give instructions to R to save the fitted model in a compressed file format (*.rda). All outputs are saved with the fold number indicated in the filenames.

In the previous example we show how to fit models to the first fold. We can ran all folds in parallel using the following bash-code

```
R CMD BATCH "--args fold=1" exBox11.R out1 &
R CMD BATCH "--args fold=2" exBox11.R out2 &
R CMD BATCH "--args fold=3" exBox11.R out3 &
R CMD BATCH "--args fold=4" exBox11.R out4 &
R CMD BATCH "--args fold=5" exBox11.R out5 &
```

Or, even simpler, using a do-loop in bash

```
#!/bin/bash
for (( fold=1; fold<=5; fold++ ))
 do
      R CMD BATCH "--args fold=$fold " exBox11.R out$fold &
 done
```

3 Discussion

We developed the first version of the Gibbs sampler in BLR in 2009 [2] and in 2010 we offered the program as an R-package [3]. In the R-package, a few steps of the Gibbs sampler were programmed in C, and this greatly increased the computational speed. We have used BLR with up to 100 thousand SNPs and close to 10 thousand records. In principle there are no limits to the number of markers or records beyond those imposed by the availability of RAM memory and processing time.

Memory requirements and run-time complexity. The incidence matrices for fixed and random effects (XF, XR, and XL) are all stored as numeric. This allows fast computations and the possibility of using the program with discrete (e.g., SNP codes) or continuous predictors. This generality comes at the price of higher memory requirements, and the user should be aware that memory requirements increase proportionally to the number of subjects and number of markers. For high density markers (e.g., hundreds of thousands of SNPs), memory requirements can be high.

In the Gibbs sampler, updates are done one effect at a time; therefore, computational time usually increases proportionally to the number of markers (p). Figure 1 displays the time in seconds that it takes for BLR to run (in an Intel Core i7 with a 2.4Gz processor) 1,000 iterations for various values of p and n. With

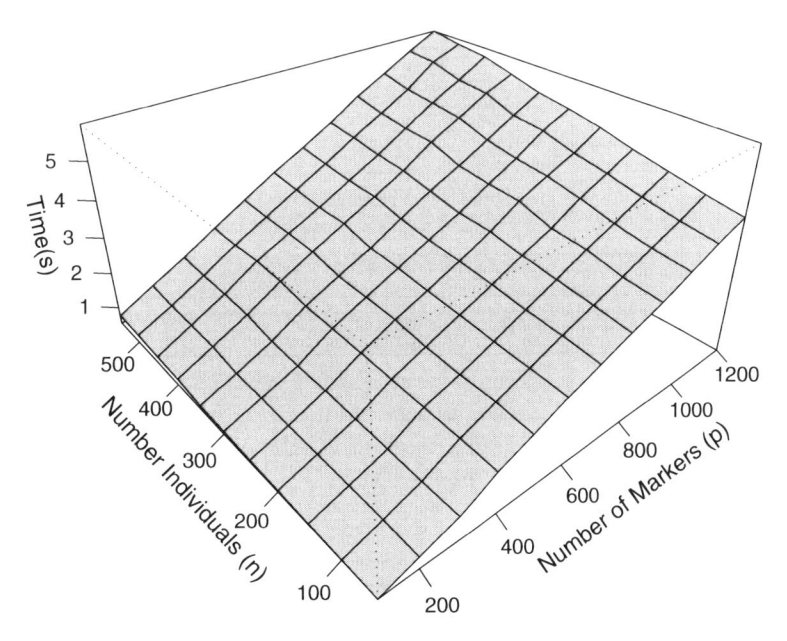

Fig. 1 Running time by sample size and number of markers (1,000 iterations run on an Intel Core i7 with a 2.4 Gz processor)

1,279 markers and 599 individuals it took approximately 6 seconds to run 1,000 iterations of the Gibbs sampler.

Influence of hyper-parameters on inferences. When $p \gg n$ marker effects cannot be uniquely estimated from the likelihood and under these conditions the influences of the prior on inferences are nontrivial. This applies to any Bayesian regression model where $p \gg n$, including the BRR and the BL. We characterized the influences of the prior on inferences about marker effects in the BL, and the results were published in de los Campos et al. [2]. In general, predictions of genetic values are quite robust within reasonable values of λ; however, estimates of marker effects are more sensitive to this parameter. Our recommendation is to first compute the value of λ that we expect using Eq. 6 and then choose values of the rate and shape parameters so that the prior has a mode that it is reasonably flat in the neighborhood of the value derived from Eq. 6.

In some instances, when p is very large and n is too small, the mixing of the MCMC algorithm may be too poor. In these cases, the only alternative is to fix λ, using `prior$lambda$type="fixed"`, and then choose the value of λ either using Eq. 6 or some type of cross-validation. Again, our experience dictates that predictions are relatively robust with respect to λ, within reasonable ranges in the neighborhood of the value given by Eq. 6. Importantly, although Bayesian methods provide a way of dealing with the "*large-p with small-n*" problem, n cannot be too small; otherwise we are at the mercy of the prior. Establishing a minimum number of records is difficult, because this will depend on factors such as number of markers, heritability and effective population size. However, the consequence of a too-small n should be evident from inspection of trace plots of the residual variance or regularization parameters.

Future developments. Over the past two years we have identified a series of limitations in the program, which we plan to address in future releases. These include:

(a) The available R-version can only handle continuous outcomes; we have already developed a beta version of BLR that handles binary and censored outcomes. This version is available upon request and we plan to incorporate this possibility in future releases.

(b) The design of the user interface and the internal structure of the program impose important restrictions. The fact than only one XL, XR, or GF is allowed, implies that we cannot estimate different regularization or variance parameters for different sets of predictors. We have developed a proposal to modify the program using an interface that will allow including varying numbers of design matrices for effects, each of which will be assigned different priors. This will give the user the possibility of generating a diverse array of models. We plan to incorporate this in future releases.

(c) Another line of development that we plan to pursue is the incorporation of alternatives for priors that induce variable selection such as those of models BayesB [4] or BayesC [12].

Releases from this project will be made available first at the R-forge site http://bglr.r-forge.r-project.org/ and, after a testing period, at the R repository.

4 Notes

1. *Use of the R-package pedigreemm for building a numerator relationship matrix and related objects.* We illustrate how to compute additive relationship matrices and functions of it using the pedigreemm package of R ([9] http://cran.r-project.org/web/packages/pedigreemm/index.html). The example uses a simple pedigree from [13].

```
##  Example 1
#############################################################
 rm(list=ls())
 library(pedigreemm)

# defines the pedigree
   id<-1:6
   sire<-c(NA,NA,1, 1,4,5)
   dam<-c(NA,NA,2,NA,3,2)

# Creating a pedigree object
   ped <- pedigree(label=id,sire=sire,dam=dam)

# Computes inbreeding coefficients
   F<-inbreeding(ped)

# Computes the additive relationship matrix (A)
   A<- getA(ped)

# The cholesky factor of A (upper-triangular)
   U<-relfactor(ped)

# T-inverse and D-inverse from A=TDT'
   TInv <- as(ped, "sparseMatrix")
   DInv <- diag(1/Dmat(ped))

# Computes the inverse of the additive relationship matrix
   AInv<-getAInv(ped)
```

Note the pedigree must be sorted so that ancestors appear before offspring and complete (i.e., any animal that appear as sire or dam must also appear in the id vector. The function `editPed()` can be used to complete and sort a pedigree before the pedigree object is created (see example below).

```
## Example 2
################################################################
  rm(list=ls()); library(pedigreemm)
# Pedigree complete and sorted (From Example 1)
    id<-1:6;
    sire<-c(NA,NA,1,1,4,5);
    dam<-c(NA,NA,2,NA,3,2)
# Now we remove sire 1 and dam 2
# and shuffle the remaining pedigree
    PED<-cbind(id,sire,dam)
    PED<-PED[-c(1,2),]      # removes 1 sire and 1 dam
    PED<-PED[4:1,]          # shuffles pedigree
# Try to build the pedigree object yields an error
    ped<- pedigree(label=PED[,1],sire=PED[,2],dam=PED[,3])

# Now we use the editPed() function to fix the problems
    PED<-editPed(label= PED[,1] ,sire=PED[,2], dam=PED[,3])
    ped<- pedigree(label=PED[,1],sire=PED[,2],dam=PED[,3])
    image(getA(ped))
```

2. *Parameterization of scaled-inverse chi-square densities.* In BLR scaled inverse chi-square densities assigned to variance parameters (σ^2) are parameterized as follows $p(\sigma^2|df, S) \propto [\sigma^2]^{-(1+(df/2))} \times \text{Exp}\{-S/2\sigma^2\}$, with this parameterization the prior expectation and mode of σ^2 are $S/(df-2)$ (for $df > 2$) and $S/(df+2)$, respectively.

3. *Gibbs Sampler.* In a Gibbs sampler [14, 15] draws from a joint density are obtained by sampling from fully conditional densities. To illustrate, let $p(\theta_1, \theta_2|\mathbf{y})$ be the joint posterior density of $\{\theta_1, \theta_2\}$, and let $p(\theta_1|\mathbf{y}, \theta_2)$ and $p(\theta_2|\mathbf{y}, \theta_1)$ denote the fully conditional densities of θ_1 and of θ_2, respectively. Using these densities one could form a Gibbs sampler by sequentially repeating the following steps:

   – Draw one sample of θ_1 from $p(\theta_1|\mathbf{y}, \theta_2)$, denote it $\theta_1^{[t]}$.
   – Set $\theta_1 = \theta_1^{[t]}$.
   – Draw one sample of θ_2 from $p(\theta_2|\mathbf{y}, \theta_1)$, denote it $\theta_2^{[t]}$.
   – Set $\theta_2 = \theta_2^{[t]}$.

   Repeat the above steps B times. Pairs $\{\theta_1^{[t]}, \theta_2^{[t]}\}$ drawn after convergence to the posterior density can be regarded as draws from the joint posterior density.

4. *Storage of samples collected by BLR.* As BLR runs, samples of fixed effects, variance and regularization parameters are stored in text files. By default, samples are saved to the working directory. However, one could use the `saveAt` argument to provide a path and prefix to be appended to filenames. For example, if one sets `saveAt="c:/myExamples/example1_"`, the samples will be saved at `c:/myExamples` and the prefix `example1_` will be added to every file that is created

by BLR. Importantly, if several runs are performed in the same folder and with the same saveAt value, samples from different runs will be stacked in the same output files. To avoid this, the user can change the path/prefix provided at saveAt (see line 6, Box 2, for an example) or remove files from previous analysis.

5. *Incidence matrices for effects.* In the example in Box 2, we use the argument XF to provide the incidence matrix for fixed effects. These should be numeric matrices, objects of other classes (e.g., data.frame) will induce errors. BLR does not do a comprehensive check of all inputs; therefore, we always recommend checking that the object type is as required (see help (BLR) for a complete description of input-types). For class-variables (e.g., sex, year-effects), the user needs to create the appropriate contrasts (dummy variables) in advance. The model.matrix() function of R can be used to create incidence matrices for class and continuous variables. If the incidence matrices have column-names, these are appended to the estimated posterior means (type names(fm$bF) in the example in Box 2 to check this).

6. We explain how to set ID and A in the GF argument for cases with and without repeated measures. Suppose we have a random effect with three levels, and assume that the A matrix has the following form:

	ID1	ID2	ID3
ID1	1.1	0.5	0.0
ID2	0.5	1.1	0.0
ID3	0.0	0.0	1.0

If we have only one record by level of the random effect (i.e., in absence of repeated measures), we sort phenotypes according to the ID variable and use the following code to set the GF argument:

```
A<-rbind(c(1.1,0.5,0),c(0.5,1.1,0),c(0,0,1))
GF<-list( ID=1:3, A=A )
```

Now, suppose that we have repeated measures. Assume, for instance that there are five phenotypic records and that the phenotypic vector (y) is sorted in a way that the first and third entry of it are associated to the first level of the random effect (ID[1]=ID[3]=1), the second entry of y is associated to the third level of the random effect (ID[2]=3) and the fourth and fifth entries are associated to the second level of the random effect (ID[4]=ID[5]=2). In this case, we set the GF argument as follows:

```
A<-rbind(c(1.1,0.5,0),c(0.5,1.1,0),c(0,0,1))
GF<-list(  ID=c(1,3,1,2,2), A=A )
```

Acknowledgments

de los Campos, Pérez, and Crossa acknowledge financial support from the International Maize and Wheat Improvement Center (CIMMYT). Pérez and de los Campos were also supported by NIH Grants R01GM101219-01 and R01GM099992-01A1.

References

1. Park T, Casella G (2008) The Bayesian lasso. J Am Stat Assoc 103(482):681–686
2. de los Campos G, Naya H, Gianola D, Crossa J, Legarra A, Manfredi E, Weigel K, Cotes JM (2009) Predicting quantitative traits with regression models for dense molecular markers and pedigree. Genetics 182(1):375–385
3. Pérez P, de los Campos G, Crossa J, Gianola D (2010) Genomic-enabled prediction based on molecular markers and pedigree using the Bayesian linear regression package in R. Plant Genome J 3(2):106–116
4. Meuwissen TH, Hayes BJ, Goddard ME (2001) Prediction of total genetic value using genome-wide dense marker maps. Genetics 157(4):1819–1829
5. de los Campos G, Hickey JM, Detwyler HD, Pong-Wong R, Calus MPL (2013) Whole genome regression and prediction methods applied to plant and animal breeding. Genetics 193:327–345
6. R Development Core Team (2010) R: a language and environment for statistical computing. R Foundation for Statistical Computing, Vienna, Austria [Internet]. R Foundation for Statistical Computing. http://www.R-project.org
7. Hoerl AE, Kennard RW (1970) Ridge regression: biased estimation for nonorthogonal problems. Technometrics 12(1):55–67
8. Vazquez AI, Bates DM, Rosa GJM, Gianola D, Weigel KA (2010) Technical note: an R package for fitting generalized linear mixed models in animal breeding1. J Anim Sci 88(2):497–504
9. Bates D, Vazquez AI (2009) Pedigreemm: pedigree-based mixed-effects models V 0.2-4 [Internet]. http://cran.r-project.org/web/packages/pedigreemm/index.html
10. McLaren CG, Bruskiewich RM, Portugal AM, Cosico AB (2005) The international rice information system. A platform for meta-analysis of rice crop data. Plant Physiol 139(2):637–642
11. Spiegelhalter DJ, Best NG, Carlin BP, van der Linde A (2002) Bayesian measures of model complexity and fit. J R Stat Soc Series B (Stat Methodol) 64(4):583–639
12. Habier D, Fernando R, Kizilkaya K, Garrick D (2011) Extension of the Bayesian alphabet for genomic selection. BMC Bioinforma 12(1):186
13. Mrode RA, Thompson R (2005) Linear models for the prediction of animal breeding values [Internet]. Cabi. [cited 15 Aug 2012] http://books.google.com/books?hl=en&lr=&id=bnewaF4Uq2wC&oi=fnd&pg=PR5&dq=mrode+animal+breeding+genetics&ots=05EITYRwMh&sig=aDDnpg69HSI4-acAmi38yTAVMjqg
14. Geman S, Geman D (1984) Stochastic relaxation, Gibbs distributions, and the Bayesian restoration of images. IEEE Trans Pattern Anal Mach Intell 6:721–741
15. Casella G, George EI (1992) Explaining the Gibbs sampler. Am Stat 46:167–174

Chapter 13

Genomic Best Linear Unbiased Prediction (gBLUP) for the Estimation of Genomic Breeding Values

Samuel A. Clark and Julius van der Werf

Abstract

Genomic best linear unbiased prediction (gBLUP) is a method that utilizes genomic relationships to estimate the genetic merit of an individual. For this purpose, a genomic relationship matrix is used, estimated from DNA marker information. The matrix defines the covariance between individuals based on observed similarity at the genomic level, rather than on expected similarity based on pedigree, so that more accurate predictions of merit can be made. gBLUP has been used for the prediction of merit in livestock breeding, may also have some applications to the prediction of disease risk, and is also useful in the estimation of variance components and genomic heritabilities.

Key words Genomic, Prediction, Genomic selection, BLUP, Genetics, Genome-wide association studies, Methods

1 Introduction

1.1 DNA Markers for the Prediction of Merit

The availability of dense DNA marker information has enabled the large-scale genotyping of individuals for prediction of an individual's genetic merit. The most common markers used for the prediction of disease risk and genetic merit are called single nucleotide polymorphisms (SNPs) and are abundant on the genome. These genetic markers have been used for various purposes in human, livestock, and plant genetics. Some uses include: the detection of areas of the genome that have a significant effect on quantitative trait variation (quantitative trait loci—QTL), the prediction of an individual's risk to disease infection, and the estimation of heritability and genetic variance components [1]. In animal and plant industries, genetic markers have also been used to determine the genetic value of individuals so that they can be selected for breeding purposes.

Numerous statistical methods have been useful in helping make these predictions more accurate. A common, simple approach is single marker linear regression, which is used to identify significant regions of the genome and has been extensively used for finding QTL (commonly referred to as association studies). However, there is an important statistical problem that the number of SNP effects is usually much larger than the number of observed phenotypes. One solution is to model SNP effects as random effects and make prior assumptions about the variance explained due to their effects. Some nonlinear methods such as Bayes A, Bayes B [2, 3], and Bayes C [4] give more emphasis to some genomic regions by allowing the variance to differ between SNP loci, whereas the genomic best linear unbiased prediction (gBLUP) method assigns the same variance to all loci and essentially treats them all as equally important.

1.2 Results Using gBLUP in Genomic Research

gBLUP has been examined in many research articles and has been shown to obtain as accurate or more accurate breeding values in livestock breeding programs than pedigree-based BLUP. VanRaden et al. [5] reported increases in breeding value accuracies of 20–50 %, similarly Harris et al. [6] used gBLUP to generate genomic predictions on 4,500 dairy cattle and found that reliabilities were 16–33 % higher than the breeding values based on parent average information for milk production traits. Moser et al. [7] also showed that there was very little difference between using gBLUP and the nonlinear models (i.e., Bayes A, B). However, Habier et al. [8] showed that these predictions quickly erode when the relationship between individuals with phenotypic information and those being predicted reduces. Clark et al. [9] showed that predictions using the Bayes B method would be more accurate if significant QTL exist, but accuracies become more equal to gBLUP when there are many QTL each with a small effect. All of these methods fit all SNP effects simultaneously in a prediction model, and in the context of a genome wide association study, Yang et al. [1] showed that in estimating the variance components and heritability of human height that common SNPs explain a larger proportion of genetic variance than the sum of significant SNPs obtained from single marker regressions.

2 Methods

2.1 Incorporating Marker Information into Best Linear Unbiased Prediction (BLUP)

The use of best linear unbiased prediction (BLUP) has enabled large amounts of genetic gain to be achieved in many livestock breeding programs. The traditional BLUP methodology relies on pedigree information to define the covariance between known relatives. This covariance can also be defined by using large amounts of DNA marker information, most commonly a large number of SNP markers. This matrix is termed the genomic relationship matrix (GRM). We will discuss two methods to incorporate genomic information into BLUP: ridge regression BLUP

(RR-BLUP) and genomic BLUP (gBLUP), and we will show how they are equivalent.

2.1.1 Ridge Regression BLUP

This method was examined by Meuwissen et al. [2] and Habier et al. [8] which assume the model

$$y = 1_n \mu + \sum_i W q_i + e$$

where μ is the mean, W is a matrix that contains genotypes coded as 0, 1, or 2, and q_i is the effect of each SNP. The elements in W in each column j have an amount $2p_j$ (where p_j is the minor allele frequency of marker j) subtracted from the genotype code to achieve that the sum of coefficients in each column is zero. Here, the SNP effects are treated as random and summed over all segments. The genetic variance explained by the SNP effects is given by $WW'\sigma_q^2$ and the residual variance is $I\sigma_e^2$, and the variance–covariance matrix among observations is therefore $WW'\sigma_q^2 + I\sigma_e^2$. The variance for each SNP can be assumed equal. This method has also been termed RR (ridge regression or random regression) BLUP [8] or SNP BLUP. Alternatively, this variance has to be estimated for each SNP, in which a prior distribution of the SNP effects has to be assumed. Bayesian methods have been proposed to achieve this (see, e.g., [2–4]).

2.1.2 Genomic Best Linear Unbiased Prediction

The second method used to combine genomic information into BLUP is using a GRM as a substitute for the numerator relationship matrix and is called gBLUP. The gBLUP method was introduced by VanRaden [10] and Habier et al. [8]. In practice, the model used to implement gBLUP is:

$$y = Xb + Zg + e$$

where y is a vector of phenotypes, X is a design matrix relating the fixed effects to each animal, b is a vector of fixed effects, Z is a design matrix allocating records to genetic values, g is a vector of additive genetic effects for an individual, and e is a vector of random normal deviates with variance σ_e^2. Furthermore $\text{var}(g) = G\sigma_g^2$ where G is the genomic relationship matrix, and σ_g^2 is the genetic variance for this model. Note that the vector g contains animals with phenotypic data but can be extended to animals with no phenotypes. The first group is then referred to as the training or reference population, whereas the latter is the test population or a set of individuals to be predicted.

gBLUP has three important features that make it more desirable to use than RR-BLUP: (1) the dimensions of the genetic effects in the mixed model equations is reduced from $m \times m$ (where m is the number of markers) in RR-BLUP to $n \times n$ (where n is the number of individuals in the population) in gBLUP, which is computationally more efficient; (2) the accuracy of an individual's genomic estimated breeding value (GEBV) can be calculated in the same way as in pedigree-based BLUP; and (3) gBLUP information can be incorporated with pedigree information in a single step method [11].

2.1.3 Equivalence Between gBLUP and RR-BLUP

Habier et al. [8] showed that gBLUP and RR-BLUP are actually equivalent models. The model for gBLUP (ignoring fixed effects) is given by:

$$y = 1_n\mu + Zg + e$$

where y is a vector of phenotypes, 1_n is a vector of ones, μ is the mean, Z is a design matrix allocating records to genetic values, g is a vector of additive genetic effects for an individual, and e is a vector of random normal deviates σ_e^2. The variance of y in this model is given by $(ZGZ'\sigma_g^2 + I\sigma_e^2)$ where σ_e^2 is the residual variance.

Similarly the model for RR-BLUP is given by:

$$y = 1_n\mu + \sum_i W q_i + e$$

where μ is the mean, W is an incidence matrix linking observations to SNP genotypes, q_i is the effect of each SNP which is treated as random, and the variance of y is $\text{var}(y) WW'\sigma_q^2 + I\sigma_e^2$ (assuming equal variance for each SNP), therefore the variance of y in gBLUP and RR-BLUP is the same (*see* **Note 1**).

2.2 Building the Genomic Relationship Matrix

Combining the information from genetic markers into a relationship matrix was first suggested by Nejati Javaremi et al. [12]. Similarly, Villaneuva et al. [13] examined the use of a GRM as a method of genomic evaluation and suggested that when genetic variation is explained by many QTL of small effect, BLUP using a GRM can be used to produce higher accuracy estimates than pedigree-based BLUP by representing additive relationships between individuals based on information using shared DNA markers. Relationship estimates in the GRM can deviate from the expected relationship given in the numerator relationship matrix **A**. For example, variation in the relationship between two full siblings may range from 0.4 to 0.6 instead of the expectation of 0.5 given in **A** [15, 16]. This exploitation of the variation in relationships is what makes the GRM a useful tool in genomic evaluations. Estimates of the GRM can be formed using different methods and various ways to make the GRM have been proposed [1, 10, 14]. Some of these will be presented in this section.

2.2.1 VanRaden [10]

The method presented by VanRaden [10] essentially develops the matrix **W** as presented in Subheading 2.1.1. He defined an incidence matrix **M**, coded as −1, 0, 1 that specifies which alleles each individual has inherited. The minor allele frequency (the SNP allele with the lowest frequency) at locus i is p_i, and the matrix **P** contains the allele frequencies expressed as a difference from 0.5 and multiplied by 2, such that column i of **P** is $2(p_i - 0.5)$. Subtraction of **P** from **M** gives exactly **W** (termed *matrix Z* in [10]). The minor allele frequency correction forces the sum of coefficients across animals to be zero for each marker. I also give more weighting to rare alleles than to common alleles when calculating genomic relationships. The GRM is calculated as $\boldsymbol{G} = \boldsymbol{WW}'/[2\sum p_i(1-p_i)]$. The division by $2\sum p_i(1-p_i)$ places **G** on the same scale to the numerator relationship matrix (**A**) which is used widely in livestock breeding; however, this is only if the allele frequencies used to scale **G** are referring to the same base population as used in **A** (*see* **Note 2**).

2.2.2 Yang et al. [1]

Other genomic matrices have been proposed, such as the one used by Yang et al. [1]. Here, they combined the information on all N SNPs (i) (coded 0, 1, 2) to calculate the relationship between individuals j and k into the GRM (G_{ijk}) using a weighting scheme based on allele frequencies similar to VanRaden [8]. Weighting the off-diagonal and diagonal elements differently, when $j \neq k$ then:

$$G_{jk} = \frac{1}{N}\sum_i G_{ijk} = \frac{1}{N}\sum_i \frac{(w_{ij}-2p_i)(w_{ik}-2p_i)}{2p_i(1-p_i)}$$

When j is equal to k (i.e., the relationship of an individual to itself), then:

$$G_{jk} = \frac{1}{N}\sum_i G_{ijk} = 1 + \frac{1}{N}\sum_i \frac{w_{ij}^2 - (1-2p_i)w_{ij} + 2p_i^2}{2p_i(1-p_i)}$$

where w_{ij} is the element of **W** pertaining to marker i and individual j. These estimates of relationship are all relative to a base population in which the average relationship between individuals is zero (all individuals are completely unrelated). Yang et al. [1] used the individuals in the sample as the base so that the average relationship between all pairs of individuals is 0 and the average relationship of an individual with him- or herself is 1. N is the number of markers.

2.2.3 Goddard et al. [14]

The matrix $\boldsymbol{G}_m$ can be constructed as $\boldsymbol{G}_m = \boldsymbol{WW}'/M$ where each element of matrix **W** is formed as in Yang et al. ([1], see above) and $M = \sum 2p_j(1-p_j)$. The matrix **A** is the matrix with expected numerator relationships as derived from the pedigree information. Then, $\hat{\boldsymbol{G}}$ can be calculated as

$$\hat{\boldsymbol{G}} = [\boldsymbol{A} + b(\boldsymbol{G}_m - \boldsymbol{A})]$$

where σ_a^2 is the additive genetic variance and σ_g^2 is the variance of each of the marker effects, $b = \sigma_g^2/\sigma_a^2$. The regression of $\hat{\boldsymbol{G}}$ back toward **A** is said to now remove some of the error associated with estimating genomic relationships from a finite number of markers, therefore acknowledging that $\hat{G}$ is an estimate of the true genomic relationship G.

2.3 How gBLUP Works

In the model used for RR-BLUP, the variance of known phenotypes can be written as $\mathbf{WW'} + \lambda \mathbf{I}$ where $\mathbf{W}$ links individual phenotypes to the marker effects, which is a matrix of animal genotypes coded 0, 1, or 2 (for the number of copies of a specific allele the animal has), and $\lambda = \sigma_e^2/\sigma_g^2$. The multiplication of $\mathbf{WW'}$ gives the correlation between the genomes of two individuals and the elements of the corresponding matrix have the same expected values as the numerator relationship matrix ($\mathbf{A}$) in the traditional BLUP equations. This is important because it further solidifies that even if there is no linkage disequilibrium (LD) then genomic estimates of merit will have a nonzero value because of the relationships between animals in $\mathbf{WW'}$. However, this also may mean that if LD is low then predictions of merit may quickly erode as the relationship between animals reduces [3, 8, 17, 18].

Using the GRM to compute genomic breeding values has simplified how genomic predictions are estimated and can be easily completed in software such as ASReml [19] (*see* **Note 3**) and R. The GRM combines data on all n animals with phenotypes and genotypes and links it to animals that have genotypes collected but no phenotypic information. ASReml [19] allows for an animal model to be fitted where the inverse of the G matrix is used to fit the covariance structure among the animal effects (*see* **Notes 4 and 5**). The mixed model looks like:

$$\begin{bmatrix} X'X & X'X & 0 \\ Z'X & Z'Z + G^{11} & G^{12} \\ 0 & G^{21} & G^{22} \end{bmatrix} \begin{bmatrix} b \\ g1 \\ g2 \end{bmatrix} = \begin{bmatrix} X'y \\ Z'y \\ 0 \end{bmatrix}$$

The positions of the GRM G^{11} is the subset of individuals that have phenotypic and genotypic information recorded on them, positions G^{12} and G^{21} pertain to the relationships between the animals with phenotypic data and those without, and G^{22} represents the animals without phenotypic measurements. The breeding values of animals without phenotypes are therefore estimated as:

$$\hat{g_2} = -(G^{22})^{-1} G^{21} \hat{g_1}$$

This is the genomic regression of breeding values of animals without data on the breeding value of animals with data or gBLUP.

2.4 A Simple Example Using Genomic BLUP

Let us assume we have five animals, four have phenotypes and we wish to use genotypic information to predict the breeding value of the fifth animal. Let us also assume that animal 1 is the parent of 2, 3, and 5, with there being no information about the ancestry of animal 4. If we assume that fixed effects are known and that each value of y is a deviation from the mean using BLUP based on pedigree, we obtain estimates of each animal as $\hat{u} = (Z'Z + A^{-1})^{-1} Z'y$. To obtain estimates using genomic information G^{-1} replaces A^{-1}.

$$A = \begin{bmatrix} 1 & 0.5 & 0.5 & 0 & 0.5 \\ 0.5 & 1 & 0.25 & 0 & 0.25 \\ 0.5 & 0.25 & 1 & 0 & 0.25 \\ 0 & 0 & 0 & 1 & 0 \\ 0.5 & 0.25 & 0.25 & 0 & 1 \end{bmatrix}$$

$$G = \begin{bmatrix} 1.0 & 0.50 & 0.50 & 0.02 & 0.50 \\ 0.50 & 1.0 & 0.20 & 0.015 & 0.20 \\ 0.5 & 0.20 & 1.0 & 0.025 & 0.30 \\ 0.02 & 0.015 & 0.025 & 1.0 & 0.20 \\ 0.5 & 0.20 & 0.30 & 0.02 & 1.0 \end{bmatrix}$$

The A matrix is derived from the path coefficients from the pedigree. Whereas the G is an arbitrary example in which animal 3 is more similar to animal 5 than animal 2 (based on the expected degrees of relationship for half siblings). In the pedigree example, animal 4 is completely unrelated to the other animals; however, now with genomic data animal 4 shares some information with the other ones.

When assuming a heritability of 0.25, the breeding value of animal 5 would be estimated as;

$\widehat{u_5} = 0.5\widehat{u_1}$ (Note that $\widehat{u_1}$ contains also phenotypic information from animals 2 and 3).

Whereas under gBLUP, the prediction would be according to:

$$\widehat{g_5} = 0.499\widehat{g_1} - 0.026\widehat{g_2} + 0.0622\widehat{g_3} + 0.0144\widehat{g_4}$$

2.5 What Information Is Used

The regression coefficients presented above may not make sense at first because it seems illogical that the weight on the breeding value of animal 2 is negative and the weight on animal 4, which is far less related, is positive. The reason for this is that most of the information in 2 is used to predict the breeding value of 1. The importance of different sources of information can also be illustrated by the regression of the breeding values on phenotypes. These can be calculated as $\widehat{u_5} = GZ'V'^{-1}y$ and for animal 5 the result shows that

$$\widehat{u_5} = 0.1136\widehat{y_1} - 0.0455\widehat{y_2} + 0.0455\widehat{y_3}$$
$$\widehat{g_5} = 0.1135\widehat{y_1} + 0.0328\widehat{y_2} + 0.0591\widehat{y_3} + 0.0519\widehat{y_4}$$

This has important consequences for interpreting the results of gBLUP. It illustrates that gBLUP and Pedigree BLUP are very similar and share similar sources of information. It also illustrates that information on unrelated individuals may now be included in an estimate of a breeding value. Given that animal 4 contributed no information in pedigree BLUP and now influences the prediction in gBLUP. In this example, this would have a small effect, as the coefficient is small, but in a larger reference population there may be thousands of records contributing a small amount of information, altogether contributing to a large increase in accuracy. Another important implication of using genomic relationships in BLUP is that now known siblings can contribute different amounts of information to breeding value estimates, as there is now some ability to differentiate between these animals and access some of the within family variation, due to Mendelian sampling.

Example 1 VanRaden 2008 implementation of the GRM and gBLUP

```
#Making the genomic relationship matrix
nmarkers=1000
data = matrix(scan("genotypes.txt"),ncol=nmarkers,byrow=TRUE);
sumpq=0
freq=dim(data)[1]
P=freq
lamda=ncol(data)
for(i in 1:ncol(data)){(freq[i]<-((mean(data[,i])/2)))
(P[i]=(2*(freq[i]-0.5)))
(sumpq=sumpq+(freq[i]*(1-freq[i])))}
Z<-data
for(i in 1:nrow(data)){
    for(j in 1:ncol(data)){(Z[i,j]<-((data[i,j]-1)-(P[j])))}
}
Zt=t(Z)
ZtZ=Z%*%Zt
G=ZtZ/(2*sumpq)
G

#gBLUP
for(i in 1:nrow(G)){
    (G[i,i]<-((G[i,i]+0.01)))}
y=matrix(scan("phen.txt"),ncol=1,byrow=TRUE);
I=matrix(1,100,1)
    EBV=(solve(1+(lamda*solve(G))))%*%y
```

3 Notes

1. The gBLUP method that uses a GRM is equivalent to the RR-BLUP method only when markers in WW' in the RR model are scaled the same as those used to calculate G, i.e., $XX' = X_{ij}/2 \sum p_i(1-p_i)$. Furthermore gBLUP and RR-BLUP are only equivalent when $G\sigma_g^2 = WW'\sigma_q^2$

2. When building the GRM, it may be important to examine the allele frequencies used to scale the matrix. In the original gBLUP method of VanRaden [10], it was proposed that frequencies should be of the base population and therefore needed to be estimated. However, recent work by Forni et al. [20] suggests that similar results can be obtained using the allele frequencies of the current population. The definition of the allele frequencies affects when the base animals are observed and is mainly important when gBLUP is used in the single step method of Misztal et al. [11]. The single step method combines pedigree and genomic information so that this information can be used to predict a single breeding value. Combining this information requires the base populations of genomic and pedigree relationship populations to be the same.

3. The ASReml software is easily used to implement gBLUP and can be downloaded from: http://www.vsni.co.uk/downloads/asreml.

4. To undertake gBLUP using ASReml, the user must provide a predefined GRM. This can be inverted and loaded into ASReml as a .giv file.

5. Often large data sets in ASReml can require more memory to be allocated to the program, this can be achieved by entering –s8 into the command line (see ASReml users guide for more details).

References

1. Yang J, Benyamin B, McEvoy BP et al (2010) Common SNPs explain a large proportion of the heritability for human height. Nat Genet 42:565–571

2. Meuwissen THE, Hayes BJ, Goddard ME (2001) Prediction of total genetic value using genome-wide dense marker maps. Genetics 157:1819–1829

3. Habier D, Tetens J, Seefried FR et al (2010) The impact of genetic relationship information on genomic breeding values in German Holstein cattle. Genet Sel Evol 42:5

4. Habier D, Fernando RL, Kizilkaya K et al (2010) Extension of the Bayesian alphabet for genomic selection. In: Proceedings of the ninth congress on genetics applied to livestock production, Leipzig, 1–6 Aug 2010

5. VanRaden PM, Van Tassell CP, Wiggans GR et al (2009) Invited review: reliability of genomic predictions for North American Holstein bulls. J Dairy Sci 92:16–24

6. Harris BL, Johnson DL, Spelman RJ (2008) Genomic selection in New Zealand and the implications for national genetic evaluation. In: Sattler JD (ed) Proceedings of the 36th ICAR session, Niagara Falls, New York, pp 325–330

7. Moser G, Tier B, Crump RE et al (2009) A comparison of five methods to predict genomic breeding values of dairy bulls from genome-wide SNP markers. Genet Sel Evol 41:56

8. Habier D, Fernando RL, Dekkers JCM (2007) The impact of genetic relationship information on genome-assisted breeding values. Genetics 177:2389–2397

9. Clark S, Hickey JM, van der Werf JHJ (2011) Different models of genetic variation and their effect on genomic evaluation. Genet Sel Evol 43:18

10. VanRaden PM (2008) Efficient methods to compute genomic predictions. J Dairy Sci 91:4414–4423

11. Misztal I, Legarra A, Aguilar I (2009) Computing procedures for genetic evaluation including phenotypic, full pedigree, and genomic information. J Dairy Sci 92:4648–4655
12. Nejati-Javaremi A, Smith C, Gibson JP (1997) Effect of total allelic relationship on accuracy of evaluation and response to selection. J Anim Sci 75:1738–1745
13. Villanueva B, Pong-Wong R, Fernandez J et al (2005) Benefits from marker-assisted selection under an additive polygenic genetic model. J Anim Sci 83:1747–1752
14. Goddard ME, Hayes BJ, Meuwissen TH (2011) Using the genomic relationship matrix to predict the accuracy of genomic selection. J Anim Breed Genet. doi:10.1111/j.1439-0388.2011.00964.x
15. Visscher PM, Medland SE, Ferreira MA et al (2006) Assumption-free estimation of heritability from genome-wide identity-by-descent sharing between full siblings. PLoS Genet 2:e41
16. Hill WG, Weir BS (2011)Variation in actual relationship as a consequence of Mendelian sampling and linkage. Genet Res doi:10.1017/S0016672310000480. http://dx.doi.org/
17. Hayes BJ, Visscher PM, Goddard ME (2009) Increased accuracy of artificial selection by using the realized relationship matrix. Genet Res 91:47–60
18. Clark SA, Hickey JM, Daetwyler H et al (2012) The importance of information on relatives for the prediction of genomic breeding values and the implications for the makeup of reference data sets in livestock breeding schemes. Genet Sel Evol 44:4
19. Gilmour AR, Gogel BJ, Cullis BR et al (2009) ASReml user guide release 30. VSN International, Hemel Hempstead
20. Forni S, Aguilar I, Misztal I (2011) Different genomic relationship matrices for single-step analysis using phenotypic, pedigree and genomic information. Genet Sel Evol 43:1

Chapter 14

Detecting Regions of Homozygosity to Map the Cause of Recessively Inherited Disease

James W. Kijas

Abstract

Homozygosity is a component of genetic patterning that can be used to search for the cause of genetic disease. In this chapter, methods are presented to analyze SNP data for the presence of homozygosity. Two exercises demonstrate methods to define runs of homozygosity, to identify shared homozygosity between individuals, and to evaluate the results in light of the expectations of a recessively inherited genetic disorder. An example dataset is used to aid in data interpretation.

Key words Recessive disease, Homozygosity mapping, Shared haplotypes, SNP genotypes, PLINK

1 Introduction

Understanding patterns of genetic variation can greatly assist in our quest to understand aspects of evolution, population history, and the basis of inherited disease. One particularly informative aspect of genetic patterning is homozygosity, whereby a genomic region lacks polymorphism. By scoring the genotypic status of loci as either homozygous or heterozygous, genomic regions within an individual can be identified where consecutive DNA markers are uniformly homozygous. These are termed runs of homozygosity (ROH) and are classified according to length. Short ROH in humans typically span tens of kilobases (Kb). Intermediate length ROH extend between hundreds of Kb up to several megabases (Mb) while long ROH may be up to tens of Mb in length [1].

1.1 Mechanisms Underpinning Runs of Homozygosity

There are at least two mechanisms that give rise to homozygous regions. The first is a high level of relatedness between individuals within a population. This may be due to a past historic bottleneck that has decreased the effective population size. In domestic animals it may be due to breed boundaries or the persistent use of a small number of high value sires. Whatever the reason for high levels of relatedness, the consequence is to increase the likelihood that an

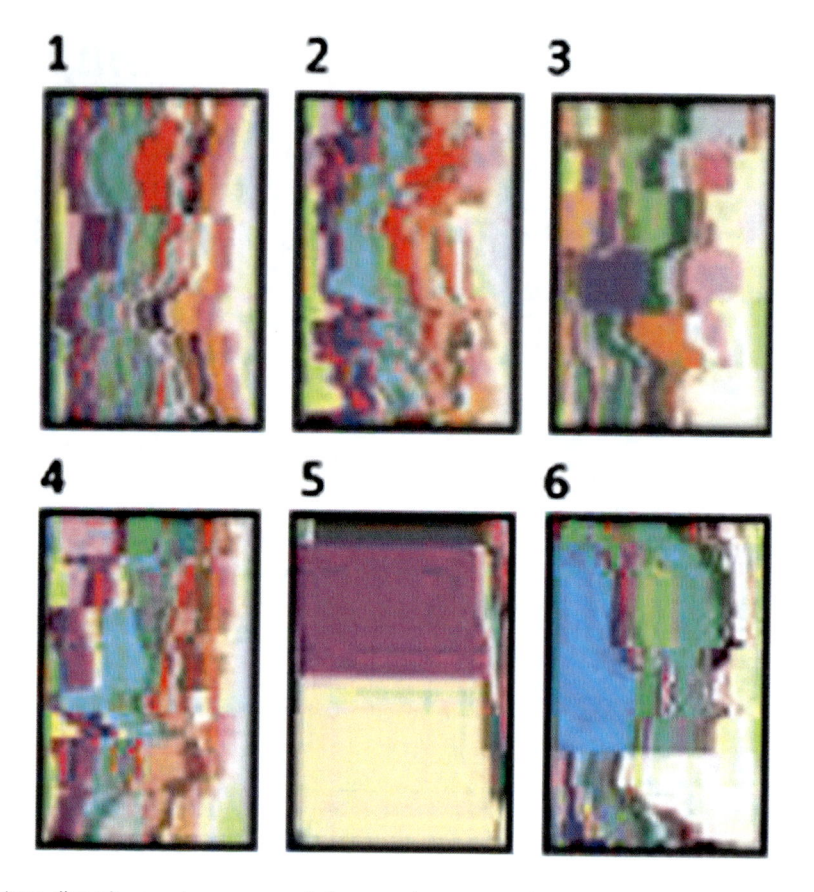

Fig. 1 Haplotype diversity on chromosome 1 for a region surrounding the *GDF8* gene in six sheep breeds (1 = Leccese, 2 = Karakas, 3 = Irish Suffolk, 4 = Norduz, 5 = Texel, 6 = Romney). Different *colors* were used to distinguish haplotypes. Positive selection has acted to dramatically increase the frequency of the haplotype that carries the c.*1232 G > A mutation responsible for muscle hypertrophy within the Texel population. Figure courtesy of Paul Scheet

individual may inherit chromosomal segments that are identical by descent (IBD) from both parents. When this occurs, the result is a long ROH where the parents are very closely related and increasingly shorter ROH as the relatedness of parents decreases.

The second mechanism is positive selection. In this scenario, the frequency of a mutation that confers some selective advantage will increase over time. Importantly, positive selection also acts on neutral loci in linkage disequilibrium with the beneficial mutation. In cases where positive selection is strong, haplotype diversity can be dramatically reduced to leave only the haplotype that carries the beneficial allele. This phenomenon is referred to as a "selection sweep" and may lead to regions of homozygosity immediately surrounding the beneficial allele [2]. Figure 1 provides a pictorial example of positive selection in sheep. In this case, selection has acted upon a mutant form (c.*1232 G > A) of the *Myostatin*

(*GDF8*) gene that results in a desirable muscle hypertrophy phenotype [3, 4]. The c.*1232 G > A allele is only present in one of the breeds and positive selection has acted to reduce diversity and increase the frequency of the associated haplotype. Such movements in allele frequency towards fixation can be used as a "signature of selection" in experiments that scan the genome for genes likely to have beneficial phenotypic consequences. A powerful illustration of this in chicken served to identify genes likely to have been directly impacted by the domestication process [5].

1.2 Homozygosity Mapping to Find Disease Causing Genes

It has long been known that high levels of relatedness and inbreeding are associated with an increased prevalence of inherited disease. This is particularly the case for recessive diseases, where affected individuals only arise where the disease allele is inherited from both parents. This means affected individuals are homozygous for the sequence variant that underpins disease. Importantly, this homozygosity extends outwards from the causal mutation and can thus be used to trace the location of the causal gene using an approach called homozygosity mapping [6]. This has proven to be a highly effective approach for identification of recessive disease genes in human and the genetic basis for thousands of disorders has been successfully elucidated [reviewed in ref. 7]. The key repository of human disease phenotypes and where known, their genetic basis, is the Online Mendelian Inheritance in Man (OMIM) (http://omim.org/). Homozygosity mapping has also proven effective in domestic species, where researchers have the ability to create research populations with tailored pedigree structures [8]. In particular, the population history and breed structure of species such as domestic dog offer some powerful advantages for disease genetics [9]. The full potential of homozygosity mapping has only been realized in the last few years through the development of high density genotyping arrays. The ability to analyze tens of thousands (cattle, sheep, chicken, goat) or even millions (human, mouse) of SNP loci has revolutionized the investigation of inherited disease. It has also posed analytical challenges for researchers who are neither bioinformaticians nor programmers. This chapter provides a practical guide for laboratory-based researchers with an interest in using SNP data to map the location of recessively inherited genetic diseases. The methods presented are based on freely available software and aim to scan genome-wide SNP data for the presence of ROH in the context of disease.

2 Materials

2.1 PLINK and Analysis of SNP Data

The protocol relies entirely on use of PLINK [10], a C/C++ program designed to manipulate and analyze large SNP datasets. It has become widely used, in part because of the extensive support material that can be found at the program website (http://pngu.mgh.harvard.edu/~purcell/plink/). Also, the program has a graphic user interface (GUI) called gPLINK that assists users handle project files and bypass the command line syntax for many common operations and analyses (http://pngu.mgh.harvard.edu/~purcell/plink/gplink.shtml). The comprehensive nature of PLINK means that a full description is beyond the scope of this chapter. Users are strongly recommended to take the PLINK tutorial that can be found at the website (http://pngu.mgh.harvard.edu/~purcell/plink/tutorial.shtml). This explains how to get started with PLINK and gives a detailed explanation of the input data files (**.ped** and **.map** files) that are an essential prerequisite to the operations described in this chapter. This chapter assumes the reader has installed PLINK and gPLINK and has some experience in manipulating **.ped** and **.map** files.

2.2 Selection of Individuals

The method can be applied where individuals have been collected in a case-control design (where samples are unrelated) or with individuals in a known pedigree. Given the length of shared homozygosity is often proportional to the relatedness of the affected individuals, it is worth giving consideration to how the samples are collected (*see* **Note 1**). The critical component is that the assignment of disease status (affected vs. unaffected) is correct for every individual. As we shall see in **step 7** below, the power of the approach stems from searching for regions of shared ROH in affected individuals that is missing in all asymptomatic controls. If the disease status is wrong for even a small number of individuals, badly misleading results may be obtained.

3 Methods

This methods section is organized into two exercises, each of which is divided into steps. In the first exercise, we will use SNP data to identify ROH within individual genomes. The user-defined parameters needed to drive the analysis are explained, along with the logic used by PLINK to call ROH. The second exercise introduces disease phenotypes (case or control) into the dataset, and then uses them to search for ROH that are shared between affected individuals. Where applicable, key additional PLINK options are given and the output files explained to assist the practitioner with interpretation.

3.1 Exercise 1: Runs of Homozygosity Within Individual Genomes

In this first exercise, we will use PLINK to search SNP genotypes and define ROH.

1. Launch gPLINK and open a project using the pull down menu option "Open." This allows the user to define a local folder

that contains the **.ped** and **.map** files containing the data. The files available for analysis are listed by gPLINK in the "Folder viewer" pane on the left hand side.

2. It is always a good idea to check that the data format is correct and to view the parameters of the dataset before starting a series of analyses. This is accomplished using the "Validate Fileset" option implemented within gPLINK. Go to the "PLINK" pull down menu, select the "Summary Statistics" category and then select "Validate Fileset." After nominating the relevant **.ped** and **.map** files and giving the output file a name, clicking "OK" will initiate the analysis within PLINK.

   The information about the dataset is contained within the **.log** file (*see* **Note 2**). If the number of SNP encountered by PLINK are different to what you expect, or the number of individuals varies from expectation it is important to find this out now. The **.log** file also contains information about sex (if known) and disease status (case or control).

3. It is often useful to have a list of the individuals that are present within the dataset, along with some basic information about the SNP. Go to the "PLINK" pull down menu, select the "Summary Statistics" category and then select "Missingness." This prompts gPLINK to construct a command line for PLINK with the *–missing* flag (*see* **Note 3**). Assume we have called the operation "indivs_and_SNP."

   ```
   plink --file mydata ---missing --out indivs
   _and_SNP
   ```

   Running the analysis creates two very useful output files. Firstly, **indivs_and_SNP.imiss** contains the FID (family ID) and IID (Individual ID) for each individual in the project. Secondly, **indivs_and_SNP.lmiss** contains a list with the SNP (SNP identifiers) and their CHR (chromosome number) for each marker. To open these, or other output files, use the "right click" mouse button after highlighting the desired file within the gPLINK folder viewing pane.

4. Before using PLINK to define ROH, it is important to understand how the analysis works. This will assist with identifying reasonable values for each of the parameters needed to analyze your dataset. The algorithm looks at the SNP data within a defined window and makes a call about if this window can be declared as homozygous or not (yes or no). The window then slides along the chromosome and the test is repeated. This generates yes/no information relating to each window of SNP, rather than to SNP individually. To assign the boundaries of the ROH, the algorithm then calculates an SNP-wise metric using the proportion of homozygous windows in which each SNP resides.

Does the window span a homozygous block?
Using the two thresholds set below, in this case YES

1) Set allowable heterozygotes using --homozyg-window-het 1
2) Set allowable missing data using --homozyg-window-missing 2

The size of the sliding window (orange box) was defined using
--homozyg-window-snp 10   or
--homozyg-window-kb 35

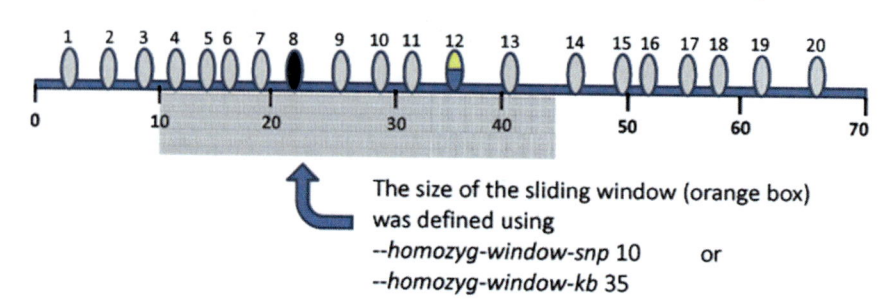

Fig. 2 The first step in homozygosity mapping involves setting the dimension of the sliding window and the parameters that direct if the window can be declared homozygous. Twenty SNP are represented (*ovals*) that are distributed across a 70 Kb region of a chromosome. Homozygous SNP are shown in *light gray* (e.g., SNP1), heterozygous SNP with mixed shades (e.g., SNP 12), and SNP with missing data in *black* (SNP 8). In this instance, the presence of one heterozygous SNP and one missing SNP is permissible. The window would therefore be declared "homozygous" before it moved along the chromosome

PLINK has commands that control these two aspects of the analysis separately (i.e., if windows are homozygous, and then the SNP-wise calculation). First, two options are available to set the size of the sliding window. To define the region in terms of SNP number, use the *–homozyg-window-snp* option. Alternatively, if you want to specify window size using a physical distance, use the *–homozyg-window-kb* flag. You also need to consider the tolerance around how PLINK will declare homozygosity within each window. To allow a certain number of heterozygous calls within regions declared homozygous, use the *–homozyg-window-het* command. Setting this to a number that represents 0.5–1 % of SNP in the window guards against a small number of genotyping errors excluding true homozygous regions. To allow a certain number of SNP to have missing data, use the *–homozyg-window-missing* command. Figure 2 provides a schematic example to illustrate how the parameters work.

The second part of the algorithm decides if a given SNP is present within an ROH segment. This is achieved by calculating the proportion of homozygous windows that overlapped the SNP under consideration. Segments are then called based on this metric. The proportion is set using *–homozyg-window-threshold* and the default value is 0.05. This is intentionally low to ensure SNP at the extreme ends of true ROH are not excluded; however, it can be raised by the user. Generally ROH analysis can be run using the default value.

To perform the ROH analysis, go to the "PLINK" pull down menu, select the "IBD Estimation" category and then select "Runs of homozygosity." gPLINK then provides the user with the opportunity to alter any of the window or threshold parameters. After giving the operation a name (e.g., ROHtest1) and clicking "OK" the analysis will commence. An example command string is given below:

```
plink --file mydata ---homozyg ---homozyg-window-snp 50 ---homozyg-window-het 1 ---homozyg-window-missing 2 --out ROHtest1
```

The key output is contained within two files.

ROHtest1.hom.indiv contains summary information for each of the individuals tested. The output has the following six columns:

FID	Family ID
IID	Individual ID
PHE	Phenotype is unaffected(1), affected (2) or missing(0)
NSEG	Number of segments for the individual declared homozygous
KB	Total number of kb contained within homozygous segments
KBAVE	Average size of homozygous segments for this individual

This is useful to compare the number and average size of ROH across the set of individuals. If no segments have been called (NSEG = 0), then a modification to the parameters is needed before the analysis is repeated (*see* **Note 4**).

ROHtest1.hom contains a line for every ROH segment assigned across individuals. The output has the following columns:

FID	Family ID
IID	Individual ID
CHR	Chromosome
SNP1	SNP at start of ROH
SNP2	SNP at end of ROH
POS1	Physical position (bp) of SNP1
POS2	Physical position (bp) of SNP2
KB	Length of region (kb)
NSNP	Number of SNPs in run
DENSITY	Average SNP density
PHOM	Proportion of sites homozygous
PHET	Proportion of sites heterozygous

The results contained can be used to start an examination of the genes that are present within each homozygous segment. However, the true power of homozygosity mapping for disease comes from the comparison of segments between individuals. This is explored in the second section.

3.2 Exercise 2: Runs of Homozygosity Within Individual Genomes

In this exercise we introduce the disease status of individuals and use this information to zero in on the chromosomal region that harbors disease causal mutations. The key here is to recall that the disease causing mutation and surrounding SNP will be homozygous within affected individuals for recessive disorders. This sets up three predictions that we will use to narrow in on the causal mutation. These predictions are that:

(a) The region immediately surrounding the causal mutation will be in an ROH.

(b) Each affected individual will be homozygous for the same shared SNP haplotype.

(c) No unaffected individuals will be homozygous for this disease carrying haplotype.

To illustrate the output and assist with interpretation, we will consider a toy dataset that contains SNP genotypes from six Angus cattle sampled from a single herd. Three of the animals (cows 1–3) are affected with a recessive disease, two are phenotypically normal (cow 4 and 6) while the last animal (cow 5) has early signs that are consistent with a number of similar types of diseases.

5. In order to introduce disease status information into the **.ped** file containing the genotypes, we first need to prepare a phenotype file. The format for this is given below:

```
FID     IID     Pheno
Angus   cow1    2
Angus   cow2    2
Angus   cow3    2
Angus   cow4    1
Angus   cow5    0
Angus   cow6    1
```

Save this as a text file without the header row (**Cow_phenos.txt**). It is worthwhile noting that homozygosity mapping relies on each of the affected animals having received the causal mutation IBD. To preserve the power of the analysis, it is therefore best to code animals that have ambiguous phenotypic status as unknown (e.g., cow 5) while keeping them in the dataset.

6. To introduce the data, use the *–pheno* command along with the name of the file that contains the disease information. Further, to rewrite a new version of the **.ped** file containing the updated information use the *–recode* flag. This can all be done in one operation using the command line below. Our new **.ped** and **.map** files will be named "cows."

```
plink --file mydata ---pheno Cow_phenos.txt ---recode --out cows
```

Once the operation has been run, check the **.log** file to ensure the data in Cow_phenos.txt has been correctly read (*see* **Note 5**). This will be indicated in the **.log** file as follows:

```
Reading alternate phenotype from [Cow_phenos.txt]
6 individuals with non-missing alternate phenotype
Missing phenotype value is also -9
3 cases, 2 controls and 1 missing
```

7. We are now ready to perform the key step in homozygosity mapping. If we add the *--homozyg-group* flag in front of *--homozyg*, PLINK prepares a detailed output file describing how homozygous segments overlap between animals. Note in the command line below we have specified that the analysis should be performed on "cows" rather than "mydata" (*see* **Note 2**).

```
plink --file cows ---homozyg ---homozyg-group --out cows_compared
```

After performing the operation PLINK generates a number of output files, the most important of which is **cows_compared.hom.overlap**. The complete list of columns in the output is given below:

```
Pool    Groups of overlapping segments are num-
        bered consecutively
FID     Family ID
IID     Individual ID
PHE     Phenotype of individual
CHR     Chromosome
SNP1    SNP at start of segment
SNP2    SNP at end of segment
BP1     Physical position of start of segment
BP2     Physical position of end of segment
KB      Physical size of segment
NSIM    Number of segments in the pool that match
        this one
GRP     Allelic-match grouping of each segment
```

The contents of the file for our toy example is reproduced below, however in the interests of space the SNP1 and SNP2 columns that provide the SNP identifiers have been removed.

```
Pool  FID    IID    PHE   CHR  BP1       BP2       KB     NSNP  NSIM  GRP
S1    Angus  cow1   2     20   56035015  61863283  5828   128   1     1
S1    Angus  cow2   2     20   50044555  62593419  12549  258   1     1*
S1    Angus  cow3   2     20   60954990  66256730  5302   128   0     2*
S1    CON    3      3:0   20   60954990  61863283  908    21    NA    NA
S1    UNION  3      3:0   20   50044555  66256730  16212  348   NA    NA

S2    Angus  cow2   2     27   20592947  26008363  5415   120   1     1
S2    Angus  cow4   1     27   8927449   28386241  19459  405   1     1*
S2    CON    2      1:1   27   20592947  26008363  5415   120   NA    NA
S2    UNION  2      1:1   27   8927449   28386241  19459  405   NA    NA
```

To assist with interpretation, the same information is also given in schematic form in Fig. 3. Focussing on the segments in Pool S1, we can see all three of the individuals share an overlapping region on cattle chromosome 20. Importantly, the output contains two lines of information to assist with interpretation called CON and UNION (Fig. 3). A ratio is provided of affected: unaffected individuals specific to each pool. In our example, the ratio for segments in Pool S1 is 3:0. The region adheres to predictions (a) and (c), i.e., the region from bp 60954990 to 61863283 (spanning 21 SNP) is both homozygous in all affected animals *and* it appears to be absent in the normal cattle tested. By contrast, the next Pool (S2) is composed of haplotypes shared by affected and normal individuals and can be discounted in the search for the disease.

The NSIM and GRP columns indicate if the homozygous segments share the same haplotype. We can use this to determine if region S1 CON conforms to the final prediction (b): each affected individual will be homozygous for the *same* shared SNP haplotype. NSIM gives the number of other animals that share the same haplotype. The NSIM value for cow3 is zero, indicating that it does not share the same allelic content across the region as cow1 or cow2. This prompted PLINK to assign it a new group number (GRP) of 2. Where analysis is performed using many more individuals than six, these outputs (NSIM and GRP) assist in classifying individuals into sets based on their haplotype similarity.

8. The final step in the exercise drills down to the genotypes. This enables the user to evaluate the SNP information immediately

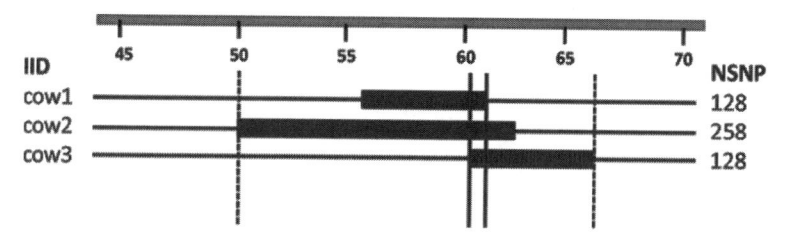

Fig. 3 Overlapping homozygous segments are shown in three cattle (cows 1–3) for a 25 Mb region of cattle chromosome 20. Homozygous segments are indicated using *black boxes* and the number of SNP within each segment is indicated at *right* (NSNP). The position of each ROH is given in base pairs in the output file, and converted to Mb in the schematic. The PLINK output file also contains information about the total distance spanned by the three homozygous regions. The *dashed vertical lines* show the total distance spanned by ROH across all animals, and this is given as UNION within the output file. Further, the minimum shared region common to each animal is indicated using *vertical solid lines*, and is described in the output as CON for the consensus region

surrounding and within regions shared between haplotypes. We will continue with our cattle example and look at the genotypes in Pool S1. This is achieved by adding the *–homozyg-verbose* flag and a new output file name to the command line we used in step 7.

```
plink --file cow ---homozyg ---homozyg-group --homozyg-verbose --out ROHtest2
```

PLINK then generates the following output files:

ROHtest2.hom contains a line for each ROH (*see* **step 4**).

ROHtest2.hom.indiv contains summary information for individuals (*see* **step 4**).

ROHtest2.hom.overlap contains segment pool information (*see* **step 7**).

ROHtest2.hom.overlap.S1.verbose contains the SNP genotypes for pool S1.

ROHtest2.hom.overlap.S2.verbose contains the SNP genotypes for pool S2 etc.

PLINK outputs a separate **.verbose** file for each of the segment pools listed in the **.overlap** file. This is because the **.verbose** files can be large where many individuals are being analyzed. The contents of **ROHtest2.hom.overlap.S1.verbose** are given below:

```
     FID     IID    GRP
1)   Angus   cow1   1
2)   Angus   cow2   1
3)   Angus   cow3   2

SNP                          1        2        3
ARS-BFGL-NGS-106134         [A/A]    [A/A]    G/A
BTB-01456930                [G/G]    [G/G]    G/G
BTB-01899482                [C/C]    [C/C]    C/C

ARS-BFGL-BAC-34303          [A/A]    [A/A]    [C/A]
ARS-BFGL-NGS-14979          [A/A]    [A/A]    [G/G]
BTA-94914-no-rs             [G/G]    [G/G]    [A/A]
ARS-BFGL-NGS-44355          [G/G]    [G/G]    [G/G]
ARS-BFGL-NGS-20185          [A/A]    [A/A]    [A/A]
ARS-BFGL-NGS-106785         [G/G]    [G/G]    [A/A]
ARS-BFGL-NGS-115561         [G/G]    [0/0]    [G/G]
BTB-00793885                [A/A]    [A/A]    [G/G]
BTB-00793877                [G/G]    [G/G]    [G/G]
ARS-BFGL-NGS-85952          [A/A]    [A/A]    [G/G]
BTB-00793802                [A/A]    [A/A]    [G/G]
ARS-BFGL-NGS-24454          [A/A]    [A/A]    [A/A]
Hapmap51796-bta-107866      [G/G]    [G/G]    [G/G]
```

```
BTA-13506-no-rs             [A/A]   [A/A]   [G/G]
ARS-BFGL-NGS-15533          [C/C]   [C/C]   [C/C]
BTB-00794397                [A/A]   [A/A]   [G/G]
BTA-50981-no-rs             [A/A]   [A/A]   [G/G]
ARS-BFGL-NGS-99194          [A/A]   [A/A]   [A/A]
ARS-BFGL-BAC-34915          [G/G]   [G/G]   [G/G]
BTB-00791947                [A/A]   [A/A]   [G/G]
Hapmap40754-bta-50982       [G/G]   [G/G]   [A/A]

ARS-BFGL-NGS-2085           G/A     [G/G]   [A/A]
ARS-BFGL-NGS-28540          G/G     [G/G]   [G/G]
Hapmap54252-rs29018304      A/G     [G/G]   [A/A]
ARS-BFGL-NGS-119910         A/A     [0/0]   [A/A]
ARS-BFGL-NGS-28540          G/G     [G/G]   [G/G]
Hapmap54252-rs29018304      A/G     [G/G]   [A/A]
ARS-BFGL-NGS-119910         A/A     [0/0]   [A/A]
```

The top four lines of the file list the animals (FID and IID) and give each a numeric identifier in the form of 1)–3). The remainder of the file is formatted differently. Each individual has their genotypes listed in a column, and the identity of the individual is given by the numeric identifier. In our example, the first column contains the SNP name and the second column contains genotypes for cow 1. PLINK uses a number of conventions to assist with interpretation. First, genotypes are enclosed by brackets where SNP are part of the defined ROH. Recall this doesn't mean every SNP genotype is homozygous, depending on the value set for –*homozyg-window-het*. The **.S1.verbose** file contains SNP data for the entire UNION region (refer to Fig. 3). Contained within this is the CON region shared between individuals (Fig. 3). An empty row is inserted by PLINK to indicate the start and end of the CON region. It can also be identified as the only region where all individuals have their genotypes enclosed within brackets. In our example the CON region spans 21 SNP (ARS-BFGL-BAC-34303 to Hapmap40754-bta-50982).

Inspection of the genotypes in the CON region confirms the haplotype carried by cow 3 is different from the haplotype present in cows 1 and 2. Recall PLINK indicated this by assignment of separate GRP numbers, and these are listed in each **.verbose** file. The biological interpretation in our example is that the region does not contain the mutation underpinning the disease, based on the assumption cow3 is affected. If cow3 is affected with a separate disorder, and therefore can be considered "normal" with respect to the disease under investigation, our interpretation would change. Assuming the phenotypic assignment for the six cows is correct, then we have to conclude that the SNP data is unable to define the critical interval housing the disease mutation. This may be due to the homozygous region surrounding the mutation being too small to detect using a given SNP density. Typically, sampling additional animals that are more closely related will assist in detecting the correct region using homozygosity mapping.

3.3 Discussion

Patterns of genetic variation can be used to trace the genetic basis of recessive disease. In this chapter, methods were explored to identify regions of homozygosity within and between individuals. The approach is particularly relevant to the search for disease genes within domestic animals where the level of relatedness between individuals may be high through inbreeding.

4 Notes

1. Where individuals are very closely related, shared ROH tends to be long. While this increases the ability to detect the shared region, it also means a large critical interval will be obtained that is likely to contain numerous positional candidate genes. Conversely, where individuals are selected that are more distantly related, recombination will have served to narrow the shared ROH. This should be considered in circumstances where the researcher has some control over the selection of affected individuals to study. In domestic animals such as dogs, the opportunity exists to perform homozygosity mapping using affected individuals drawn from multiple breeds. Where the mutation is common across breeds, homozygosity mapping has proven highly successful in quickly defining very small critical intervals.

2. A **.log** file is generated at the completion of each analysis performed by PLINK. It is good practice to check this *every time*. It lists the name of the **.ped** and **.map** file used for the analysis. This is important where multiple **.ped** files are present within the working directory, as a check to ensure you have used the correct **.ped** file that can avoid time-consuming mistakes. Similarly, the **.log** file lists information about the number of SNP and the status of the individuals. Where you have filtered a dataset, it is worthwhile checking the **.log** file to ensure the change you intended to make has indeed been correctly implemented.

3. Many commands, such as - - *missing*, have been built into gPLINK so users don't need to alter the command line issued from gPLINK to drive PLINK. Many commands, however, are not accessible via a pull down menu in gPLINK and therefore have to be manually inserted onto the command line. To open

a blank command line, go to the "Advanced" pull down menu and select "Create PLINK command." After defining the relevant **.ped** and **.map** file, the user is asked to give the operation a name (e.g., SNP_pruned). This opens a command line with the following form:

```
plink --file {insert PLINK command here} --out SNP_pruned --gplink
```

The user then replaces with text `{insert PLINK command here}` with the desired command. For example

```
plink --file ---extract mysnps.txt --out SNP_pruned --gplink
```

Once complete, hitting "OK" sends the command to PLINK and the operation will appear within the gPLINK Operations Viewer.

4. The parameters need to be tailored to the SNP set available. One of the infrequently used commands allows the user to specify the minimum allowable density of SNP in Kb using –*homozyg-density*. The default value is 50, meaning data from SNP pairs separated by more than 50 Kb are unavailable for analysis. This is particularly important for datasets derived using SNP arrays available for a number of livestock species. These typically have between 50,000 and 60,000 SNP with an average spacing of approximately 50 Kb. Using the default setting in these instances will mean the analysis fails. Users simply need to include the flag –*homozyg-density* 100 to increase the value well above the average spacing of the SNP.

5. If the recoded **.ped** and **.map** files do not contain the updated phenotypic information (i.e., if the **.log** file indicates 0 cases, and 0 controls are present after the operation), then the user can combine both the –*pheno* flag (step 6) and homozygosity mapping commands (step 7) in the same operation. The command line would be:

```
plink --file mydata ---pheno Cow_phenos.txt ---homozyg ---homozyg-group --out cows_compared
```

References

1. Pemberton TJ, Absher D, Feldman MW, Myers RM, Rosenberg NA, Li JZ (2012) Genomic patterns of homozygosity in worldwide human populations. Am J Hum Genet 91(2):275–292
2. Sabeti PC, Schaffner SF, Fry B, Lohmueller J, Varilly P, Shamovsky O, Palma A, Mikkelsen TS, Altshuler D, Lander ES (2006) Positive natural selection in the human lineage. Science 312(5780):1614–1620
3. Clop A, Marcq F, Takeda H, Pirottin D, Tordoir X, Bibé B, Bouix J, Caiment F, Elsen JM, Eychenne F, Larzul C, Laville E, Meish F, Milenkovic D, Tobin J, Charlier C, Georges M (2006) A mutation creating a potential illegitimate microRNA target site in the myostatin gene affects muscularity in sheep. Nat Genet 38(7):813–818
4. Kijas JW, McCulloch R, Edwards JE, Oddy VH, Lee SH, van der Werf J (2007) Evidence

for multiple alleles effecting muscling and fatness at the ovine GDF8 locus. BMC Genet 8:80

5. Rubin CJ, Zody MC, Eriksson J, Meadows JR, Sherwood E, Webster MT, Jiang L, Ingman M, Sharpe T, Ka S, Hallböök F, Besnier F, Carlborg O, Bed'hom B, Tixier-Boichard M, Jensen P, Siegel P, Lindblad-Toh K, Andersson L (2010) Whole-genome resequencing reveals loci under selection during chicken domestication. Nature 464 (7288):587–591

6. Lander ES, Botstein D (1987) Homozygosity mapping: a way to map human recessive traits with the DNA of inbred children. Science 236:1567–1570

7. Alkuraya FS (2010) Autozygome decoded. Genet Med 12(12):765–771

8. Charlier C, Coppieters W, Rollin F, Desmecht D, Agerholm JS, Cambisano N, Carta E, Dardano S, Dive M, Fasquelle C, Frennet JC, Hanset R, Hubin X, Jorgensen C, Karim L, Kent M, Harvey K, Pearce BR, Simon P, Tama N, Nie H, Vandeputte S, Lien S, Longeri M, Fredholm M, Harvey RJ, Georges M (2008) Highly effective SNP-based association mapping and management of recessive defects in livestock. Nat Genet 40(4):449–454

9. Rimbault M, Ostrander EA (2012) So many doggone traits: mapping genetics of multiple phenotypes in the domestic dog. Hum Mol Genet 21(R1):R52–R57

10. Purcell S et al (2007) PLINK: a tool set for whole-genome association and population-based linkage analyses. Am J Hum Genet 81(3):559–575

Chapter 15

Use of Ancestral Haplotypes in Genome-Wide Association Studies

Tom Druet and Frédéric Farnir

Abstract

We herein present a haplotype-based method to perform genome-wide association studies. The method relies on hidden Markov models to describe haplotypes from a population as a mosaic of a set of ancestral haplotypes. For a given position in the genome, haplotypes deriving from the same ancestral haplotype are also likely to carry the same risk alleles. Therefore, the model can be used in several applications such as haplotype reconstruction, imputation, association studies or genomic predictions. We illustrate then the model with two applications: the fine-mapping of a QTL affecting live weight in cattle and association studies in a stratified cattle population. Both applications show the potential of the method and the high linkage disequilibrium between ancestral haplotypes and causative variants.

Key words Association studies, QTL mapping, Haplotypes, Hidden Markov models, Ancestral haplotypes

1 Introduction

In GWAS, DNA sequence variants (DSV) (such as SNP, CNV, deletions, ...) influencing a phenotype of interest (disease, height, ...) are identified using statistical tests for association between markers and the phenotype. These DSV arise randomly in the population on a specific haplotype. *Haplotypes* are an observed specific combination of successive marker alleles on a given chromosomal segment (*see* **Note 1**). The piece of DNA carrying the variant can be transmitted to the next generations but the recombination process breaks it into smaller parts. Therefore, in the current generation, the length of the haplotype fragment common to all individuals carrying the DSV is a function of the number of generations since the variant appeared in the population.

The resulting association between haplotypes and DSV can be used to identify causal DSV underlying phenotypes. Most often this is performed in a two-step approach. First, genotypes from a reference panel are imputed in samples from the design.

Imputation (*see* **Note 2**) consists in predicting ungenotyped marker alleles observed in a so-called reference panel, a group of individuals genotyped for a larger number of markers (e.g., HapMap individuals or sequenced individuals from the 1000 genomes project). Thanks to this method, all individuals are genotyped in silico for the large number of markers. Haplotype-based techniques are used to perform the imputation, stressing the good predictive properties of the haplotypes (superior to single SNP). Association is then performed between the SNPs in the reference set and the phenotype. This first strategy works well if some of the markers in the reference set are in high linkage disequilibrium (LD) with the causal DSV. A second strategy would be to directly perform association between the haplotypes and the phenotypes. In comparison with single marker-based tests, haplotype-based tests require estimation of more variables (with a potential loss of power) but can present higher LD (depending on the marker density). Therefore, they are recommended when the causal DSV are in weak LD with surrounding markers, which is a function of the marker density and of the population history. A potential additional benefit of using haplotypes is that they can be associated with interacting alleles or with multiple closely linked alleles. Many different haplotype-based approaches have been developed, each relying on different definitions of the haplotypes. In this chapter, we present a method relying on the idea that in all populations, haplotypes from the current population trace back to a finite set of ancestral haplotypes, present a given number of generations ago. We want to group those haplotypes deriving from the same ancestral haplotype because we assume they carry the same DSV affecting traits and the same marker alleles. Therefore this information could be used in applications such as GWAS, phasing (haplotype reconstruction), prediction of ungenotyped markers (imputation) or prediction of genetic merit (or risk disease). To group haplotypes deriving from the same ancestral haplotypes, we will model them with hidden Markov models (HMM).

We will start this chapter with an example describing how haplotypes in an experimental line can be modeled with such hidden Markov chains. This first example is simple since the history of the population (the pedigree and the number of generations of crossing) and founders (their number and their marker alleles) are known. In addition, equal contribution of founder lines is assumed and haplotypes are known in both founder and terminal lines (the final lines obtained after several generations of crossing). Then we will extend this first case to more complex situations where haplotypes are not observed (but genotypes are known) and to more general situations in populations of unrelated individuals (such as in case/control studies) where the founder haplotypes and the history of the population are not known. These situations correspond to GWAS in human or to outbred livestock populations.

We will also show that in particular cases found in agronomical species, haplotypes can be modeled as a combination of the haplotypes of the parents. This ancestral haplotypes model will then be used to fine-map a QTL affecting live weight in cattle and to perform association studies in cattle pedigrees.

2 Hidden Markov Models for Haplotype Modeling

2.1 Modeling Haplotypes in Recombinant Inbred Lines

We start to explain our model with a specific example from experimental lines as those used in mice or *Arabidopsis Thaliana* for instance [1]. Recombinant inbred lines (RILs) are created to dissect the genetic architecture of complex traits (including identifying QTLs). These lines are constructed by first creating a set of lines based on the cross of L (two or more) inbred lines, and second, by taking the obtained individuals through several rounds of selfing (or close relationship mating, such as brother–sister mating, if selfing is not possible). After G generations, the resulting lines show essentially no variation within line—they are almost homozygous for the autosomes, while variation between lines is large, since the various terminal lines each present a different mosaic of the founder lines, with the switches between the various lines occurring randomly in each specific terminal line. Furthermore, since the RILs are fixed, the genotypes correspond to two identical haplotypes, eliminating the need for phasing (Fig. 1).

Chromosomal fragments cannot be directly observed, and the mosaic structure of the individuals in the terminal lines remains unknown. Haplotypes origins can nevertheless be studied using a set of M genetic markers, that we will initially assume equally spaced (this condition will be relaxed later). Haplotypes can be described by modeling markers sequentially according to a genetic map; at the first marker, the modeled chromosome fragment descends from one of the original lines. The probability π_i for the first marker region to originate from line i is assumed randomly distributed with probability $\pi_i = 1/L$ (where L is the number of founder lines). For the subsequent markers, the origin can be modeled as a function of the origin at the previous marker in the sequence. This leads us to define the probability to observe a jump—a "transition"—from a region descending from one original line to a region potentially originating from another line (Fig. 2). Noting θ the (constant) recombination rate between two markers, the probabilities τ_{ij} to switch from origin i (or founder i) to origin j between two markers can be computed as:

$$\tau_{ij} = \begin{cases} (1-(1-\theta)^G)\frac{1}{L} + (1-\theta)^G & \text{if } i = j \\ (1-(1-\theta)^G)\frac{1}{L} & \text{if } i \neq j \end{cases} \quad (1)$$

Indeed, transmitted fragments can either be transmitted without recombination during G generations between two

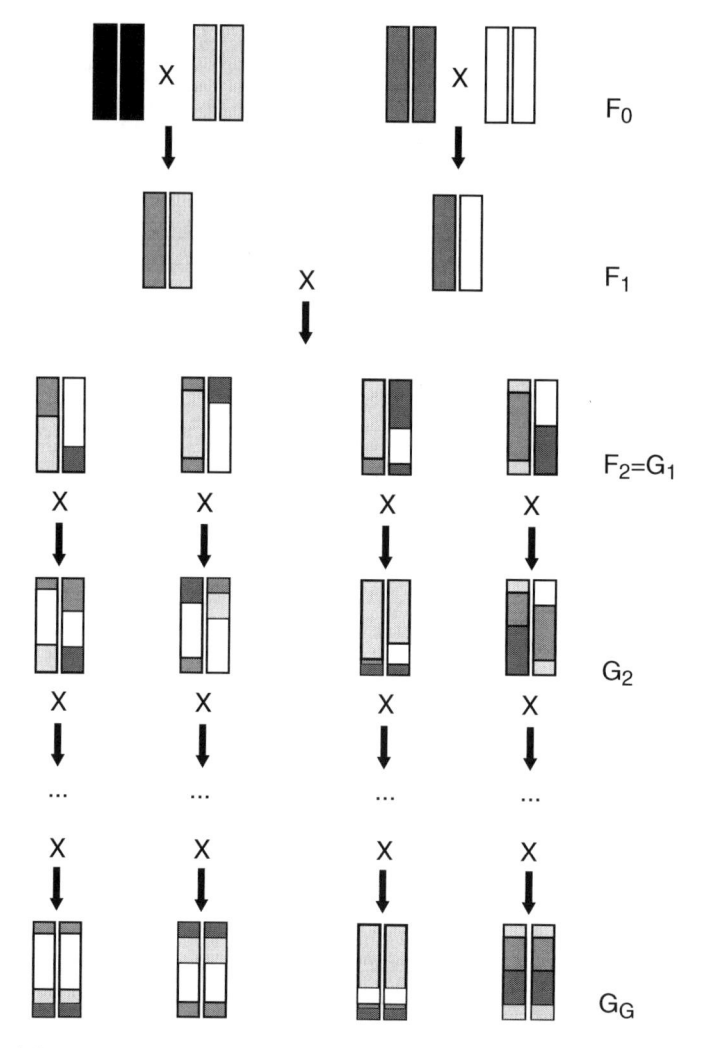

Fig. 1 An example of RILs, with $L = 4$ original lines, represented through four different colors. A F1 generation is obtained by crossing the two first lines on one hand and two last lines on the other hand. Then, several individuals are obtained by crossing individuals from the two distinct crosses. The next $(G - 1)$ generations are obtained by selfing, leading to (quasi) homozygous terminal lines, that can be seen as mosaics of the original lines

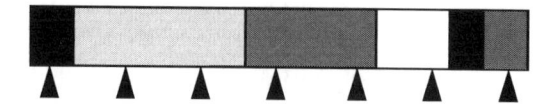

Fig. 2 An experimental line haplotype, represented as a mosaic of the original inbred lines regions (various colors patterns indicate various original inbred lines). *Black* triangles indicate the positions of a set of equally spaced markers. The transition from marker 2 to marker 3 is not associated to a change of line origin, while the transition from marker 3 to marker 4 is

markers—this situation has a probability equal to $(1-\theta)^G$ and the founder origin remains unchanged ($i = j$)—or with a recombination with the complementary probability $1-(1-\theta)^G$; in that case we can jump to any other founder line (including i) with equal probability $1/L$. Therefore, transitions between lines, implying a recombination, have a probability equal to $[1-(1-\theta)^G]/L$, while transitions to the same origin can result from events with or without recombinations. For small distances between markers, $(1-\theta)^G$ is often approximated as $e^{-\theta G}$.

With nowadays technologies, the number of markers can be very high, introducing the need for appropriate models for which efficient computational algorithms are available. To that end, we will now define hidden Markov models (*see* **Note 3**). For more details on HMM, we invite the reader to consult textbooks or a tutorial from Rabiner [2].

2.2 Hidden Markov Models

2.2.1 Definition

A Markov chain is a sequence of states taken in a possible set $S = \{S_1, S_2, \ldots, S_L\}$ of L states observed at regular intervals (distances such as time for example) noted $t = 1, 2, \ldots$ (and loosely referred to as time in the following). The chain can start in any state j with a probability π_j. The vector $\boldsymbol{\pi} = (\pi_1, \pi_2, \ldots, \pi_L)$ represents the initial states probabilities. Transitions between successive states depend on a set of transition probabilities which can be summarized in a transition matrix $\mathbf{T} = (\tau_{ij})$ where τ_{ij} is the probability to observe a transition from state i to state j. Note that the system is memoryless: the probability to reach state j at time t is only dependent on the state of the system at time $(t-1)$, and conditionally independent of previous states. This is called the Markov property.

This description of a Markov chain closely matches the modeling of the RILs. Indeed, we described these as a succession (a mosaic) of L founder lines. The origin at the first marker is obtained by a set of initial states probabilities described above and transition probabilities are obtained from Eq. 1. As for a Markov chain, we assumed that transition probabilities (switches between founder lines) were a function only of the origin at the previous and the current markers. A Markov chain is graphically represented in Fig. 3.

As in our RILs example, in HMM the states are not directly observed (they are "hidden") but a set of symbols $O = \{o_1, o_2, \ldots, o_N\}$ can be emitted from each state. Each state i has its own emission probabilities ε_{ij} (the probabilities to emit the symbol j) described with the matrix $\mathbf{E} = (\varepsilon_{ij})$ where $i = 1, 2, \ldots, L$ and $j = 1, 2, \ldots, N$. The emission probabilities are only a function of the underlying state at that time. Similarly, in the RILs example, marker alleles (two alleles in case of SNPs) are emitted as a function of the underlying origin (see Fig. 4).

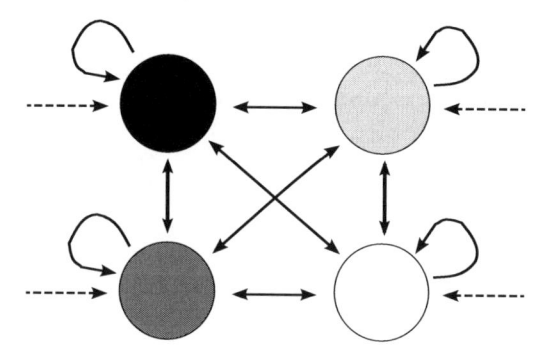

Fig. 3 A Markov chain that could be used to represent the cross described in figures 1 and 2. *Colored circles* stand for states and represent the corresponding parental lines. *Full arrows* represent transition probabilities, while *dashed arrows* indicate initial probabilities

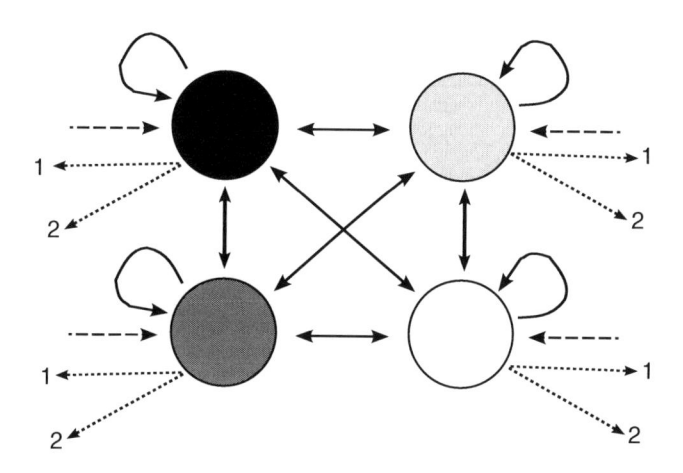

Fig. 4 A Markov chain that could be used to represent the cross described in figures 1 and 2. *Colored circles* stand for states and represent the corresponding parental lines. *Full arrows* represent transition probabilities, while *dashed arrows* indicate initial probabilities. *Dotted arrows* indicate emission probabilities and emitted symbols are reported (assuming that the used markers are SNP)

To summarize, we can describe a hidden Markov model as a set of five elements:

1. A set of L states, noted $S = \{S_1, \ldots, S_L\}$
2. A vector $\boldsymbol{\pi}_{L \times 1}$ of initial state probabilities $\boldsymbol{\pi}' = (\pi_1, \pi_2, \ldots, \pi_L)$
3. A matrix $\mathbf{T}_{L \times L}$ of transition probabilities $\boldsymbol{\tau} = (\tau_{ij})$, with $i, j = 1, 2, \ldots, L$
4. A set of N emitted symbols $O = \{O_1, O_2, \ldots, O_N\}$
5. A matrix $\mathbf{E}_{L \times N}$ of emission probabilities $\boldsymbol{\varepsilon} = (\varepsilon_{ij})$, with $i = 1, 2, \ldots, L$ and $j = 1, 2, \ldots, N$

HMM are often represented synthetically as a triplet $\lambda = (\tau, \varepsilon, \pi)$. Before detailing how to perform efficient inference with these models, a few remarks should be made on their use for modeling RILs:

- In general HMM, transition probabilities remain constant. This would not be appropriate to model haplotypes when distance between markers varies. In that case, we must define a "non-homogenous" HMM where the transition probabilities change at each transition (time) t. In our example, we should replace θ by $\theta(t)$, the recombination rate between the current marker and the next in the formulae given above.

- Similarly, emission probabilities remain constant in general HMM whereas for haplotypes modeling, emission probabilities vary with marker position. Consider markers 2 and 3 in Fig. 2. Although the terminal line descends from the same founder line at these markers, there is no reason that the founder line carries identical alleles for these two markers. If they are different, the emission probabilities will also be different, while still being in the same state. To include that feature in the model, we replace $\mathbf{E}$ by $\mathbf{E}(t)$ (a marker dependent emission probability matrix).

- When markers are SNPs, the same alphabet $O = \{1, 2\}$ is used for all states and positions. If other types of markers (microsatellites, CNV, ...) are included in the analysis, then the alphabet would need to be adapted accordingly at each position, which could be formally represented as $O = O(t)$.

Another representation can be provided, with $L \times M$ states instead of L states, where transition and emission probabilities are fixed (Fig. 5). In that case, a state would correspond to a combination of a position and a founder line. Transition probabilities would be such that transitions are only possible from the L states corresponding to one position to the L states corresponding to the next position, and initial states probabilities would be non zero only for states corresponding to the first position.

2.2.2 Inference in Hidden Markov Models

After describing the model, we now turn to the inferences that can be conducted using these models. Various questions can be addressed, with some of importance to genetic problems.

Computing the probability of an observed sequence of symbols. Assume that we observed a sequence of symbols (a haplotype in our case) and that we want to obtain the probability of this event. In the example of the previous section, where we have seven markers, and assuming that all the markers are SNP with two alleles noted "1" and "2," we could try to compute the probability to obtain a sequence Q, such as "1211221," using the described model. If the sequence R of states was known, this would be a trivial task (Fig. 6). Writing r_i the ith state in the states sequence (of length M) and q_i the ith observation in the observations sequence, the

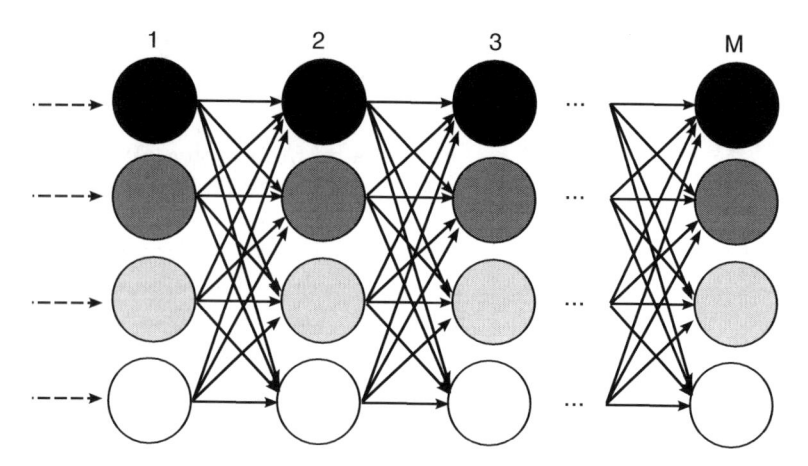

Fig. 5 Another representation of the Markov chain of figure 4. *Circles* stand for states and each color represents a parental line. *Full arrows* represent transition probabilities, while *dashed arrows* indicate initial probabilities. Note that, although L × M states are present, only L are available from any state (except the L states on the right of the graph, which are states for which no transition will occur). Emission probabilities are not represented for clarity. Numbers represent the marker reached in the chain

probability would be the product of corresponding emission probabilities:

$$P(Q|R,\lambda) = \prod_{i=1}^{M} \varepsilon_{r_i q_i}$$

The probability of any particular states sequence R is easy to compute: it is the probability to start with r_1, multiplied by the successive transition probabilities corresponding to jumps from state r_i to state r_{i+1}:

$$P(R|\lambda) = \pi_{r_1} \prod_{i=2}^{M} \tau_{r_{i-1}, r_i}$$

Since the actual states sequence is unknown, we have to compute the needed probability, $P(Q|\lambda)$, using all possible states sequences:

$$P(Q|\lambda) = \sum_R P(Q, R|\lambda) = \sum_R P(Q|R, \lambda) P(R|\lambda)$$

The sum involving all possible states sequences R is potentially hard to compute: since we have M positions (i.e., markers) in the sequence (i.e., the haplotype), and L possible states for every position, the brute force computation of the sum involves L^M states. In the small example in the Figs. 2, 3, 4, 5, and 6, this would lead to $4^7 = 16,384$ distinct states sequence. In practical

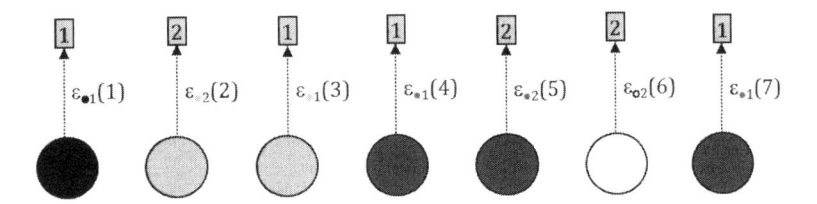

Fig. 6 Illustration of the computation of the probability of Q = "1211221" assuming that the sequence of states is known

situations (see applications below), we could have $L = 10$ or more and $M = 1,000$ or more, which would lead to more than 10^{1000} sequences, a number which does not allow exhaustive computation of $P(Q|\lambda)$. Fortunately, an efficient algorithm, called "forward–backward" algorithm exists, which can drastically reduce the computations. The principle is as follows. Forward variables are defined as the probability to emit the t first symbols of the sequence and to be in state j at time t. Writing $\alpha_j(t)$ that probability, we have:

$$\alpha_j(t) = P(q_1 q_2 \ldots q_t, r_t = S_j | \lambda)$$

Using that definition, three properties are easily demonstrated:

1. "Initialization" property:

$$\alpha_j(1) = P(q_1, r_1 = S_j | \lambda) = P(r_1 = S_j | \lambda) P(q_1 | \lambda, r_1 = S_j) = \pi_j \varepsilon_{j\, q_1}$$

2. "Induction" property: starting from forward variables at time t, forward variables at time $(t + 1)$ can be inferred by observing that the latter can be obtained using only the variables at time t and transitions from states at time t to state at time $(t + 1)$. More formally:

$$\alpha_j(t+1) = P(q_1 q_2 \ldots q_t q_{t+1}, r_{t+1} = S_j | \lambda)$$
$$= \sum_{i=1}^{K} P(q_1 q_2 \ldots q_t q_{t+1}, r_t = S_i, r_{t+1} = S_j | \lambda)$$
$$= \sum_{i=1}^{K} P(q_1 q_2 \ldots q_t, r_t = S_i | \lambda)$$
$$\quad \times P(q_{t+1}, r_{t+1} = S_j | q_1 q_2 \ldots q_t, r_t = S_i, \lambda)$$
$$= \sum_{i=1}^{K} \alpha_i(t) P(r_{t+1} = S_j | r_t = S_i, \lambda) P(q_{t+1} | r_{t+1} = S_j, \lambda)$$
$$= \sum_{i=1}^{K} \alpha_i(t) \tau_{ij} \varepsilon_{j\, q_{t+1}}$$

3. "Termination" property: when a sequence of M symbols is observed, the corresponding probability can easily be obtained from forward variables:

$$P(Q|\lambda) = \sum_{i=1}^{K} P(Q, r_M = S_i|\lambda) = \sum_{i=1}^{K} \alpha_i(M)$$

The computation of $P(Q|\lambda)$ can be performed using these three properties. Using this procedure, called "forward algorithm," the number of operations to achieve the computation of $P(Q|\lambda)$ is $L^2 \times M$, which is reasonable to compute. In the preceding example, we would go from 10^{1000} computations using a brute force approach to $10^2 \times 1,000 = 10^5$ using the forward algorithm. A similar procedure, called "backward algorithm," can also be used to achieve the same goal. We define backward variables $\beta_i(t)$ as the probability to emit the observed symbols sequence from time $(t + 1)$ to time M while being in state i at time t:

$$\beta_i(t) = P(q_{t+1} q_{t+2} \ldots q_M | r_t = S_i, \lambda)$$

Similarly to the forward procedure, the backward procedure is then:

1. Initialization: $\beta_i(M) = 1$, arbitrarily for $1 \leq i \leq L$
2. Induction: $\beta_i(t-1) = \sum_{j=1}^{K} (\beta_j(t)\tau_{ij})\varepsilon_{j\ q_t}$ for $t = M$ to 2 and for $1 \leq i \leq L$
3. Termination: $P(Q|\lambda) = \sum_{i=1}^{K} \beta_i(1)$

Computing the probability to be in a given state at a given time. Another question of interest when studying genetic problems using crosses is how to know which of the ancestral haplotypes is underlying the specific position we are interested in? Using our model, this corresponds to asking which is the current state at a given position, given the observed sequence. Of course, most of the time, several states are possible, meaning that the answer will be probabilistic. Formally, the probability to be in state i at time t is:

$$\gamma_i(t) = P(r_t = i | Q, \lambda)$$

This probability can be computed using the variables $\alpha_i(t)$ and $\beta_i(t)$ described above as follows. First, the probability can be written as a ratio:

$$\gamma_i(t) = \frac{P(r_t = i, Q|\lambda)}{P(Q|\lambda)}$$

The denominator is the sum over i of the numerator probabilities. Now, the numerator probability can be computed by developing the expression as follows:

$$P(r_t = i, Q|\lambda)$$
$$= P(r_t = i, q_1, q_2, \ldots, q_t|\lambda) \times P(q_{t+1}, \ldots, q_M|r_t = i, q_1, q_2, \ldots, q_t, \lambda)$$
$$= \alpha_i(t) \times P(q_{t+1}, \ldots, q_M|r_t = i, \lambda) = \alpha_i(t) \times \beta_i(t)$$

Consequently, the requested probability is computed using the equation:

$$\gamma_i(t) = \frac{\alpha_i(t) \times \beta_i(t)}{\sum_i [\alpha_i(t) \times \beta_i(t)]}$$

Identifying the states sequence underlying an observed sequence. In the genetic problem described in the previous section, this would correspond to determining the inherited parental lines at any given marker position in the RILs. A large number of states sequences could give rise to the observed sequence, and some criterion is needed to infer the states sequence that is optimal to explain the observations. The optimality criterion we will consider (and which is very largely used in practice) can be defined as finding the single sequence of states R that maximizes $P(R|Q,\lambda)$. The algorithm solving that problem is called the Viterbi algorithm. It can be explained as a recursive procedure. Assume we define $\delta_j(t)$ as the maximum probability obtained for observing the $(t-1)$ first symbols of the observed sequence with any states sequence $r_1, r_2, \ldots, r_{t-1}, r_t = j$ ending in state j at the time t:

$$\delta_j(t) = \max_{r_1 r_2 \ldots r_{t-1}} P(q_1 q_2 \ldots q_{t-1}, r_t = j|\lambda)$$

We can initialize the computations using:

$$\delta_j(1) = \pi_j \varepsilon_{jq_1}$$

And we can progress along the observed sequence using the recurrence relationship:

$$\delta_j(t+1) = [\max(\delta_i(t) \times \tau_{ij})] \times \varepsilon_{jq_{t+1}}$$

This procedure leads to the best (i.e., largest) probability and, keeping track of the states at time t that maximizes the probability for each of the L states at time $t+1$ while performing the computations, to the optimal states sequence corresponding to the observations.

To clarify the procedure, we end up this section showing a small example: assume we have a three states model, with the following parameters:

$$\tau = \begin{pmatrix} 0.6 & 0.2 & 0.2 \\ 0.1 & 0.7 & 0.2 \\ 0.5 & 0.3 & 0.2 \end{pmatrix}, \varepsilon = \begin{pmatrix} 0.6 & 0.4 \\ 0.7 & 0.3 \\ 0.2 & 0.8 \end{pmatrix}, \pi = \begin{pmatrix} 0.5 \\ 0.2 \\ 0.3 \end{pmatrix}$$

We can compute the most likely states sequence leading to an observed sequence $Q = (1, 1, 2, 1)$ as follows (the sequence within square brackets corresponds to the current best sequence):

$\delta_1(1) = 0.5*0.6 = 0.30$ [1]
$\delta_2(1) = 0.2*0.7 = 0.14$ [2]
$\delta_3(1) = 0.3*0.2 = 0.06$ [3]

$\delta_1(2) = \max(\underline{0.30*0.6};0.14*0.1;0.06*0.5)*0.6 = 0.30*0.6*0.6 = 0.1080$ [1-1]
$\delta_2(2) = \max(0.30*0.2;\underline{0.14*0.7};0.06*0.3)*0.7 = 0.14*0.7*0.7 = 0.0686$ [2-2]
$\delta_3(2) = \max(\underline{0.30*0.2};0.14*0.2;0.06*0.2)*0.2 = 0.30*0.2*0.2 = 0.0120$ [1-3]
$\delta_1(3) = \max(\underline{0.1080*0.6};0.0686*0.1;0.0120*0.5)*0.4 = 0.1080*0.6*0.4 = 0.025920$ [1-1-1]
$\delta_2(3) = \max(0.1080*0.2;\underline{0.0686*0.7};0.0120*0.3)*0.3 = 0.0686*0.7*0.3 = 0.014406$ [2-2-2]
$\delta_3(3) = \max(\underline{0.1080*0.2};0.0686*0.2;0.0120*0.2)*0.8 = 0.1080*0.2*0.8 = 0.017280$ [1-1-3]
$\delta_1(4) = \max(\underline{0.02592*0.6}; 0.014406*0.1; 0.01728*0.5)*0.6 = 0.00933120$ [1-1-1-1]
$\delta_2(4) = \max(0.02592*0.2; \underline{0.014406*0.7}; 0.01728*0.3)*0.7 = 0.00705894$ [2-2-2-2]
$\delta_3(4) = \max(\underline{0.02592*0.2}; 0.014406*0.2; 0.01728*0.2)*0.2 = 0.00103680$ [1-1-1-3]

So the best states sequence is [1-1-1-1] and the corresponding probability of observing the sequence Q is 0.0093312.

Estimating the parameters of the model based on observed sequences. The idea here is to learn from an observed sequence the values of the parameters of $\lambda = (\pi, \tau, \varepsilon)$. There is no optimal way of estimating the parameters, but a choice can be made such that $P(Q|\lambda)$ is locally maximized. A procedure to achieve this goal, known as Baum–Welsch procedure, will be given here. The algorithm can be described defining one additional probability:

1. The conditional (given the observed sequence Q and the parameters λ) probability of being in state i at time t and in state j at time $(t + 1)$ is written $\xi_{ij}(t)$. It can be computed as:

$$P(r_t = i, r_{t+1} = j|Q, \lambda) = P(r_t = i, r_{t+1} = j, Q|\lambda)/P(Q|\lambda)$$

The numerator of the right-hand side can be written as the product of four probabilities:

$P(r_t = i, q_1 \ldots q_t|\lambda) = \alpha_i(t)$
$P(r_{t+1} = j|r_t = i, q_1 \ldots q_t, \lambda) = \tau_{ij}$
$P(q_{t+1}|r_{t+1} = j, r_t = i, q_1 \ldots q_t, \lambda) = \varepsilon_{jq_{t+1}}$
$P(q_{t+2} \ldots q_M|r_{t+1} = j, r_t = i, q_1 \ldots q_t, \lambda) = \beta_j(t+1)$

The denominator can be seen as the marginalization over all pairs of states of the numerator probability. Consequently, the probability can be written as:

$$\xi_{ij}(t) = \frac{\alpha_i(t) \times \tau_{ij} \times \varepsilon_{j\,q_{t+1}} \times \beta_j(t+1)}{\sum_i \sum_j \alpha_i(t) \times \tau_{ij} \times \varepsilon_{j\,q_{t+1}} \times \beta_j(t+1)}$$

2. The conditional probability to be in state i at time t can be defined using the probability $\xi_{ij}(t)$:

$$\gamma_i(t) = \sum_j \xi_{ij}(t)$$

This is the same probability as the one defined above.

$\sum_{t=1}^{M-1} \xi_{ij}(t)$ provides the expected number of transitions from state i to state j and $\sum_{t=1}^{M-1} \gamma_i(t)$ provides the expected number of transitions from state i (and the expected number of times the chain is in state i). Using these probabilities, the parameters of λ can be reestimated by using constrained maximization:

- $\hat{\pi}_i = \gamma_i(1)$

- $\hat{\tau}_{ij} = \dfrac{\sum\limits_{t=1}^{M-1} \xi_{ij}(t)}{\sum\limits_{t=1}^{M-1} \gamma_i(t)}$

- $\hat{\varepsilon}_{ij} = \dfrac{\sum\limits_{i=1}^{M-1} \gamma_i(t)\delta(q_t = o_j)}{\sum\limits_{i=1}^{M-1} \gamma_i(t)}$

In these equations, the hat sign is used to denote estimators, and $\delta(i = j)$ is equal to 1 if $i = j$, and to 0 otherwise. These equations have a simple interpretation, they correspond to expected counts: initial state probabilities are estimated as the expected number of times the chain started in a given state, transition probabilities are estimated as the expected number of transitions from state i to j divided by the total transitions from i and emission probabilities are estimated as the expected number of times symbol j was emitted from state i divided by the total number of emissions from state i. Baum and Welsch proved that these equations lead to values that are either equal or better than the initial λ values in the sense that $P(Q|\hat{\lambda}) > P(Q|\lambda)$. This means that, using this estimation procedure repeatedly, maximum likelihood solutions can be obtained, i.e., values of the parameters that make the observations more likely become available. It should nevertheless be commented that the obtained solutions provide a local maximum, but, to our knowledge, no procedure has been devised yet to obtain the global maximum.

2.3 The Ancestral Haplotypes Model

2.3.1 Introduction

The concept of "ancestral haplotype" is used for more general situations than the example of RILs. However, the idea behind the concept is similar to that underlying the approach followed for the experimental lines: each today individual's chromosomes can be regarded as a mosaic of a set of ancestral haplotypes, with conserved ancestral regions separated by breakpoints corresponding to ancestral recombinations. Although seemingly similar, notable differences exist in the definition of the parameter from the two models.

2.3.2 The Ancestral Haplotypes (Hidden States)

The hidden states of the model will be L ancestral haplotypes. One major difference with the situation relative to RILs is that these ancestral haplotypes are not observed, as was the case for experimental crosses of inbred lines. Actually, the *ancestral haplotypes* are rather a conceptual view (*see* **Note 4**): they can be regarded as a set of haplotypes from which all today individuals are assumed to have descended. The ancestral haplotypes can correspond to haplotypes from physical ancestors. The set of these physical ancestors can change at every position and the physical ancestors can belong to different generations. In that case and for a long chromosomal fragment, one ancestral haplotype would be a mosaic of real haplotypes from different physical ancestors (eventually belonging to different generations). Ancestral haplotypes can also correspond to mixtures of haplotypes from physical ancestors. Indeed, the model might group haplotypes from the current population which derived from several ancestors. In some situations, ancestral haplotypes might even not correspond to physical ancestors and be purely conceptual, obtained from the inference process (*see* **Note 5**).

Since the ancestral haplotypes are not known and rather conceptual, their number is not known either. The optimal number of states depends on different factors:

- The demographic history of the population (the effective population size): a relatively isolated population expanding from a small set of individuals can probably be described using a fairly small value of L. For example, in the bovine applications given below, the intensive use of artificial insemination has led to the large diffusion of a small set of chromosomes, leading to very small effective population sizes, even when the populations themselves are very large. In these populations, values of $L = 10–20$ have been shown to work well (see below).

- The age of the modeled ancestral haplotypes. For a given population, reducing L (the pool of ancestral haplotypes) would result in modeling elder ancestral haplotypes. This is directly related to the number of generations between the ancestral haplotypes and the current population and has therefore an impact on transition probabilities too.

- The age and the frequency of the DSV of interest (for mapping, imputation or genomic selection) also have an impact. We want to obtain the best possible LD between the ancestral haplotypes and the DSV. A recent and frequent DSV will be associated to a long and unique haplotype that will be relatively easy to isolate. Older DSV will be linked to shortened haplotypes with possibly a few differences due to more recent mutations. Most often this information is unknown and we must test which L performs best. For instance, for imputation techniques, the L maximizing imputation accuracy in a cross-validation experiment can be chosen as is done in fastPHASE [3].

- In addition to high LD between ancestral haplotypes and DSV, we want methods to remain powerful (keeping the number of ancestral haplotypes low). This objective will be reached if we are able to isolate into one cluster all haplotypes carrying a mutation with a directional effect on the scrutinized trait, and in another cluster, all other haplotypes. This would lead us ideally to use a two clusters model. Unfortunately, reducing L can cause haplotypes carrying the mutation of interest to be grouped with other haplotypes (breaking the desired LD between ancestral haplotypes and DSV). In addition, several haplotypes might carry DSV of interest. Therefore, L must be chosen to maintain a balance between high LD and low number of ancestral haplotypes.

- Computational considerations must also be taken into account. Increasing the value of L also potentially increases the computational burden, which could become a problem when large marker sets, as those available today, are used.

Clearly, the value of L is the result of a compromise that has to take into account the biological reality and modeling constraints and limitations.

2.3.3 Initial State and Transition Probabilities

In the ancestral haplotypes model, initial state and transition probabilities rely on several assumptions:

- The ancestral haplotypes do not contribute equally to the current generation. Indeed, due to demographic forces such as genetic drift or selection, some ancestral haplotypes might be more frequent. Another feature of ancestral haplotypes is that their frequencies generally vary along the chromosome: for example, since selection only affects specific regions, the haplotypes bearing favorable loci could be overrepresented in these.

- In case of switch, the transition probability is independent of the current state (as in the RILs example).

- The number G of generations between ancestral haplotypes and the current population is unknown and might vary among haplotypes and among regions. The recombination rate to be used to compute the transition probability is not obtained from an analytical formula based on biological data but is estimated from the data. A global parameter, including the effects of the distance between markers and of the number of generations—respectively θ and G in the RILs applications—called the jump probability, is used and assumed constant for all states (thus, G is assumed identical for all ancestral haplotypes, although, as mentioned above, the generational gap between today and ancestral haplotypes might vary).

As a result of these assumptions, initial state probabilities are not uniform and must be estimated. Another consequence is that transition probabilities are position specific:

$$\tau_{ij}(t) = \begin{cases} J(t)f_i(t) + (1 - J(t)) & \text{if } i = j \\ J(t)f_j(t) & \text{if } i \neq j \end{cases}$$

where $J(t)$ is the jump probability at position t, $f_i(t)$ is the frequency of ancestral haplotype i at position t. These parameters must be estimated with a procedure similar to the Baum–Welsch procedure described above. $J(t)$ is estimated as the expected number of jumps at time t divided by the expected number of observations at time t (equal to the number of modeled haplotypes) whereas $f_i(t)$ are estimated as the expected number of jumps to state i at time t divided by the expected number of jumps at time t (only jumps provide information to estimate $\mathbf{f}(t)$). Similarly, $\pi_i(t)$ is estimated as the expected number of chains (haplotypes) starting in state i divided by the total number of chains.

2.3.4 Emission Probabilities

In this model, emission probabilities $\varepsilon_{ij}(t)$ correspond to the probability to observe marker allele j in ancestral haplotype i at position (or marker) t. These correspond to allele frequencies for each ancestral haplotype. Ideally, an ancestral haplotype should be labeled with a unique marker allele and the emission probabilities should be either 0 or 1. However, due to mutations, genotyping errors, grouping of different haplotypes and imperfect inference, the haplotype of an observed individual is not an exact copy of the ancestral haplotype from which it has descended and the emission probabilities can differ from 0 or 1 and must be estimated. Emission probabilities are estimated as the expected number of alleles j observed on ancestral haplotype i at marker t divided by the expected number of haplotypes deriving from ancestral haplotype i at marker t.

In general, haplotypes are unknown and the marker information consists in genotypes (two marker alleles for diploid species). In that case, two haplotypes must be modeled simultaneously. One solution is to define L^2 states (a combination of two ancestral haplotypes). Alternatively, we model this as two independent hidden chains emitting jointly a genotype. Emission probabilities are then changed to $\varepsilon'_{ii'jj'}(t)$, the probability to observe genotype with alleles j and j' in ancestral haplotypes i and i' at marker t:

$$\varepsilon'_{ii'jj'}(t) = \begin{cases} \varepsilon_{ij}(t)\varepsilon_{i'j'}(t) + \varepsilon_{ij'}(t)\varepsilon_{i'j}(t) & \text{if } j \neq j' \\ \varepsilon_{ij}(t)\varepsilon_{i'j}(t) & \text{if } j = j' \end{cases}$$

The emission probabilities can be modeled with flexibility. For example, the model could be set up to accept both genotypes and haplotypes (if haplotypes are only partially known). In that case, emission probability would be the product of emission probabilities from each haplotype (*see* **Note 6**).

2.4 Extension to Pedigreed Data: The Example of Half-Sib Families

The ancestral haplotypes model assumes unrelated individuals but in many situations relatives are genotyped. For instance, it happens that haplotypes from a parent are included in the set of haplotypes to be modeled. This has two consequences. First, information from relatives can be used to reconstruct haplotypes. For instance, if an individual is heterozygous whereas the father is homozygous for a given marker, it is trivial to determine which allele was inherited from the father (and from the mother). The second consequence is that haplotypes known to be identical (based on pedigree relationship and marker information) should not be modeled as two independent haplotypes.

To illustrate these considerations, we now turn to large half-sibs families (as often met in livestock species or crops). In these families, haplotypes from parents with many genotyped offspring can be reconstructed with high accuracy. Indeed, marker alleles from close markers are most often transmitted without recombination. With many offspring, it is easy to determine which marker alleles cosegregate together and hence, to reconstruct the haplotypes from the parent.

Therefore, we propose to extend the model to particular situations where the parent is genotyped. In that case, the corresponding haplotype will be modeled as a mosaic of the two haplotypes of the parent ($L = 2$ for that chain). This is a top-down approach (*see* **Note 7**), transferring information from parents to offspring which is particularly useful in livestock species or crops where parents can have many offspring. In that case, haplotypes of parents are reconstructed with high accuracy and modeling haplotypes from offspring as a mosaic of haplotypes from the common parent increases phasing and imputation accuracy [4]. This particular model can also be described with an HMM:

- With two states (the two haplotypes from the genotyped parent).
- With initial state probabilities equal to 0.5.
- With transition probabilities equal to $1 - \theta(t)$ (no switch) and $\theta(t)$ (a switch).
- With two possible emitted symbols: the marker alleles carried by the parent.
- With emission probabilities equal to $1 - e$ and e if the marker allele is identical or not to the marker allele in the corresponding haplotype from the parent (e is a small value used to allow for genotyping errors).

3 Applications

The model designed for half-sibs families will now be applied to several real-life problems to illustrate how this model can be used in mapping applications (in GWAS).

3.1 Introduction

Clustering haplotypes into groups of haplotypes derived from a set of L ancestral haplotypes as described earlier in this chapter can be used for several purposes such as haplotype phasing, missing marker prediction (imputation), association studies (including QTL fine-mapping) or for phenotype prediction (disease risk prediction, marker-assisted selection or genomic selection). Indeed, haplotypes within a cluster are likely to descend from the same ancestral haplotype and to carry the same DSV and combination of alleles. Here, our main focus is on association studies, but the efficiency of this model for haplotype reconstruction, imputation or genomic selection has already been proven, stressing the high predictive ability of the model (e.g., [3–6]).

For association studies, the model can be applied in a wide-range of designs. First, it has been used in experimental designs resulting from the crosses of a set of founder lines to obtain a terminal line, these lines being inbred or not. Different variants of the method can be applied, according to the knowledge (or not) we have of the founder or terminal lines haplotypes. We showed earlier that the model can be used in (multiple) (RILs) [1]. A very similar application is the MAGIC design such as described in Kover et al. [7]. Although the model fits better for inbred lines, the original idea came from association studies in outbred animal stocks [8]. The model has also been applied in heterogeneous stock mice [9, 10]. In these experimental lines, the general idea is to use the forward-backward algorithm to estimate at each tested position the probability that each haplotype in the mapping population descends from each of the founder lines. A design matrix containing probabilities to descend from the founder lines (where each line corresponds to an individual and each column a founder line) can then be used with the phenotypes to perform an ANOVA test.

The model has also been proposed for more general association studies for unrelated individuals [10]. In this second application, haplotypes of cases and controls are assigned to a set of L ancestral haplotypes with the Viterbi algorithm. The assumption is that one of these ancestral haplotypes is carrying a risk variant. Therefore K tests are performed where one of the K ancestral haplotypes is contrasted with the $K - 1$ others using the Armitage trend tests.

Now we will illustrate application of the general model using ancestral haplotypes to other types of association studies: one example of QTL mapping for live weight in cattle (complex trait) and then a set of examples of disease or coat patterns mapping (binary traits) in cattle populations presenting stratification.

3.2 Use of Ancestral Haplotypes to Fine-Map a QTL Affecting Live Weight in Cattle

The application of the method to the mapping of a QTL affecting live weight in cattle has been described in Karim et al. [11]. Here we present a small part of this study performed by a much larger team [11]. The QTL was first detected in a Holstein × Jersey line cross, two dairy cattle breeds with relatively different zootechnical performances. By mating Holstein and Jersey F0 individuals, six F1 sires and 796 F1 dams were obtained. Further crossing of F1 individuals yielded 745 F2 cows which were phenotyped for a broad range of traits (more than 500 traits). A genome scan relying on genotypes for 294 microsatellites detected a very significant QTL affecting live weight on chromosome BTA14.

Additional genotyping of F1 sires and of the 745 F2 individuals for 48 new microsatellites and 925 BTA14 SNPs (resulting in a total of 981 markers on BTA14) was realized. The HMM described in Subheadings 2.3 and 2.4 was then used to reconstruct haplotypes and assign them to $L = 20$ ancestral haplotypes. Haplotypes from the six F1 sires were first reconstructed very accurately by observing marker alleles cosegregating in their numerous (>100) genotyped offspring (close marker alleles which are frequently transmitted jointly belong to the same sire haplotype). Combining almost perfect knowledge of F1 sire haplotypes and LD information (modeled through ancestral haplotypes), the haplotypes from the F2 individuals were reconstructed accurately with DualPHASE [4]. This procedure provides better results than haplotype reconstruction based on LD alone [4]. Maternal haplotypes from the F2 individuals are directly assigned through the model to the 20 ancestral haplotypes (using the Viterbi algorithm) whereas the paternal haplotypes from the F2 individuals are indirectly related to the ancestral haplotypes: the model links these paternal haplotypes to haplotypes from the corresponding F1 sire and these are each assigned to one of the 20 ancestral haplotypes. As a result, each F2 individual is associated to two (eventually identical) ancestral haplotypes and we can use that information to construct a design matrix $\mathbf{W} = (w_{ij})$ relating individuals to the ancestral haplotypes. Element w_{ij} is equal to the number of copies (from 0 to 2) of ancestral haplotype j carried by individual i.

QTL mapping is performed using a standard linear mixed model (LMM). The vector of records (live weight) is modeled as:

$$\mathbf{y} = \mathbf{Xb} + \mathbf{Zu} + \mathbf{Wh} + \mathbf{e}$$

where $\mathbf{y}$ is a $n \times 1$ vector of live weight measurements, $\mathbf{b}$ is a vector containing effects of covariates of the model (the mean and the cohort), $\mathbf{u}$ is a vector containing random polygenic effects (the effect of all the variants affecting live weight except the tested QTL—*see* **Note 8**) and $\mathbf{h}$ is a vector containing 20 effects of the ancestral haplotypes. Effects are associated to records with incidence matrices ($\mathbf{X}$, $\mathbf{Z}$, and $\mathbf{W}$). (Co)variance matrices of residual and ancestral haplotypes effects are diagonal, $\mathbf{I}_n \sigma_e^2$ and $\mathbf{I}_{20} \sigma_h^2$ where $\mathbf{I}_j$ is an identity matrix $j \times j$, σ_e^2 and σ_h^2 are the residual and ancestral haplotypes variances respectively. For the polygenic terms, the (co)variance matrix is $\mathbf{A}\,\sigma_a^2$ where $\mathbf{A}$ is the additive relationship matrix estimated from pedigree relationships and σ_a^2 the polygenic variance.

Testing for the presence of a QTL is performed at each marker position by a likelihood ratio test (LRT) comparing the likelihood of two models: a model where the variance of the QTL is assumed to be 0 (H0: no QTL effect) and a model where the variance of the QTL is estimated (H1). When the QTL variance is null, the likelihoods computed under H0 and H1 should be identical except for stochastic differences, and the LRT has then been shown to be distributed as a chi-square distribution with between 1 and 2 degrees of freedom. Using that distribution allows to determine whether the QTL variance is significantly different from 0 at a given threshold α. When significant values of the LRT have been obtained, a QTL is detected and a confidence interval of this QTL can then be specified around the position at which the LRT curve maximizes. The LRT curve obtained using the F2 individuals and the dense marker set is presented in Fig. 7 (two curves are

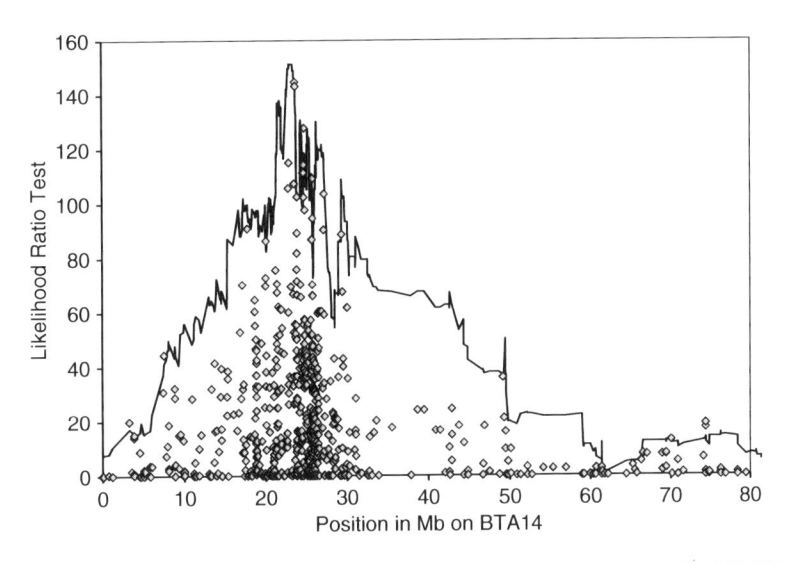

Fig. 7 LRT curve along BTA14 testing for a QTL affecting live weight in 745 F2 individuals genotyped for 981 markers with ancestral haplotypes (*black curve*) or single point tests (*grey squares*)

presented, using either ancestral haplotypes or SNPs as covariates, in the latter case, instead of the 20 ancestral haplotypes, the two possible SNP alleles are contrasted). The QTL is highly significant and the signal maximizes between positions 22,750,000 and 23,750,000 (Baylor 4.0/bosTau4). Use of ancestral haplotypes as covariates achieved higher significance than single point analyzes.

Solutions for the ancestral haplotypes effects obtained at the maximal location are represented in Fig. 8. Distribution of the ancestral haplotypes effects is bimodal, suggesting a biallelic QTL. Table 1 gives the frequency of each of the ancestral haplotypes and their effect. Four ancestral haplotypes seem to capture a variant associated to higher live weight. Frequencies of ancestral haplotypes range from 0 to 20 % (note that 3 ancestral haplotypes are not used). Some haplotypes are associated to rare ancestral haplotypes (with a frequency below 2 %). For these ancestral haplotypes, we don't have enough records to estimate their effect accurately. In this case the mixed models solutions are "regressed" towards the mean. Based on frequent ancestral haplotypes, the contrast between the two QTL alleles would approximately be equal to 20 kg live weight.

To confirm the localization of the QTL, we also used 3548 Holstein, Jersey and crossbred progeny-tested bulls genotyped for genomic selection with the Illumina BovineSNP50 BeadChip. We used 293 markers on chromosome BTA14 (approximately from positions 15,000,000 to 30,000,000 Baylor 4.0/bosTau4). Phenotypes were breeding values for live weight and the applied mixed model was almost the same. Fixed effects included

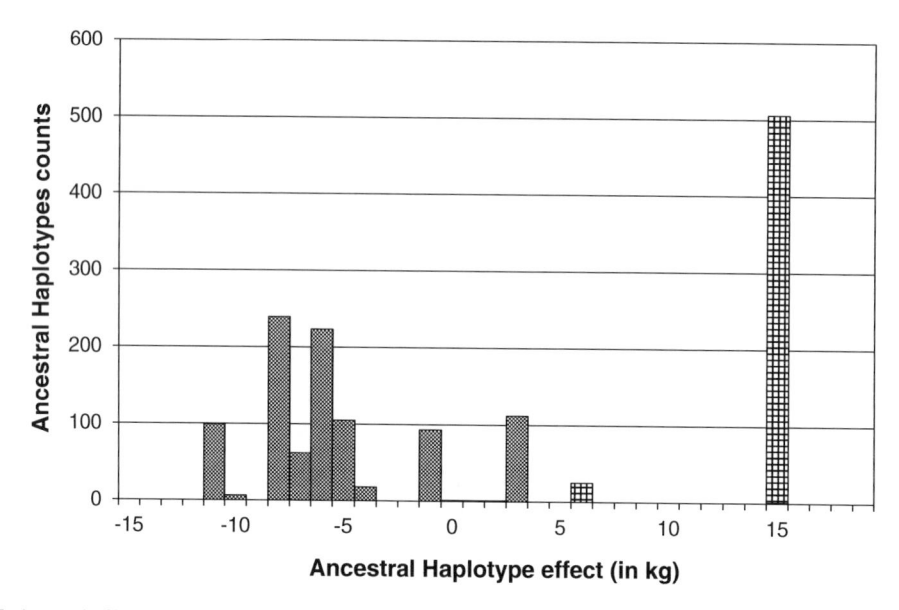

Fig. 8 Estimated effect on live weight and frequencies of ancestral haplotypes in the F2 population. Ancestral haplotypes having similar effects were pilled up. Ancestral haplotypes associated with the q and Q allele of the later identified candidate QTN variant are represented in *dark* and *light grey*, respectively

regressions on the percentage of Holstein or Jersey bloods. LRT curves obtained in the F2 population and with the progeny-tested bulls are presented in Fig. 9; both curves maximize in a <1 Mb chromosomal segment. The analysis on progeny-tested bulls confirms the QTL position with the use of a different set of individuals and markers.

Solutions of ancestral haplotypes are presented in Fig. 10, presenting also a bimodal distribution. Each haplotype was associated to a color according to the proportion of Holstein (light grey) and Jersey (black) blood of the bulls. Ancestral haplotypes with large effects were essentially associated to individuals with Holstein blood. Ancestral haplotypes with small effects were mainly linked to individuals with some Jersey blood but also to a few pure Holstein individuals. As expected, the ancestral haplotypes capture breed origin (some haplotypes being Jersey haplotypes and other Holstein haplotypes) and the q allele is from Jersey origin whereas the Q allele in from Holstein origin. In New-Zealand, even in individuals declared as 100 % Holstein, there is still presence of some Jersey haplotypes (in Dutch Holstein, Q seems to be fixed).

Within the confidence interval, the 6 F1 sires were resequenced in addition to one Holstein and one Jersey individual. Among identified DSV (9,572 in total), only those matching the sire QTL genotypes were kept for further analysis: 14 such candidate DSV (13 SNPs and one VNTR) were found and genotyped in the F2 population. The QTL mapping method was repeated and the LRT signal found for some of the selected SNPs in very high

Table 1
Association between ancestral haplotypes and candidate QTN or splice-site variant alleles and estimated effect of the 20 ancestral haplotypes in the F2 population

		Candidate QTN		Splice-site variant	
Ancestral haplotype number	Estimated effect	Number of alleles q	Number of alleles Q	Number of alleles T	Number of alleles A
1	3.1	1	0	0	1
2	15.4	0	99	0	99
3	−6.5	62	0	0	62
4	−7.2	229	0	0	229
5	15.6	1	289	1	289
6	6.9	0	25	0	25
7	0	0	0	0	0
8	15.8	0	117	0	117
9	0	0	0	0	0
10	−5.3	122	0	0	122
11	−5.5	101	0	0	101
12	−7.9	10	0	0	10
13	−9.9	6	0	0	6
14	−4.6	105	0	0	105
15	0	0	0	0	0
16	−3.6	18	0	0	18
17	1.2	1	0	0	1
18	3.4	111	0	0	111
19	−0.4	93	0	0	93
20	−10.4	99	1	99	1

LD exceeded previous signals (Fig. 11). This illustrates that when a SNP is in high LD with the causal DSV, a single point test has higher power than haplotype-based tests.

Based on results from other breeds, five of these candidate SNPs could be excluded. In addition to the candidate QTN, a T to A donor splice-site variant affecting expression of CHCHD7 was found in the region. Table 1 presents the association between ancestral haplotypes and alleles at the candidate QTN and the splice-site variant. The LD is almost perfect with only 2 misclassified alleles out of 1,490 for the candidate QTN and 1 out of 1,490

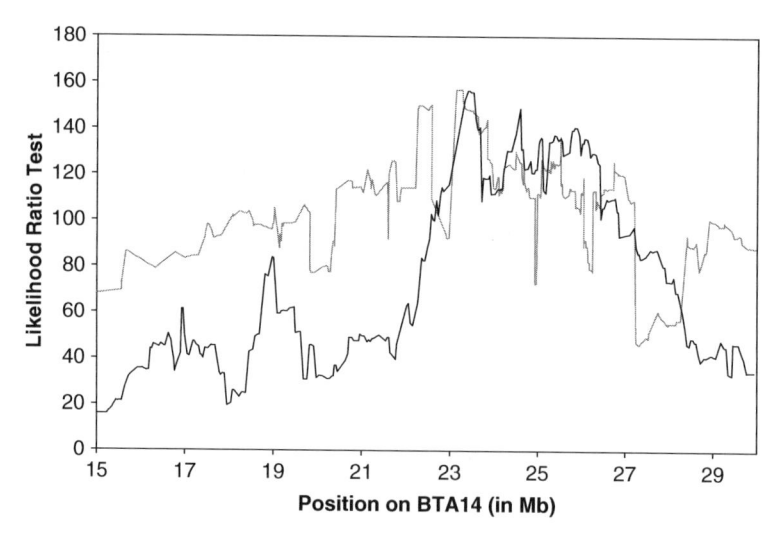

Fig. 9 LRT curve on BTA14 testing for a QTL affecting live weight in 745 F2 individuals (*black curve*) or in 3,548 progeny tested-bulls (*grey curve*) with ancestral haplotypes

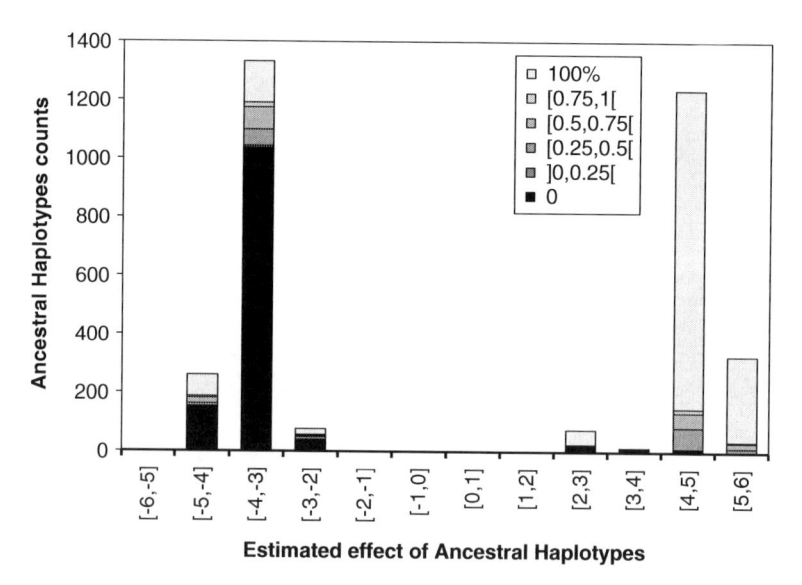

Fig. 10 Estimated effect on live weight and frequencies of ancestral haplotypes in the progeny-tested bulls population. Ancestral haplotypes having equal effect were pilled up. The intensity of *grey* reflects the percentage of Jersey blood; bulls were divided into six categories ranging from 100 % Holstein in *light grey* to 100 % Jersey in *Black*

for the splice-site variant (differences might also be due to genotyping errors) indicating the good predictive ability of the ancestral haplotypes. The table shows also the good association between ancestral haplotypes estimated effects and candidate QTN alleles.

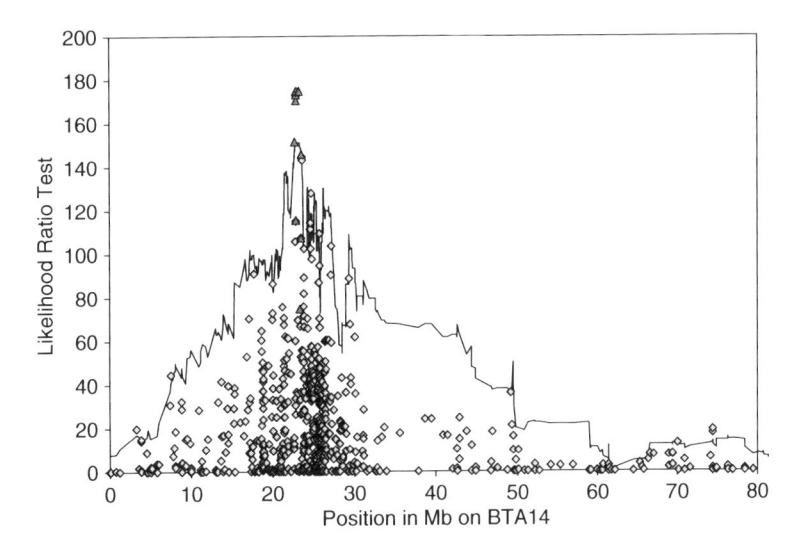

Fig. 11 LRT curve along BTA14 testing for a QTL affecting live weight in 745 F2 individuals with ancestral haplotypes (*black curve*), single point tests (*grey squares*) and genotyped candidates QTN (*dark grey triangles*)

This first illustration shows that, in the context of QTL mapping for complex traits in outbred livestock species, methods relying on ancestral haplotypes can be used to refine the localisation of a QTL and that the ancestral haplotypes are in high LD with DSV in the studied region.

3.3 Association Mapping in Structured Populations

In the second application, association is performed with ancestral haplotypes for discrete (binary) traits. This illustration is based on a study by Zhang et al. [12] and data from Durkin et al. [13] and Sartelet et al. [14, 15]. The data consists of a sample of Belgian Blue cattle genotyped with a custom made 50K bovine chip. In addition to 301 controls, 79 individuals present particular phenotypes: 33 are affected by gingival hamartoma, 24 by prolonged gestation, 13 by arthrogryposis and eight present a particular coat with a white-stripe on the back (the phenotype is called color-sideness [13]). The three first phenotypes are monogenic recessive diseases.

47,574 SNPs mapping to autosomal chromosomes and satisfying quality controls were used. Haplotypes were reconstructed using Beagle [16] and assigned to 10 ancestral haplotypes with DualPHASE [4]. Ancestral haplotypes-based association study was then performed using a generalized linear mixed model (GLMM). A linear predictor η is first modeled with a model similar to the one used for complex traits:

$$\eta = \mathbf{Xb} + \mathbf{Zu} + \mathbf{Wh}$$

This linear predictor is linked to the observed phenotype with a logit link function:

$$\mu_i = h(\eta_i) = \frac{e^{\eta_i}}{1 + e^{\eta_i}}$$

where μ_i is the probability to express the phenotype (for instance coding "normal" phenotypes as 0 and affected or color-sided individuals as "1"). The solutions of this GLMM can be obtained using an iterative procedure based on the Laplacian approximation of the likelihood. At each iteration, current estimates of **b**, **u** and **h** are used to approximate the GLMM as an equivalent LMM which is then solved to obtain new estimates of the parameters. This procedure is repeated until convergence.

In this application, the additive relationships were not estimated from the pedigree but from the markers. A genomic relationship matrix (*see* **Note 9**) was built based on similarity between haplotypes at each position; the values were then averaged over the whole genome. Haplotype similarity was based on ancestral haplotypes with the following similarity score between individuals x and y at marker m:

$$S_{xy,m} = 0.25[I_{11} + I_{12} + I_{21} + I_{22}]$$

where I_{ij} is equal to 1 if the ith and jth haplotypes of individuals x and y are assigned to the same ancestral haplotype, and 0 else (if not). The values are standardized to obtain values between 0 (unrelated) and 1 (completely inbred). The inclusion of this polygenic effect corrects for population and/or family structure (stratification) as demonstrated by Yu et al. [17].

Testing for the presence of a QTL is based on a score test. The idea is to test whether the variance of **h** is significantly different from 0 (H1) by estimating the slope of the first derivative of the log-likelihood when $\sigma_h^2 = 0$. Positive value of the slope then indicates that the most likely value of the variance is greater than 0. Tzeng and Zhang [18] derived a test statistic T for haplotype-based models for GLMM. Applied to the logit function, the test statistic is equal to:

$$T = 0.5(\mathbf{W}'(\mathbf{y} - \mathbf{Xb} - \mathbf{Zu}))' \times \mathbf{W}'(\mathbf{y} - \mathbf{Xb} - \mathbf{Zu})$$

It is based on residuals $(\mathbf{y} - \mathbf{Xb} - \mathbf{Zu})$ obtained by first solving the GLMM without the ancestral haplotypes effects. The successive positions to be scanned can then be tested using the value of T in the above equation, where the **W** matrix is position dependent.

The significance of the values of T can be obtained using the distribution of T under the null hypothesis ($\sigma_h^2 = 0$). Although analytically difficult, this distribution can be approximated using a gamma distribution. Since the parameters of the distribution are local (different at every tested position), we obtain them through permutations: shuffling the phenotypes breaks the potential

Table 2
Association between disease and ancestral haplotypes for three monogenic recessive disorders

	Associated ancestral haplotype	Number of homozygous cases	Number of homozygous controls
Harmatoma	#9	33/33	0/346
Arthrogryposis	#8	13/13	0/366
Prolonged gestation	#7	24/24	2/356

relationship between phenotypes and haplotypes, leading to situations corresponding to H0, and consequently to T values sampled from the distribution of T (or approximately from the sought gamma distribution). Repeating this procedure a large number of times allows having an accurate estimation of the parameters of the gamma distribution.

As a side remark, the T score can be interpreted as a sum over the ancestral haplotypes groups of the sum of the squared differences between observed (i.e., **y**) and expected (i.e., **Xb** + **Zu**) values of the phenotypes for each group.

For the monogenic recessive diseases, the signal is clear and identifies unambiguously the region harboring the causal variant (we don't present Manhattan plots for these). In each case, a specific ancestral haplotype is associated with the causal variant (see Table 2). The association is almost perfect, all cases and very few controls are homozygous for the identified ancestral haplotype; indicating their usefulness for association studies.

For color-sideness, we selected only colored individuals as controls. We excluded white individuals because the phenotype (color-sideness) is not clearly visible for these animals. Color-sideness is due to a CNV [13] mapping to chromosome BTA29. The Manhattan plot is presented in Fig. 12 and the region harboring the CNV is identified with high significance (below 10^{-9}).

Each of the eight color-sided individuals carries at least one copy of the same ancestral haplotype associated to the CNV (which has a dominant behavior). This illustrates that haplotype-based studies can also capture variants associated to CNV.

This second set of illustrations confirms that ancestral haplotypes present a strong association with ungenotyped variants and are useful in association and fine-mapping studies (see **Note 10**). This association is expected to be stronger with recent haplotypes which are longer. Therefore, they would be more efficient for variants under recent selection or for mapping diseases for which the frequency increased recently in the population.

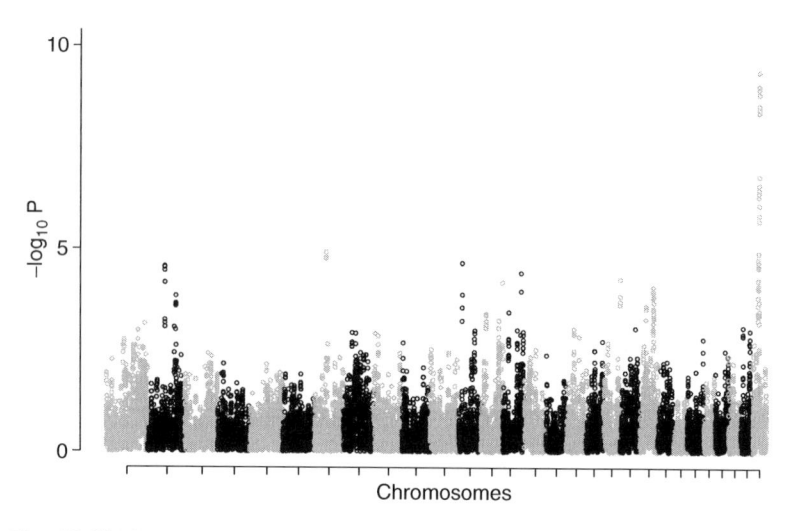

Fig. 12 Manhattan plot for ancestral-haplotyped based association with color-sideness in Belgian Blue cattle. The *Y*-axis expressed the level of signifance and the *X*-axis the position in the cattle genome (chromosomes are alternatively represented in *grey* and *black*)

3.4 Further Analysis of the Identified Region

In both application sections, it was illustrated that LMM or GLMM associated with ancestral haplotypes can help to determine chromosomal regions harboring interesting variants. However, these regions remain large and identifying the causative variants requires other techniques which will not be discussed in this chapter.

A confidence interval spanning the region detected in the fine-mapping study must be determined. LOD drop-off techniques are often used in that situation but are not extremely precise for LRT tests associated with mixed models whereas boot-strapping techniques are difficult in structured populations (with pedigrees and with relationships among individuals). Confirmation with a second cohort, as has been done for the live weight QTL, is more meaningful since different samples can help us to discard different regions.

After a significant signal has been detected and refined with models using ancestral haplotypes, use of identity-by-descent (IBD) sharing techniques is helpful to refine the causal DSV location (*see* **Note 11**). These techniques compare haplotypes from individuals known to carry identical QTL or risk alleles and assume that around the variant, these haplotypes must be identical. For monogenic recessive diseases, we would compare all haplotypes from all cases; for color-sideness (which was dominant), we would compare the common haplotype of color-sided individuals and for the QTL mapping experiment we would compare sire haplotypes carrying Q and q alleles. This is an advantage of designs where parents have many offspring (as in agronomical or model species); in that case,

QTL genotypes from the parents can be determined with high confidence based on statistical techniques and linkage information. With IBD sharing methods, the confidence interval would then exclude all regions where at least one haplotype is distinct.

DSV within the confidence interval must be found through resequencing. Candidate DSV should be filtered according to their biological relation with the phenotype (for instance, gene function), to their expected impact on the normal biological process (changing level of expression, changing amino acids, stop codon, etc.) and to their consistency with known genotypes from all individuals. For Mendelian traits, the genotypes at the causal variant are known from the phenotype. As mentioned above, QTL genotypes of parents with many offspring can be determined statistically with high confidence from linkage analysis. In other cases, ancestral haplotypes might still be useful. If association with ancestral haplotypes and underlying DSV is high, we could also search DSV consistent with haplotype groups with extreme phenotypic effects (which increase our confidence on the allele carried by the haplotype). This approach has been successfully applied to find DSV affecting recombination rate in cattle [19].

4 Discussion

4.1 General Versus Biological Model

The general modeling of ancestral haplotypes is not directly related to biological processes such as evolution but is rather predictive. The ancestral haplotypes do not necessarily correspond to real ancestors of individuals in the sample. Although the model fits well to the biological data, it does not make use of knowledge of the sample parameters such as effective population size, physical distance between markers, history, etc. Neither does it rely on a coalescent model or on demographic or evolutionary processes. A few attempts have been made to link model parameters (such as jump probability) to biological properties (such as local recombination rates) without clear success, even for more sophisticated models using Bayesian priors to influence parameter distribution. Interaction between parameters such as mutation rate, recombination rate and number of founders also makes their interpretation more difficult.

Nevertheless the HMM fits the data particularly well without relying on assumptions, which makes the model robust. The model can adjust itself to the data whereas models based on precise parameters would be sensitive to wrong priors and parametric assumptions. Although not equivalent, parameters remain vaguely related to biological parameters: jump probability is linked to recombination rates (including physical distance, number of generations to ancestors, recombination hotspots), the number of ancestral haplotypes would be linked to their age (more ancestral haplotypes correspond to more recent ancestors), etc. It is also interesting to note that the model groups haplotypes with a recent common ancestor (or with high IBD probability) almost as well [4] as more complex models based on the coalescent theory and using parameters such as effective population size, allele frequencies and physical distance between markers. In summary, without trying to translate all biological processes into realistic details, the model nevertheless performs well.

4.2 Predictive Properties of Ancestral Haplotypes

The model was first designed to perform haplotyping and to impute ungenotyped marker alleles [3–5]. In these studies, it was shown that the model was efficient for these applications stressing the high linkage disequilibrium between ancestral haplotypes and underlying DSV. Use of ancestral haplotypes in association studies confirmed that ancestral haplotypes are tightly associated with underlying causal variants and are a useful tool for GWAS. More on the use of ancestral haplotypes in association studies can be found in Su et al. [10], Druet and Georges [4] and Zhang et al. [12]. Software have been developed to assign haplotypes to ancestral haplotypes [4, 10] and to perform association studies accounting for stratification with GLMM [12]. Examples of successful applications for fine-mapping are Karim et al. [11], Durkin et al. [13], Sandor et al. [19], Sartelet et al. [14, 15]. These various examples illustrate that the ancestral haplotypes can capture different types of DSV, including CNV, insertions and deletions.

4.3 Flexibility of the Ancestral Haplotype Model

There is a large variety of haplotype-based methods and we will not present them. Each method relies on a different definition of haplotypes. The present method shows a relatively high flexibility since jumps can occur between any pair of markers: it does not need the definition of arbitrary window size for haplotype definition or definition of haplotype block boundaries. Recombinant haplotypes do not create additional (or new) haplotypes but are dynamically assigned to different ancestral haplotypes before and after crossing-over. Keeping the number of haplotypes low is important to maintain power in association studies. Similarly, haplotypes with small differences (genotyping errors or mutations) can still be associated to the same ancestral haplotypes without creating new rare haplotypes. Finally, since the number of ancestral haplotypes is defined, it remains under control.

4.4 Use in Association Studies

The method we presented can be compared to a so-called LD/LA fine-mapping method. Within the pedigree (mainly large families), linkage information is used whereas across pedigrees LD information is used by grouping similar haplotypes in clusters of related haplotypes. Through LD, information from more distant information is exploited and historical recombinations (*see* **Note 12**) are used (the recombinations from the common ancestor to the current population which cut the shared fragment into smaller pieces). Indeed, within the pedigree, shared segments are easily identified but extremely long. Therefore, the confidence interval of the region harboring the QTL is too large. When using ancestral haplotypes, shared segments are much smaller because it traces back to more distant ancestors. This allows obtaining smaller confidence intervals for the QTL localization. In addition, keeping the number of ancestral haplotypes low is important to correctly estimate their effect and detect potentially segregating QTL.

5 Notes

1. Haplotypes are specific combinations of successive marker alleles on a chromosome segment. We can compare them to different words we could use at a particular position in a text. If some alleles are missing in the sequence, they can be guessed similarly to missing letters that can be guessed from the other letters in the words or in surrounding words. Therefore, haplotype information can be used to predict ungenotyped alleles or to capture underlying variants of interests (i.e., having an effect on the studied phenotype).

2. Imputation consists in the prediction of missing genotypes based most often on haplotypes information (predicting missing letters based on observed letters). Thanks to this method, marker alleles observed in a reference sample can be genotyped in silico in another sample of the population. This technique is often used in GWAS studies. Since they rely on haplotype modeling, imputation methods are closely related to haplotype-based GWAS methods.

3. HMM are statistical tools used to describe sequences of observations. Parameters of the model are automatically learned from the observed data and the model can be used to make predictions. Furthermore, HMM are computationally efficient. These tools have been used for applications such as speech recognition. In genetics, they are well suited to haplotype modeling. Consequently, they have already been used to describe sequences of marker alleles, to learn patterns from genomic data, to predict sequences of unordered genotypes (phasing) and missing genotypes (imputation) and to link phenotypes with unobserved variants (association studies).

4. In the ancestral haplotype model, founder haplotypes are unknown and learned from the data. They are purely conceptual—the model is not expected to describe explicitly the underlying biology—and correspond to groups of similar haplotypes probably derived from a common ancestor. They do not correspond necessarily to physical ancestors. Approaches relying

on this model are quite flexible and nonparametric—they do not rely on assumptions, which makes them relatively robust.

5. Nevertheless, the ancestral haplotypes can capture real genealogical processes. For a better understanding we can compare ancestral haplotypes to the notion of most recent common ancestor (MRCA). The ancestral haplotype would correspond to the MRCA of a set of haplotypes. At one position, several ancestral haplotypes are defined and hence, several MRCA. Along the chromosome, the set of MRCA can change. Similarly, each MRCA do not trace back exactly the same number of generations backwards in the pedigree.

6. In many species, such as humans, individuals are diploid. Therefore haplotypes are not observed and the observation units are genotypes. HMM must then be extended and changed accordingly. One solution is to model two chains (i.e., haplotypes) simultaneously, making then emission probabilities a bit more complex. The HMM can easily be extended to different levels of ploidy (but with additional computation costs when the ploidy increases).

7. The extension of the model to include pedigree information is a top-down approach: information is transferred vertically from the informative parent to its offspring. It is only recommended when the parent has more precise information than the offspring, in terms of accuracy of haplotypes or in terms of number of genotyped markers. For instance, it can be used with sires having many offspring (for which accurate haplotypes are consequently available) or with parents genotyped at higher density. The model would not be recommended if a parent has a single progeny or if the parent is genotyped at lower density.

8. In general, phenotypes are affected by multiple genes, most of them having a small effect. Still, the combination of all small effects can result in a large genetic effect. Relatives share common polygenic effects (they are more similar). When performing association studies or QTL mapping, sharing of involved loci can be confounded with sharing of other genomic regions between relatives. This should be taken into account by including such polygenic effects in the model to obtain a cleaner statistic. The correlation between the polygenic effects of relatives can be modeled with a so-called relationship matrix, which can be obtained from the known genealogy (pedigree) of the individuals.

9. When the pedigree data is not known, then a genomic relationship matrix can be obtained from marker data by observing the number of shared alleles, and used as a surrogate of the relationship matrix. It could work as well, or even better, than the pedigree relationship matrix, since real segregation of the

parental chromosomes can indirectly be observed. For that reason, this matrix is also able to correct for polygenic links between individuals and for potential stratification of the data (mixture of sub-populations with varying allelic contents), and is thus a precious tool to avoid spurious signal in associations studies.

10. The examples and our experience show that ancestral haplotypes are in high LD with underlying DSV. Ancestral haplotypes proved efficient in imputation studies, haplotype reconstruction programs and in association studies, genomic selection and QTL fine-mapping. The association will be better for recent variants which will be easier to identify.

11. Ancestral haplotypes can also be useful in IBD sharing methods which identify common segments shared by individuals carrying the same risk alleles. Alleles can be determined based on segregation in offspring (QTL mapping in large half-sib families), phenotype (in the case of monogenic recessive or dominance) or eventually on solutions of ancestral haplotype effects. The solutions can be used to identify extreme groups. We can then sequence individuals carrying different mutations to identify them.

12. Linkage analysis is powerful to detect QTL in large sire families because the power to estimate differences between alleles of sires is large. Indeed, each sire allelic effect is estimated based on a large number of offspring. However, resolution is low since progeny share long haplotype fragments in common. Only those recombinations occurring in the last generation (on the sire chromosome) help to determine the confidence interval. With the ancestral haplotype model, shared segments around the QTL are small and comparison can be performed across families. Power is maintained if haplotypes are grouped in large clusters. The resolution is much improved because we use historical recombinations (those from the ancestral haplotypes to the current population) which break haplotypes into very small pieces.

References

1. Broman KW (2005) The genomes of recombinant inbred lines. Genetics 169:1133–1146
2. Rabiner LR (1989) A tutorial on hidden Markov chains and selected applications in speech recognition. Proc IEEE 77:257–286
3. Scheet P, Stephens M (2006) A fast and flexible statistical model for a large-scale population genotype data applications to inferring missing genotypes and haplotype phase. Am J Hum Genet 78:629–644
4. Druet T, Georges M (2010) A hidden Markov model combining linkage and linkage disequilibrium information for haplotype reconstruction and quantitative trait locus fine mapping. Genetics 184:789–798
5. Sun S, Greenwood CMT, Neal RM (2007) Haplotype inference using a Bayesian hidden Markov model. Genet Epidemiol 31:937–948
6. de Roos APW, Schrooten C, Druet T (2011) Genomic breeding value estimation using

genetic markers, inferred ancestral haplotypes, and the genomic relationship matrix. J Dairy Sci 94:4708–4714

7. Kover PX, Valdar W, Trakalo N, Scarcelli I, Ehrenreich M, Purugganan MD et al (2009) A multiparent advanced generation inter-cross to finemap quantitative traits in Arabidopsis thaliana. PLoS Genet 5(7):e1000551

8. Mott R, Talbot CJ, Turri MG, Collins AC, Flint J (2000) A method for fine mapping quantitative trait loci in outbred animal stocks. PNAS 97(23):12649–12654

9. Valdar W, Solberg LC, Gauguier D, Burnett S, Klenerman P, Cookson WO et al (2006) Genome-wide genetic association of complex traits in heterogeneous stock mice. Nat Genet 38:879–887

10. Su SY, Balding DJ, Coin LJ (2008) Disease association tests by inferring ancestral haplotypes using a hidden Markov model. Bioinformatics 24:972–978

11. Karim L, Takeda H, Lin L, Druet T, Ariaz JAC, Baurain D et al (2011) Variants modulating the expression of a chromosome domain encompassing PLAG1 influence bovine stature. Nat Genet 43:405–413

12. Zhang Z, Guillaume F, Sartelet A, Charlier C, Georges M, Farnir F, Druet T (2012) Ancestral haplotype-based association mapping with generalized linear mixed models accounting for stratification. Bioinformatics 28(19): 2467–2473. doi:10.1093/bioinformatics/bts348

13. Durkin K, Coppieters W, Drögmüller C, Ahariz N, Cambisano N, Druet T et al (2012) Serial translocation via circular intermediates underlies colorsidedness in cattle. Nature 482:81–84

14. Sartelet A, Druet T, Michaux C, Fasquelle C, Géron S, Tamma N (2012) A splice site variant in the bovine RNF11 gene compromises growth and regulation of the inflammatory response. PLoS Genet 8(3):e1002581

15. Sartelet A, Stauber T, Coppieters W, Fasquelle C, Druet T, Zhang Z (2012) A missense mutation in the ClC-7 chloride channel causes hamartomas with osteopetrosis in cattle. Submitted

16. Browning SR, Browning BL (2007) Rapid and accurate haplotype phasing and missing data inference for whole-genome association studies by use of localized-haplotype clustering. Am J Hum Genet 81:1084–1097

17. Yu J et al (2006) A unified mixed-model method for association mapping that accounts for multiple levels of relatedness. Nat Genet 38:203–208

18. Tzeng JU, Zhang D (2007) Haplotype-based association analysis via variance-components score test. Am J Hum Genet 81:927–938

19. Sandor C, Li W, Coppieters W, Druet T, Charlier C, Georges M (2012) Genetic variants in REC8, RNF212 and PRDM9 influence male recombination in cattle. PLoS Genet 8(7):e1002854. doi:10.1371/journal.pgen.1002854

Chapter 16

Genotype Phasing in Populations of Closely Related Individuals

John M. Hickey

Abstract

Knowledge of phase has many potential applications for empowering genomic information. For example, phase can facilitate the identification of identical by descent sharing between pairs of individuals, as part of the process of genotype imputation, or to facilitate parent of origin of allele modeling in order to quantify the effect of parental imprinting.

Long-range phasing and haplotype library imputation are a fast heuristic phasing method that is particularly suited to application in isolated populations, such as those that prevail in animal and plant breeding programs or human populations in remote locations.

This chapter describes the general principle of long-range phasing, the specifics of the long-range phasing, and haplotype imputation algorithm implemented in the software AlphaPhase1.1 (Hickey et al., Genet Sel Evol 43:12, 2011). Examples of how to apply AlphaPhase1.1 to different data sets are described with particular reference to the avoidance of common pitfalls. Finally, some notes describing some of the key terms are given.

Key words Phasing, Haplotyping, GWAS, AlphaPhase, Imputation, IBD

1 Introduction

Knowledge of the phase of genotype data can be useful in genome-wide association studies (GWAS). Phase information allows the use of analysis frameworks that account for identity by descent (IBD) or parent of origin of allele [2] and it can lead to a large increase in data quantities via genotype or sequence imputation e.g., [3, 4]. Phasing entire genomes in GWAS data sets has been a computational bottleneck due to the unavailability of robust heuristic phasing methods when a family structure does not exist and due to the computationally intensive nature of statistically based phasing methods, e.g., fastPHASE [5] and Beagle [6]. Long-range phasing and haplotype library imputation (LRPHLI; [1, 7]) are a fast and accurate heuristic method for phasing, which uses information from both related and seemingly unrelated individuals by invoking the concepts of surrogate parents and Erdös numbers [7] (*see* **Note 1**) and haplotype libraries [1]. In comparison to phasing methods based on statistical inference (e.g.,fastPHASE; [5]), this paradigm is much faster and much more accurate [1, 7]. While in theory LRPHLI methods could work well in any population (including the global human population) if enough data were available, practical implementation of the method requires that the data comprise individuals that are reasonably related to each other so that haplotypes of moderate size (e.g., 10 cM) are likely to be shared by individuals. This scenario prevails in isolated human populations such as those on islands (e.g., Iceland, Orkney), and in livestock and plant breeding populations. This chapter is divided into six sections. Firstly, the general principle of long-range phasing is described; secondly, the LRPHLI algorithm implemented in AlphaPhase1.1 [1] is described; thirdly, some results relating to the phasing performance of AlphaPhase1.1 are given; fourthly, examples of how to apply AlphaPhase1.1 to different data sets are described; fifthly, some potential applications of the resulting data and pitfalls of using AlphaPhase1.1 are discussed; finally, some notes are given which define some of the important terms and give more specific advice on the setting of parameter values of AlphaPhase1.1.

1.1 Long-Range Phasing

The LRP method of Kong et al. [7] is illustrated in Fig. 1. These authors have suggested to phase a string of consecutive SNP in a single genome region (termed a core in [1]) by first identifying surrogate parents of each proband. Surrogate parents are individuals who share a haplotype with the proband (*see* **Note 2**) and are identified as those individuals that do not have any opposing homozygote genotypes with the proband. These surrogates are termed Erdös 1 surrogates, meaning that they are one degree removed from the proband on the basis of haplotype identity.

The Erdös 1 surrogates of the proband are partitioned into surrogates of the paternal and maternal haplotypes. For the proband, inference of the phase at each locus within the paternal/maternal haplotype is attempted by stepping through the paternal/maternal surrogates until a surrogate is found that is homozygous at that locus and thus can be used to declare the phase. This is termed accessing Erdös 1 information. If a homozygote is not found at Erdös 1, the algorithm proceeds to information from surrogates at the Erdös 2 layer. Erdös 2 surrogates of a proband are surrogates who do not share a haplotype with the proband but do share a haplotype with Erdös 1 surrogates of the proband. The algorithm can continue like this for as many Erdös layers as contained within the data (Fig. 1). Errors created due to incorrect surrogate identification can be partially resolved by pruning from the surrogate list those surrogate parents whose haplotypes (phased by an earlier round of LRP) do not agree with the genotype of the

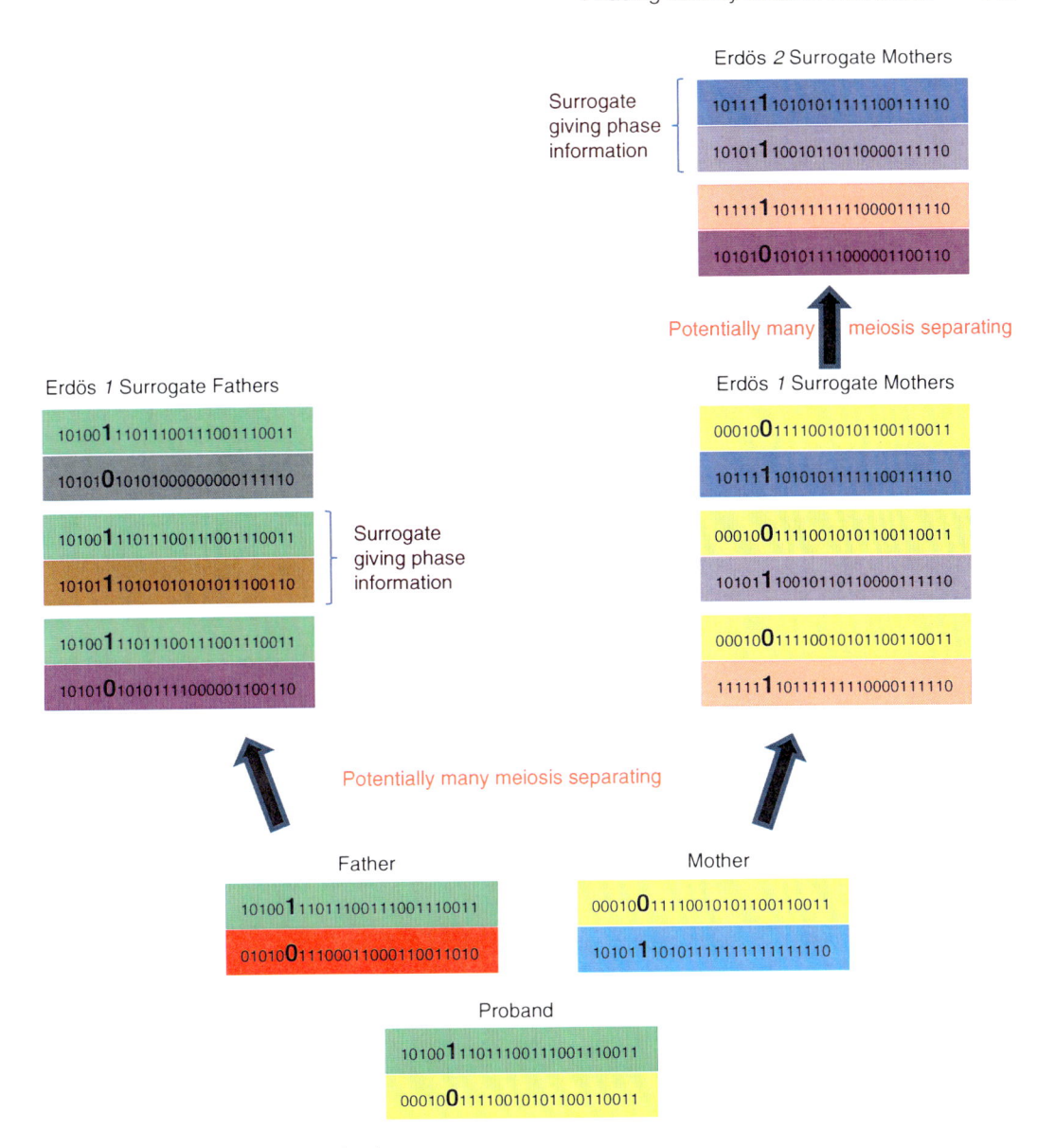

Fig. 1 Illustration of long-range phasing process

proband (because of genotyping errors or insufficient combinatorial power to eliminate or partition surrogate parents properly), and then re-phasing all individuals using the pruned list of surrogates.

2 LRPHLI Algorithm Implemented in AlphaPhase1.1

AlphaPhase1.1 uses 2 component algorithms concurrently for the purpose of resolving phase. It uses a modified long-range phasing algorithm and a haplotype library imputation algorithm both of which are extensively described in Hickey et al. [1].

A brief description of the important features and modifications from the Kong et al. [7] algorithm is given here.

In the definition of surrogate parents in AlphaPhase1.1, information is used on the markers in the core and its adjacent tails (*see* **Note 3**). A core is defined as a consecutive string of SNP loci for which phasing is being attempted. Tails are defined as consecutive strings of SNP loci immediately adjacent to either end of a core. Information on homozygous loci across both core and tails is used to define surrogates. Specifically, opposing homozygotes between two individuals illustrates lack of IBD and surrogacy in that region. Tails provide additional information thus reducing the risk of false surrogate definition, especially near the ends of the core region. Without using tails, combinatorial power may be insufficient at the ends of cores and this would result in failure to eliminate individuals that are not surrogate parents due to recombination events in these regions.

In AlphaPhase1.1, the partitioning of surrogate parents into surrogates of the paternal and maternal haplotypes is done in two ways using pedigree information if it is available and using a k-medoids clustering algorithm if it is not. Partitioning of surrogates if both parents are genotyped involves dividing the surrogate parents into two groups, those that are only surrogates of the proband and its sire (i.e., not the dam) and those that are only surrogates of the proband and its dam (i.e., not the sire). The k-medoids algorithm works by making two mutually exclusive clusters of the surrogate parents and arbitrarily labeling one the "paternal" and the other the "maternal."

For each core, each individual in a data set has potentially many surrogate parents at each Erdös level and potentially many Erdös levels (Fig. 1). In AlphaPhase1.1, this information is collapsed into a single square matrix of order equal to the number of individuals in the data set with genotype information. For each individual, only information on Erdös 1 surrogate parents is stored explicitly with this information identifying whether a surrogate parent is of the paternal haplotype, maternal haplotype, or both. Surrogate parents at higher Erdös layers are thus stored implicitly. Storing an indicator as to whether a surrogate parent is paternal, maternal, or both facilitates the use of the partitioned surrogates at Erdös layer 1 while allowing all information to be used at higher Erdös layers. It is not necessary to partition the surrogates into maternal/paternal at higher Erdös layers, giving greater flexibility and power.

AlphaPhase1.1 attempts to phase each locus for each proband by sequentially stepping through each Erdös layer and accessing the surrogate parents at each of these layers until a target number of surrogate parents is accessed. Once this target number of surrogate parents is accessed, the phase is determined if a proportion of these greater than a small error threshold (i.e., <10 %) agrees on what the phase is. If there is disagreement among the surrogate parents on what is the phase, a statistical significance test can be performed. The sequential approach ensures that even though individual surrogate parents can appear at several different Erdös levels, they are only used at their lowest Erdös level because it steps through the Erdös levels from the lowest to the highest. A restriction is included to ensure that the route to a homozygous surrogate parent can only pass through heterozygous surrogate parents. Phasing using consensus information from many surrogates reduces the impact of any individual surrogate. If information is used only from the first found homozygote, surrogate phasing error can result when false surrogates have been identified or a surrogate contains a genotype error. Using information from multiple surrogates alleviates the need to carry out the surrogate pruning step of the [7] algorithm.

In AlphaPhase1.1, genotyping errors are handled at the step that identifies the surrogates of a proband. Individuals who show no opposing homozygotes with that proband across long strings of consecutive loci are easily disrupted by both genotype and mapping errors. This can be overcome by allowing a small percentage of opposing homozygotes without rejecting surrogacy. However, this creates a new problem of increased numbers of individuals being identified as surrogates that are not in fact true surrogates. Several steps are taken to deal with this problem, including the haplotype library imputation (described below), using information from multiple surrogates, and the removal of surrogates breaking certain rules (described in Appendix A of [1]).

Finally, long-range phasing may not work in some individuals for which there is insufficient surrogate information (e.g., due to a recombination in a gamete), or for which surrogate information is inconsistent. Some of these problems can be overcome at the end of the long-range phasing step by building a library of all unique haplotypes that the long-range phasing step has found in the data set, and sequentially imputing phase for unphased individuals from this library. At each round, individuals that have one of their pair of gametes unphased can have it phased as the complement of its phased gamete via the genotype. At each round, new haplotypes that have been created through recombination can be detected and added to the library. A number of steps (described in Appendix A of [1]) are invoked during HLI to determine if a suitable haplotype exists in the library and for it to be declared as phase for the unphased gamete of the proband. The entire LRPHLI algorithm is described in Box 1.

Box 1
The Entire LRPHLI Algorithm

Step 1: Define start point and end point of the cores and tails.

Step 2: Loop across the cores and complete the following steps for each core.

(continued)

> **Box 1**
> **(continued)**
>
> Step 2a: Identify Erdös 1 surrogate parents for each genotyped individual by looping across all SNP in the core and tails and counting the numbers of homozygote genotypes in agreement and in disagreement between the individual and all other genotyped individuals. If the count of the loci in disagreement is less than a determined threshold (e.g., 2 %), an individual is taken to be a surrogate parent.
>
> Step 2b: Partition Erdös 1 surrogates into surrogates of the paternal or maternal gamete.
>
> Step 2c: Loop across the genotyped individuals and phase their loci based on information from the various Erdös levels as required.
>
> Step 2d: Build a haplotype library containing all completely phased haplotypes found in the data set.
>
> Step 2e: Impute the phase for gametes that are not completely phased by LRP by matching their phased loci to haplotypes in the haplotype.
>
> Step 2f: In each proband, all phased loci are paired to create a genotype and this genotype is checked for compatibility with the genotype and if across a whole core more than 10 % disagreement is found, the heterozygous loci have their phase call removed for this proband. In this case, homozygous loci are assumed to have no genotyping error and are thus phased de facto for this core of this proband.
>
> Steps 2e and 2f are iterated until no new haplotypes are added to the library.

3 Phasing Performance of AlphaPhase1.1

The performance of AlphaPhase1.1 was extensively tested in Hickey et al. [1] using both simulated and real data. Effective population size, SNP density, pedigree structure, and species were varied, and a wide variety of program parameters were used. AlphaPhase1.1 performed well in both simulated and real livestock and human data sets in terms of both phasing accuracy and computation efficiency. The percentage of alleles phased in both simulated and real data sets of varying size generally exceeded 98 %, whereas the percentage of alleles incorrectly phased in simulated data was generally less than 0.5 %. The phasing accuracy was affected by data set size, with lower yields for data set sizes less than 1,000, but was not affected by effective population size, family data structure, presence or absence of pedigree information, and SNP density. The method was computationally fast but required appropriate tuning of the program parameters in order to obtain fast and accurate phasing for large data sets. The reader is referred to Hickey et al. [1] for more details.

4 Applying AlphaPhase1.1 in Practice

4.1 Input Files

Two files are required when running the program. A third (Pedigree file) can also be supplied. A fourth (the true phase) can also be supplied if simulated data are being used and performance is required to be tested. The two primary input files are AlphaPhaseSpec.txt (a parameter file used to control the program) and a genotype file.

AlphaPhaseSpec.txt. An example of AlphaPhaseSpec.txt is shown in Box 2. Everything to the left of the comma should not be changed.

Box 2
Example of AlphaPhaseSpec.txt

PedigreeFile	,PreSpecifiedPedigree.txt
GenotypeFile	,60kGenotypeGF.txt,GenotypeFormat
NumberOfSnp	,2000
GeneralCoreAndTailLength	,300
GeneralCoreLength	,100, NotOffset
UseThisNumberOfSurrogates	,10
PercentageSurrDisagree	,10.00
PercentageGenoHaploDisagree	,1.00
GenotypeMissingErrorPercentage	,1.00
NrmThresh	,0.00
FullOutput	,1
Graphics	,0
Simulation	,0
TruePhaseFile	,TruePhaseFile.txt

The program is controlled by changing the input top the right of the comma.

Below is a description of what each line is for. It is important to note that *AlphaPhaseSpec.txt* is case sensitive.

PedigreeFile gives the name of the file containing the pedigree information. If there is no pedigree file available, include *NoPedigree* in place of a pedigree file name.

GenotypeFile gives the name of the file containing the genotypes, followed by a comma, followed by the format of the genotype file. There are three possible formats, *GenotypeFormat* (where the genotypes are coded as 0, 1, and 2, with missing markers coded as 3) and *UnorderedFormat* (with the genotypes as unordered alleles coded as 1's and 2's). Further details are given in the Genotype File format description given below.

NumberOfSnp gives the number of SNP in the genotype file.

GeneralCoreAndTailLength gives the overall length in terms of numbers of SNPs of the core and its adjacent tails. For example, if the *GeneralCoreLength* (described below) is 100 and the *GeneralCoreAndTailLength* is 300, this means that the core is 100 SNPs long and the tails are the 100 SNPs adjacent to each end of the core, thus the length of the core and tail is 300 SNPs long. At the end of a chromosome, the tail can only extend in one direction. Thus in this case, the core and tail length would be only 200 SNPs, the 100 SNPs in the core, and the 100 SNPs adjacent to the core.

GeneralCoreLength gives the overall length in terms of numbers of SNPs of the core. The *GeneralCoreLength* can never be longer than the *GeneralCoreAndTailLength*. The mode is also set at the end of this line. The two options are *Offset* and *NotOffset*. The *Offset/NotOffset* mode allows the program to be run in an *Offset* mode or a *NotOffset* mode. The *NotOffset* mode starts at the first core at the first SNP and defines all subsequent cores from this point. The *Offset* mode is designed to create overlaps between cores. First running the program in *NotOffset* phases several cores, then re-running the program in *Offset* mode moves the start of the cores to halfway along the first core, thereby creating 50 % overlaps between cores for the *NotOffset* mode and the *Offset* mode. The mode option is specified at the end of the *GeneralCoreLength* line.

UseThisNumberOfSurrogates gives the number of surrogates across which information pertaining to phase must be accumulated before phase can be declared. This number specifies how many homozygote surrogates the program needs to find before the locus can be declared phased in the proband.

PercentageSurrDisagree gives the percentage of surrogates that are allowed to conflict with the majority of the surrogates and still have phase declared. For example, a 10.00 (10 %) value means that if information about phase is accumulated across ten surrogates and nine of them indicate phase is in one direction and one indicates it is in the other, phase is declared to be in the direction of the nine. But if these counts are eight in one direction and two in the other, phase is undeclared (i.e., the minority is more than 10 %).

PercentageGenoHaploDisagree gives the percentage of disagreement across all SNPs in a core which are allowed to disagree between the true genotype and the genotype suggested the candidate pair of haplotypes for the candidate haplotypes. For example, a 1.00 (1 %) value means that across a core of 100 SNPs, 1 SNP is allowed to conflict between its actual genotype and the genotype comprised of the sum of the alleles of the candidate haplotypes.

GenotypeMissingErrorPercentage gives the percentage of SNPs that are allowed to be missing or in conflict across the entire core and tail length during surrogate definition. A 1.00 (1 %) value means that across a *GeneralCoreAndTailLength* of 300 SNPs, 3 of these SNPs are allowed to be missing or in disagreement between two otherwise compatible surrogate parents. Thus these two individuals are allowed to be surrogate parents of each other in spite of the fact that 1 % of their genotypes are missing or are in conflict (i.e., opposing homozygotes).

NrmThresh gives the maximum value (between 0.00 and 1.00) that the coefficient of relationship can take between a dummy sire and the true dam when pedigree information is used to partition surrogates in situations where parents are not genotyped. Section 2b (iv.) of Appendix A of Hickey et al. [1] gives more details.

FullOutput determines whether the extra output files are suppressed or not. A value of "1" gives the full output. A value of "0" suppresses the full output. Full output includes many files that can be used as diagnostics or to understand the data set, such as the matrix of surrogates. The suppressed output only includes the final-phased information.

Graphics determines whether the graphical output is invoked or not. The graphical components are not yet functional hence a value of "0" is required here.

Simulation determines whether the analysis involves simulated data where the true phase is known and performance measurement is required or not. A value of "1" indicates that it is a simulation. A value of "0" indicates that it is not a simulation.

TruePhaseFile gives the name of the file containing the true phase when working with simulations. The program does not read this line when the value of the line above is set to "0"; hence, it is irrelevant when working with real data.

Some general advice on setting suitable parameters is given in *see* **Note 4**.

4.2 Data Format

Pedigree File. The pedigree file should have three columns: individual, father, and mother. It should be space or comma separated with missing parents coded as 0. No header line should be included in the pedigree file, and both numeric and alphanumeric formats are acceptable. The pedigree does not have to be sorted in any way as the program automatically does this. If no pedigree file is available, *NoPedigree* should be given in place of a pedigree file name in *AlphaPhaseSpec.txt*.

Genotype File. The genotype information should be contained in a single file containing one line for each individual. The first column of this file should contain the individual's identifier with numeric and alphanumeric formats being acceptable. The next columns should contain the SNP information with two formats being acceptable, *GenotypeFormat* and *UnorderedFormat*. *GenotypeFormat* has a single column for each SNP where the genotypes are coded as 0, 1, and 2 and missing genotypes are coded as 3, with 0 being homozygous AA, 1 being heterozygous AB or BA, and 2 being homozygous BB. *UnorderedFormat* has two consecutive columns for each SNP, with AA being coded as 1 1, AB and BA being coded as 1 2 or 2 1, and BB being coded as 2 2. Missing genotypes can take any other numeric format (e.g., 3 3). Examples of these formats are included in the examples subdirectory of the program. The genotype file should not have a header line.

All markers in the genotype file should be ordered by chromosome position and the program can only operate on one chromosome at a time.

4.3 Output

The output of AlphaPhase is organized into a number of subdirectories (*PhasingResults*, *Miscellaneous*, and *Simulation*, when simulated data are used). A description of what is contained within these folders is given below.

4.4 PhasingResults

PhasingResults contains the primary results file and an index file with its coordinates. FinalPhase.txt contains the final-phased output for each individual. It has two rows for each individual and a column for each locus. The first column contains the individual's identification, followed by the phased information for the SNPs in the same order as the input genotype file. The coordinates of *FinalPhase.txt* are contained within *CoreIndex.txt*. By the coordinates what is meant is the start point and end point of each core (i.e., where a haplotype begins and ends). Cores are unaligned. Three columns exist in *CoreIndex.txt*. Column 1 is the core identifier, column 2 is the starting SNP of the core, and column 3 is the ending SNP of the core. As the algorithm phases the data on a core by core basis, it is important to recognize that these cores are unaligned. Therefore the subsequent analysis of the phased data needs to account for this.

IndivPhaseRate.txt contains the percentage of alleles phased in each of the cores for each individual, with columns being Id, percentage phased core 1, percentage of phased core 2, etc.

SnpPhaseRate.txt contains the percentage of individuals phased for each SNP, with the columns being SNP ordered number and percentage of individuals phased for that SNP.

PhasingYield.txt contains the average percentage of phased across all the individuals and all the SNPs for each core. It is a handy file for checking the performance for each core.

The directory HaplotypeLibrary contains the library of haplotypes (e.g., HapLib1.txt is the library for core 1) for each core and the directory Extras. In the first column of HapLibX.txt is the haplotype Id, then its frequency, then the haplotype. Extras contains files called HapCommonalityX.txt which contain matrices of relationships between the haplotypes within a core. These relationships are calculated as the count of alleles which match each pair of haplotypes divided by the total number of SNPs in a core.

Miscellaneous contains files which summarize the data. The genomic relationship matrix is contained within GenotypedMarkerNRM.txt. The pedigree-derived numerator relationship matrix between the genotyped individuals is contained within GenotypedNRM.txt, a pseudo version of this showing relationships which are above the NrmThresh as 1 and below it as 0 is given in GenotypedPseudoNRM.txt. This matrix identifies pairs of relationships between individuals that exceed a threshold.

SurrogatesX.txt contains a matrix of how animals are surrogate of each other for core X. A one means it is a surrogate of one of the clusters (i.e., paternal/maternal) and a two means it is surrogate of the other. The labeling paternal/maternal is arbitrary.

SurrogatesSummaryX.txt contains six columns. Column 1 is the Id, column 2 is the count of cluster 1 surrogates (e.g., paternal), column 3 is the count of cluster 2 surrogates (e.g., maternal), column 4 is the count of surrogates that are in both clusters (e.g., paternal and maternal), column 5 is the count of all surrogates, and column 6 is a code for how the surrogates were partitioned (1 = both parents genotyped, 2 = sire genotyped and used for partitioning, 3 = dam genotyped and used for partitioning, 4 = pseudo NRM partitioning, 5 = progeny genotyped and used for partitioning, 6 = k-medoids partitioning). Details on these partitioning strategies are given in Hickey et al. [1].

Timer.txt contains the time taken to complete the task.

Simulation contains files summarizing the comparisons between the simulated data and the phased output. CoreMistakesPercent.txt has a row for each core, followed by an empty row followed by a row containing the average across each of the cores. The columns are percentage of all alleles phased correctly within a core, percentage of all heterozygous alleles phased correctly within a core, percentage of all alleles not phased, percentage of heterozygous alleles not phased, percentage of all alleles incorrectly phased, and percentage of heterozygous alleles incorrectly phased. In IndivMistakesPercentX.txt, column 1 is the Id, column 2 is the count of cluster 1 surrogates (e.g., paternal), column 3 is the count of cluster 2 surrogates (e.g., maternal), and column 4 is the count of all surrogates for each individual for core X.

Columns 5 and 6 are the percentage of all alleles correctly phased within a core for the paternal and maternal alleles. Columns 7 and 8 are the percentage of all alleles not correctly phased within a core for the paternal and maternal alleles. Columns 9 and 10 are the percentage of all alleles not phased within a core for the paternal and maternal alleles. The next six columns are the same as the previous six except that they refer to the heterozygous SNPs. The next six columns are also the same except that they refer to the missing SNPs, while the final six columns refer to the SNPs simulated to have genotype error (must be identified in the program (contact John Hickey)). IndivMistakesX.txt contains the raw counts of what IndivMistakesPercentX.txt contains as percentages. MistakesX.txt contains the raw individual by SNP mistakes, with alleles phased correctly coded as 1, not phased coded as 9, and incorrectly phased coded as 5.

4.5 Examples

A full set of worked examples are available for download from http://sites.google.com/site/hickeyjohn/ and the reader is referred to that source. In this chapter, a single worked example is described involving a phasing scenario where pedigree information is available, the genotypes are stored in *GenotypeFormat*, and a true phase file is available. This example is contained in the folder *PhasingWithPedigreeInformation* of the AlphaPhase1.1 download and the AlphaPhaseSpec.txt file is shown above in Box 2. Explicitly, the name of the pedigree file is PreSpecifiedPedigree.txt, the name of the genotype file is 60kGenotypeGF.txt, and the name of the true phase file is 60kPhase.txt. There are 2000 SNPs (equivalent to 60,000 SNPs across the whole genome because the chromosome was simulated to be 1 Morgan in length) to be phased and a core length of 100 is set here. This is a fairly small core length and if the data set contained many closely related animals, such a short core length would be suboptimal (for both speed and accuracy). The core and tail length is 300 SNPs meaning that the tails themselves involve 100 SNPs on each side of the core. Again this is probably a small core length for most data sets. The UseThisNumberOfSurrogates is specified to 10 meaning that information from ten homozygote surrogates is used to derive a consensus phase for each data set. Ten percent of the surrogates are allowed to disagree which means that one out of the ten surrogates can propose a conflicting phase for the allele in question and phase will be declared as the phase proposed by the other nine. PercentGenoHaploDisagree allows 1 % of the alleles in a haplotype to disagree with the genotypes of the proband and the haplotype still accepted as being the haplotype that the proband carries. GenotypeMissingErrorPercentage controls the percentage of missing or erroneous genotypes that are allowed during surrogate diagnosis. Because these are simulated data, the parameter is set to 0 %; however, in real data a small percentage (e.g., 0.5 % or 1 %) of errors should be allowed. The remaining parameters are not very important.

5 Notes

1. **Erdös numbers** Erdös numbers are a system to describe the shortest route linking a pair of things. In the context of long-range phasing, Erdös 1 surrogates are surrogates that share a haplotype with the proband, meaning that they are one degree removed from the proband on the basis of haplotype identity. Erdös 2 surrogates do not share a haplotype with the proband but share a haplotype with an individual that shares a haplotype with the proband.

2. **Surrogate parents** are individuals who share a haplotype with the proband and are identified as those individuals that do not have any opposing homozygote genotypes with the proband.

3. **Core and tail** A core is defined as a consecutive string of SNP loci for which phasing is being attempted. Tails are defined as consecutive strings of SNP loci immediately adjacent to either end of a core.

4. **Advice on values for parameters**

    GeneralCoreLength and *GeneralCoreAndTailLength* Short cores and intermediate core and tail lengths give the best results. However, the algorithm is robust to small variations about what the optimal is likely to be. For 60,000 SNP density, a core length of 100 SNPs and a core and tail length of 300–500 SNPs are advisable. For 300,000 SNP density, a core length of 400 SNPs and a core and tail length of 1,200–2,000 SNPs are advisable.

    UseThisNumberOfSurrogates and *PercentageSurrDisagree* Good results were obtained using values of 10 for *UseThisNumberOfSurrogates* and 10.00 (10 %) for *PercentageSurrDisagree*.

PercentageGenoHaploDisagree and *GenotypeMissingErrorPercentage* It is best to be stringent with the editing of data (i.e., remove animals with large numbers of missing or poorly called SNP and remove SNPs with large numbers of or poorly called missing individuals) and then use low values for these parameters (e.g., 0.00 (0 %) or 1.00 (1 %)).

It is advisable to play with all of these parameters to fine tune them for a particular data set. Making *GeneralCoreAndTailLength* too short and *GenotypeMissingErrorPercentage* too high can increase the computational time considerably and can give poorer phasing performance. For large data sets with lots of related individuals, it is best to use longer *GeneralCoreAndTailLength* and *GeneralCoreLength* as such a data set is likely to have lots of individuals that share long haplotypes and by using long haplotypes reduces the numbers of surrogates an individual has and thus reduces the computational cost of partitioning these surrogates. Partitioning surrogates is the biggest computational task in AlphaPhase1.1. The trends in Hickey et al. [1] can be used to give a feel for what is sensible.

References

1. Hickey JM, Kinghorn BP, Tier B, Wilson JF, Dunstan N, van der Werf JHJ (2011) A combined long-range phasing and long haplotype imputation method to impute phase for SNP genotypes. Genet Sel Evol 43:12
2. Kong A, Steinthorsdottir V, Masson G, Thorleifsson G, Sulem P, Besenbacher S, Jonasdottir A, Sigurdsson A, Kristinsson KT, Jonasdottir A, Frigge ML, Gylfason A, Olason PI, Gudjonsson SA, Sverrisson S, Stacey SN, Sigurgeirsson B, Benediktsdottir KR, Sigurdsson H, Jonsson T, Benediktsson R, Olafsson JH, Johannsson OT, Hreidarsson AB, Sigurdsson G, DIAGRAM Consortium, Ferguson-Smith AC, Gudbjartsson DF, Thorsteinsdottir U, Stefansson K (2009) Parental origin of sequence variants associated with complex diseases. Nature 462:868–874
3. Howie BN, Donnelly P, Marchini J (2009) A flexible and accurate genotype imputation method for the next generation of genome-wide association studies. PLoS Genet 5:e1000529
4. Hickey JM, Kinghorn BP, Tier B, van der Werf JHJ, Cleveland MA (2012) A phasing and imputation method for pedigreed populations that results in a single-stage genomic evaluation. Genet Sel Evol 44:9
5. Scheet P, Stephens M (2006) A fast and flexible statistical model for large-scale population genotype data: applications to inferring missing genotypes and haplotypic phase. Am J Hum Genet 78:629–644
6. Browning SR, Browning BL (2007) Rapid and accurate haplotype phasing and missing-data inference for whole-genome association studies by use of localized haplotype clustering. Am J Hum Genet 81:1084–1097
7. Kong A, Masson G, Frigge ML, Gylfason A, Zusmanovich P, Thorleifsson G, Olason PI, Ingason A, Steinberg S, Rafnar T, Sulem P, Mouy M, Jonsson F, Thorsteinsdottir U, Gudbjartsson DF, Stefansson H, Stefansson K (2008) Detection of sharing by descent, long-range phasing and haplotype imputation. Nat Genet 40:1068–1075

Chapter 17

Genotype Imputation to Increase Sample Size in Pedigreed Populations

John M. Hickey, Matthew A. Cleveland, Christian Maltecca, Gregor Gorjanc, Birgit Gredler, and Andreas Kranis

Abstract

Genotype imputation is a cost-effective way to increase the power of genomic selection or genome-wide association studies. While several genotype imputation algorithms are available, this chapter focuses on a heuristic algorithm, as implemented in the AlphaImpute software. This algorithm combines long-range phasing, haplotype library imputation, and segregation analysis and it is specifically designed to work with pedigreed populations.

The chapter is organized in different sections. First the challenges related to genotype imputation in pedigreed populations are described, along with the specifics of the imputation algorithm used in AlphaImpute. In the second section, factors affecting the accuracy of genotype imputation using this algorithm are discussed. The different parameters that control AlphaImpute are detailed and examples of how to apply AlphaImpute are given.

Key words Imputation, Phasing, Genomic selection

1 Introduction

The success of genome-wide association studies and genetic prediction systems using genomic information depends on the ability to accurately estimate individual marker effects or the sum of these marker effects. Large numbers of individuals with genotypic and phenotypic information are required to obtain accurate estimates of these parameters. Furthermore, in order to disseminate the resulting marker estimates, genotypes of a large number of individuals, for which we wish to predict either breeding values or disease risk, are needed. Densely genotyping of these individuals is currently expensive. Therefore, much research effort has been devoted to methods and strategies to impute genotypes from cheaper low-density to high-density platforms [e.g., 1–8]. Although the general aim remains to generate highly accurate genotypes for a set of target individuals, the proposed approaches differ in how they account for population structure, alternative levels of genotyping in the population, and in the ability to handle sparse genotype and pedigree information.

The aim of this chapter is to give a brief introduction to genotype imputation, with a more specific discussion on the use of pedigree in imputation algorithms, and to describe factors influencing imputation accuracy. A specific approach for pedigree-based imputation is outlined using the software AlphaImpute [8], and detailed instructions for using this software to obtain imputed genotypes are discussed.

2 Imputation Overview (*See* Box 1)

Imputation to recover genotype information of individuals genotyped at low-density involves two steps. First, the haplotypes carried by the high-density genotyped individuals need to be resolved by a process known as phasing. Subsequently, low-density genotypes are used, in conjunction with pedigree, linkage, and/or linkage disequilibrium information, to determine which combination of these haplotypes are carried by animals that are genotyped at low-density. In some situations the imputation can also be extended to completely ungenotyped individuals.

Imputation methods can be divided into two main categories, those using linkage disequilibrium information (e.g., fastPHASE [9]; MaCH [3]; Beagle [1]; IMPUTE2 [2]) and those using pedigree and linkage information (e.g., PHASEBOOK [4]; Chromophase [6]; findhap.f90 [7]; AlphaImpute [8]; FImpute [10]). Use of pedigree information generally provides more power than using only linkage disequilibrium information [8]. There are at least two reasons for this. Firstly, the phase of high-density genotyped individuals can generally be resolved more accurately using pedigree rules compared to linkage disequilibrium-based phasing algorithms. Secondly, pedigree information can be used during imputation to significantly reduce the number of plausible haplotypes that can be carried by an individual.

Box 1
Genotype coding

Throughout this chapter it is assumed that single-nucleotide polymorphism (SNP) alleles are coded as 0 or 1, and genotypes are coded as the numbers of copies of allele 1, so therefore take the values 0, 1, and 2, where 0 is homozygous for allele 0, 1 is heterozygous, and 2 is homozygous for allele 1.

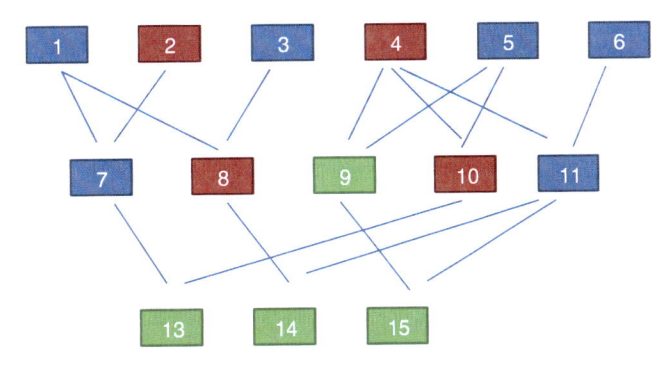

Fig. 1 Example pedigree with individuals genotyped at high-density (1, 3, 5, 6, 7, 11), low-density (9, 13, 14, 15), and not genotyped at all (2, 4, 8, 10)

The pedigree depicted in Fig. 1 illustrates the information available and the challenge of genotype imputation in a pedigreed population. In the example, some of the individuals are genotyped at high density, others at low density, with the remainder not being genotyped. Pedigree-based algorithms often use pedigree coupled with Mendelian rules of inheritance to resolve the phase of haplotypes that are carried by the relatives of an individual to be imputed. Then the missing genotypes are imputed by using the same information to determine which of its relative's haplotypes an individual carries at each of its genomic regions.

3 Factors Affecting the Accuracy of Imputation

The accuracy of imputation (*see* **Note 1**) is influenced by several factors, including the number and distribution of markers on the low-density genotyping panel, the number of individuals genotyped at high-density and their relationships to the individuals to be imputed, allele frequencies, and finally the local linkage disequilibrium between each low-density genotype and its surrounding high-density genotypes [8, 11, 12]. While several researchers have explored alternative genotyping strategies to maximize imputation accuracies, Huang et al. [12] explored the two major determinants of imputation accuracy in pedigreed populations: (1) the genotyping status of the immediate ancestors of individuals to be imputed and (2) the number of markers on the low-density panel.

Alternative options with respect to these two factors have different costs and imputation accuracies. A conservative approach would involve genotyping the eight great-grandparents, the four grandparents, and the two parents at high-density. This approach is likely to ensure that the phase of the parents is resolved for almost all markers, therefore reducing the task of imputation to the choice of the gamete passed to the offspring and the modelling of any recombination events that occur at meiosis. Furthermore, increasing the density of the low-density genotyping panel reduces the length of the regions for which recombination has to be modelled, resulting in higher imputation accuracy. However, such a conservative strategy can be costly, especially considering that for most breeding programs, individual female parents make a relatively small genetic contribution to the next generation. Alternative genotyping strategies can be far less expensive. For example, only male ancestors could be genotyped at high-density and female ancestors at low- or intermediate-density or not be genotyped at all. However, these cheaper alternatives may lead to a sizeable reduction in imputation accuracy.

Huang et al. [12] showed that the conservative approach involving high-density genotyping of all parents and grandparents in a pig population resulted in average imputation accuracy (as measured by the R-squared between the true and imputed genotypes which is a measure of imputation accuracy that is not affected by allele frequency) of 0.967, 0.990, and 0.996, costing $34.84, $49.84, and $62.84 (prices based on 2011 costs of different SNP chips) when genotyping the low-density individuals with 384, 3,000, and 6,000 marker density, respectively. By only genotyping the male ancestors at high-density the accuracies reduced to 0.888, 0.968, and 0.981. However much of this loss of accuracy could be recovered at a much reduced cost by utilizing strategies that involved genotyping the female ancestors at low or intermediate marker density. For example, genotyping the candidates and their female ancestors at 384 marker density cost $20.58 and gave an accuracy of 0.935. Increasing the marker density on the female ancestors to 3,000 or 6,000 markers while maintaining the 384 marker density on the candidates increased the accuracy to 0.955 or 0.956 while increasing the costs to $24.74 or $26.28. While the optimal strategy for different situations depends on the specifics of the individual breeding program, Huang et al. [12] showed that much of the imputation accuracy achievable when using a conservative approach could be recovered at a considerably lower cost.

4 AlphaImpute Software for Pedigree-Based Imputation

4.1 The AlphaImpute Algorithm

AlphaImpute combines simple phasing and imputation rules, long-range phasing, haplotype libraries, and segregation analysis to impute genotypes of all animals in a pedigree for all genotyped SNP. Imputed genotypes are constructed as the sum of the alleles or allele probabilities, which is termed allele dosage. The algorithm imputes alleles, when it has information to do this. It imputes allele dosage when the information is incomplete based on either the algorithm of Kerr and Kinghorn [13], which uses single locus information, or in the regions surrounding detected recombination locations allele dosage based on physical position relative to the informative flanking markers.

As described by Hickey et al. [8] AlphaImpute proceeds by first separating out a set of animals that are genotyped at high-density. These animals have their SNP phased using long-range phasing and haplotype library imputation, generating haplotype libraries in the process [14]. Allele probabilities are calculated for all SNP for all animals in the pedigree and alleles are imputed when these probabilities are >0.99. This can be thought of as single-locus phasing. Once the single-locus phasing and the long-range phasing are completed, the haplotypes identified in the long-range phasing are matched to alleles phased by single-locus phasing. This matching step begins with parental and other ancestral haplotypes by processing the data from the oldest to the youngest individual in the pedigree. The second part of the matching step involves processing the haplotype libraries to see if the haplotypes that an animal carries exist in the library in animals that are not identified as ancestors within the available pedigree information. The haplotype libraries are updated with any new haplotypes that are identified during this process. These steps are iterated a number of times (user defined). Finally, using the updated dataset, which now includes a large number of imputed SNP, allele probabilities are recalculated for all SNP for all animals in the pedigree. The complete algorithm is outlined in [8].

Pedigree-based imputation is generally not useful in human data sets due to pedigree information often not being available. However if pedigree information is available for data sets from isolated populations, AlphaImpute is expected to perform well. AlphaImpute will not work in human populations that are not isolated because the phasing mechanisms of long-range phasing require relatively related individuals to work.

4.2 Software Overview

AlphaImpute consists of a single program however it calls both AlphaPhase1.1 [14] and GeneProbForAlphaImpute [13]. All information on the model of analysis, input files and their layout, is specified in a single parameter file. The program has been designed for large scale and routine implementation of imputation in breeding programs that have high quality pedigree information available. Several options are included in order to reduce the computational requirements. Before describing these, we first outline the input files and their formats. We then describe strategies to reduce the computational load when carrying out routine use. The program is available for download from http://sites.google.com/site/hickeyjohn/alphaimpute

4.3 Input Files

A single specification file called AlphaImputeSpec.txt controls the program. An example of AlphaImputeSpec.txt is shown in Box 2. Changing the input to the right of the comma controls the program. Everything to the left of the comma should not be changed and is included for the purposes of identification.

Box 2
Example of AlphaImputeSpec.txt to Run the Program to Impute Genotypes Involving the Full Phasing of the High-Density Genotypes

PedigreeFile	,MyPedigree.txt
GenotypeFile	,MyGenosFile.txt
SexChrom	,No,None,None
NumberSnp	,3129
InternalEdit	,Yes
EditingParameters	,95.0,2.0,98.0,AllSnpOut
NumberOfPairsOfPhasingRounds	,10
CoreAndTailLengths	,200,300,400,500,600,250,325, 410,290,700
CoreLengths	,100,200,300,400,500,150,225, 310,190,600
PedigreeFreePhasing	,No
GenotypeErrorPercentage	,0.00
NumberOfProcessorsAvailable	,20
InternalIterations	,3
PreprocessDataOnly	,No
UserDefinedAlphaPhaseAnimalsFile	,None
PrePhasedFile	,None
TrueGenotypeFile	,MyTrueGenos.txt

Below is a description of what each line does in AlphaImputeSpec.txt. It is important to note that AlphaImputeSpec.txt is case-sensitive. Before proceeding, it is worth pointing out that internally AlphaImpute divides all the animals in the pedigree into two groups, one called a high-density group and the other the low-density group. The high-density group is the group of animals that are genotyped for enough SNP that they can have their haplotypes resolved by AlphaPhase1.1. The low-density group are all remaining animals in the pedigree and comprise animals that are not genotyped at all, are genotyped at low-density, or are genotyped at high-density but have a proportion (greater than a threshold the user can set) of their SNP missing (e.g., not called by the genotype calling algorithm). This partitioning is done because placing animals with too many SNP missing into AlphaPhase1.1 can result in dramatic increases in computational time and dramatic reduction in the accuracy of phasing.

PedigreeFile gives the name of the file containing the pedigree information. Details on the format are below. The number of generations required to obtain good imputation results are discussed in *see* **Note 2**.

GenotypeFile gives the name of the file containing the genotypes. Details on the format are below.

SexChrom is a command that specifies whether a sex chromosome is to be imputed or not. For autosomes the commands should be given as shown in Box 2. For sex chromosomes a file with the gender of each animal in the pedigree needs to be supplied, and the gender of the heterogametic individuals needs to be supplied on this line of AlphaImputeSpec.txt.

NumberOfSnp gives the number of SNP in the genotype file.

InternalEdit specifies whether the program should edit the data internally or not. The two options are ***Yes*** or ***No*** (note these are case-sensitive). Editing the data allows the program to remove SNP that are missing in too many animals/remove animals from the high-density group that have too many SNP that are missing. Editing the data may increase the speed and accuracy of the imputation. It is particularly important not to allow too many missing genotypes to enter the phasing step in AlphaPhase1.1 as this can dramatically increase the time required to complete the phasing and reduce the phasing accuracy.

EditingParameters controls the internal editing that is invoked by the *InternalEdit* option described above. The three numerical parameters control the internal editing while the case-sensitive qualifier that controls the final output of the results with regard to the editing. The internal editing involves three steps run in sequence (Step 1, Step 2, and Step 3).

The first numerical parameter controls Step 1, which divides the animals in the data into two initial groups, the high-density group, and the low-density group. Animals in the data set that are genotyped for more than XX.X% (In Box 2 this figure is 95.0 %) of the SNP enter the high-density group, with the remainder entering the low-density group.

The second numerical parameter controls Step 2, which removes some SNP from the analysis. SNP that are missing in more than XX.X % (In Box 2 this figure is 2.0) of the animals that have been placed in the high-density set by the previous parameter are removed.

The third numerical parameter controls Step 3, which finalizes the animals in the high-density group. It is similar to that of the first numerical parameter in that it divides the data into two groups, the finalized high-density group and low-density group. The animals in the data set that are genotyped for more than XX.X% (In Box 2 this figure is 98.0) of the SNP that remain after Step 2 has removed some SNP enter the finalized high-density set. The remaining animals enter the finalized low-density set. The final high-density group is passed to AlphaPhase1.1 to be phased.

The case-sensitive qualifier controls the SNP for which results are outputted and it has two options *AllSnpOut* or *EditedSnpOut* (note that these are case-sensitive). *AllSnpOut* produces output for all the SNP that are inputted and ignores the numerical parameters in Step 1, 2, and 3. *EditedSnpOut* produces output only for the SNP that survive the internal editing. The SNP that survive the internal editing are outlined in the output file ***EditingSnpSummary.txt*** that is described below.

NumberOfPairsOfPhasingRounds has two alternatives.

Alternative 1 (Box 2) controls the number of pairs of phasing rounds that are performed by AlphaPhase1.1 on the high-density group. The minimum for this number is 2 while the maximum is 30. It is worth pointing out that a pair of rounds comprises one round with AlphaPhase1.1 in Offset mode and the other in NotOffset mode. Different phasing rounds are required so that each SNP is phased multiple times as a part of cores that span different SNP. Additionally the different core spans and Offset/NotOffset modes create overlaps between cores. This helps to partially remove the small percentages of phasing errors that AlphaPhase1.1 makes. The concept of cores (and their tails) is outlined in [13] and briefly in *see* **Note 3**. Offset/NotOffset mode is described below.

Alternative 2 (Box 3) can be used to read in data sets that have been previously phased by AlphaPhase1.1. This allows users to read in results of previous phasing work. Three parameters are required here.

The first is the case-sensitive qualifier *PhaseDone*. This specifies that the phasing rounds have been done previously.

The second is the complete path to where these phasing rounds are stored. This path must be surrounded by quotations (e.g., "/here/is/the/full/path/").

The third is the number of phasing jobs that are to be read from the folder. The folders containing each of the phasing rounds must be labelled Phase1, Phase2, ..., Phase *N*, where *N* is the number of phasing rounds. It is important to realize that Alternative 1 (described above) for *NumberOfPhasingRounds* sets a number that is half the actual number of phasing rounds carried out (because of it specifies the number of pairs of rounds rather than rounds). Therefore it is good to check how many phasing rounds are actually in the folder you are reading in.

The second alternative can be used in conjunction with ***PreProcessDataOnly*** (described below) to give greater control on the computational time required to perform the phasing.

CoreAndTailLengths gives the overall length in terms of numbers of SNP of the core and its adjacent tails for each of the phasing runs. For example, if the *CoreLengths* (described below) value is 100 and the *CoreAndTailLengths* is 300, this means that the core is 100 SNP long and the tails are the 100 SNP adjacent to each end of

Box 3
Example of AlphaImputeSpec.txt Showing How to Read in Previously Phased Data Sets

PedigreeFile	,MyPedigree.txt
GenotypeFile	,MyGenosFile.txt
SexChrom	,No,None,None
NumberSnp	,3129
InternalEdit	,Yes
EditingParameters	,95.0,2.0,98.0,AllSnpOut
NumberOfPairsOfPhasingRounds	,PhaseDone,"/Users/john/PhaseOld/",20
CoreAndTailLengths	,200,300,400,500,600,250,325,410,290,700
CoreLengths	,100,200,300,400,500,150,225,310,190,600
PedigreeFreePhasing	,No
GenotypeErrorPercentage	,0.00
NumberOfProcessorsAvailable	,20
InternalIterations	,3
PreprocessDataOnly	,No
UserDefinedAlphaPhaseAnimalsFile	,None
PrePhasedFile	,None
TrueGenotypeFile	,MyTrueGenos.txt

the core, thus the length of the core and tail is 300 SNP. At the end of a chromosome, the tail can only extend in one direction. Thus in this case the core and tail length would only be 200 SNP, the 100 SNP in the core, and the 100 SNP adjacent to the one end of the core. The total number of *CoreAndTailLengths* specified must equal the number specified for *NumberOfPairsOfPhasingRounds* (i.e., in Box 2 there are 10 rounds of phasing specified and there are 10 *CoreAndTailLengths* specified).

CoreLengths gives the overall length in terms of numbers of SNPs of each core. The *CoreLengths* can never be longer than its corresponding *CoreAndTailLengths*. The total number of *CoreLengths* specified must equal the number specified for *NumberOfPairsOfPhasingRounds* (i.e., in Box 2 there are 10 rounds of phasing specified and there are 10 *CoreLengths* specified).

The order of the *CoreAndTailLengths* must correspond to the order of the *CoreLengths* (i.e., in Box 3 the *CoreAndTailLenghts* 200 is for the first pair of phasing runs and corresponds to the *CoreLenths* 100).

PedigreeFreePhasing tells the program to perform the long-range phasing step of AlphaPhase1.1 without using pedigree information. In some cases this may be quicker and more accurate, but it is not likely to be commonly applicable (the command options to the right of the comma are a case-sensitive "*No*" or "*Yes*").

GenotypeErrorPercentage gives the percentage of SNP that are allowed to be missing or in conflict across the entire core and tail length during the surrogate definition in AlphaPhase1.1. A value of 1.00 (i.e., 1 %) means that across a *CoreAndTailLengths* of 300 SNP, 3 of these SNP are allowed to be missing or in disagreement between two otherwise compatible surrogate parents. Thus these two individuals are allowed to be surrogate parents of each other in spite of the fact that 1 % of their genotypes are missing or are in conflict (i.e., opposing homozygotes). Small values are better (e.g., <1.0 %). See the manual for AlphaPhase1.1 for more details.

NumberOfProcessorsAvailable sets the number of processors you want to use to complete the tasks. Setting more processors will reduce the computational time because the program will parallelize the calculation of the genotype probabilities and the phasing rounds.

UserDefinedAlphaPhaseAnimalsFile gives the user an option to read in a list of individuals that are phased using long-range phasing in AlphaPhase1.1. Specify "*None*" to the right of the comma if no file is to be read in, specify the name of the file to the right of the comma if a file is to be read in. The file to be read in should contain a single column of the ID's of the individuals to be sent to AlphaPhase1.1. This option is useful for routine runs involving large data sets.

PrePhasedFile gives the option to read in pre-phased data (e.g., phased by a previous round of AlphaImpute or by another program such as a half-sib haplotyping program). Specify "*None*" to the right of the comma if no file is to be read in, specify the name of the file to the right of the comma if a file is to be read in. The file to be read in should contain a two lines for each individual, the first line being its phased paternal gamete (alleles coded as 0 or 1 or another integer (e.g., 3) for missing alleles) and the second line being the phased maternal gamete. The first column should be the ID's of the individuals. The file takes the same format as *ImputePhase.txt* in the *Results* section of AlphaImpute. Care must be taken here to ensure that only reliable phased individuals are included when using this option.

TrueGenotypeFile gives the name of the file containing the true genotypes if you want to test the program. For example, this file could contain the true genotypes of a set of animals that have a proportion of their genotypes masked. If no such file is available, you can set the parameter to "*None*". Testing the program can be useful when applying the program to a new population; perhaps you should mask some SNP in a small percentage of your animals and see how it performs imputing them.

InternalIterations controls the number of iterations of the internal haplotype matching and imputation steps. A good number for this parameter is 3.

PreProcessDataOnly has the two options "*Yes*" or "*No*".

Yes sets the program so that it stops after it has preprocessed the data and set up the files for the analysis. *No* sets the program to do a complete imputation run. The Yes option is useful for getting to know your data set. The different data *EditingParameters* alter the number of SNP to be included in the analysis, and alter the numbers of animals that are included in the high-density group that is passed to AlphaPhase1.1. These numbers are printed to the screen. It is best to try different editing options to tune to each data set. Preprocessing the data creates the files for the phasing rounds. The phasing rounds can then be run externally to AlphaImpute to see if the phasing parameters (*CoreLengths*, *CoreAndTailLengths*, *GenotypeErrorPercentage*) are appropriate in terms of speed and phasing yield for the *EditingParameters* used on the data set.

The phasing rounds can be then run directly by the user by first running the program with *PreProcessDataOnly* set to *Yes*, then renaming the folder *Phase* to something else (e.g., PhasePreProcess because the folder Phase gets deleted each time you run the program) and then the program can be rerun with *PreProcessDataOnly* set to *No* and having the *NumberOfPhasingRuns* altered so that it reads the Phasing rounds in the PhasePreProcess folder (N.B. Check the number of folders in this folder, you don't want to leave phase rounds behind!!!!!). This option allows the user to tweak the phasing parameters.

4.4 Data Format

The program generally requires two input files, a pedigree file and a genotype file.

4.4.1 Pedigree File

The pedigree file should have three columns: individual, father, and mother. It should be space or comma separated with for missing parents coded as 0. No header line should be included in the pedigree file both numeric and alphanumeric formats are acceptable. The pedigree does not have to be sorted in any way as the program automatically does this.

4.4.2 Genotype File

The genotype information should be contained in a single file containing 1 line for each individual. The first column of this file should contain the individual's identifier with numeric and alphanumeric formats being acceptable. The next columns should contain the SNP information with a single column for each SNP where the genotypes are coded as 0's, 1's, and 2's and missing genotypes are coded as another integer between 3 and 9 (e.g., 3), with 0 being homozygous AA, 1 being heterozygous AB or BA, and 2 being homozygous BB. The genotype file should not have a header line.

4.5 Output

The output of AlphaImpute is organized into a number sub directories (*Results and Miscellaneous*, and in the case of when a true genotype data file is supplied *TestAlphaImpute*). A description of what is contained within these folders is given below.

The folder *Results* contains four files.

4.5.1 Genotype Data

ImputeGenotypeProbabilities.txt is the primary genotype output file. It contains, for each SNP and each animal in the pedigree, a real number, the genotype probability, which is the sum of the two allele probabilities (i.e., the genotype) at that locus. Therefore genotypes are coded as real numbers between 0 and 2. The first column is the animal ID, with the subsequent columns being for each SNP.

ImputeGenotypes.txt is the secondary genotype output file. It contains a genotype for each SNP and each animal in the pedigree where it was possible to match it to a haplotype or was already genotyped. SNP that could not be matched or were not genotyped are denoted as being missing by a 9 (in the previous file these missing values were replaced with genotype probabilities). The first column is the animal ID, with the subsequent columns being for each SNP.

4.5.2 Phased Data

ImputePhaseProbabilities.txt is the primary output file containing phased data. It contains an allele probability for each of the two alleles of each SNP and each animal in the pedigree. The first column is the animal ID, with the subsequent columns being for each SNP. Each animal has two rows, with the first of these being for the paternal gamete and the second being for the maternal gamete. Alleles are coded as real numbers between 0 and 1 (i.e., probability of allele being a 1).

ImputePhase.txt is the secondary output file containing phased data. It contains an allele for each of the two alleles of each SNP and each animal in the pedigree where it was possible to match it to a haplotype. Alleles that could not be matched are denoted by a 9 as being missing. The first column is the animal ID, with the subsequent columns being for each SNP. Each animal has two rows, with the first of these being for the paternal gamete and the second being for the maternal gamete. Alleles are coded as integers either 0 or 1 with missing alleles set to 9 (in the previous file these missing values were replaced with allele probabilities).

The folder ***Miscellaneous*** contains files that summarize the editing of the data. *EditingSnpSummary.txt* contains three columns, the first being the sequential number of the SNP, the second being the count of animals that are missing each SNP in the high-density set, and the third being an indicator of whether the SNP was included in the analysis or not (1 = included/0 = excluded). *Timer.txt* contains the time needed to complete the task.

TestAlphaImpute is only invoked if a TrueGenotypeFile is supplied. The resulting folder contains four files.

IndividualAnimalAccuracy.txt contains a row for each animal in the test file. The first column is the animals ID, the second a classifier as to what genotyping status its ancestors had (1 being both parents genotyped, 2 being sire and maternal grandsire genotyped, 3 being dam and paternal grandsire genotyped, 4 being sire genotyped, 5 being dam genotyped, and 6 being any other scenario; An ancestor is considered genotyped if it was genotyped for more than 50 % of the SNP), and the next columns are for each of the SNP, with the coding 1 being for SNP correctly imputed, 2 being for SNP incorrectly imputed, 3 being for SNP not imputed, and 4 being for SNP that were already genotyped.

IndividualSummaryAccuracy.txt summarizes the information in *IndividualAnimalAccuracy.txt*. Columns 1 and 2 were as for the previous file, column 3 is the percentage of SNP to be imputed that were imputed correctly for this animal, column 4 is the percentage imputed incorrectly, column 5 is the percentage not imputed, column 6 is the percentage of paternal alleles that were imputed or phased, and column 7 is the percentage of maternal alleles that were imputed or phased.

IndividualSummaryYield.txt summarizes the yield in terms of the percentage of paternal/maternal alleles that have been imputed or phased for all animals in the pedigree. Column 1 is the ID, column 2 is an indicator as to whether it was genotyped for more than 50 % of the SNP or not (1 = was genotyped, 0 = was not genotyped), column 3 is the percentage of paternal alleles imputed or phased, and column 4 is the percentage of maternal alleles imputed or phased.

4.6 Advice on Values for Parameters

For a data set that comprises 10,000 animals, of which 3,000 animals are genotyped for 3129 SNP (on chromosome 1, thus equivalent to 50 k density) and 1,000 animals are genotyped for (180 SNP on chromosome 1, thus equivalent to some low-density chip) a good way to proceed would be with the parameters outlined in Box 2. Another example is given in the examples section of this chapter. Further examples are given in the user manual of AlphaImpute, which is available from this link: http://sites.google.com/site/hickeyjohn/alphaimpute

5 Routine Use of AlphaImpute

AlphaImpute can be used for the routine imputation of genotypes (e.g., weekly use). In such circumstances a number of things can be tweaked to make the program computationally much faster. Gametes of individuals (e.g., key sires) that have been phased externally by other software or that have been previously phased by AlphaImpute can be read into the program directly by supplying a pre-phased data file. Phasing large numbers of individuals with AlphaPhase1.1 takes a lot of computational time. Haplotype library and pedigree-based phasing internal to AlphaImpute is much less computationally demanding. Therefore when large numbers of high-density genotyped individuals are to be phased, it is best to restrict the individuals phased by AlphaPhase1.1 to a subset comprising only the key individuals and their close relatives. This can be implemented via the UserDefinedAlphaPhaseAnimalsFile option.

6 Examples

Three example scenarios will be outlined

Example 1: Run the program to impute genotypes involving the full phasing of the high-density genotypes. The AlphaImputeSpec.txt file containing suitable parameters for this example is shown in Box 2.

Example 2: Run the program to impute genotypes involving when the high-density genotypes have been previously phased. The AlphaImputeSpec.txt file containing suitable parameters for this example is shown in Box 3.

Data sets are available for download from http://sites.google.com/site/hickeyjohn/alphaimpute

7 Notes

1. Imputation accuracy

    Imputation accuracy can be measured in many ways including the percentage of genotypes imputed correctly, percentage of alleles imputed correctly, correlation between true and imputed genotype, and the proportion of variation contained within the true high-density genotypes that is explained by the imputed genotypes (R^2). The correlation and the R^2 are more appropriate measures of imputation accuracy than the percentage of genotypes imputed correctly. Firstly they contain a measure of the accuracy of imputation and the yield of imputation in a single measure, and secondly they are not affected by allele frequency. A clear description of the effect of allele frequency on different measures of imputation accuracy is given in Hickey et al. [8].

2. Effect of the depth of pedigree
   The number of generations of pedigree may affect the imputation accuracy when using AlphaImpute. When most of the imputation power comes from having genotypes on the immediate ancestors (e.g., sire and dam), this power can only be accessed if the gametes of these ancestors are reasonably well phased for the whole chromosome. Phasing whole chromosomes generally requires having genotype information on ancestors or information from descendants of ancestors that can be used in its place. Therefore AlphaImpute works best with at least two generations of animals that are genotyped and a few more generations of ancestors which have pedigree relationships recorded. However as Huang et al. [12] have shown good imputation results can be obtained when having the sire and maternal grandsire genotyped and having a few ancestral generations of ungenotyped animals included in the pedigree.

3. Core and tail lengths
   A full description of core and tail lengths is given in the chapter of this book describing long-range phasing (*see* also **Note 4**). Users should note that the core and tail lengths should be constructed with reference to the expected degree of relationships in the data set (i.e., how many centi-Morgans of genome is likely to be shared between pairs of surrogate parents) and the marker density. Data sets of highly related animals should have long cores so as to maximize the phasing accuracy and minimize the computational time.

4. Long-range phasing
   Long-range phasing is a heuristic phasing mechanism that can operate either with or without pedigree information. Full details are given in the chapter of this book describing long-range phasing.

References

1. Browning SR, Browning BL (2007) Rapid and accurate haplotype phasing and missing-data inference for whole-genome association studies by use of localized haplotype clustering. Am J Hum Genet 81:1084–1097
2. Howie BN, Donnelly P, Marchini J (2009) A flexible and accurate genotype imputation method for the next generation of genome-wide association studies. PLoS Genet 5: e1000529
3. Li Y, Willer CJ, Ding J, Scheet P, Abecasis GR (2010) MaCH: using sequence and genotype data to estimate haplotypes and unobserved genotypes. Genet Epidemiol 34 (8):816–834
4. Druet T, Georges M (2010) A hidden Markov model combining linkage and linkage disequilibrium information for haplotype reconstruction and quantitative trait locus fine mapping. Genetics 184:789–798
5. Habier D, Fernando RL, Garrick DJ (2010) A combined strategy to infer high-density SNP haplotypes in large pedigrees. Proceedings of the 9th World Congress on genetics applied to livestock production, Leipzig, 1–6 August 2010. pdf 09–15

6. Daetwyler HD, Wiggans GR, Hayes BJ, Woolliams JA, Goddard ME (2011) Imputation of missing genotypes from sparse to high density using long-range phasing. Genetics 189:317–327
7. VanRaden PM, O'Connell JR, Wiggans GR, Weigel KA (2011) Genomic evaluations with many more genotypes. Genet Sel Evol 243:10
8. Hickey JM, Kinghorn BP, Tier B, van der Werf JHJ, Cleveland MA (2012) A phasing and imputation method for pedigreed populations that results in a single-stage genomic evaluation. Genet Sel Evol 44:9
9. Scheet P, Stephens M (2006) A fast and flexible statistical model for large-scale population genotype data: applications to inferring missing genotypes and haplotypic phase. Am J Hum Genet 78:629–644
10. Sargolzaei M, Chesnais JP, Schenkel F (2011) FImpute—an efficient imputation algorithm for dairy cattle populations. J Dairy Sci 94(1 (E-Suppl 1)):421, Abstract
11. Zhang Z, Druet T (2010) Marker imputation with low-density marker panels in Dutch Holstein cattle. J Dairy Sci 93:5487–5494
12. Huang Y, Hickey JM, Cleveland MA, Maltecca C (2012) Assessment of alternative genotyping strategies to maximize imputation accuracy at minimal cost. Genet Sel Evol 44:25
13. Kerr RJ, Kinghorn BP (1996) An efficient algorithm for segregation analysis in large populations. J Anim Breed Genet 113:457–469
14. Hickey JM, Kinghorn BP, Tier B, Wilson JF, Dunstan N, van der Werf JHJ (2011) A combined long-range phasing and long haplotype imputation method to impute phase for SNP genotypes. Genet Sel Evol 43:12

Chapter 18

Validation of Genome-Wide Association Studies (GWAS) Results

John M. Henshall

Abstract

Validation of the results of genome-wide association studies or genomic selection studies is an essential component of the experimental program. Validation allows users to quantify the benefit of applying gene tests or genomic prediction, relative to the costs of implementing the program. Further, if implemented, an appropriate weight in a selection index can only be derived if estimates of the accuracy of genomic predictions are available. In this chapter the reasons for validation are explored, and a range of commonly encountered scenarios described. General principles are stated, and options for performing validation discussed. Designs for validation are heavily dependent on the availability of phenotyped animals, and also on the pedigree structures that characterize the breeding program. Consequently, there is no single plan that is always applicable, and a custom plan often must be developed.

Key words Validation, Genome-wide association studies, Genomic selection, Cross validation

1 Introduction

When genomic regions are identified as affecting a trait in a genome-wide association study (GWAS), it is important to verify that the association can be generalized to other populations. With high dimensional genomic markers such as SNP, tens or hundreds of thousands of effects can be estimated, and the estimated effects that are most significant are likely to be overestimates [1–3]. Unbiased estimates of effect can only be made from a data set that was not used in the discovery process.

Where the purpose of the study is genomic selection, estimates of effect are not of intrinsic interest, but may still be used in prediction equations. The purpose of genomic selection is ultimately prediction, and once a study to estimate associations between the genome and phenotype is complete it is appropriate to validate the predictive power of the estimated associations. Features of the design of the experiment, in particular the heritabilities of the traits and the number of animals used, provide an indication of the likely power, but the influence of other factors on predictive power is difficult to quantify. Examples are: the unknown distribution of genomic effects; population structure; the selection history of the traits of interest; and confounding of fixed effects and subpopulations. It is only by comparing breeding values estimated from genomic data with breeding values estimated from an independent data set that the predictive power of the genomic breeding values can be confirmed.

For genomic selection, it is convenient to think of the animals in the study as belonging either to a training set, which is used to parameterize the model used for prediction, and a validation set, which is used to confirm the predictive power of the parameterized model. Sometimes there is an obvious distinction between the training set and the validation set, such as a training set comprising an experimental population and a validation set comprising commercial sires. At other times there may be one large population, and there is no clear reason to choose any particular animal to be in either set. In these circumstances cross validation from within the population may be applied.

For the purposes of this chapter, it doesn't matter what methods are used to estimate genomic breeding values. The principles and methods described here would be equally applicable if breeding values were estimated using phenotypes and pedigree alone. All that is required is a method for estimating breeding values for the validation set of individuals, and an independent estimate of the breeding values of those individuals. Given this starting point, this chapter will focus on the range of scenarios that might be encountered, highlighting issues and suggesting approaches for dealing with them.

2 Why Validate?

If the intent of a GWAS is to discover causal mutations for genetic effects of known effect (such as the cause of polled in cattle or sheep), then the purpose of validation is to confirm that the discovery is indeed the causal mutation, and that the results therefore generalize across other unrelated populations. Where the intent of the GWAS is to discover genes affecting quantitative traits, as noted in the introduction, the estimates of effect that achieved most significance are likely to be overestimates [1–3]. Corrections to the bias are available, but are conditional on assumptions regarding the distributions of effects [e.g., 4–6]. If the purpose of the study is to report on the size of the effect of genomic regions, then without validation in an independent population the estimated effects would be either knowingly biased or conditional on model assumptions (*see* **Notes 1** and **2**).

With genomic selection, the degree of validation required is dependent on the novelty of the data and process being applied, and on the degree of trust held by the intended end-user of the predictions. For example, the accuracy of a breeding value estimated as the mean of progeny trait records is a function of the heritability and the number of progeny. If these are both known, and if the end-user of the estimated breeding value believes in the relationship between heritability, number of recorded progeny, and breeding value accuracy, then there is little need for validation. But it is not so long ago that EBV not too different from progeny means were applied in demonstration flocks and herds, with the dual aims of validating the underlying genetic model and related parameter estimates, and of increasing the degree of trust held by the stud industry [e.g., 7–10]. Each of these dual aims will be explored further.

With response to selection widely accepted, validation no longer requires breeding selected animals and measuring the response, but there are aspects of the genetic model and its parameterization that may require validation. Currently, our understanding of the relationship between DNA sequence and phenotype is increasing for some model organisms [e.g., 11], but limited for livestock species. Different methods for genomic prediction make different assumptions about the underlying genetic architecture, and evidence in support of these models is scanty. Until such time, as the weight of historical evidence shows that a model and associated methods are robust across species, populations, and traits, it will be necessary to validate genomic predictions that apply new methods, or that apply established methods to new species, populations, or traits.

As with GWAS, the feature of the genomics data that makes validation of new methods particularly important is the risk of overfitting. With SNP or sequence data there are far more genetic data points per individual than there are individuals with trait records. The number of parameters (p) available to include in the model is far greater than the number of records available to parameterize the model (n). When $p >> n$, it is trivial to build a model that perfectly predicts all records in the training population, but that has no utility at all to predict records outside the training population. This has been well understood since the beginnings of genomic selection [12], and much research has been directed at finding methods that do not over-fit, methods that exploit all of the genetic information in the data but that do not confuse environmental noise for genetic effects, as detailed in previous chapters in the book. Validation of predictions from new methods ensures that performance in the training set is due to genetics and not due to inadvertent overexploitation of the abundance of genomic parameters available.

The second aim of validation, to increase the level of trust held by animal breeders in genetic improvement, is as important as confirming scientific hypotheses. It is trust in the worth of a genetic improvement tool that drives adoption. The level of trust in genetic improvement varies considerably. In some industries, poultry breeding for example, response to selection has been so dramatic that it would be difficult if not impossible to profitably produce unimproved stocks for commercial sale. In other industries, standard quantitative genetics practices, such as using BLUP to estimate breeding values from pedigree and trait data, are not widely used. Paradoxically, it is in the first of these cases that validation to increase the level of trust may be more important. In an intensive industry such as poultry, trust in the power of genetic improvement is high, but breeding programs are already highly tuned. The investment in new breeding procedures may have to be argued before company owners who are not geneticists, and the economics closely scrutinized. Trust in proven genetic models is coupled with healthy skepticism in new, unproven genetic models. Fortunately, it is in this environment that it is easiest to perform a meaningful validation trial, as data and resources are more likely to be available. Conversely, in an industry where little genetic improvement currently takes place, any genetic gain is an improvement, and quantification of the exact amount is less important, provided that increases in costs are negligible. In these industries data for validation may be difficult and costly to obtain, and trust in genetic improvement is lacking. This places a significant responsibility on the advocates of genomic selection programs to ensure that the lack of potential for validation is not exploited, and that adequate validation in other populations has taken place.

Unlike GWAS, where validation would almost always be required, there is an implication in the above that validation of genomic selection may not always be necessary. If the predictive power of a particular genomic breeding product has been validated across a range of populations, then perhaps validation in a new but similar population is unnecessary. If a model and method for parameterization of the model have been validated in a number of populations for a range of traits, then perhaps the model and method could be applied without validation in the same populations for new traits, or in new populations. In deciding when further validation is unnecessary, or not cost-effective, many factors need to be taken into account, and some (such as population structure) are difficult to summarize for use in tools to assist in making the decision.

3 Independence and Choice of the Validation Population

3.1 Genomic Selection

Using a model parameterized from phenotype and genotype data on a training set, breeding values are estimated on a validation set. Typically these breeding values will be based on genotype data from the validation set, and here these are referred to as GEBV. To validate the GEBV, they are compared to breeding values for the validation set estimated from an independent source of data. Often these EBV will be based on phenotype or progeny phenotype of the validation set, so these will be referred to as PEBV. It is most important that any data contributing to the choice of model for the GEBV or its parameterization are not included in the PEBV. For example, it is not appropriate to use the phenotypes of animals in the validation set both in estimating the GEBV and in estimating the PEBV. This is not to say that phenotypes for the validation set can't ever be used in estimating the GEBV. A GEBV might be estimated using genomic and phenotypic data on selection candidates, if that is what would commonly be available at the time of selection. However, in this case the PEBV must be estimated without the phenotypes from the selection candidates, using instead, for example, phenotypes on progeny (*see* **Note 3**).

Any data contributing to the choice of method and model used to produce the GEBV must also be excluded from the estimation of the PEBV. It is not uncommon to use a set of PEBV to compare several sets of GEBV, each estimated using the same phenotype and genotype data on the training set, but estimated using different models or methods [e.g., 13–16]. However, one can't then choose the model or method that produced the highest correlation between the GEBV and PEBV, and assume that the correlation obtained is what would be expected if another, independent set of PEBV were tested. The PEBV have clearly influenced the final set of GEBV and validation will require a second, independent set of PEBV (*see* **Note 4**).

Ideally, the validation population should be chosen to be representative of animals in which the GEBV will be used. If a genomic prediction product is developed for use within a flock or herd, then the animals in the validation set should be representative of that flock or herd. If the product is intended for use within a breed, then the animals in the validation set should be representative of the breed (*see* **Note 5**). Sometimes choosing the validation set amounts to deciding how to split a larger population into a training and validation set. In this case the validation set should be those animals most similar to animals in which the GEBV will be used in practice—typically this is the youngest group of animals [e.g., 17, 18].

The accuracy of the GEBV is assessed by the strength of the association between the GEBV and the PEBV, such as estimated by the correlation. The accuracy of the PEBV therefore places an upper bound on the measurable accuracy of the GEBV. For example, if the PEBV are records on a trait with heritability of 40 %, then even if the GEBV are estimated with no error, the correlation between the GEBV and the PEBV cannot exceed 0.63, the square root of the heritability. For this reason PEBV should be sought that are as accurate as is possible. Ideally, we would measure the accuracy of the GEBV as the correlation between the GEBV and the true, but unknown breeding value of each individual. In industries where accurate breeding values are available for sires with many trait recorded progeny, such as dairy, a validation population of older sires might be appropriate (provided that only records for progeny of even older sires are included in the training population), as even though the target market for the genomic prediction might be younger sires, the lower accuracy of their EBV makes them less useful for validation purposes (*see* **Note 6**).

For traits which are not routinely recorded by industry, where an experiment is performed to record traits specifically to parameterize a genomic prediction model, it is necessary to balance resources between the training population and the validation population. It may be necessary to restrict the population in which the product is to be applied, as validating in all breeds may be too expensive.

In all cases, planning for validation from the outset will help to avoid problems in finding the resources to conduct an appropriate validation. The validation population must be identified early, and quarantined from any involvement in model development or parameterization.

3.2 GWAS

The same principle of an independent validation population applies to GWAS, but here, independent has a different meaning. Genomic selection often deliberately exploits information on related individuals, and as such related individuals may not only be appropriate, but also necessary in the validation population. However, GWAS results are usually expected to apply across a breed, or even across a species, and to eliminate linkage disequilibrium as the cause of an observed effect it is necessary to reestimate the effect in populations unrelated to the training population. Hence, animals for the validation population should be representative of the breed or species, but should exclude animals closely related to animals in the training set.

4 Statistics and Methods

4.1 GWAS

For GWAS, a gene or locus effect is validated when the gene or locus is shown to have a significant effect in the independent validation population. If the validation population is sufficiently representative of the diversity of a breed or species, then the association could be considered to be validated for that breed or species. The claim that a DNA variant is causal and that the association is not due to linkage would be supported by the effect having a consistent linkage phase across diverse populations, but an association study on a validation population does not constitute proof. The requirements of the leading genetics journals for reporting GWAS results are strict [e.g., 19], and these guidelines are worthy aspirations for livestock research programs, but they may exceed what is necessary to provide a useful diagnostic test. For example, there are commercial tests currently available for poll in cattle that acknowledge that the test is based on linkage [e.g., 20]. When this is the case, it is necessary for the validation study to provide estimates of the type I and type II error rates associated with the test. These are commonly reported as the sensitivity (proportion of positive test results that are true positives) and specificity (proportion of negative test results that are true negatives). In addition, if some test results are inconclusive, such as when one marker allele or haplotype allele is associated with multiple phenotypes, then an estimate of the frequency of inconclusive results must also be provided.

4.2 Genomic Selection

The validation study for genomic selection reports on the strength of the association between GEBV and PEBV on the validation population. The actual measure reported (e.g., Pearson correlation, rank correlation) is less important than the choice of individuals to include in the comparison, and the interpretation of the strength of the observed association.

Correlation estimates are often sensitive to the range of values included in the dataset. With respect to genomic selection, these values are phenotypes or estimated breeding values, and a correlation estimated across a wide range of phenotypes, such as across a breed, is likely to be higher than one estimated from a narrow range of phenotypes. For example, Daetwyler et al. [14] found that correlations between GEBV and PEBV for wool traits in Merino sheep were lower within wool types (superfine, fine, and strong wool) than when estimated using all Merinos in the validation population. As breeders would usually be selecting animals from within a wool type, it is the within wool type correlations that best describe the likely accuracy of the GEBV as it will be applied commercially. While correlations for a range of subpopulations might be estimated and reported, it is important that those that are most relevant to the commercial application of genomic selection are included and highlighted. This may be an across stud correlation if GEBV will be used to choose between studs, within stud correlations if the GEBV will be used to rank animals within studs, or even within family correlations if the GEBV are to be used to rank animals within families.

To interpret the strength of the observed correlation, it is necessary to refer to the theoretical maximum correlation achievable, which is the accuracy of the PEBV available in the validation population. One alternative is to report the correlation as a proportion of the PEBV accuracy, but this may give the casual reader the impression that the GEBV are better than they really are. More commonly, raw correlations are given, along with estimates of bias from a regression of PEBV on GEBV [e.g., 17]. Plots of PEBV on GEBV may also assist with the interpretation of the strength of the correlations, especially if information on subpopulation such as stud or family can be illustrated in the plot.

4.3 Cross Validation   When there is no obvious distinct validation population, for example, when phenotypes or genotypes are available only on a single generation, one approach to validation would be to choose in advance a random set of individuals to keep aside as a validation population. Cross validation differs from this in that multiple random sets are chosen, and each set is in turn considered to be the validation population. This has the advantage of providing multiple estimates of the correlation between the GEBV and PEBV of a randomly chosen validation population. Cross validation is commonly used for model selection or model tuning [e.g., 16, 21–24] but also for estimating the accuracy or reliability of genomic breeding values for a given parameterized model and population [e.g., 25, 26]. As noted earlier, where cross validation is used for model selection or tuning, then that does not constitute validation of the selected, tuned model. However, the principles of cross validation are the same regardless of whether for model selection or model validation (*see* **Notes 7–9**).

The repeated validation populations can be chosen in a number of ways. One approach is to repeatedly sample with replacement from the whole population [e.g., 22, 24]. Alternatively, in K fold cross validation the population is split into K pairwise disjoint sets, with each set in turn used as the validation population [e.g., 16, 23, 25]. Hybrid designs are also possible, for example, sampling validation sets without replacement, but not requiring all animals to occur at least once in a validation set [e.g., 21].

Perhaps more important than the method (K fold or repeated sampling with replacement) is accounting for pedigree, such as by applying constraints to the random sampling process. Clearly, it would seldom be appropriate to allow, for example, a son to occur in the training set and his sire in the validation set, but often whether a constraint is appropriate will be situation-dependent. Whether or not siblings (for example) are allowed to be split over training and validation sets can potentially have a big impact on the validation population correlations between GEBV and PEBV. In breeding systems where animals are selected based on sibling performance, such as where families are split between hatchery broodstock and the commercial grow-out environment in aquaculture, then including siblings in both training and validation sets is not only appropriate but necessary. In other systems, where selections are made before sibling performance data is available, it would not be appropriate and constraints to prevent it should be applied.

As demonstrated by [21], geneticists need not limit themselves to published methods such as K fold cross validation. Pedigree structures are highly regular and for the cross validation to be meaningful the sampled validation sets should be subjected to the same pedigree constraints as the populations where genomic selection will be applied. This may require a custom design and custom software.

5 Notes

1. Plan early—the validation population should be identified as part of the design of the overall experiment.

2. Document the validation plan in advance of commencing the experiment. Variations to the plan on the basis of experimental outcomes could indicate deficiencies in the validation process, and must be documented and justified.

3. Choose animals for the validation population that have PEBV that are as accurate as possible, as the accuracy of the PEBV places an upper bound on the measurable accuracy of the GEBV.

4. Once identified, the validation population must be quarantined from development of the GEBV. PEBV or data contributing to PEBV the validation set should not be used in the training sets. Further, PEBV in the validation sets should not be used in any analysis used to select or parameterize the model used in the estimation of GEBV.

5. The validation population should be representative of the populations in which the GEBV will be used commercially. Typically this would be either the youngest generation, or a population with a particular pedigree relationship to animals in the training set.

6. Evaluate the accuracy of GEBV within groups that are relevant to groups in which GEBV will be used commercially. If GEBV are to be used to select between full-sibs, then report within family correlations between GEBV and PEBV. If GEBV are to be used to select animals across studs, then report across stud correlations between GEBV and PEBV.

7. Cross validation can be used to select between models and to tune models, but in that case the selected model cannot be considered to be validated. A further independent validation population is required.

8. Cross validation can be used to gain multiple estimates the accuracy of GEBV from a given tuned model, provided that the validation data did not contribute to the model selection or tuning process.

9. In cross validation, population structures should be considered when sampling validation populations, just as they are when performing a true validation (*see* **Note 5** above).

References

1. Beavis WD (1994) The power and deceit of QTL experiments: lessons from comparative QTL studies. 49th Annual Corn and Sorghum Industry Research Conference Chicago, 1994. pp 250–266
2. Xu SZ (2003) Theoretical basis of the Beavis effect. Genetics 165(4):2259–2268
3. Lande R, Thompson R (1990) Efficiency of marker-assisted selection in the improvement of quantitative traits. Genetics 124(3):743–756
4. Bogdan M, Doerge RW (2005) Biased estimators of quantitative trait locus heritability and location in interval mapping. Heredity 95(6):476–484. doi:10.1038/sj.hdy.6800747
5. Goddard ME, Wray NR, Verbyla K, Visscher PM (2009) Estimating Effects and Making Predictions from Genome-Wide Marker Data. Stat Sci 24(4):517–529. doi:10.1214/09-sts306
6. Weller JI, Shlezinger M, Ron M (2005) Correcting for bias in estimation of quantitative trait loci effects. Genet Sel Evol 37(5):501–522. doi:10.1051/gse:2005013 g0515 [pii]
7. Woolaston RR, Piper LR (1996) Selection of Merino sheep for resistance to Haemonchus contortus: Genetic variation. Anim Sci 62:451–460
8. Parnell PF, Arthur PF, Barlow R (1997) Direct response to divergent selection for yearling growth rate in Angus cattle. Livestock Product Sci 49(3):297–304. doi:10.1016/S0301-6226(97)00045-6
9. Pereira MC, Mercadante MEZ, Razook AG, Figueiredo LA, Albuquerque LG (2008) Results of 23 years of selection for postweaning weight in a Caracu beef herd. South Afr J Anim Sci 38(2):136–144
10. Koch RM, Cundiff LV, Gregory KE (1994) Cumulative selection and genetic change for weaning or yearling weight or for yearling weight plus muscle score in Hereford cattle. J Anim Sci 72(4):864–885
11. Valdar W, Solberg LC, Gauguier D, Burnett S, Klenerman P, O Cookson W, Taylor MS, Rawlins JNP, Mott R, Flint J (2006) Genome-wide genetic association of complex traits in heterogeneous stock mice. Nat Genet 38(8):879–887. doi:10.1038/Ng1840
12. Meuwissen THE, Hayes BJ, Goddard ME (2001) Prediction of total genetic value using genome-wide dense marker maps. Genetics 157(4):1819–1829
13. Hayes BJ, Bowman PJ, Chamberlain AC, Verbyla K, Goddard ME (2009) Accuracy of genomic breeding values in multi-breed dairy cattle populations. Genet Sel Evol 41:51. doi:1297-9686-41-51 [pii] 10.1186/1297-9686-41-51
14. Daetwyler HD, Hickey JM, Henshall JM, Dominik S, Gredler B, van der Werf JHJ, Hayes BJ (2010) Accuracy of estimated genomic breeding values for wool and meat traits in a multi-breed sheep population. Anim Product Sci 50(11–12):1004–1010. doi:10.1071/An10096
15. Moser G, Khatkar MS, Hayes BJ, Raadsma HW (2010) Accuracy of direct genomic values in Holstein bulls and cows using subsets of SNP markers. Genet Sel Evol 42:37. doi:1297-9686-42-37 [pii] 10.1186/1297-9686-42-37
16. Moser G, Tier B, Crump RE, Khatkar MS, Raadsma HW (2009) A comparison of five methods to predict genomic breeding values of dairy bulls from genome-wide SNP markers. Genet Sel Evol 41:56. doi:1297-9686-41-56 [pii] 10.1186/1297-9686-41-56
17. Mäntysaari E, Liu Z, VanRaden P (2010) Interbull validation test for genomic evaluations. Interbull Bull 41:17–21
18. Su G, Brondum RF, Ma P, Guldbrandtsen B, Aamand GP, Lund MS (2012) Comparison of genomic predictions using medium-density (approximately 54,000) and high-density (approximately 777,000) single nucleotide polymorphism marker panels in Nordic Holstein and Red Dairy Cattle populations. J Dairy Sci 95(8):4657–4665. doi:S0022-0302(12)00455-9 [pii] 10.3168/jds.2012-5379
19. Framework for a fully powered risk engine (2005) Nat Genet 37(11):1153–1153
20. Mariasegaram M, Harrison BE, Bolton JA, Tier B, Henshall JM, Barendse W, Prayaga KC (2012) Fine-mapping the POLL locus in Brahman cattle yields the diagnostic marker CSAFG29. Anim Genet 43(6):683–688. doi:10.1111/j.1365-2052.2012.02336.x
21. Habier D, Tetens J, Seefried FR, Lichtner P, Thaller G (2010) The impact of genetic relationship information on genomic breeding values in German Holstein cattle. Genet Sel Evol 42:5. doi:1297-9686-42-5 [pii] 10.1186/1297-9686-42-5
22. Legarra A, Robert-Granie C, Manfredi E, Elsen JM (2008) Performance of genomic selection in mice. Genetics 180(1):611–618. doi:genetics.108.088575 [pii] 10.1534/genetics.108.088575

23. Luan T, Woolliams JA, Lien S, Kent M, Svendsen M, Meuwissen THE (2009) The accuracy of genomic selection in Norwegian Red cattle assessed by cross-validation. Genetics 183(3):1119–1126. doi:10.1534/genetics.109.107391
24. Mujibi FD, Nkrumah JD, Durunna ON, Stothard P, Mah J, Wang Z, Basarab J, Plastow G, Crews DH Jr, Moore SS (2011) Accuracy of genomic breeding values for residual feed intake in crossbred beef cattle. J Anim Sci 89(11):3353–3361. doi:jas.2010-3361 [pii] 10.2527/jas.2010-3361
25. Saatchi M, McClure MC, McKay SD, Rolf MM, Kim J, Decker JE, Taxis TM, Chapple RH, Ramey HR, Northcutt SL, Bauck S, Woodward B, Dekkers JCM, Fernando RL, Schnabel RD, Garrick DJ, Taylor JF (2011) Accuracies of genomic breeding values in American Angus beef cattle using K-means clustering for cross-validation. Genet Sel Evol 43:40. doi:Artn 40 doi:10.1186/1297-9686-43-40
26. Su G, Guldbrandtsen B, Gregersen VR, Lund MS (2010) Preliminary investigation on reliability of genomic estimated breeding values in the Danish Holstein population. J Dairy Sci 93(3):1175–1183. doi:S0022-0302(10)00085-8 [pii] 10.3168/jds.2009-2192

Chapter 19

Detection of Signatures of Selection Using F_{ST}

Laercio R. Porto-Neto, Seung Hwan Lee, Hak Kyo Lee, and Cedric Gondro

Abstract

Natural selection has molded the evolution of species across all taxa. Much more recently, on an evolutionary scale, human-oriented selection started to play an important role in shaping organisms, markedly so after the domestication of animals and plants. These selection processes have left traceable marks in the genome. Following from the recent advances in molecular genetics technologies, a number of methods have been developed to detect such signals, termed *genomic signatures of selection*. In this chapter we discuss a straightforward protocol based on the F_{ST} statistic to identify genomic regions that exhibit high variation in allelic frequency between groups, which is a characteristic of genomic regions that have gone through differential selection. How to define the borders of these regions and further explore its genetic content is then discussed.

Key words SNP, F-statistics, Population genetics, Evolution, Divergence, Speciation, Adaptation, R statistics

1 Introduction

Positive selection, either natural or human-oriented, has shaped all species in the world. It exposes an advantageous trait and favors its increase in prevalence in a population. Beneficial traits are often different between species, in humans likely traits include bipedalism and speech, in cattle docility and increased milk and beef production, in plants increased crop yield and early flowering, but beneficial traits also include traits such as increased disease resistance which is of major importance across many species [1].

Positive selection for a particular allele (genetic variant) at a locus leads to an increase of its prevalence in a population and in the process leaves unique genetic patterns or signatures in the DNA sequence. To illustrate, natural selection that occurred during human adaptation to live at high altitudes [2], exemplifies selection on genes associated to environmental conditions, in this case on loci related to "cellular oxygen sensing," among others; and also demonstrates that populations in different localities might go through different adaptation processes, even if the selective pressure is similar in both places [2]. In domesticated animals, natural factors that would be able to impose selective pressure are often overshadowed by human intervention. This was observed in the development of poll-sheep breeds, the hornless sire would be disadvantaged in the wild, but remained in the population due to human intervention and, in fact, was actively selected for during the development of some breeds [3]. Similarly in dogs, the wrinkle Shar-Pei, which predisposes the animal to a number of skin conditions, is selected for and the phenotype is highly prevalent in the population [4]. In cattle, intensive human-oriented selection for increased milk production [5] and adaptation to the natural environment [6], also defined regions that were under selection. Importantly, both natural and human-oriented selection can lead to genomic changes that are called signatures of selection.

Many statistical approaches have been proposed to detect genetic signatures of selection. Several of them are based broadly on two concepts, the identification of allelic frequency differences between populations, and the detection of long haplotypes with or without differentiation of the ancestral allele.

Direct observation of differences in allelic frequencies between populations is a rough way to identify loci showing differentiation. A more descriptive statistic and measure of genetic differentiation is F_{ST}, it is the most commonly used method since it is directly related to the variance in allele frequency between populations [7]. The term F_{ST} originated with Sewall Wright in 1951 [8] described as "the correlation between random gametes, drawn from the same subpopulation, relative to the total," this imprecise definition allowed for different interpretations, and nowadays a number of derived statistics, all often called F_{ST}, are in use. Generally speaking, the approaches to estimate F_{ST} can be grouped into method-of-moments, maximum-likelihood and Bayesian methods [7]. Method-of-moments estimates are based on analysis of variance of allelic frequencies between two or more groups (populations) to test whether the mean values are equal and to what extent there is a differentiation between the groups [9, 10]. Differently from the method-of-moments, maximum-likelihood requires the definition of a probability distribution from which samples are drawn beforehand. It is harder to implement and generally biased, however it has smaller variance and is the method of choice in many applications [7]. Bayesian estimates share the likelihood concept with the maximum-likelihood approach, but differ in the sense that they take into account a prior distribution of the unknown parameters and base the estimates on the posterior distribution that are related to the likelihood of the prior. The estimates are calculated via a computational method that uses a Markov chain Monte Carlo method (MCMC). Bayesian methods allow estimation of very complex models that would be hard or impossible using the maximum-likelihood approach [7, 11, 12].

The detection of long haplotypes is based on the relationship between an allele's frequency and the linkage disequilibrium surrounding it [13]. In other words, it relies on the assumption that, under positive selection, a favorable allele raises in frequency fast enough so that long-range associations with other polymorphisms in the region are not eliminated by recombination, hence defining long-range haplotypes (LRH) [14]. Examples of methods based on this principle includes the LRH or EHH method [13], integrated Haplotype Score (iHS) which is based on the ancestral allele differentiation [15], Cross Population Extended Haplotype Homozygosity (XP-EHH) [14], and contrasting EHH between populations [16].

All proposed methods have limitations regardless of whether they are based on the identification of allelic frequency differences or the detection of long haplotypes. Methods based on the first concept require at least partial reproductive isolation between the populations, and the effects of demographic history, especially bottlenecks can be an important confounding factor [1]. On the other hand, most of the methods for the detection of long haplotypes require accurate phasing of chromosomes and identification of the ancestral allele which sometimes are challenging to obtain. Trying to avoid the limitations and extract the most from the distinct information that each of the methods can offer, a composite of multiple signals (CMS) test was developed [17], potentially increasing in 100-fold the detection resolution.

In this chapter we describe the detection of signatures of selection using the F_{ST} statistic based on the pure drift model proposed by Nicholson et al. [18]. This method follows from the concept of identification of allelic frequency differences between populations and derives from a *model-based* interpretation of the firstly described F_{ST}, and is, in many cases, highly correlated to the F_{ST} proposed by Cockerham and Weir [9]. This derivation tries to reflect an evolutionary model in which new lineages emerged and evolved in isolation after an ancestral population split.

The individual SNP values of F_{ST} are then grouped within genomic windows (contiguous regions), using a kernel regression smoothing algorithm [19], to define smoothed F_{ST} values that facilitate the identification of genomic regions containing more extreme F_{ST} values. In other words, this method uses a local averaging of the observations (F_{ST}) when estimating the regression function. An important point in this step is the size of the local bandwidth that is used, which refers to the number of SNP that are included in each window for the smoothing. Small bandwidths will use less SNP and will potentially lead to highly wiggly curves, while large bandwidths might smooth the data too much resulting in loss of important peaks [19, *R package Lokern*]. The signals identified can be used independently or further complemented with other methodologies [3, 5, 20]. The output of the analyses is a score for each individual locus (F_{ST}) or genomic region (smoothed F_{ST}), in which the higher the value the stronger is the signal of differentiation or selection.

2 Materials

- SNP genotypes from a number of individuals in each population.
- SNP map. A file describing the genomic position of each marker.

SNP data can come in various formats—a common one is the PLINK [21] format. PLINK is a widely used and freely available software that implements several analyses for SNP data. Moreover, conveniently, commercially available SNP genotyping platforms can export directly to this format, including a genotype and an SNP map file.

A dense coverage of SNP markers is required for these analyses. Different types of molecular markers can be used, but the development of commercially available SNP genotyping platforms for many species has made SNP the markers of choice.

3 Methods

3.1 F_{ST} and Smoothed F_{ST}

This section will guide you through the steps for the identification of signatures of selection using F_{ST} and smoothed F_{ST} based on SNP-allele frequencies. We will use the statistical programming language R [22] for illustration purposes. The sample data and code used in this chapter can be downloaded from the publisher's website (http://extras.springer.com/). The steps are:

- Clearly define the scientific question. F_{ST} will identify loci with divergent allelic frequencies between two or more populations; therefore, a precise definition of what differences are of interest is fundamental so as to correctly define (allocate individuals) the populations relevant for the analyses (*see* **Notes 1** and **2**).
- Select a representative sample for each population. In general, the bigger the sample size for each population, the more reliable the allelic frequency estimates will be (*see* **Notes 3** and **4**).
- To calculate F_{ST}, we will need the allelic frequencies in each population. Depending on the format of the genotypes, different approaches can be taken to convert genotypes into frequencies but this will be rather data-/problem-specific. To illustrate, we will use three very distinct cattle breeds (Hanwoo, Angus, and Brahman) with 25 animals genotyped per breed on a single

chromosome (~8,000 SNP). The genotypes are in a plain text file ("*3breeds.txt*" available from the publisher's website: http://extras.springer.com/) coded as 0/1/2 for, respectively, *AA*, *AB*, and *BB* alleles. The files are organized one SNP per line and one animal per column. To read in the data in R [22]:

```
samples=read.table("3breeds.txt",header=F,
sep=" ")
dim(samples)
> 7763 75
```

- There are 7,763 SNP and 75 samples. The data is already organized so that the SNP are ordered based on their physical location on the chromosome and there is no missing data. Ideally no SNP that failed should be included in the analyses (*see* **Note 5**). For the samples; every 25 are animals from one of the breeds (Hanwoo—1:25; Angus—26:50; Brahman—51:75). Note that the genotypes used herein are for illustration purposes only; they are a small subset from the bovine 700 K and in high linkage disequilibrium which is not desirable for these kinds of studies. The first 5 SNP/samples look like:

```
2 2 0 1 1
0 0 2 1 1
2 2 0 1 1
2 2 0 1 1
0 0 1 1 1
```

- To calculate the frequencies in each population (*see* **Note 6**):

```
#create matrix to store freqs of each SNP in each pop
M=matrix(NA,7763,3)
colnames(M)=c(Hanwoo,Angus,Brahman)

# hanwoo
M[,1]=apply (samples[,1:25],1,function(x) sum
(x)/(length(x)*2))

# angus
M[,2]=apply (samples[,26:50],1,function(x) sum
(x)/(length(x)*2))

# brahman
M[,3]=apply (samples[,51:75],1,function(x) sum
(x)/(length(x)*2))
```

- And the first lines of the matrix of frequencies:

```
Hanwoo  Angus  Brahman
0.56    0.40   0.56
0.44    0.94   0.02
0.56    0.60   0.94
0.56    0.72   0.96
0.36    0.44   0.04
0.36    0.90   0.02
```

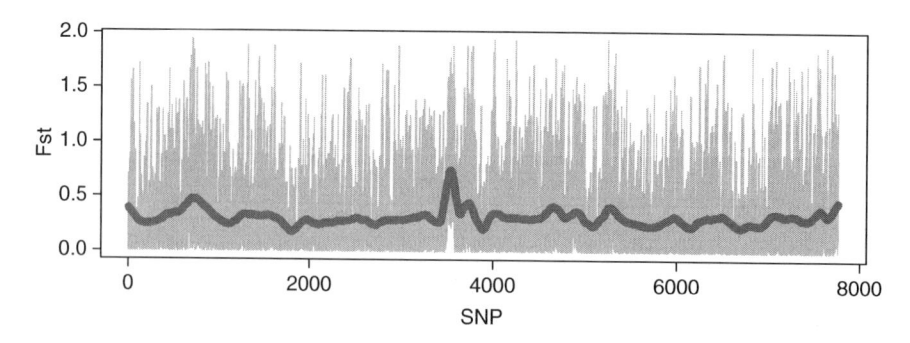

Fig. 1 Genome-wide plot of F_{ST} and smoothed F_{ST} (*thick line*). Notice how the smoothed F_{ST} highlights a region under potential selection around SNP 3,500—it is less clear just from the F_{ST} due to the strong fluctuations from SNP to SNP

- Now, to calculate F_{ST} (*see* **Note 7**):

  ```
  # average allele frequency across populations
  meansB=rowMeans(M)
  ```

  ```
  alleleVar=meansB*(1-meansB) # p*q variance
  ```

  ```
  # deviation of each population from mean
  meanDevB=M-meansB
  ```

  ```
  # deviation squared divided by var
  FST=meanDevB^2/alleleVar
  ```

- And the first few lines of F_{ST}:

  ```
       Hanwoo       Angus      Brahman
  0.011379801  0.045519203   0.0113798
  0.002857143  0.900178571   0.8016071
  0.093333333  0.047619048   0.2742857
  0.184210526  0.003759398   0.2406015
  0.031746032  0.126984127   0.2857143
  0.018168605  0.915879360   0.6760538
  ```

- There is a lot of noise in the F_{ST} values we just calculated (Fig. 1). A smoothing algorithm can help filter this out to identify more robust trends in the data. A useful R package for smoothing is *lokern* which can be used to calculate smoothed F_{ST} values for each population

  ```
  Library(lokern)
  smoothHanwoo = lokerns(FST[,1], n.out=77)
  smoothAngus = lokerns(FST[,2], n.out=77)
  smoothBrahman = lokerns(FST[,3], n.out=77)
  ```

- The SNP data is already ordered based on chromosomal location but if it was not, an additional parameter could be added to *lokern* as a vector of indexes for reordering the SNP (this could be based on the SNP map information). The *n.out* parameter defines the number of points to which the dataset will be smoothed to; in this case we used 77 which is approximately one point for every 100 SNP (*see* **Note 8**). The output from the smoothing process has a few items; the most important at this point is *est*. These are the SNP smoothed values per se.

- For a quick visualization of the Hanwoo F_{ST} and smoothed F_{ST}, you can plot the analysis output with (Fig. 1):

  ```
  plot(FST[,1],type="l",xlab="SNP",ylab="Fst",
  col="gray")
  lines(smoothHanwoo$est,type="l",col="red",
  lwd=6)
  ```

- Save the results to disk. If the SNP are not ordered, the parameter *x.out* stores the indexes of SNP across the genomes (*see* **Notes 9** and **10**).

  ```
  write.table(FST,"FST3breeds.txt",quote=F,row.
  names=F)
  ```

  ```
  smoothed=data.frame(SNPindex=smoothHanwoo$x.
  out,
  Hanwoo=smoothHanwoo$est
  Angus=smoothAngus$est,
  Brahman=smoothBrahman$est)
  ```

  ```
  write.table(smoothed,"FSTsmooth3breeds.txt",
  quote=F,row.names=F)
  ```

- Identify F_{ST} and smoothed F_{ST} peaks. Highly divergent loci (F_{ST}) or genomic regions (smoothed F_{ST}) between populations have high F_{ST} and smoothed F_{ST} values, and those are, generally, associated with either natural or artificial selection. The method here is applied to the smoothed F_{ST} but could similarly be applied to F_{ST} values. A straightforward approach to identify the loci or regions under selection is to calculate the average and standard deviation of the smoothed F_{ST} values for each population, and identify the regions that have values greater than the average plus three standard deviations (~extreme 1 % values) (*see* **Note 11**).

- Once the smoothed F_{ST} peaks are identified, the next step is to define the borders for each of these genomic regions, the density of the SNP in the dataset will define the size of the window to be considered (*see* **Note 12**). Once the regions are defined, a genome browser for the species under investigation can be used to identify the genes under the smoothed F_{ST} peaks (*see* **Note 13**).

- Explore the genes in the selection regions. Do you recognize any gene? What are their functions? Are they involved with any genetic condition? Have they been identified in previous studies? Is there an overrepresentation of any particular ontology pathway?

3.2 Genomic Relationship Matrix

Another useful way of looking at the data is with a genomic relationship matrix (GRM). Other chapters in this volume discuss the GRM in details but within the context of genetic variability between populations, a simple plot of the principal components of the GRM can be very revealing. Let's return to the original genotype data coded as 0/1/2 and use van Raden's [23] formulation to build the GRM:

```
M=as.matrix(samples)
# frequency of second allele = num alleles / total num alleles
p=apply(M,1,function(x) sum(x)/(length(x)*2))
# deviation from 0.5 - P should use base population frequencies not the data itself
M=M-1 # genotypes as -1 / 0 / 1
P=2*(p-0.5)
Z=M-P
ZtZ = t(Z) %*% Z
d=2*sum(p*(1-p)) # scalar
G=ZtZ/d # GRM
```

We can visualize the relationships in the data with a *heatmap* plot (Fig. 2). The *heatmap* helps to highlight the connection between individuals of the same population and across populations. In Fig. 2 we see that all individuals clustered perfectly into their own breed group. But further, in this example, the Hanwoo and Angus are *Bos taurus* animals and the Brahman, *Bos indicus*. The first two are closely related while the common ancestor between *Taurine* and *Indicine* cattle is much more ancestral; this is clear from the two *block* structures in Fig. 2.

```
# color coding for each breed
cols=c(rep("red",25),rep("blue",25),rep("green",25))

heatmap(G,symm=T,col=gray.colors(16,start=0,end=1),
RowSideColors=cols,ColSideColors=cols)

legend("topright",c("Hanwoo","Angus","Brahman"),
fil=c("red","blue","green"),cex=1)
```

Similarly, a singular value decomposition of the GRM can be used to visualize the relationships between samples (Fig. 3).

```
SVD=svd(G)
plot(SVD$v[,2],-1*SVD$v[,1],cex.main=0.9,main="singular value decomposition",xlab="PC1",
ylab="PC2",col=cols,pch=20)
legend("topright",c("Hanwoo","Angus","Brahman"),
fil=c("red","blue","green"),cex=1)
```

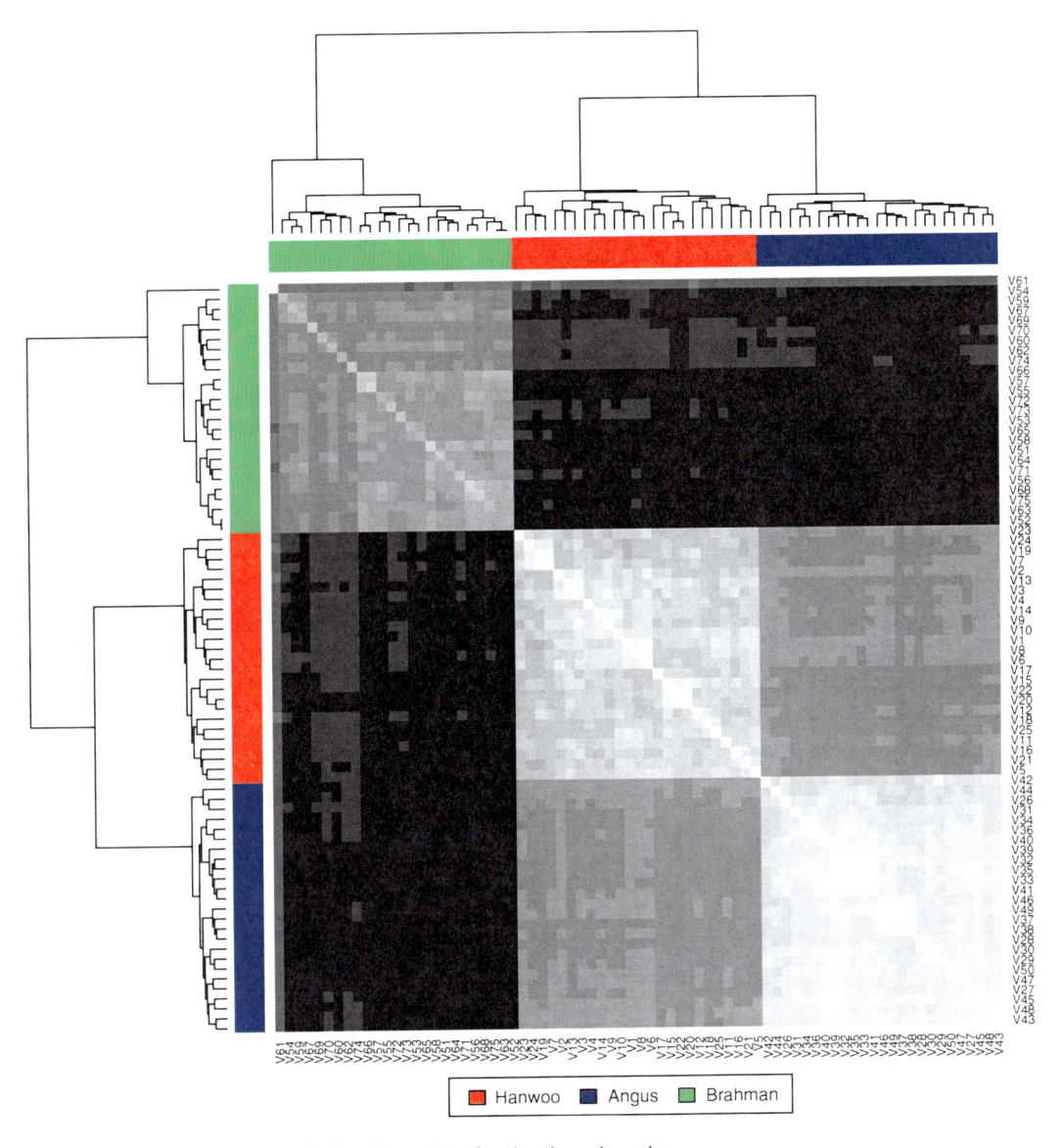

Fig. 2 Heatmap of genomic relationship matrix for the three breeds

4 Notes

1. The F_{ST} and smoothed F_{ST} can be used to explore genetic differences between very contrasting and distantly related populations, and also to detect much more subtle differences between defined genetic groups. It is expected that as relatedness between populations reduce, more extreme F_{ST} and smoothed F_{ST} signals can be detected.

2. The F_{ST} and smoothed F_{ST} are used to detect regions along the genome that have divergent allelic frequencies between

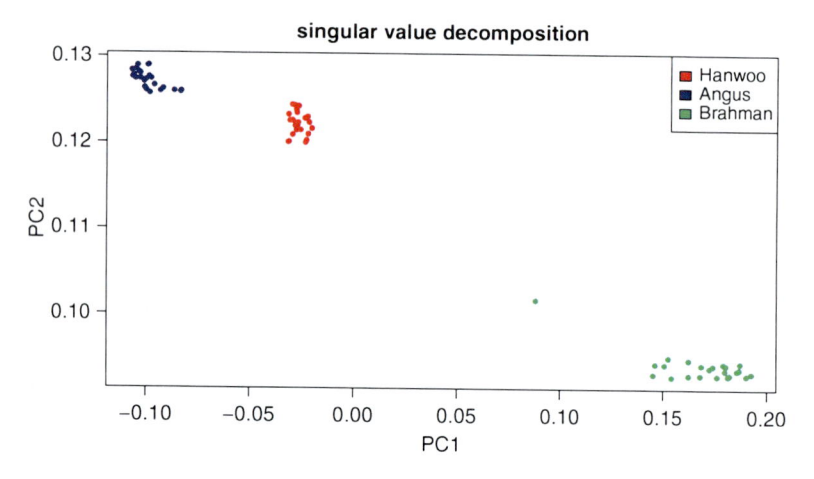

Fig. 3 Plot of first and second principal components of the GRM

populations. The use of closely related populations that have few differences between them could potentially help the identification of candidate genes, in a later step. For example, if studying human adaptation to high altitudes, a valuable comparison would be a study based only on Bolivian populations that live at low and high altitudes or, if studying cattle horn development, by identifying a horned and polled population in the same breed.

3. Since allelic frequencies can be biased by pedigree structure, each population should include individuals as unrelated as possible.

4. The method described in this chapter does not include a sample/population size correction, so a balanced number of individuals across the different populations would be preferred.

5. Before starting out, sort your genotype files in genomic order and keep them like this. The F_{ST} and smoothed F_{ST} calculations are done without SNP ids and genomic positions, and these are not included in the output, so it is essential that the genotype files and result files are in the same order. Moreover for a correct data smoothing the SNP need to be in genomic order (although an index of SNP ordering can be passed to the *lokern* smoothing algorithm as an additional parameter).

6. The calculations of the chromosome and minor allele counts can also be done using the commonly used, freely available, software PLINK [21]. This software can assist in putting the markers in order and, using the options "–freq –within" (see PLINK manual), it returns allelic frequencies of a reference allele for all populations, the reference allele count (MAC), and chromosome count (NCHR) for each population.

The reference allele is the minor allele in most of the samples, but it can be the major allele in one of the groups or breeds.

7. Due to advances of genotyping technologies, an ever-larger number of SNP are included in commercial genotyping platforms, reaching millions of SNP in human platforms, for instance. Availability of computational resources is not necessarily a constraint to explore these large datasets, at least in the case of signatures of selection. Since F_{ST} is calculated individually for each locus across the different populations, if the computer cannot handle the whole dataset, it can simply be split into two or three files and run separately. Then merge the results after the calculations. In doing so it is possible to get the exact same result without high-end computational resources.

8. In simple terms, the smoothing process tries to reduce the number of *points* in the data by calculating a representative value for each *window* of data, which is defined using the *n.out*=X parameter, where X is the number of output values that the data will be reduced to. So if you have 1000 SNP and n.out=100, a window of 10 SNP will be used and a representative value calculated for that window, reducing the data to 100 points. This is not entirely realistic since no data is actually *dropped*, but rather the transition between one SNP and the next is smoothed.

9. There are a number of ways to export your data from R. The *write.table* function is convenient to export data in tabular format but can be rather slow. For large datasets the *write* function is faster but less flexible. If all analyses will be conducted in R, binary versions can be stored using save—much faster and smaller in size but will require advanced programming skills to access out of the R environment (check the R help files for usage).

10. Use the *x.out* values to cross reference smoothed values to a reference SNP map and its genetic position if the data was originally not in order (and you supplied the correct order as a parameter to *lokerns*). As common practice, just routinely add this information to the file.

11. There are a number of ways to identify and select those loci or genomic regions; the method of choice is really dependent on the researcher and how stringent he/she wants to be. A researcher might want to highlight only the top 10 regions for each population, or the extremes identified by regions with F_{ST} or smoothed F_{ST} values greater than the average values plus three standard deviations; on the other hand if one intends to identify genes under the selection regions and build gene networks, for instance, he/she might want to be more flexible

and instead consider all regions with average values plus two standard deviations instead. Other more formal approaches are to calculate q-values based on smoothed F_{ST}, as used by Flori et al. (2009) or use a resampling method to obtain p-values from the empirical distributions.

12. Generally low-density SNP panels will define larger genomic regions in contrast to denser-SNP panels that will zoom in on narrower regions. The number of F_{ST} values that are transformed into a smoothed F_{ST} needs to be taken into account. A simplified way to do this assumes that all markers are equally spaced along the genome, and then the average inter-marker distance is calculated and multiplied by the number of SNP fitted into each smoothed F_{ST}. For instance, if the average inter-marker distance is 2 kbp, and for every 50 SNP one smoothed F_{ST} value was generated, this smoothed F_{ST} value represents the interval under those 50 SNP, so the observed window would be 50 × 2 kbp or 100 kbp.

13. There are a number of web browsers available, which include genomic information of several different species (e.g., http://genome.ucsc.edu/cgi-bin/hgGateway, www.livestockgenomics.csiro.au).

Acknowledgment

This work was supported by a grant from the Next-Generation BioGreen 21 Program (No. PJ008196), Rural Development Administration, Republic of Korea.

References

1. Sabeti PC, Schaffner SF, Fry B, Lohmueller J, Varilly P, Shamovsky O, Palma A, Mikkelsen TS, Altshuler D, Lander ES (2006) Positive natural selection in the human lineage. Science 312:1614–1620
2. Bigham A, Bauchet M, Pinto D, Mao XY, Akey JM, Mei R, Scherer SW, Julian CG, Wilson MJ, Herraez DL, Brutsaert T, Parra EJ, Moore LG, Shriver MD (2010) Identifying signatures of natural selection in Tibetan and Andean populations using dense genome scan data. PLoS Genet 6
3. Kijas J, Lenstra JA, Hayes B, Boitard S, Porto Neto LR, San Cristobal M, Servin B, McCulloch R, Whan V, Gietzen K, Paiva S, Barendse W, Ciani E, Raadsma H, McEwan J, Dalrymple B, The International Sheep Genomics Consortium (2012) Genome wide analysis of the World's sheep breeds reveals high levels of historic mixture and strong recent selection. PLoS Biol 10(2):e1001258
4. Akey JM, Ruhe AL, Akey DT, Wong AK, Connelly CF, Madeoy J, Nicholas TJ, Neff MW (2010) Tracking footprints of artificial selection in the dog genome. Proc Natl Acad Sci USA 107:1160–1165
5. Flori L, Fritz S, Jaffrezic F, Boussaha M, Gut I, Heath S, Foulley JL, Gautier M (2009) The genome response to artificial selection: a case study in dairy cattle. PLoS One 4:e6595
6. Gautier M, Naves M (2011) Footprints of selection in the ancestral admixture of a New World Creole cattle breed. Mol Ecol 20:3128–3143
7. Holsinger KE, Weir BS (2009) Fundamental Concepts in Genetics. Genetics in geographically structured populations: defining, estimating and interpreting FST. Nat Rev Genet 10:639–650

8. Wright S (1951) The genetical structure of populations. Ann Eugen 15:323–354
9. Weir BS, Cockerham CC (1984) Estimating F-statistics for the analysis of population-structure. Evolution 38:1358–1370
10. Nicholson G, Smith AV, Jonsson F, Gustafsson O, Stefansson K, Donnelly P (2002) Assessing population differentiation and isolation from single-nucleotide polymorphism data. J Royal Stat Soc Series B-Stat Methodol 64:695–715
11. Excoffier L (2001) Analysis of population subdivision. In: Balding DJ, Bishop M, Cannings C (eds) Handbook of statistical genetics. Wiley, Chichester, pp 271–307
12. Robert CP (2001) The Bayesian choice: from decision-theoretic foundations to computational implementation, 2nd edn. Springer, New York, NY
13. Sabeti PC, Reich DE, Higgins JM, Levine HZP, Richter DJ, Schaffner SF, Gabriel SB, Platko JV, Patterson NJ, McDonald GJ, Ackerman HC, Campbell SJ, Altshuler D, Cooper R, Kwiatkowski D, Ward R, Lander ES (2002) Detecting recent positive selection in the human genome from haplotype structure. Nature 419:832–837
14. Sabeti PC, Varilly P, Fry B, Lohmueller J, Hostetter E, Cotsapas C, Xie X, Byrne EH, McCarroll SA, Gaudet R, Schaffner SF, Lander ES, Frazer KA, Ballinger DG, Cox DR, Hinds DA, Stuve LL, Gibbs RA, Belmont JW, Boudreau A, Hardenbol P, Leal SM, Pasternak S, Wheeler DA, Willis TD, Yu F, Yang H, Zeng C, Gao Y, Hu H, Hu W, Li C, Lin W, Liu S, Pan H, Tang X, Wang J, Wang W, Yu J, Zhang B, Zhang Q, Zhao H, Zhou J, Gabriel SB, Barry R, Blumenstiel B, Camargo A, Defelice M, Faggart M, Goyette M, Gupta S, Moore J, Nguyen H, Onofrio RC, Parkin M, Roy J, Stahl E, Winchester E, Ziaugra L, Altshuler D, Shen Y, Yao Z, Huang W, Chu X, He Y, Jin L, Liu Y, Sun W, Wang H, Wang Y, Xiong X, Xu L, Waye MM, Tsui SK, Xue H, Wong JT, Galver LM, Fan JB, Gunderson K, Murray SS, Oliphant AR, Chee MS, Montpetit A, Chagnon F, Ferretti V, Leboeuf M, Olivier JF, Phillips MS, Roumy S, Sallee C, Verner A, Hudson TJ, Kwok PY, Cai D, Koboldt DC, Miller RD, Pawlikowska L, Taillon-Miller P, Xiao M, Tsui LC, Mak W, Song YQ, Tam PK, Nakamura Y, Kawaguchi T, Kitamoto T, Morizono T, Nagashima A, Ohnishi Y, Sekine A, Tanaka T, Tsunoda T, Deloukas P, Bird CP, Delgado M, Dermitzakis ET, Gwilliam R, Hunt S, Morrison J, Powell D, Stranger BE, Whittaker P, Bentley DR, Daly MJ, de Bakker PI, Barrett J, Chretien YR, Maller J, McCarroll S, Patterson N, Pe'er I, Price A, Purcell S, Richter DJ, Sabeti P, Saxena R, Sham PC, Stein LD, Krishnan L, Smith AV, Tello-Ruiz MK, Thorisson GA, Chakravarti A, Chen PE, Cutler DJ, Kashuk CS, Lin S, Abecasis GR, Guan W, Li Y, Munro HM, Qin ZS, Thomas DJ, McVean G, Auton A, Bottolo L, Cardin N, Eyheramendy S, Freeman C, Marchini J, Myers S, Spencer C, Stephens M, Donnelly P, Cardon LR, Clarke G, Evans DM, Morris AP, Weir BS, Johnson TA, Mullikin JC, Sherry ST, Feolo M, Skol A, Zhang H, Matsuda I, Fukushima Y, Macer DR, Suda E, Rotimi CN, Adebamowo CA, Ajayi I, Aniagwu T, Marshall PA, Nkwodimmah C, Royal CD, Leppert MF, Dixon M, Peiffer A, Qiu R, Kent A, Kato K, Niikawa N, Adewole IF, Knoppers BM, Foster MW, Clayton EW, Watkin J, Muzny D, Nazareth L, Sodergren E, Weinstock GM, Yakub I, Birren BW, Wilson RK, Fulton LL, Rogers J, Burton J, Carter NP, Clee CM, Griffiths M, Jones MC, McLay K, Plumb RW, Ross MT, Sims SK, Willey DL, Chen Z, Han H, Kang L, Godbout M, Wallenburg JC, L'Archeveque P, Bellemare G, Saeki K, An D, Fu H, Li Q, Wang Z, Wang R, Holden AL, Brooks LD, McEwen JE, Guyer MS, Wang VO, Peterson JL, Shi M, Spiegel J, Sung LM, Zacharia LF, Collins FS, Kennedy K, Jamieson R, Stewart J (2007) Genome-wide detection and characterization of positive selection in human populations. Nature 449:913–918
15. Voight BF, Kudaravalli S, Wen X, Pritchard JK (2006) A map of recent positive selection in the human genome. PLoS Biol 4:e72
16. Tang K, Thornton KR, Stoneking M (2007) A new approach for using genome scans to detect recent positive selection in the human genome. PLoS Biol 5:1587–1602
17. Grossman SR, Shylakhter I, Karlsson EK, Byrne EH, Morales S, Frieden G, Hostetter E, Angelino E, Garber M, Zuk O, Lander ES, Schaffner SF, Sabeti PC (2010) A composite of multiple signals distinguishes causal variants in regions of positive selection. Science 327:883–886
18. Nicholson G, Smith AV, Jonsson F, Gustafsson O, Stefansson K, Donnelly P (2002) Assessing population differentiation and isolation from single-nucleotide polymorphism data. *Annual meeting of the Royal-Statistical-Society*, London, England, pp 695–715
19. Gasser T, Kneip A, Kohler W (1991) A flexible and fast method for automatic smoothing. J Am Stat Assoc 86:643–652
20. van Heerwaarden J, Doebley J, Briggs WH, Glaubitz JC, Goodman MM, de Jesus Sanchez Gonzalez J, Ross-Ibarra J (2011) Genetic signals of origin, spread, and introgression in a

large sample of maize landraces. Proc Natl Acad Sci USA 108:1088–1092
21. Purcell S, Neale B, Todd-Brown K, Thomas L, Ferreira MAR, Bender D, Maller J, Sklar P, de Bakker PIW, Daly MJ, Sham PC (2007) PLINK: a tool set for whole-genome association and population-based linkage analyses. Am J Hum Genet 81:559–575
22. R Development Core Team (2012) R: a language and environment for statistical computing. R Foundation for Statistical Computing, Vienna, Austria
23. VanRaden P (2008) Efficient methods to compute genomic predictions. J Dairy Sci 91:4414–4423

Chapter 20

Association Weight Matrix: A Network-Based Approach Towards Functional Genome-Wide Association Studies

Antonio Reverter and Marina R.S. Fortes

Abstract

In this chapter we describe the Association Weight Matrix (AWM), a novel procedure to exploit the results from genome-wide association studies (GWAS) and, in combination with network inference algorithms, generate gene networks with regulatory and functional significance. In simple terms, the AWM is a matrix with rows represented by genes and columns represented by phenotypes. Individual $\{i, j\}$th elements in the AWM correspond to the association of the SNP in the ith gene to the jth phenotype. While our main objective is to provide a recipe-like tutorial on how to build and use AWM, we also take the opportunity to briefly reason the logic behind each step in the process. To conclude, we discuss the impact on AWM of issues like the number of phenotypes under scrutiny, the density of the SNP chip and the choice of contrast upon which to infer the cause–effect regulatory interactions.

Key words Genome-wide association studies, Systems biology, Gene network, Transcription factor analysis, Pathway analysis, Complex multivariate phenotypes

1 Introduction

We describe the Association Weight Matrix (AWM), a novel procedure to exploit the results from genome-wide association studies (GWAS) and, in combination with network inference algorithms, generate gene networks that hopefully have regulatory and functional significance. Our motivation to develop AWM was twofold: on the one hand, the desire to treat the results from GWAS beyond a simple enumeration of the association of (potentially very many) genetic markers; on the other hand, the availability of a series of phenotypes of interest in a (not necessarily) livestock production system. This latter justification contrasts with the often dichotomous outcome (i.e., Yes/No) and from a single phenotype (e.g., disease) that are found in studies involving model organisms and humans.

In the spirit of multivariate analyses, while a single phenotype may have the primary focus in a given study, we believe that having and formally exploiting a set of phenotypes that inevitably will be somewhat correlated with the primary focus, is prone to increase the accuracy of any statistical inference.

First reported by Fortes et al. [1] on a single population of beef cattle, the AWM was then further implemented to consider a second population [2] as well as the two populations simultaneously [3]. This latter example was enriched by the added ingredient of regulatory impact factors (RIF), a metric designed to identify the regulatory molecules responsible for phenotypic differences [4, 5].

In spite of this prior work, no publication exists explicitly detailing and discussing how the AWM operates. To address this deficiency, in the present chapter we provide a recipe-like tutorial on how to build and use AWM. We also take the opportunity to briefly reason the logic behind each step in the process. We discuss the impact on AWM of issues like the number of phenotypes under scrutiny, the density of the SNP chip, and the choice of contrast upon which to infer the cause–effect regulatory interactions.

2 Materials

The only necessary and also sufficient condition to build the AWM is the availability of the results from GWAS of a series of phenotypes. For convenience, it is assumed that standard quality control filters (such as the removal of SNP with low minor allele frequency, with poor genotype calling rate or in gross violation of the Hardy–Weinberg equilibrium) have been applied prior to obtaining the GWAS results. These results should include for each SNP and phenotype, the estimate of its additive effect along with the P-value associated with this estimate being different from zero. For example, if one is studying diabetes the series of phenotypes would include fasting plasma glucose, estimates of insulin release and insulin sensitivity, body weight, body mass index, waist circumference, blood pressure, cholesterol levels, diabetes duration, and so on; and GWAS results for each of these phenotypes would be required. Details regarding the population under study are not needed, nor are the models for the analysis of the phenotypes in question. The AWM is flexible and adaptable to a range of GWAS methodologies. However, issues like population sample size, SNP filtering criteria, and model goodness of fit will impact on the quality of the GWAS which in turn will affect the quality of the AWM.

The utility of the AWM will be influenced by the extent and accuracy of the annotation of the genome of the species under study. In this respect, unmapped SNPs won't contribute to the generation of the gene network unless some specific criteria are defined for them during the SNP selection criteria (more on this in the Methods section).

When considering a series of phenotypes to be included in the AWM, the actual number of phenotypes under study and the strength to which they relate to each other are important. By and large, the pivotal purpose of AWM is the generation of gene networks based on the co-association of pair-wise SNPs across phenotypes. The basic metric measuring the strength of such co-association is the correlation coefficient. As the standard error of the estimate of a correlation coefficient is proportional to the inverse of the square root of the number of phenotypes, only when a large number of phenotypes is available will a seemingly small estimate of correlation of ~0.5, in absolute value, have a reasonable chance to be deemed significant and hence establish an edge in the generation of the network. We recommend having at least 10 phenotypes (*see* **Note 1**). In the diabetes example, a case–control study would be the main phenotype of interest, while the other phenotypes mentioned (body mass index, insulin sensitivity, etc.) would be investigated as physiologically related traits that complete the scenario. In short, the required materials are three input files as follows (in brackets, file names as those used in the R scripts provided in the Methods section): (1) a file with SNP *P*-values (AWM_P); (2) a file with additive effects estimated for each SNP (AWM_A); and (3) a file containing a list of SNP with corresponding mapped chromosomal position and distance to the nearest gene (GenomeMap).

3 Methods

In what follows, we provide a "Step by Step" tutorial on how to build the AWM and how to use it, in combination with algorithms such PCIT [6] and RIF [4, 5], to infer gene co-association and regulatory networks. Figure 1 provides a schematic representation of the flow chart involved for the development of the AWM/PCIT strategy.

The AWM is a matrix of reals with rows represented by genes and columns represented by phenotypes (body mass index, insulin sensitivity, diabetes, etc.). Individual $\{i, j\}$th elements in the AWM correspond to the association of the SNP in the ith gene to the jth phenotype. The building of the AWM requires a series of six steps as follows:

Step 1: *Primary SNP Selection*: Select the SNPs that are associated with the key phenotype of interest (in the diabetes example, this would be the presence or absence of clinical diabetes, case–control). At this stage, the strength of the statistical significance for the selection of SNPs can be kept at not too stringent levels (e.g., raw *P*-values <0.05) because the power of the AWM-PCIT approach steams from the integration of a collection of seemingly

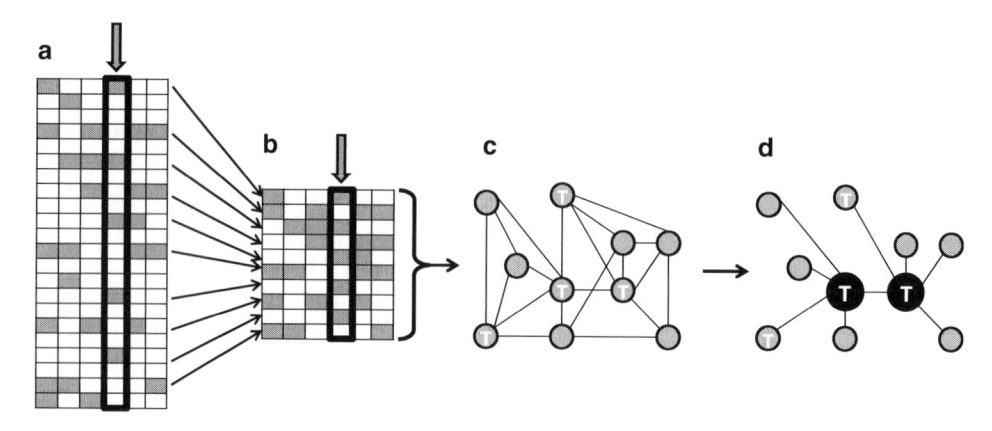

Fig. 1 *Flowchart of the association weight matrix:* (**a**) The association of 20 SNPs (arranged in rows) on 6 phenotypes (arranged in columns) is represented by a matrix with cells occupied by the additive effects. Significant effects are indicated by shaded cells and the key phenotype is highlighted in the fourth column. (**b**) A total of 10 SNPs associated with the key phenotype and/or with a large number of phenotypes are selected to generate a subset of the original matrix. (**c**) A co-association analysis is performed to generate a gene network with 10 genes (one for each SNP) linked by 20 significant connections. Four of the 10 genes are annotated as transcription factors (indicated here by a circumscribed "T"). (**d**) A final analysis reveals that the original network can be further pruned to reveal two master transcription factors connected to themselves and their respective targets by nine connections

independent actions each of which has an associated significance (*see* **Note 2**). The R script corresponding to this step is as follows:

```
AWM_P <- read.table("AWM_Pvalue.txt", header=T, row.names=1)
AWM_A <- read.table("AWM_AdditiveEffects.txt", header=T, row.names=1)
Step1SNPs <- subset(AWM_P, KeyPhenotype <= 0.05)
Step1SNP_IDs <- as.list(rownames(Step1SNPs))
```

Step 2: *Exploring the dependency among phenotypes:* For the SNPs selected above, register the average number of non-key phenotypes to which they are associated. For notation, let A_P denote this average. For consistency, we suggest to use the same significance as before (e.g., *P*-values <0.05) when counting the number of phenotypes to which a given SNP is associated to. As implied, A_P is a function of the amount of dependency existing among the available phenotypes. Clearly, when the phenotypes are highly related to each other, an SNP associated with the key phenotype is also likely to be associated with many of the other phenotypes. The R script corresponding to this step is as follows:

```
Step2Ap <- t(apply(Step1SNPs, 1, function(Step1SNPs)
table(Step1SNPs<=0.05)))
Step2Ap <- as.data.frame(Step2Ap)
Ap <- mean(Step2Ap[,2])
Step2SNP_IDs <- as.list(rownames(Step2Ap))
```

Step 3: *Secondary SNP Selection*: Select the SNPs that are associated with at least A_P phenotypes. In this step, we make proper use of the dependency existing among phenotypes by further capturing those SNPs that, while not associated with the key phenotype of interest, they are associated to the remaining phenotypes in a larger than average fashion. The R script corresponding to this step is as follows:

```
Step3_select <- function(AWM_P){ AWM_P<=0.05 }
Step3Ap <- t(apply(AWM_P, 1, Step3_select))
SignificantPhenotypes <- apply(Step3Ap, 1, sum)
Step3Ap <- subset(SignificantPhenotypes, SignificantPhenotypes>=Ap)
Step3Ap <- as.data.frame(Step3Ap)
Step3SNP_IDs <- as.list(rownames(Step3Ap))
Step3SNPs <- merge(Step1SNP_IDs, Step3SNP_IDs)
Step3SNPs <- t(as.data.frame(Step3SNPs))
rownames(Step3SNPs) <- Step3SNPs[,1]
```

Step 4: *Exploiting the genome map*: From the SNPs captured in the Primary and Secondary selection steps, identify those that (*i*) map to coding regions, (*ii*) are located very close to known genes, or (*iii*) are located substantially far from any coding region. This step is where we first attempt to bring some "*genomic functionality*" to the overall process. Conditions (*i*) and (*iii*) are straightforward and depend only on the density of the SNP chip used for the GWAS and on the quality of the assembly of the genome of the species under consideration. Condition (*iii*) can be expanded by also including unmapped SNPs. Condition (*ii*) has a more subtle connotation as it depends on the extent of the linkage disequilibrium (LD) and the size of the likely size of the promoter region of a given gene. Promoter regions are essential for gene expression [7]. If the expression of a particular gene is to be modulated (i.e., activated or inhibited), then its regulator must bind to its promoter region. Mutations in a promoter region that impact on the expression of the neighboring gene are termed as having a cis-acting effect. This is in contrast with the term trans-acting, which refers to mutations that impact on the expression of a gene located farther away on the genome, even on a different chromosome. In bovine, we have used 2,500 bp as the threshold distance from an SNP to a gene to define condition (*ii*) and we have used unmapped or a distance greater than 1.5 Mb to define condition (*iii*). Any SNP-to-gene distance that fell between 2,500 bp and 1.5 Mb is annotated as far. In this step, the remaining far or unmapped SNPs are discarded from further analysis and we continued to select from the close and very far groups only. The R script corresponding to this step is as follows:

```
GenomeMap <- read.table("GenomeMapFile.txt", header=T, row.names=1)
SNPsofInterest <- GenomeMap[rownames(Step3SNPs),]
Step4SNPclose <- subset(SNPsofInterest, Distance2Gene <= 2500)
Step4SNPfar <- subset(SNPsofInterest, Distance2Gene > 1500000)
Step4SNPclose_IDs <- as.list(rownames(Step4SNPclose))
Step4SNPfar_IDs <- as.list(rownames(Step4SNPfar))
```

Step 5: *Mapping the "One SNP—One Gene" relationship*: One final SNP selection criterion is imposed to those genes represented by more than one SNP. This applies to SNP in conditions (*i*) and (*ii*) of Step 4. In these cases, the SNP associated with the highest number of phenotypes is selected as the representative for that gene. If still there is more than one SNP per gene, then the SNP with the lowest *P*-value average across all phenotypes will be selected. Similarly, for sets of SNP in strong LD and located in regions far from any gene (cf. Condition (*iii*) of Step 4 above) it is recommended to select one SNP as the representative of that LD block and the selection based on the strength of the association averaged across all phenotypes. The R script corresponding to this step is as follows:

```
SignificantPhenotypes <- as.data.frame(SignificantPhenotypes)
Step5SNPclose <- as.data.frame(SignificantPhenotypes[rownames(Step4SNPclose),])
rownames(Step5SNPclose) <- rownames(Step4SNPclose)
Step5Genes <- data.frame(Step4SNPclose$Gene_ID, Step5SNPclose[,1])
colnames(Step5Genes) <- c("Gene_ID", "SignifPhenos")
Step5Sorted <- Step5Genes[order(Step5Genes$SignifPhenos, decreasing = TRUE),]
Step5Unique <- subset( Step5Sorted, !duplicated(Step5Sorted$Gene_ID))
Unique_SNP_IDs <- as.list(rownames(Step5Unique))
Step5SNPs <- merge(Unique_SNP_IDs, Step4SNPfar_IDs)
Step5SNPs <- t(as.data.frame(Step5SNPs))
rownames(Step5SNPs) <- Step5SNPs[,1]
```

Step 6: *Populating the AWM*: Using the results from the GWAS, in each {i, j} cell of the AWM place the association of the ith gene with the jth phenotype. The association of the ith gene with the jth phenotype is measured by the additive effect of the SNP assigned to that gene as per the previous mapping steps for SNP in conditions (i) and (ii) of Step 4. For SNP in condition (iii) simply label the AWM row with the SNP code instead of an official gene symbol. In order to allow direct comparisons across phenotypes, SNP additive effects should be standardized by dividing each individual effect by the standard deviation of the effect of all SNP in the jth column phenotype in question. To avoid selection bias, the standard deviation of the SNP effects should be the one obtained from all the SNP used in the entire GWAS as opposed to the one resulting from the set selected to be included in the AWM. The R script corresponding to this step follows:

```
Step6SNP_ID <- AWM_A[rownames(Step5SNPs),]
rownames(Step5Unique) <- Step5Unique[,1]
Step5Unique <- as.list(rownames(Step5Unique))
Step6SNPs <- merge(Step5Unique, Step4SNPfar_Ids)
Step6SNPs <- t(as.data.frame(Step6SNPs))
rownames(Step6SNPs) <- Step6SNPs[,1]
write.table(Step6SNP_ID,file="PCIT.txt",row.name=    rownames
   (Step6SNPs))
```

The table-like structure of AWM renders itself to be used as the input of most cluster analysis software programs. To this effect, we recommend PermutMatrix [8] (available at http://www.lirmm.fr/~caraux/PermutMatrix/) a user-friendly program to perform graphical analysis of a dataset. It accommodates several types of cluster analysis and a variety of methods for statistical seriation.

In our context, results from these cluster analyses, the way in which columns (phenotypes) cluster and, perhaps to a lesser extent unless strong prior knowledge exists, the way in which rows (genes) cluster together allow for an initial validation of the quality of the data included in the AWM and hence provide confidence in the results from subsequent analyses of AWM (*see* **Note 3**).

The most important aspect of such subsequent analyses is the generation of a gene network with predictive regulatory power. However, prior to generating a gene network, we recommend other analyses. Operating on the columns, we recommend to compute the correlations obtained from pair-wise correlation of columns and to compare these values with estimates of genetic correlation obtained from traditional quantitative genetic approaches such as REML approaches. With the possible limitation of AWM-based estimates of genetic correlations not being free from the impact of environmental covariances, these estimates should show a great similarity with the equivalent estimates resulting from multivariate REML-based approaches.

Operating on the rows, the correlation between all pair-wise genes (and/or SNP) can be used to reconstruct a gene network. In essence, gene pairs with an extreme correlation estimate (either close to −1.0 or +1.0) will be connected by a direct link in the network. The problem reduces to identify which correlation estimates are deemed to be significant and hence establish an edge in the reconstruction of the network.

Reflecting the enormous interest in the area of gene networks coupled with the richness of computational mathematics, multivariate statistics, and computer science, many methods have been proposed to generate gene networks. Broadly speaking, these methods can be classified into two categories: unsupervised and supervised. In the former, networks are inferred exclusively from the data, whereas supervised methods require the additional knowledge of known regulatory interactions as a training set. Unsupervised methods in particular are algorithmically diverse and tend to be computationally faster than supervised methods. Among unsupervised methods, we recommend the PCIT algorithm [6, 9] that uses the partial correlation (PC) in an information theory (IT) framework to ascertain which correlations between gene pairs are significant. Nonetheless, we accept that other methods to generate gene networks would also be appropriate for AWM and this remains to be tested.

One appealing feature of PCIT is that, unlike hard-threshold methods where each correlation estimate is compared against a single threshold, in PCIT each correlation between two genes is compared relative to the correlation between each of these two genes and any other gene in the dataset. This feature implies that a seemingly small correlation between two genes of say $|r| < 0.4$ can be deemed to be significant if no other gene has a correlation with any of those two genes that is larger than 0.4 in absolute value. On the other extreme, an estimated correlation coefficient very close to +1 or −1 could be considered nonsignificant by PCIT.

The output from PCIT can be used as the input for Cytoscape [10] (available at http://www.cytoscape.org), a software program to visualize and analyze the resulting network. A number of Cytoscape plug-ins are available to characterize various features of the networks. Among these, we recommend MCODE [11] (http://baderlab.org/Software/MCODE) to identify clusters of highly interconnected genes that hopefully will be consistent with the known functional and mechanistic behavior of the genes in the cluster and have biological significance for the key phenotype upon which the AWM-PCIT was derived.

Prominent among gene connections with mechanistic significance are regulatory interactions between a transcription factor (TF) and their target genes. Connections of this nature predicted in the

AWM-PCIT gene network can be verified by mining databases of experimentally validated TF–target interactions such as TRANSFAC (http://www.biobase-international.com/gene-regulation) or Genomatix (http://www.genomatix.de). Furthermore, information derived from these databases can be used to formally examine the confidence on the derived network when this network is compared with a control network that could be built after permuting the rows and columns of the AWM.

4 Discussion

One final subject to consider is how to implement the results from the AWM-PCIT approach. Because it does not relate in a direct manner with standard methodologies such as QTL detection or prediction of breeding values, it is not immediately obvious how to apply the findings. To this question, our most unambiguous answer is that SNPs anchored to genes of known biological relevance are more likely to shed biological knowledge and hence be of relevance across populations. Indeed, other authors have shown that integrating information about gene function and regulation with statistical associations from GWAS might enhance knowledge of genomic mechanisms affecting complex phenotypes [12, 13].

A second, perhaps less obvious answer, is that the (hopefully small) minimal set of gene interactions that through their connections spanned most of the topology of the entire AWM-PCIT gene network could reveal a set of SNPs that could be used to exploit their apparent epistatic interaction. This and other value propositions remain to be explored.

5 Notes

1. An issue worth noting when implementing the AWM is that SNPs that are "minimally" associated with a large number of phenotypes will be included in the AWM regardless of which phenotype is the focus or key phenotype of interest. We see this phenomenon as the result of two not necessarily exclusive scenarios: either the set of phenotypes under scrutiny are highly related to each other and hence do not sufficiently span the entire biology underlying the focus phenotype, or these "always associated" SNPs correspond to genes of ubiquitous importance in a wide range of biological processes such as master regulators. While we have not encountered examples of this second possibility, we anticipate it emerging when dealing with tissue differentiation and development where master regulators of cell cycle are prone to impact on many of the phenotypes studied. Regardless of the scenario, we suggest to proceed with the entire AWM-PCIT as usual while keeping a cautious eye on these SNP and the genes they represent.

2. Quite possibly, one of the most anticipated critiques of AWM is that of dealing with the problem of multiple testing. As thousands of SNPs are being tested in any given GWAS, prudence dictates that only those SNPs with a very stringent statistical significance (i.e., very small P-value) should be considered. This argument has its formal justification in the control of false discovery rates (FDR). However, it also seems to clash with the seemingly relaxed criteria for selecting SNPs to be included in the AWM. As mentioned earlier, the strength of the statistical significance for the selection of SNPs can be kept at a low stringency level (e.g., raw P-values <0.05) because the power of the AWM-PCIT approach steams from the integration of a collection of independent actions each of which has an associated significance which in themselves amount at controlling the FDR. Clearly an SNP that is minimally associated with the focus phenotype (i.e., P-value less than but close to 0.05) but not with any of the other phenotypes included in the AWM will have a flat profile of its association across phenotypes. A flat profile won't have the power to generate the co-association with the profile of any other SNP needed to establish an edge in the generation of the AWM-PCIT gene network. Furthermore, when analyzing the network, we encourage the user to visualize it in sequence at various levels of resolution, starting from all significant correlations (edges) to those edges corresponding to successive increases in the co-association correlation they originated from.

3. If traits known to be genetically correlated, or even very closely aligned physiologically, do not appear close to each other in the initial cluster analysis, then the quality of the data should be questioned. For instance, in the diabetes example, one should be concerned with the results if body mass index and body weight do not cluster together. Our suggestion in this case would be to take a step back and, if at all possible, review the GWAS results, including exploring the fixed effects and covariates fitted in the GWAS models.

Acknowledgments

We are grateful to Yuliaxis Ramayo-Caldas for assistance composing the R scripts included in this work.

References

1. Fortes MRS et al (2010) Association weight matrix for the genetic dissection of puberty in beef cattle. Proc Natl Acad Sci USA 107:13642–13647
2. Fortes MRS et al (2010) A new method for exploring genome-wide associations applied to cattle puberty. 9th World Congress on genetics applied to livestock production. German Society for Animal Science, Leipzig, Germany, pp 4–166. ISBN 978-3-00-031608-1.
3. Fortes MRS et al (2011) A single nucleotide polymorphism-derived regulatory gene network underlying puberty in 2 tropical breeds of beef cattle. J Anim Sci 89:1669–1683
4. Hudson NJ, Reverter A, Dalrymple BP (2009) A differential wiring analysis of expression data correctly identifies the gene containing the causal mutation. PLoS Comput Biol 5: e1000382
5. Reverter A et al (2010) Regulatory impact factors: unraveling the transcriptional regulation of complex traits from expression data. Bioinformatics 26:896–904

6. Reverter A, Chan EKF (2008) Combining partial correlation and an information theory approach to the reversed engineering of gene co-expression networks. Bioinformatics 24:2491–2497
7. Cookson W et al (2009) Mapping complex disease traits with global gene expression. Nat Rev Genet 10:184–194
8. Caraux G, Pinloche S (2005) Permutmatrix: a graphical environment to arrange gene expression profiles in optimal linear order. Bioinformatics 21:1280–1281
9. Watson-Haigh NS, Kadarmideen HN, Reverter A (2010) PCIT: an R package for weighted gene co-expression networks based on partial correlation and information theory approaches. Bioinformatics 26:411–413
10. Shannon P et al (2003) Cytoscape: a software environment for integrated models of biomolecular interaction networks. Genome Res 13:2498–2504
11. Bader GD, Hogue CW (2003) An automated method for finding molecular complexes in large protein interaction networks. BMC Bioinformatics 4:2
12. Snelling WM et al (2012) How SNP chips will advance our knowledge of factors controlling puberty and aid in selecting replacement beef females. J Anim Sci 90(4):1152–1165. doi:10.2527/jas.2011-4581
13. Weller JI, Ron M (2011) Invited review: quantitative trait nucleotide determination in the era of genomic selection. J Dairy Sci 94:1082–1090

Chapter 21

Mixed Effects Structural Equation Models and Phenotypic Causal Networks

Bruno Dourado Valente and Guilherme Jordão de Magalhães Rosa

Abstract

Complex networks with causal relationships among variables are pervasive in biology. Their study, however, requires special modeling approaches. Structural equation models (SEM) allow the representation of causal mechanisms among phenotypic traits and inferring the magnitude of causal relationships. This information is important not only in understanding how variables relate to each other in a biological system, but also to predict how this system reacts under external interventions which are common in fields related to health and food production. Nevertheless, fitting a SEM requires defining a priori the causal structure among traits, which is the qualitative information that describes how traits are causally related to each other. Here, we present directions for the applications of SEM to investigate a system of phenotypic traits after searching for causal structures among them. The search may be performed under confounding effects exerted by genetic correlations.

Key words Causal structure search, Inductive causation algorithm, Multiple-trait models, Phenotype networks, Structural equation models

1 Introduction

Causal relationships between variables are pervasive in biology, such as genetic and phenotypic networks. For example, disease status may affect body weight, which may be also controlled by genetic effects. Gene expression may affect some phenotypic traits, but may also be affected by other phenotypes, as well as by eQTL (i.e., QTL that affects the magnitude of the expression of genes). These networks with complex relationships may be expressed by structural equation models, or SEM [1, 2], which are multivariate models whose equations convey functional links among traits, and may be interpreted as causal mechanisms [3]. Unlike the information provided by standard multivariate models, the causal information carried by SEM allows understanding how the values of some traits are affected by (and not only associated with) the values of other traits. This enables the prediction of the effect of external interventions in the system. For example, a causal network containing three variables, such as A→B→C, informs that any external physical intervention on B would change the value of C, but would not modify A. This information could not be provided by a covariance matrix or a joint distribution. Therefore, such modeling approach may be very important not only for understanding relationships between variables in a biological system, but also in fields where external interventions in such systems are common and predicting their effects is useful, such as in pharmacology, medicine, veterinary, and agriculture.

Gianola and Sorensen [4] adapted SEM to quantitative genetics mixed effects settings, where they simultaneously represent the causal network involving phenotypic traits as well as genetic effects affecting each trait. The genetic effects may be inferred on the basis of pedigree or genomic information. These models may be written as:

$$\mathbf{y}_i = \Lambda \mathbf{y}_i + \mathbf{X}_i \boldsymbol{\beta} + \mathbf{u}_i + \mathbf{e}_i \quad (1)$$

where $\mathbf{y}_i$ is a vector containing observations on t traits measured in subject i, Λ is a $t \times t$ matrix containing structural coefficients (parameters that express the magnitude of causal associations) and zeroes. The decision of which entries of Λ are coerced to 0 and which ones are not depends on the causal structure describing the functional relationship among phenotypic traits. Such structure is the qualitative information (generally represented by a directed graph) that describes, for each trait, the subset of remaining traits that have a causal influence on it. Moreover, the vector $\boldsymbol{\beta}$ contains "fixed" effects associated with exogenous covariables in $\mathbf{X}_i$, $\mathbf{u}_i$ is a vector of genetic effects, and $\mathbf{e}_i$ is a vector of model residuals. The vectors $\mathbf{u}_i$ and $\mathbf{e}_i$ are assumed to have the following joint distribution:

$$\begin{bmatrix} \mathbf{u}_i \\ \mathbf{e}_i \end{bmatrix} \sim N \left\{ \begin{bmatrix} \mathbf{0} \\ \mathbf{0} \end{bmatrix}, \begin{bmatrix} \mathbf{G}_0 & \mathbf{0} \\ \mathbf{0} & \Psi_0 \end{bmatrix} \right\},$$

where $\mathbf{G}_0$ and Ψ_0 are the additive genetic and residual covariance matrices, respectively.

A relevant feature of SEM is that it can be reduced by solving it for the terms containing $\mathbf{y}_i$ in the right-hand side [4, 5]. This can be performed by transforming the model to $(\mathbf{I} - \Lambda)\mathbf{y}_i = \mathbf{X}_i \boldsymbol{\beta} + \mathbf{u}_i + \mathbf{e}_i$, such that:

$$\begin{aligned} \mathbf{y}_i &= (\mathbf{I} - \Lambda)^{-1} \mathbf{X}_i \boldsymbol{\beta} + (\mathbf{I} - \Lambda)^{-1} \mathbf{u}_i + (\mathbf{I} - \Lambda)^{-1} \mathbf{e}_i \\ &= \boldsymbol{\theta}_i^* + \mathbf{u}_i^* + \mathbf{e}_i^* \end{aligned} \quad (2)$$

where:

$$\begin{bmatrix} \mathbf{u}_i^* \\ \mathbf{e}_i^* \end{bmatrix} \sim N \left\{ \begin{bmatrix} \mathbf{0} \\ \mathbf{0} \end{bmatrix}, \begin{bmatrix} \mathbf{G}_0^* & \mathbf{0} \\ \mathbf{0} & \mathbf{R}_0^* \end{bmatrix} \right\}, \quad (3)$$

with $\mathbf{G}_0^* = (\mathbf{I} - \Lambda)^{-1}\mathbf{G}_0(\mathbf{I} - \Lambda)^{-1'}$ and $\mathbf{R}_0^* = (\mathbf{I} - \Lambda)^{-1}\Psi_0(\mathbf{I} - \Lambda)^{-1'}$. Note that the model reduction essentially transforms the SEM into a classical mixed effects multiple-trait model, or MTM [6], that results in the same joint distribution of data.

As SEM, in principle, can present more unknowns than what can be identifiable from its likelihood function [5, 7], inferences generally require imposing parametric restrictions. An alternative for achieving that is combining constraints on dispersion parameters with the choice of the causal structure among traits. For example, if residuals are considered as independent, all the model parameters become identifiable if the chosen causal structure among traits is acyclic. However, if these models are used to express causal relationships, restrictions must reflect causal assumptions. For example, assuming acyclic causal structures means assuming that no trait has indirect causal effects on itself, and the premise of independent residuals mirrors the assumption that no hidden variables have causal effects on two or more traits explicitly modeled. Such an approach for fitting mixed effects SEM was followed, for example, by [8–12]. One challenge for selecting a causal structure is that the number of possible structures is generally huge, even for acyclic models encompassing few traits, which makes it unfeasible to compare them exhaustively on the basis of standard model comparison criteria such as AIC [13], BIC [14], or DIC [15]. However, there are search algorithms that allow for data driven exploration of spaces of acyclic causal structures. Without using these algorithms, one may rely completely on prior knowledge to select causal structures.

Graphs representing causal structures may present features that are important for the search algorithms, such as d-separations and unshielded colliders, which are briefly described next. In a graph, sequences of connected variables (nodes), regardless of the direction of the arrows that connect them, are called paths. A collider is a node in a path for which both edges are directed towards it (e.g., the variable B in path A→B←C). Flows of information through paths are blocked by the presence of a collider, or by conditioning on a non-collider. On the other hand, conditioning on a collider allows flows of dependence, as does a non-collider unconditionally. In a graph, two nodes are d-separated conditionally on a specific subset of remaining nodes if, under this condition, there are no paths that allow flows of dependence between them. Additionally, an unshielded collider consists of a collider in some path and its neighboring nodes, if the latter nodes are not connected in the fully directed graph. The features described may be present on the causal structure of a causal model and, under some assumptions, they leave marks on the joint distribution of the generated data. Therefore, it is possible to learn about the underlying causal structure from the distribution, allowing some algorithms to explore the space of causal structures and to search for points of that space that are compatible with that joint distribution [3, 16].

The Inductive Causation (IC) algorithm [3, 17] is one of the so-called *constrained based algorithms*, which performs the search on the basis of some assumptions, from which the causal sufficiency assumption [16] is normally deemed as the strongest one. Nevertheless, this premise is equivalent to assuming independent residuals, which is frequently used in SEM applications, as presented above [8–12]. For this reason, the use of search methods actually allows for the relaxation of modeling assumptions, since following that approach does not require assuming the causal structure as fully known. One challenge in quantitative genetics mixed effects settings is that, even assuming independent residuals, the search of causal structures is confounded by unobserved and correlated genetic effects. This is corrected by conducting the search of causal structures on the basis of $p(\mathbf{y}|\mathbf{u})$ instead of $p(\mathbf{y})$. As demonstrated by [18, 19], the residual covariance matrix of a MTM (i.e., $\mathbf{R}_0^*$) has sufficient information for the search in this setting.

In this chapter, we present an approach for SEM implementation combined with search for causal structures. For this search, we apply an adaptation of the IC algorithm to a scenario where correlated genetic effects may act as confounders in the search, so it can be implemented in a quantitative genetics mixed effects setting. Following [18], the modeling approach is divided into three main parts: (1) obtaining inferences for $\mathbf{R}_0^*$, (2) search for causal structures with the IC algorithm, and (3) fit SEM with the chosen causal structure.

2 Materials: Codes and Files

The code and examples described herein are for the statistical programming language R. Code and data files can be downloaded from the publisher's website (http://extras.springer.com/). Prepare data frames containing the observations for the set of traits for which causal inference is sought, as well as information regarding the exogenous covariables (e.g., environmental effects) associated to each observation. The formatting of these objects is described in the next section.

The first part of the analysis requires a program capable of fitting mixed effects MTM with a specific covariance structure describing kernel distances among levels of the random genetic effects. This program should then be able to construct a matrix representing such structure on the basis of information given by the user, or to allow a covariance structure (i.e., kernel distance matrix) to be inserted and accounted for when fitting the mixed model. Traditionally, this distance matrix is constructed as a function of pedigree information, e.g., the numerator relationship matrix commonly used in GWAS or association mapping to account for population structure and relatedness among individuals, as for example in [20]; or in the prediction of infinitesimal genetic effects in animal breeding [21]. Conversely, the genetic distance matrix can be constructed on the basis of genomic information [22–24]. Additionally, random effects with different kernel distances may be accounted for in the model if effects that are distinct from additive genetic effects are considered, as for example in [25].

For causal structure selection, a program that performs a search algorithm is necessary. Here, we propose an approach based on the IC algorithm. However, instead of implementing the required statistical decisions using point estimates of partial correlations and z tests [16], we adopt the approach of [18], based on samples of the posterior distributions of these correlations. A program that executes this task is provided by the authors, and is referred to as ICPS.

Additional software is also provided to fit SEM with independent residuals for the final part of the analysis, here referred to as gibbsREC. The likelihood equivalence between fully recursive SEM and MTM may be explored, so that one can fit the former model using gibbsREC and obtain posterior samples of the parameters of the latter model (*see* **Note 1**). As a result, the initial part of the analysis may also be performed on the basis of gibbsREC, so that there is no need of distinct programs for fitting MTM and SEM. This program explicitly models functional connections among traits, and the architecture describing how they are related to each other (i.e., the causal structure) must be defined a priori through a matrix with logical variables provided by the user. However, similar results for SEM could be obtained by employing any other program that fits MTM with linear continuous covariates (so that one may declare causal parents as covariates in the equations for each trait) and diagonal residual covariance matrix.

To illustrate the approach and its implementation, a very simple example is presented, which can be run in a matter of minutes. Here, we apply the method to four traits evaluated in 300 individuals, and use a genetic distance matrix representing a group of 20 full sibs with 15 individuals each. Additionally, no exogenous covariates are considered, so that each trait is directly affected only by other traits, direct genetic effects, and residuals. In this case, these are the only variables that are placed in the right-hand side of the SEM. However, applications to real data may get more complicated (with more data points and traits to be analyzed, or with a more complex genetic relationship structure among subjects), requiring more computing time.

3 Methods

Applying the approach proposed by [18] consists of three steps, which are described next:

3.1 Fitting MTM and Obtaining Posterior Samples of R_0^*

1. For the application of gibbsREC in the example, the function can be called as:

```
> fm<-gibbsREC(Y=Y,M=M, FELevels=rep(1,times=
  nrow(Y)),
  AInv=GInv,ID=c(1:nrow(Y)),
  prior=prior,nIter=4000,thin=1,burnIn=2000,
  oneBlock =FALSE)
```

where the arguments carry the following information:

Y—a matrix of phenotypes, where each row is associated to one of n subjects, and each column associated to one of t traits (*see* **Note 2**).

M—a square $t \times t$ matrix with logic entries, indicating with "TRUE" the causal coefficients in Λ allowed to be free structural coefficients (*see* **Note 3**). For a fully recursive model (*see* **Note 1**), a matrix with lower triangular TRUE values may be provided irrespective of the order of the traits in the Y matrix. For the example, the matrix M can be defined as:

```
> M<-matrix(nrow=4,ncol=4,FALSE)
> M[2,1]<-TRUE
> M[3,c(1,2)]<-TRUE
> M[4,c(1:3)]<-TRUE

> M
      [,1]   [,2]   [,3]   [,4]
[1,]  FALSE  FALSE  FALSE  FALSE
[2,]  TRUE   FALSE  FALSE  FALSE
[3,]  TRUE   TRUE   FALSE  FALSE
[4,]  TRUE   TRUE   TRUE   FALSE
```

FELevels—numerical vector of dimension n identifying fixed effect levels assigned to each row of Y (*see* **Note 4**). As there are no exogenous variables affecting the traits in this example, a vector of 1's is provided.

AInv—$n \times n$ inverse of the genetic distance matrix (*see* **Note 5**).

ID—vector of numeric values associating rows of Y and levels of additive genetic effects (*see* **Note 6**).

prior—a list of hyperparameters (*see* **Note 7**) assigned to the prior distribution of residual variances (degrees of freedom and scale parameter of scaled inverted χ^2 distribution) and of additive genetic covariance matrix (degrees of freedom and scale matrix of inverted Wishart distribution). In the example, we used the following prior specifications:

```
> prior<-list(varE=list(df=rep(3,traits),s=rep
  (250,traits)),
  varG=list(df=6,S=diag(150,traits)))
```

nIter—total number of MCMC samples to be drawn from the posterior distribution of the parameters (*see* **Note 8**).

burnIn—number of initial MCMC samples discarded as burn-in (*see* **Note 9**).

thin—thinning interval (*see* **Note 10**).

oneBlock—TRUE value for this argument prompts the program to sample location parameters and genetic effects predictors from one single multivariate distribution at each Gibbs sampling iteration. The default value is FALSE, so that location parameters and genetic effects are sampled in scalar mode from their fully conditional distributions.

2. The program gibbsREC produces a number of output files that will be more thoroughly described later, when the code is used to fit a SEM with the chosen causal structure. At this point, we are particularly interested in one output file, namely the resCov.dat, which contains posterior samples of the MTM residual covariance matrix. After a representative sample has been obtained from the posterior distribution (*see* **Note 11**), it can be used as input for the IC algorithm.

    For the presented example, the program was set to draw 4000 posterior samples of the parameters. Such a small sample is applied here just to assure that the code will complete the task in a matter of minutes and still return a sample carrying enough information to allow the recovery of the causal structure properly. Notice, however that, in real applications, a longer chain might be necessary to reduce Monte Carlo error and improve results for statistical decisions. A comprehensive discussion on MCMC implementations including issues related to convergence and serial correlation can be found, for example, in [26].

3.2 The IC Algorithm

The IC algorithm is detailed below. Considering a set V of random variables, the IC algorithm consists of the following steps:

Step 1: For each pair of variables A and B in V, search for a set of variables S_{AB} such that A is independent of B given S_{AB}. If A and B are dependent for every possible conditioning set, connect A and B with an undirected edge. This step results in an undirected graph U. Connected variables in U are called adjacent.

Step 2: For each pair of nonadjacent variables A and B with a common adjacent variable C in U (i.e., A–C–B), search for a set S_{AB} that contains C such that A is independent of B given S_{AB}. If this set does not exist, then add arrowheads pointing at C ($A \rightarrow C \leftarrow B$). If this set exists, then continue.

Step 3: In the resulting partially oriented graph, orient as many undirected edges as possible in such a way that it does not result in new colliders or in cycles (*see* **Note 12**).

As previously specified, for the methodology proposed by [18], the statistical decisions required by the IC algorithm must be made on the basis of the posterior samples of $\mathbf{R}_0^*$. For each query about the statistical independence between traits A and B given a set of traits S and, implicitly, the genetic effects, the ICPS performs the following tasks:

(a) Obtain the posterior distribution of residual partial correlation $\rho_{A,B|S}$. These partial correlations are functions of $\mathbf{R}_0^*$. Therefore, a sample of its posterior distribution can be obtained by computing this partial correlation at each sample drawn from the posterior distribution of $\mathbf{R}_0^*$.

(b) Compute the highest posterior density (HPD) interval with some specified probability content for the posterior distribution of $\rho_{A,B|S}$ (*see* **Note 13**).

(c) If the computed HPD interval contains 0, declare $\rho_{A,B|S}$ as null. Otherwise, declare A and B as conditionally dependent.

Following that approach, the ICPS is provided as an R function, which could be called as:

```
> r<-read.table('resCov.dat') #loading output file
  from the first method
> outIC<-ICPS(r=r,burnIn=2000,thinning=1,hpd
  Content=.95,
nodeNames=c('A','B','C','D'))
```

The following arguments may be given to that function:

r—a matrix containing posterior samples of $\mathbf{R}_0^*$, where the number of rows is the size of its MCMC sample, and each row refers to a half-vectorization of each $\mathbf{R}_0^*$ sample.

burnIn—number of initial MCMC samples discarded, default value is 0 (*see* **Note 14**).

thinning—thinning of the MCMC samples for the statistical decisions, default value is 1 (*see* **Note 14**).

hpdContent—probability content of HPD interval used for the statistical decisions. Default value is 0.95 (*see* **Note 13**).

nodeNames—a vector of characters containing the names of the traits to be displayed in the plotted partially oriented graph. Default is assigning integers to identify each trait.

The method assigns to object outIC a list containing the pairs of connected traits, a list of unshielded colliders and its parents, and a matrix representation of the graph, which is a $t \times t$ matrix where 1's in outIC[[3]][i,j] represent directed edges from trait j to trait i, unless:

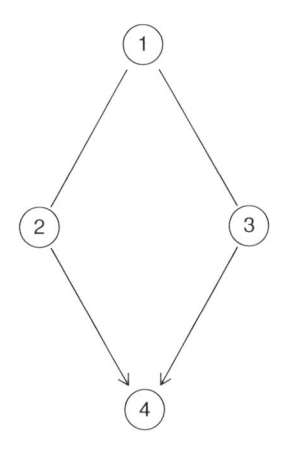

Fig. 1 Partially oriented graph returned by ICPS

```
> outIC[[3]][i,j]==outIC[[3]][j,i] & outIC[[3]]
```
[i,j]==1, which describes an undirected edge between traits i and j (*see* **Note 15**). For the example, ICPS results in:

```
> outIC
[[1]]
     [,1] [,2]
[1,]   1    2
[2,]   1    3
[3,]   2    4
[4,]   3    4

[[2]]
[[2]][[1]]
[[2]][[1]][[1]]
[1] 4

[[2]][[1]][[2]]
[1] 3 2

[[3]]
  1 2 3 4
1 0 1 1 0
2 1 0 0 0
3 1 0 0 0
4 0 1 1 0
```

The code provided also displays a partially oriented graph as a plotted object (*see* **Note 16**), as presented in Fig. 1.

3.3 Fitting SEM with the Selected Causal Structure

Lastly, a SEM using the selected causal structure (or one member within the class of observationally equivalent structures retrieved by the IC algorithm, *see* **Note 17**) is fitted, as in [4], such that causal relationships (i.e., recursive effects) can be estimated. The following two steps are needed for this task:

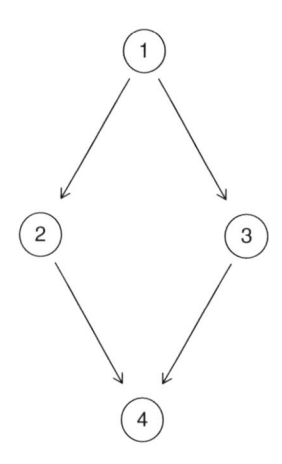

Fig. 2 Causal structure selected from the class of equivalent structures represented in Fig. 1 from temporal information (trait 1 is expressed in the individual before traits 2 and 3)

1. Choose the causal structure and construct a matrix M accordingly, to inform the chosen structure to the code that fits the final SEM. This matrix may be constructed on the basis of the output returned by the ICPS function, replacing the zeroes by "FALSE" and the ones by "TRUE". For the edges left undirected, one of the entries will have to be replaced by a FALSE value.

   In the present case, consider that trait A is expressed in the individual before traits B and C, and such knowledge forbids orienting edges from B and C toward A. Therefore, the only member of the class of DAG's returned by the IC algorithm which agrees with this temporal sequence information is the structure shown in Fig. 2.

   In the example, such matrix would be:

```
> M<-outIC[[3]]
> M[1,2:3]=0
> mode(M)<-"logical"
> M
      1      2      3      4
1 FALSE  FALSE  FALSE  FALSE
2 TRUE   FALSE  FALSE  FALSE
3 TRUE   FALSE  FALSE  FALSE
4 FALSE  TRUE   TRUE   FALSE
```

2. After choosing the causal structure, fit the SEM. Here, we use the same code and arguments applied in (Subheading 3.1, item 1). The only change is in the inserted causal structure matrix, which was previously designed for a fully recursive model.

   The following output files are created after the model is fitted:

B.dat—posterior samples of the entries of Λ. Columns containing values not restrained to 0 represent the posterior samples of structural coefficients.

G.dat—posterior samples of SEM genetic covariance matrix.

genCov.dat—posterior samples of MTM genetic covariance matrix.

psi.dat—posterior samples of residual variances pertaining to a SEM.

resCov.dat—posterior samples of MTM residual covariance matrix.

post_FE.txt—posterior means of exogenous fixed effects.

Dj.dat and DTetaBar.dat—output files that are used for evaluating the Deviance Information Criterion, or DIC [15], where the former contains $D(\theta) = -2\log(p(\mathbf{y}|\theta))$ evaluated at each Gibbs sampler iteration (i.e., conditionally on the parameter values drawn for each posterior sample), allowing Monte Carlo estimation of $\bar{D} = E_{\theta|\mathbf{y}} D(\theta)$; and the latter contain $D(\bar{\theta}) = -2\log(p(\mathbf{y}|\bar{\theta}))$, based on the posterior means of the parameters. The target quantity, obtained by $\text{DIC} = 2\bar{D} - D(\bar{\theta})$, can be computed as:

```
> burnin = 2000
> thinning = 1
> Dj<-read.table('Dj.dat')
> DTB<-read.table('DTetaBar.dat')
> indexT<-c(1:((nrow(Dj)-burnin)/thinning))*
  thinning
> Dj<-Dj[burnin:nrow(Dj),]
> Dj<-Dj[indexT]
> DIC<-2*(sum(Dj)/length(Dj)) - DTB
> DIC
     V1
1 10046.61
```

For interpretation of results, *see* **Note 18**.

4 Notes

1. Using gibbsREC to obtain inferences for MTM dispersion parameters explores the relationship $\mathbf{R}_0^* = (\mathbf{I} - \Lambda_{\text{fr}})^{-1} \Psi_{\text{fr}} (\mathbf{I} - \Lambda_{\text{fr}})'^{-1}$ [4, 5]. In this expression, $\mathbf{R}_0^*$ is the residual covariance matrix of a MTM, while Λ_{fr} and Ψ_{fr} are, respectively, structural coefficients matrix and diagonal residual covariance matrix from a fully recursive SEM. For a model with a fully recursive causal structure, Λ_{fr} is constructed following a structure that is as less sparse as possible while being devoid of cycles, which may be more easily attained by considering all entries below the diagonal as free parameters.

2. For gibbsREC, Y must provide phenotypes without missing records. The number of traits affects the time required for the causal structure search, by affecting the time required for obtaining each joint posterior sample of parameters, and the number of queries performed by the IC algorithm.

3. Assigning TRUE to entry M[i,j] indicates that in the causal structure of the SEM, trait j is allowed to have a causal effect on trait i. The SEM to be fitted by gibbsREC must be acyclic, which implies that the matrix inserted here must not have causal cycles, and TRUE value must not be assigned simultaneously to M[i,j] and M[j,i], for any i and j smaller or equal to the number of traits.

4. The program gibbsREC requires exogenous effects to be categorical. Additionally, this code does not support more than one exogenous variable affecting the traits. It is also assumed that the exogenous covariate affects all traits.

5. Different methods may be used to obtain the genetic distance matrix. From genomic markers data, one may construct this matrix on the basis of methods presented by [22–24]. Inverting such a matrix may be computationally intensive depending on its size.

6. Multiple observations per level of genetic effect are allowed, e.g., with multiple subjects within inbred lines, as in [18]. Assigning one level per observation would result in fitting an animal model [27]. Animal model applications, depending on the size of data set and population structure (and connectness), may present identifiability issues, making it more difficult to estimate genetic and residual (co)variance components. If gibbsREC is applied to obtain posterior samples of the MTM residual covariance matrix, such issues could harm the quality of the posterior samples of the residual covariance matrix, requiring longer MCMC for parameter inferences and IC algorithm decisions.

7. The list prior given to gibbsREC contains SEM dispersion hyperparameters, so that prior scale parameters inserted here do not reflect prior expectations of MTM dispersion parameters. For example, for the additive genetic dispersion, we should input a scale matrix that reflects our prior expectation about the dispersion of direct genetic effects, and not of the overall genetic effects [18]. However, when the purpose is fitting a MTM through a fully recursive model, parameterization of prior distributions gets awkward since we are interested in $\mathbf{R}_0^* = (\mathbf{I} - \Lambda_{\text{fr}})^{-1} \Psi_{\text{fr}} (\mathbf{I} - \Lambda_{\text{fr}})'^{-1}$ and not in Ψ_{fr}. Assessing the prior expectation of $\mathbf{R}_0^*$ is cumbersome because it is a function of structural coefficients, for which improper priors are assigned in gibbsREC. Here, we try to minimize the

impact of this prior by using the minimum value for the degrees of freedom that guarantees that prior expectations are within parameter spaces (i.e., number of traits plus 2). As data get more informative, the influence of the prior distributions on the results of the search tends to vanish.

8. The number of iterations may impact not only the time required to draw the posterior samples but also the time required to execute ICPS, which performs multiple partial correlation calculations for each posterior sample provided. The latter impact may be alleviated by thinning the posterior sample when applying the ICPS.

9. The values assigned to burnIn only affect the computations of posterior expectations retrieved by the code, e.g., the posterior expectations retrieved for the fixed effects and posterior means of parameters that are necessary to obtain $D = E_{\theta|y}[-2\log(p(\mathbf{y}|\theta))]$ (*see* Subheading 3.3, step 2), where θ is a vector containing model parameters. This quantity is saved in output file dTetaBar.dat, and used for obtaining the DIC for the model. Regardless of the amount of burn-in required, the complete MCMC samples are withheld for the dispersion parameters and structural coefficients. For inferences regarding these parameters, a suitable burn-in should be applied *a posteriori* to their MCMC samples.

10. As with the burnIn specification, thinning is applied only to the samples used to compute the posterior expectations, as well as $\bar{D}$.

11. As with any application of Gibbs sampling, one should evaluate MCMC convergence and obtain a reasonable sample from the posterior distribution, in terms of effective sample size and Monte Carlo error. Performing this evaluation ranges from simply relying on visual inspection of trace plots to using R packages such as BOA [28], which allows applying more formal convergence diagnostic criteria and detailed analysis of posterior samples. Insufficient burn-in or effective sample size may lead to decreasing search quality, especially when the zero value is located in the vicinities of the HPD interval of partial correlations considered for some of the statistical decisions. Generating larger chains and discarding more initial samples are a conservative approach that makes the decisions more reliable, but it also increases the computational burden, especially for analyses involving a large number of traits.

12. Search methods such as the IC algorithm rely on a series of assumptions, as discussed by [16]. However, there are sets of vanishing and nonvanishing partial correlations that cannot be reflected by any Markovian acyclic model [3, 16]. Results of this kind indicate that one or more assumptions do not hold.

Nevertheless, the results may still carry interesting features that should be considered in the exploration of causal structures spaces.

13. The default HPD interval considered by the ICPS for the statistical decisions is 0.95, but this value may not be suitable depending on the setting. Using intervals with high probabilities content (such as the default) may be critical if the posterior distributions of the correlations are not too sharp, for which they may become too wide, making it difficult for the algorithm to declare small but nonnull partial correlations as nonnull. In the decision, two types of mistakes can be made: declaring null correlations as nonnull and declaring nonnull correlations as null. Following [29], as exploring causal structures spaces does not involve protecting more against one type of mistake than the other, it may be suitable to apply smaller probability contents for the HPD interval when posterior densities are not too sharp. Whenever that is the case, the approach suggested by [18] was to apply the ICPS for a reasonable grid of different probabilities contents and then, if it results in the selection of different causal structures, to compare models with the selected distinct structures using DIC.

14. When applying the IPCS to a posterior sample of $\mathbf{R}_0^*$, failure in removing samples drawn before convergence may reduce the quality of the decisions, particularly if one of the extremes of the computed HPD interval is located close to 0. Thinning the samples may reduce drastically the execution time of the IC algorithm, but it may increase Monte Carlo error and damage the quality of the decisions.

15. Although not graphical, `outIC[[3]]` is sufficient to completely characterize the algorithm output.

16. Rgraphviz package [30] is necessary to display the partially oriented graphs.

17. Most of the times, the output of the IC algorithm will not be a completely directed graph, but a partially oriented graph representing a class of statistically equivalent structures. In this case, some non-statistical criteria must be used to choose the remaining directions, since a fully directed structure is necessary to fit a SEM. Alternatives are either to arbitrarily choose a direction and ignore the causal interpretation of the structural coefficient associated to these edges, and to use prior causal knowledge to choose edge directions. In some situations, considering such prior knowledge (due to temporal sequence information, for example) for one single edge may aid in directing several other undirected edges, as can be seen in the simulated example given by [18].

18. With these files at hand, it is interesting to make some comparisons with results produced by the MTM. For example, it is good if posterior means obtained from genCov.dat and resCov.dat with the fitted SEM do not deviate too much from the posterior means obtained from the same files produced by a fully recursive model. This is so because the IC algorithm searches for a causal structure that is compatible with the data distribution. Therefore, it is expected that restrictions imposed on the joint distribution by the causal structure of the fitted model reflect the conditional independencies found by the IC algorithm in the observed distribution of data. In other words, SEM with the selected causal structure is expected to keep the equality $\mathbf{R}_0^* = (\mathbf{I} - \Lambda)^{-1}\Psi(\mathbf{I} - \Lambda)'^{-1}$, even though SEM may induce partial correlation restrictions that are not induced by MTM. Furthermore, since the SEM with selected causal structures is a less flexible model that essentially represents the same distribution obtained from fitting a MTM, then it is also presumed that the fitted SEM presents smaller DIC when compared to the MTM's (except if no edges are removed from the IC algorithm). In this example, we computed DIC of the MTM as 10113.85, while for the selected SEM it was 10046.61. The reader may have slightly different results due to Monte Carlo error.

Acknowledgments

The authors thank Gustavo de los Campos for his contribution on the development of the function gibbsREC. This work was supported by the Agriculture and Food Research Initiative Competitive Grant no. 2011-67015-30219 from the USDA National Institute of Food and Agriculture.

References

1. Wright S (1921) Correlation and causation. J Agric Res 201:557–585
2. Haavelmo T (1943) The statistical implications of a system of simultaneous equations. Econometrica 11:12
3. Pearl J (2000) Causality: models, reasoning and inference. Cambridge University Press, Cambridge, UK
4. Gianola D, Sorensen D (2004) Quantitative genetic models for describing simultaneous and recursive relationships between phenotypes. Genetics 167:1407–1424
5. Varona L, Sorensen D, Thompson R (2007) Analysis of litter size and average litter weight in pigs using a recursive model. Genetics 177:1791–1799
6. Henderson CR, Quaas RL (1976) Multiple trait evaluation using relative records. J Anim Sci 43:1188–1197
7. Wu XL, Heringstad B, Gianola D (2010) Bayesian structural equation models for inferring relationships between phenotypes: a review of methodology, identifiability, and applications. J Anim Breed Genet 127:3–15
8. de los Campos G, Gianola D, Boettcher P, Moroni P (2006) A structural equation model for describing relationships between somatic cell score and milk yield in dairy goats. J Anim Sci 84:2934–2941

9. Heringstad B, Wu XL, Gianola D (2009) Inferring relationships between health and fertility in Norwegian Red cows using recursive models. J Dairy Sci 92:1778–1784
10. Maturana EL, Wu XL, Gianola D, Weigel KA, Rosa GJM (2009) Exploring biological relationships between calving traits in primiparous cattle with a Bayesian recursive model. Genetics 181:277–287
11. Ibanez-Escriche N, de Maturana EL, Noguera JL, Varona L (2010) An application of change-point recursive models to the relationship between litter size and number of stillborns in pigs. J Anim Sci 88:3493–3503
12. Jamrozik J, Schaeffer LR (2010) Recursive relationships between milk yield and somatic cell score of Canadian Holsteins from finite mixture random regression models. J Dairy Sci 93:5474–5486
13. Akaike H (1973) Information theory and an extension of the maximum likelihood principle. In: Petrov BN, Csaki F (eds) Second International symposium on information theory. Publishing House of the Hungarian Academy of Sciences, Budapest
14. Schwarz G (1978) Estimating dimension of a model. Ann Stat 6:461–464
15. Spiegelhalter DJ, Best NG, Carlin BR, van der Linde A (2002) Bayesian measures of model complexity and fit. J R Stat Soc Ser B-Stat Methodol 64:583–616
16. Spirtes P, Glymour C, Scheines R (2000) Causation, prediction and search, 2nd edn. MIT Press, Cambridge, MA
17. Verma T, Pearl J (1990) Equivalence and synthesis of causal models. Proceedings of the 6th conference on uncertainty in artificial intelligence 1990, Cambridge, MA
18. Valente BD, Rosa GJM, de los Campos G, Gianola D, Silva MA (2010) Searching for recursive causal structures in multivariate quantitative genetics mixed models. Genetics 185:633–644
19. Rosa GJM, Valente BD, de los Campos G, Wu XL, Gianola D, Silva MA (2011) Inferring causal phenotype networks using structural equation models. Genet Sel Evol 43:6
20. Yu JM, Pressoir G, Briggs WH, Bi IV, Yamasaki M, Doebley JF, McMullen MD, Gaut BS, Nielsen DM, Holland JB, Kresovich S, Buckler ES (2006) A unified mixed-model method for association mapping that accounts for multiple levels of relatedness. Nat Genet 38:203–208
21. Mrode RA, Thompson R (2005) Linear models for the prediction of animal breeding values, 2nd edn. Cabi Publishing-C a B Int, Wallingford
22. VanRaden PM (2008) Efficient methods to compute genomic predictions. J Dairy Sci 91:4414–4423
23. Gianola D, de los Campos G, Hill WG, Manfredi E, Fernando R (2009) Additive genetic variability and the Bayesian alphabet. Genetics 183:347–363
24. Forni S, Aguilar I, Misztal I (2011) Different genomic relationship matrices for single-step analysis using phenotypic, pedigree and genomic information. Genet Sel Evol 43
25. Gianola D, de los Campos G (2008) Inferring genetic values for quantitative traits non-parametrically. Genet Res 90:525–540
26. Gelman A, Carlin JB, Stern HS, Rubin DB (2004) Bayesian data analysis, 2nd edn. Chapman & Hall, New York, NY
27. Mrode MA (1996) Linear models for the prediction of animal breeding values. CAB International, Wallingford
28. Smith BJ (2008) Bayesian Output Analysis Program (BOA) for MCMC. http://cran.r-project.org/web/packages/boa/citation.html
29. Shipley B (2002) Cause and correlation in biology. Cambridge University Press, Cambridge
30. Gentry J, Long L, Gentleman R, Falcon S, Hahne F, Sarkar D, Hansen K (2012) Rgraphviz: provides plotting capabilities for R graph objects package version 2.2.1

Chapter 22

Epistasis, Complexity, and Multifactor Dimensionality Reduction

Qinxin Pan, Ting Hu, and Jason H. Moore

Abstract

Genome-wide association studies (GWASs) and other high-throughput initiatives have led to an information explosion in human genetics and genetic epidemiology. Conversion of this wealth of new information about genomic variation to knowledge about public health and human biology will depend critically on the complexity of the genotype to phenotype mapping relationship. We review here computational approaches to genetic analysis that embrace, rather than ignore, the complexity of human health. We focus on multifactor dimensionality reduction (MDR) as an approach for modeling one of these complexities: epistasis or gene–gene interaction.

Key words GWAS, MDR, Filter approach, Stochastic search, GPUMDR, Network approach

1 Introduction

The current strategy for studying the genetic basis of disease susceptibility is to measure millions of single-nucleotide polymorphisms (SNPs) across the human genome and test each individually for association [1, 2]. This genome-wide association study (GWAS) approach is based on the idea that genetic variations with alleles that are common in the population will additively explain much of the heritability of common diseases. As the price of genome-wide genotyping has dropped, the number of studies utilizing GWAS has increased dramatically, and this approach is now relatively common. The GWAS approach has been successful in that hundreds of new associations have been reported using rigorous statistical significance and replication criteria [3]. It is anticipated that these SNPs will reveal new pathobiology that will in turn lead to new treatments. While this may be true, a few of the loci identified are associated with a moderate or large increase in disease risk, and some well-known genetic risk factors have been missed [4]. At best, perhaps 20 % of the total genetic variance has been explained for a few select common diseases such as Crohn's disease [5]. As a result, many have asked where the missing heritability is [6]. One possibility is that complexities such as locus heterogeneity, gene–environment interaction, and gene–gene interactions or epistasis can limit the power of analysis approaches that only consider one SNP at a time [7–9]. In this chapter, we focus on epistasis as an important component of genetic architecture that should be considered when approaching the genetic analysis of GWAS data.

Epistasis is an old idea that has been discussed since the early 1900s [10]. Epistasis has been discussed in the literature in two different forms, biological and statistical. At the cellular level in an individual, biological epistasis is observed as the physical interactions between biomolecules that are influenced by genetic variation at multiple different loci; while at the population level, statistical epistasis is observed as the nonadditive mathematical relationship among multiple genetic variants [11–15]. The relationship between biological and statistical epistasis is confusing but will be important to understand if we are to make biological inferences from statistical results [16, 17]. Although the biological mechanism that forms epistasis is not clearly understood yet, epistasis is likely due, in part, to canalization or mechanisms of stabilizing selection that evolve robust (i.e., redundant) gene networks [8, 18, 19]. We expect that epistasis will be recognized as a ubiquitous phenomenon in complex human diseases once it is properly investigated [20].

One of the primary reasons that epistasis is not more commonly investigated in genetics studies of human disease is because its detection and characterization are statistically and computationally complex. One issue is the dimensionality of putting multiple SNPs together. Parametric statistical methods such as logistic regression often fail to converge on accurate parameter estimates when multiple variables and their interactions are being modeled. Another issue is that there are a large number of hypotheses involved in modeling interactions, even when considering only pairwise interactions [7, 21, 22]. Finally, the exhaustive evaluation of all possible combinations of genetic variants is at best computationally expensive. Specifically, when studying k-way interaction among n variables, the computational cost is on the order of n^k. The high computational cost does not allow us to investigate large values of k yet [7, 23, 24]. All together, these issues can be significant barriers to the routine modeling of epistasis.

Several statistical and computational techniques have been proposed to detect epistasis in human populations [13, 25]. As a staple in the data mining and machine learning community, decision trees have been widely used [26–28]. Here we use attribute to mean a variable such as a SNP or a demographic variable such as gender, that is, used to make a prediction [8]. A decision tree classified subjects as case or control by sorting them through a tree from node to node where each node is an attribute with a decision rule that guides that subject through different branches of the tree to a leaf that provides its classification [29, 30]. A RF is a collection of individual decision tree classifiers, where each tree in the forest has been trained using a bootstrap sample of instances (i.e., subjects) from the data, and each attribute in the tree is chosen from a random subset of attributes. Classification of instances is based upon aggregate voting over all trees in the forest [31, 32]. Although easy to interpret, tree-based methods have one major limitation that the standard implementations condition on marginal effects. In other words, the algorithm finds the best single variable for the root node before adding additional variables as nodes in the model [7, 14, 33, 34]. This can preclude the detection of epistasis in the absence of significant single-SNP effects.

Multifactor dimensionality reduction (MDR) was developed as a novel nonparametric data mining and machine learning strategy for identifying combinations of discrete genetic and environmental factors that are predictive of a discrete clinical endpoint [35–41]. An advantage of MDR is that it assumes no particular model and that it can be included in other methods such as decision trees or logistic regression.

At the heart of the MDR approach is a feature or attribute construction algorithm that creates a new variable or attribute by pooling genotypes from multiple SNPs. The general process of defining a new attribute as a function of two or more other attributes is referred to as constructive induction, or attribute construction. Constructive induction, using the MDR kernel, is accomplished in the following way. Given a threshold T, a multilocus genotype combination is considered high risk if the ratio of cases (subjects with disease) to controls (healthy subjects) exceeds T, otherwise it is considered low risk. Genotype combinations considered to be high risk are labeled G1 while those considered low risk are labeled G0. This process constructs a new one-dimensional attribute with values of G0 and G1. Later this new single variable can be assessed, using any classification method [42]. Statistical significance is determined by a combination of cross-validation and permutation testing [43]. The MDR algorithm has reasonable power to detect epistasis but carries out knowledge discovery in a very specific manner using an exhaustive search and a single classifier to identify the optimal combination of polymorphisms for predicting a discrete disease endpoint. There have been numerous methodological extensions to MDR including recent approaches such as robust MDR [44], survival MDR [45], covariate MDR [46], and model-based MDR [47–49]. We discuss extensions to GWAS data below. A user-friendly and open-source MDR software package written in Java is freely available from www.epistasis.org or by request.

2 Materials

2.1 Minimum System Requirements

- Java Runtime Environment, version 5.0 or higher
- 1 GHz Processor
- 256 MB Ram
- 800 × 600 screen resolution

2.2 Data Format

Genotypes should be encoded as 0–2. Each column should represent one attribute except that the last column represents the values of the trait.

3 Methods

As discussed above, GWASs have generated a number of important bioinformatics challenges including the modeling of complex genotype–phenotype relationships using data mining and machine learning methods, the use of biological knowledge databases to help guide and interpret genetic association studies, and the development of powerful and user-friendly software [7]. The flowchart in Fig. 1 represents the relationship between several approaches to scaling MDR to GWAS data. Below we review statistical and biological filter approaches, stochastic search approaches, high-performance computing approaches, and network approaches (*see* **Notes 1–4**).

3.1 Filter Approach

To overcome the computational cost for combinatorial exploration of all interactions in GWAS, one natural approach is to filter out a subset of variations that can then be efficiently analyzed using a method such as MDR. The filter criteria can be either statistical or biological. We review below a powerful filter method based on the ReliefF algorithm and then discuss prospects for using biological knowledge to filter genetic variations.

3.1.1 Statistical Filter

Kira and Rendell developed an algorithm called ReliefF, that is, capable of detecting attribute dependencies or interactions [50]. ReliefF estimates the quality of attributes through a type of nearest neighbor algorithm that selects neighbors (instances) from the same class and from the different class based on the vector of values across attributes. Weights (W) or quality estimates for each attribute (A) are estimated based on whether the nearest neighbor (nearest hit, H) of a randomly selected instance (R) from the same class and the nearest neighbor from the other class (nearest miss, M) have the same or different values. This process of adjusting weights is repeated for m instances. The algorithm produces weights for each attribute ranging from -1 (worst) to $+1$ (best). The pseudo code is outlined below:

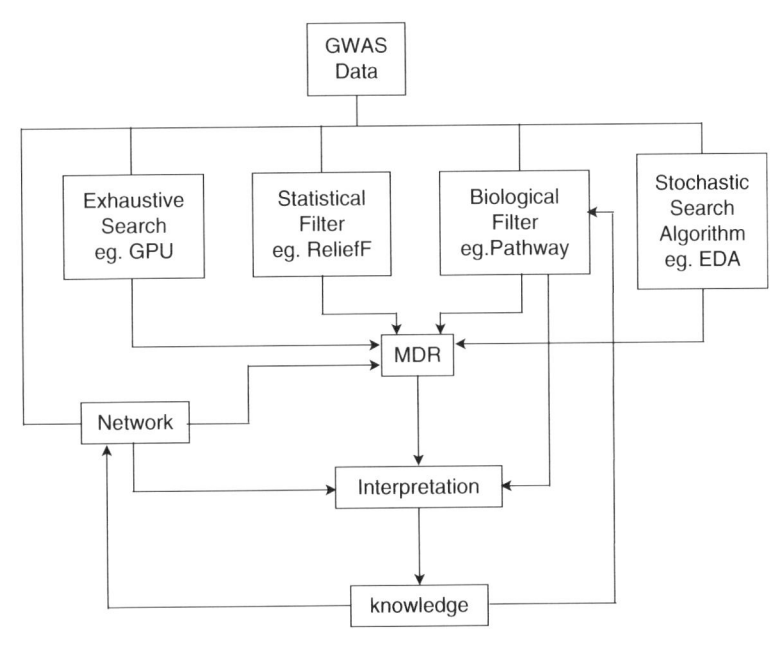

Fig. 1 Flowchart for MDR based bioinformatics analyses of GWAS data. The use of statistical filter, biological filter, and stochastic search algorithm efficiently reduce the number of SNPs to be tested by MDR. GPU based MDR provides an exhaustive search for all two-way interactions. Differently, a network built from either biological knowledge or statistical epistasis captures the entire pairwise interaction space first and then provides guidance for other computational modeling. The interpretation of MDR results produces new biological knowledge that plays a very important role at all levels of the analysis and interpretation

Set all weights $W[A] = 0$
For $i = 1$ to m do begin
    randomly select an instance R_i
    find nearest hit H and nearest miss M
    for $A = 1$ to a do
        $W[A] = W[A] - \text{diff}(A, R_i, H)/m + \text{diff}(A, R_i, M)/m$
end
The function $\text{diff}(A, I_1, I_2) = 0$ if $\text{genotype}(A, I_1) = \text{genotype}(A, I_2)$,
1 otherwise.

Kononenko improved upon Relief by choosing n nearest neighbors instead of just one [51–53]. This new ReliefF is more robust to noisy attributes and is widely used in data mining applications now. Several modifications and extensions of the ReliefF algorithm have been proposed for genome-wide analysis. Moore and White proposed a "tuned" ReliefF algorithm (TuRF) that systematically removed attributes that have low quality estimates so that the ReliefF values in the remaining attributes can be reestimated [54]. McKinney et al. proposed a hybrid ReliefF algorithm that uses measures of entropy to boost performance [55]. Greene et al. developed a spatially uniform ReliefF (SURF) algorithm that uses a fixed distance for picking neighbors [56]. Greene et al.

proposed a SURF* algorithm that extends SURF by also using information from the furthest neighbors [57]. All of these methods have demonstrated the ability to successfully filter interacting SNPs in an efficient way. The ReliefF algorithms described here are available in the open-source MDR software package.

3.1.2 Biological Filter

ReliefF and other measures such as interaction information are likely to be very useful for providing an analytical means for filtering genetic variations prior to epistasis analysis using MDR. However, the wealth of accumulated knowledge about gene function can also be useful to prioritize which genetic variations should be analyzed for gene–gene interactions [39]. For any given diseases there are often multiple biochemical pathways, for instance, that have been experimentally confirmed to play an important role in the etiology of the disease. Genes in these pathways can be selected for gene–gene interaction analysis thus significantly reducing the number of gene–gene interaction tests that need to be performed. Gene Ontology (GO), chromosomal location, and protein–protein interactions are all example sources of expert knowledge that can be used in a similar manner. For instance, Pattin and Moore have specifically reviewed protein–protein interaction databases as a source of expert knowledge that can be used to guide genome-wide association studies of epistasis [58]. Bush et al. have developed a Biofilter algorithm for integrating expert knowledge for gene–gene interaction analysis [59]. Askland et al. have shown how the exploratory visual analysis (EVA) method can be used to select SNPs in specific pathways and GO groups [60].

3.2 Stochastic Search Approach

Stochastic search or wrapper methods may be more powerful than filter approaches because no attributes are discarded in the process. Instead, every attribute retains some probability of being selected for evaluation by the classifier. Although there are many different stochastic wrapper algorithms that can be applied to this problem [61], when interactions are present in the absence of marginal effects, there is no reason to expect that any wrapper method would perform better than a random search because there are no "building blocks" for this problem when accuracy is used as the fitness measure. That is, the fitness of any given classifier would look no better than any other with just one of the correct SNPs in the MDR model.

We review below a stochastic search method based on estimation of distribution algorithms (EDAs). The general idea of EDAs is that solutions to a problem are statistically modeled and the ensuing probability distribution function is used to generate new models. Greene et al. compared an EDA that utilizes only accuracy from an MDR classifier to update the probabilities with which SNPs are selected with an EDA that uses both accuracy and preprocessed TuRF scores [62]. The results indicate that the EDA approach using TuRF scores as expert knowledge significantly outperformed the approach that just used MDR accuracies. Here, TuRF scores provide the building blocks that are needed to point the search algorithm in the right direction. More recent studies have focused on the optimal use of expert knowledge in the EDA algorithm [62]. The EDA algorithms described here are available in the open-source MDR software package.

3.3 Exhaustive Search Approach: GPU Based MDR

Filter and wrapper approaches provide an algorithmic way to apply MDR to genome-wide scale data via feature selection; however, they do not achieve exhaustive search on all possible combinations since this is computationally intractable. An exhaustive search of even only pairwise interactions in a 550,000 SNP dataset would require the analysis of a maximum of 1.5×10^{11} combinations. While an analysis of this scale is approachable with modern cluster computing an analysis that includes permutation testing to assess the statistical significance of results remains infeasible with CPU-based approaches. As an alternative, Sinnott-Armstrong et al. examined the use of commodity hardware designed for computer graphics [63]. In modern computers Graphics Processing Units (GPU) have more memory, bandwidth, and computational capability than Central Processing Units (CPU) and are well-suited for this problem [64]. After showing that Graphics hardware based computing provides a cost effective means to perform genetic analysis of epistasis using MDR by benchmarking, Greene et al. developed the software package MDRGPU to implement GPU based MDR and applied it to a genome-wide analysis of epistasis of sporadic amyotrophic lateral sclerosis (ALS) [65]. The MDRGPU software described here are available in the open-source MDR software package.

3.4 Network Approach

Albeit GPU based MDR is capable of exhaustively exploring all two-way interactions it cannot effectively test all three way or higher order interactions due to the computational costs. Tools that capture the whole interaction space for GWAS are needed. Network science has emerged as a particularly intuitive and promising approach to exploring the entire pairwise interaction space and, consequently, the underlying genetic architecture of complex phenotypic traits. A network is generally defined as a collection of vertices joined in pairs by edges and is often used to represent and study complex interaction systems [66, 67]. It allows for a structured representation of entities and their relationships, which provides a well-suited framework for the study of epistasis.

In epistasis studies, most existing network models are used to represent biological interactions and dependencies using gene expression and regulatory data [68–70]. However for statistical epistasis, networks also have the potential to systematically infer the statistical dependencies between genetic variants and the underlying genetic architecture that predict the phenotypic status. Below we highlight statistical epistasis networks (SEN) proposed by Hu et al. [71] and discuss potential use of SEN.

SEN exploits information-theoretical measures to quantify the pairwise synergetic interactions. In information theoretic terms, the main and interaction effects correspond to the so-called mutual information and information gain [72]. Specifically, $I(A;C)$ denotes the mutual information of variable A's genotype and C, usually the class variable with status *case* or *control* in population-based disease association studies. Intuitively, $I(A;C)$ is the reduction in the uncertainty of the class C due to knowledge about variable A's genotype. Its precise definition is

$$I(A;C) = H(C) - H(C|A),$$

where $H(C)$ is the entropy of C and $H(C|A)$ is the conditional entropy of C given knowledge of A. Entropy and conditional entropy are defined by

$$H(C) = \sum_c p(c) \log \frac{1}{p(c)} \quad \text{and} \quad H(C|A) = \sum_{a,c} p(a,c) \log \frac{1}{p(c|a)},$$

where $p(C)$ is the probability that an individual has class c. $p(a,c)$ is that of having genotype a and class c, and $p(c|a)$ is that of having class c given the occurrence of genotype a. Mutual information $I(A;C)$ takes only nonnegative values. If the class C is independent of a variable A's genotype, $I(A;C) = 0$, i.e., variable A does not predict the disease status. Given the pair of variables A and B, its synergetic interaction strength is the information gain $IG(A;B;C)$, defined as

$$IG(A;B;C) = I(A;B;C) - I(A;C) - I(B;C).$$

As such, $IG(A;B;C)$ is the reduction in the uncertainty, or the information gained, about class C from the genotypes of variables A and B considered together minus each of these variables considered separately. A higher value indicates a stronger synergetic interaction. Note that $IG(A;B;C)$ can take non-positive values. A negative value indicates that the genotypes of two variables tend to vary together (redundant information), while a value of zero indicates either that the genotypes of the two variables are independent or, more likely, that they interact with a mixture of synergy and redundancy. The synergetic part of the mix tends to make the information gain positive while the redundant part lowers the information gain.

Hu et al. use SEN to characterize the space of pairwise interactions among 1,422 SNPs in a large population-based bladder cancer study [71]. In the networks, each vertex corresponds to a SNP, and v_A is used to denote the vertex corresponding to SNP A. An edge linking a pair of vertices, for instance, v_A and v_B, corresponds to an interaction between SNPs *A* and *B*. Weights assigned to each SNP and each pair of SNPs quantify how much of the disease status the corresponding SNP and SNP pair can explain. These weights correspond to the mutual information $I(A;C)$, $I(B;C)$ and information gain $IG(A; B; C)$, where *C* is the class variable with status *case* or *control*. A series of statistical epistasis networks were constructed by incrementally adding edges. These networks were denoted by G_t, where edges between SNPs were added only if their pair weights were greater than or equal to a threshold *t*. The networks G_t grew as the threshold *t* decreased. Statistical significances were determined by comparing networks observed from real data to null networks built from permutated datasets. The derived networks showed interesting properties including a large connected component and an approximately scale-free topology. A scale-free network has many vertices with few neighbors and a few vertices with lots of neighbors, also known as hubs [66]. The existence of hubs in a scale-free network implies strong robustness against failure. Because random vertex removal is very unlikely to affect hubs, the connectivity of the network most likely remains intact. In biological networks, this robustness translates into the resilience of organisms to intrinsic and environmental perturbations. For instance, in protein–protein interaction networks [73], most proteins interact with only one or two other proteins but a few are able to interact with a large number. Such hub proteins are rarely affected by mutations and organisms can remain functional under most perturbations. We also expect transcriptional networks defined by the interaction of regulatory variants to be a source of epistasis [74]. The simultaneous emergence of scale-free topologies in many biological networks suggests that evolution has favored such a structure in natural systems.

Although SEN is developed based on statistical rather than real biochemical interactions, it is encouraging to observe similar topologies between biological and statistical networks. More importantly, the scale-free topology of SEN well supported the evolution perspective of epistasis and indicated the complex yet robust genetic architecture of bladder cancer. Due to these reasons, network science opens up a very promising research direction for modeling epistasis. We expect to see many more future applications of networks to GWAS as these methods can efficiently and systematically analyze high-throughput data. Future extensions include using SEN as a guide for machine learning algorithms such as RF or MDR. The information provided by a network's structure, such as degree, modularity and motifs, and the quantitative weights, will provide a better understanding of the true underlying biological epistasis.

4 Notes

1. For all MDR test, especially when the dataset is big, users may want to run in batch mode with no graphical user interface. MDR can be started by command line:

   java –Xmx1024M –jar mdr.jar

   All available options can be viewed by using command line:

   java -mdr.jar –help.

2. To do permutation test in the GUI mode, a separate program called MDRPT needs to be downloaded. To do permutation test on a cluster, a python script permutations_on_cluster.py is provided in the distribution file.

3. Depending on the size of the data, running MDR on a cluster may be more efficient. A python script called large_file_on_cluster.py is provided in the distribution file to set up MDR runs on a cluster.

4. To interpret MDR results, an interaction dendrogram is provided in the software [27]. This is based on measures of interaction information and information gain. The shorter the line connecting two attributes the stronger the interaction is. The color of a line indicates the type of interaction: red/orange suggests synergistic relationship (i.e., epistasis), yellow suggests independence, and green/blue suggests redundancy or correlation (i.e., due to linkage disequilibrium).

References

1. Hirschhorn JN, Daly MJ (2005) Genome-wide association studies for common diseases and complex traits. Nat Rev Genet 6:95–108
2. Wang WY, Barratt BJ, Clayton DG, Todd JA (2005) Genome-wide association studies: theoretical and practical concerns. Nat Rev Genet 6:109–118
3. Manolio TA (2010) Genome-wide association studies and assessment of the risk of disease. N Engl J Med 363(2):166–176
4. Ritchie MD, Hahn LW, Roodi N, Bailey LR, Dupont WD, Parl FF, Moore JH (2001) Multifactor dimensionality reduction reveals high-order interactions among estrogen metabolism genes in sporadic breast cancer. Am J Hum Genet 69(1):138–147
5. Franke A et al (2010) Genome-wide meta-analysis increases to 71 the number of confirmed Crohn's disease susceptibility loci. Nat Genet 42:1118–1125
6. Eichler EE et al (2010) Missing heritability and strategies for finding the underlying causes of complex disease. Nat Rev Genet 11:446–450
7. Williams SM, Canter JA, Crawford DC, Moore JH, Ritchie MD, Haines JL (2007) Problems with genome-wide association studies. Science 316:1840–1842
8. Moore JH, Williams SM (2009) Epistasis and its implications for personal genetics. Am J Hum Genet 85(3):309–320
9. Moore JH (2010) Bioinformatics challenges for genome-wide association studies. Bioinformatics 26(4):445–455
10. Bateson W, Saunders ER, Punnett RC, Hurst CC (1905) Reports to the Evolution Committee of the Royal Society, report II. Harrison and Sons, London
11. Thornton-Wells TA, Moore JH, Haines JL (2004) Genetics, statistics and human disease: analytical retooling for complexity. Trends Genet 20(12):640–647

12. Phillips PC (2008) Epistasis—the essential role of gene interactions in the structure and evolution of genetic systems. Nat Rev Genet 9:855–867
13. Cordell HJ (2002) Epistasis: what it means, what it doesn't mean, and statistical methods to detect it in humans. Hum Mol Genet 11 (20):2463–2468
14. Cordell HJ (2009) Detecting gene–gene interactions that underlie human diseases. Nat Rev Genet 10:392–404
15. Phillips PC (1998) The language of gene interaction. Genetics 149(3):1167–1171
16. Moore JH, Williams SW (2005) Traversing the conceptual divide between biological and statistical epistasis: systems biology and a more modern synthesis. Bioessays 27(6):637–646
17. Tyler AL, Asselbergs FW, Williams SM, Moore JH (2009) Shadows of complexity: what biological networks reveal about epistasis and pleiotropy. Bioessays 31(2):220–227
18. Gibson G (2009) Decanalization and the origin of complex disease. Nat Rev Genet 10:134–140
19. Moore JH (2005) A global view of epistasis. Nat Genet 37(1):13–14
20. Moore JH (2003) The ubiquitous nature of epistasis in determining susceptibility to common human diseases. Hum Hered 56 (1–3):73–82
21. Teare MD, Barrett JH (2005) Genetic linkage studies. Lancet 336(9940):1036–1044
22. Cordell HJ, Clayton DG (2005) Genetic association studies. Lancet 336(9491):1121–1131
23. Moore JH, Ritchie MD (2004) The challenges of whole-genome approaches to common diseases. J Am Med Assoc 291(13):1642–1643
24. Clark AG, Boerwinkle E, Hixson J, Sing CF (2005) Determinants of the success of whole-genome association testing. Genome Res 15:1463–1467
25. McKinney BA, Reif DM, Ritchie MD, Moore JH (2006) Machine learning for detecting gene–gene interactions: a review. Appl Bioinformatics 5(2):77–88
26. Jiang R, Tang W, Wu X, Fu W (2009) A random forest approach to the detection of epistatic interactions in case–control studies. BMC Bioinformatics 10(Suppl 1):S65
27. Lunetta KL, Hayward LB, Segal J, Eerdewegh PV (2004) Screening large-scale association study data: exploiting interactions using random forest. BMC Genet 5:32
28. Bureau A, Dupuis J, Falls K, Lunetta KL, Hayward B, Keith TP, Eerdewegh PV (2005) Identifying SNPs predictive of phenotype using random forest. Genet Epidemiol 28(2):171–182

29. Breiman L, Friedman JH, Olshen RA, Stone CJ (1984) Classification and regression trees. Chapman and Hall, New York
30. Mitchell T (1997) Machine learning. McGraw-Hill, New York
31. Breiman L (2001) Random Forests. Machine Learning 45(1):5–32
32. Cook NR, Zee RY, Ridker PM (2004) Tree and spline based association analysis of gene–gene interaction models for ischemic stroke. Stat Med 23(9):1439–1453
33. McKinney BA, Crowe JE, Guo J, Tian D (2009) Capturing the spectrum of interaction effects in genetic association studies by simulated evaporative cooling network analysis. PLoS Genet 5:e1000432
34. Strobl C, Boulesteix A, Zeileis A, Hothorn T (2007) Bias in random forest variable importance measures: illustrations, sources and a solution. BMC Bioinformatics 8:25
35. Hahn LW, Ritchie MD, Moore JH (2003) Multifactor dimensionality reduction software for detecting gene–gene and gene–environment interactions. Bioinformatics 19(3):376–382
36. Ritchie MD, Hahn LW, Moore JH (2003) Power of multifactor dimensionality reduction for detecting gene–gene interactions in the presence of genotyping error, phenocopy, and genetic heterogeneity. Genet Epidemiol 24 (2):150–157
37. Hahn LW, Moore JH (2004) Ideal discrimination of discrete clinical endpoints using multilocus genotypes. In Silico Biol 4:183–194
38. Moore JH (2004) Computational analysis of gene–gene interactions in common human diseases using multifactor dimensionality reduction. Expert Rev Mol Diagn 4(6):795–803
39. Moore JH et al (2006) A flexible computational framework for detecting characterizing, and interpreting statistical patterns of epistasis in genetic studies of human disease susceptibility. J Theor Biol 241:252–261
40. Moore JH et al (2007) Genome-wide analysis of epistasis using multifactor dimensionality reduction: feature selection and construction in domain of human genetics. In: Zhu X, Davidson I (eds) Knowledge Discovery and Data Mining: Challenges and Realities, IGI Global 17–30
41. Moore JH (2010) Detecting, characterizing, and interpreting nonlinear gene–gene interactions using multifactor dimensionality reduction. Adv Genet 72:101–116
42. Velez DR, White BC, Motsinger AA, Bush WS, Ritchie MD, Williams SM, Moore JH (2007) A balanced accuracy function for epistasis modeling in imbalanced datasets using multifactor

dimensionality reduction. Genet Epidemiol 31(4):306–315
43. Greene CS, Himmelstein DS, Nelson HH, Kelsey KT, Williams SM, Andrew AS, Karagas MR, Moore JH (2010) Enabling personal genomics with an explicit test of epistasis. Pac Symp Biocomput 2010:327–336
44. Gui J, Andrew AS, Andrews P, Nelson HM, Kelsey KT, Karagas MR, Moore JH (2011) A robust multifactor dimensionality reduction method for detecting gene–gene interactions with application to the genetic analysis of bladder cancer susceptibility. Ann Hum Genet 75(1):20–28
45. Gui J, Moore JH, Kelsey KT, Marsit CJ, Karagas MR, Andrew AS (2011) A novel survival multifactor dimensionality reduction method for detecting gene–gene interactions with application to bladder cancer prognosis. Hum Genet 129(1):101–110
46. Gui J, Andrew AS, Andrews P, Nelson HM, Kelsey KT, Karagas MR, Moore JH (2010) A simple and computationally efficient sampling approach to covariate adjustment for multifactor dimensionality reduction analysis of epistasis. Hum Hered 70(3):219–225
47. Calle ML, Urrea V, Malats N, Van Steen K (2010) mbmdr: an R package for exploring gene–gene interactions associated with binary or quantitative traits. Bioinformatics 26(17):2198–2199
48. Cattaert T, Calle ML, Dudek SM, Mahachie John JM, Van Lishout F, Urrea V, Ritchie MD, Van Steen K (2011) Model-based multifactor dimensionality reduction for detecting epistasis in case–control data in the presence of noise. Ann Hum Genet 75(1):78–89
49. Lou XY, Chen GB, Yan L, Ma JZ, Zhou J, Elston RC, Li MD (2007) A generalized combinatorial approach for detecting gene-by-gene and gene-by-environment interactions with application to nicotine dependence. Am J Hum Genet 80(6):1125–1137
50. Kira K, Rendell LA (1992) A practical approach to feature selection. Proceedings of the ninth international workshop on machine learning, pp 249–256
51. Kononenko I (1994). Estimating attributes: analysis and extension of Relief. Proceedings of the European conference on machine learning, pp 171–182
52. Robnik-Siknja M, Kononenko I (2003) Theoretical and empirical analysis of ReliefF and RReliefF. Machine Learning 53:23–69
53. Robnik-Sikonja M, Kononenko I (2001) Comprehensible interpretation of Relief's estimates. Proceedings of the eighteenth international conference on machine learning, pp 433–440
54. Moore JH, White BC (2007) Tuning ReliefF for genome-wide genetic analysis. Lect Notes Comput Sci 4447:166–175
55. McKinney BA, Reif DM, White BC, Crowe JE Jr, Moore JH (2007) Evaporative cooling feature selection for genotypic data involving interactions. Bioinformatics 23(16):2113–2120
56. Greene CS et al (2008) Spatially uniform ReliefF (SURF) for computationally-efficient filtering of gene–gene interactions. BioData Min 2:5
57. Greene CS, Himmelstein DS, Kiralis J, Moore JH (2010) The informative extremes: using both nearest and farthest individuals can improve Relief algorithms in the domain of human genetics. Lect Notes Comput Sci 6023:182–193
58. Pattin KA, Moore JH (2008) Exploiting the proteome to improve the genome-wide genetic analysis of epistasis in common human diseases. Hum Genet 124:19–29
59. Bush WS, Dudek SM, Ritchie MD (2009) Biofilter: a knowledge-integration system for the multi-locus analysis of genome-wide association studies. Pac Symp Biocomput 368–379
60. Askland K, Read C, Moore J (2009) Pathways-based analyses of whole-genome association study data in bipolar disorder reveal genes mediating ion channel activity and synaptic neurotransmission. Hum Genet 125:63–79
61. Michalewicz Z, Fogel DB (2004) How to solve it: modern heuristics. Springer, New York
62. Greene CS et al (2009) Optimal use of expert knowledge in ant colony optimization for the analysis of epistasis in human disease. Lect Notes Comput Sci 5483:92–103
63. Sinnott-Armstrong NA, Green CS, Cancare F, Moore JH (2009) Accelerating epistasis analysis in human genetics with consumer graphics hardware. BMC Res Notes 2:149
64. Payne JL, Sinnott-Armstrong NA, Moore JH (2010) Exploiting graphics processing units for computational biology and bioinformatics. Interdiscip Sci 2(3):213–220
65. Greene CS et al (2010) Multifactor dimensionality reduction for graphics processing units enables genome-wide testing of epistasis in sporadic ALS. Bioinformatics 26:694–695
66. Newman MEJ (2010) Networks: an introduction. Oxford University Press, New York
67. Strogatz SH (2001) Exploring complex networks. Nature 410:268–276

68. Andrei A, Kendziorski C (2009) An efficient method for identifying statistical interactors in gene association networks. Biostatistics 10:706–718
69. Chu JH et al (2009) A graphical model approach for inferring large-scale networks integrating gene expression and genetic polymorphism. BMC Syst Biol 3:55
70. Schafer J, Strimmer K (2005) An empirical Bayes approach to inferring large-scale gene association. Bioinformatics 21(6):754–764
71. Hu T, Sinnott-Armstrong NA, Kiralis JW, Andrew AS, Karagas MR, Moore JH (2011) Characterizing genetic interactions in human disease association studies using statistical epistasis networks. BMC Bioinformatics 12:364
72. Cover TM, Thomas JA (2006) Elements of information theory, 2nd edn. Wiley, New York
73. Jeong H et al (2001) Lethality and centrality in protein networks. Nature 411:41–42
74. Cowper-Sal lari R, Cole MD, Karagas MR, Lupien M, Moore JH (2011) Layers of epistasis: genome-wide regulatory networks and network approaches to genome-wide association studies. Wiley Interdiscip Rev Syst Biol Med 3(5):513–526

Chapter 23

Applications of Multifactor Dimensionality Reduction to Genome-Wide Data Using the R Package 'MDR'

Stacey Winham

Abstract

This chapter describes how to use the R package 'MDR' to search and identify gene–gene interactions in high-dimensional data and illustrates applications for exploratory analysis of multi-locus models by providing specific examples.

Key words Multifactor dimensionality reduction, R Statistical Software, Gene–gene interactions, Epistasis, Variable selection, High-dimensional data, Data-mining, Machine learning

1 Introduction

Detecting genetic variation associated with complex human traits is a primary objective in studies of genetic epidemiology, which is currently addressed with genome-wide association studies that produce high-dimensional data. Complex human diseases such as breast cancer and heart disease have already been determined to be associated with a number of clinical, genetic, and environmental risk factors but these known risk factors do not explain the full etiology of the diseases. Biological systems are collections of complex networks and pathways that interact in complicated ways and many common complex diseases are thought to have complex genetic architectures involving large numbers of variants associated with a number of genes and biological pathways. Gene–gene, and particularly SNP–SNP, interactions are one type of complex genetic model which may be an important component of understanding the etiology of common complex diseases [1–3].

In a general sense, a gene–gene interaction (epistasis) occurs when the effect of one genetic variant on the phenotype depends on the values of one or more other genetic variants; more specifically, statistical epistasis is defined as a deviation from an additive allelic effect on the phenotype at different loci. Currently many analysis approaches for genome-wide data focus on variable selection of genetic risk factors associated with disease by considering only univariate associations, and ignore potentially complex joint models such as gene–gene interactions. The high-dimensionality of the data poses difficult computational and statistical challenges, which are exacerbated in the case of interactions by exponential increases in the search space of potential variables and variable combinations.

One popular class of methods is combinatorial techniques, which exhaustively consider all combinations of variables. Although these approaches are thorough in their consideration of joint effects, they can be time-consuming and in high-dimensional genome-wide data may be intractable. A frequently used alternative is to consider filtering or variable screening techniques based on these approaches. Multifactor dimensionality reduction (MDR) is one such method, which examines all combinations of potentially interacting loci after implementing a data-reduction step to reduce dimensionality and classifies individuals to disease status based on their genetic information [4]. MDR has frequently been used in real data applications, and many software packages are available, including an R package 'MDR' (available at http://cran.r-project.org/web/packages/MDR/) [5]. In this chapter we will outline how the MDR method can be applied to high-dimensional data to identify complex models, and how to use the R package using an example dataset. We will utilize MDR as a filtering technique and also apply MDR after a previous filtering/screening step, and demonstrate how to use the 'MDR' package for exploratory data analysis.

1.1 Multifactor Dimensionality Reduction

MDR is used to investigate the relationship between a binary phenotype (such as disease status) and the joint categorical genotype across a set of SNPs. MDR reduces the dimensionality of the data by first exhaustively considering a set of combinations of loci that may interact, and viewing these combinations as a series of multifactorial genotype classes. For instance, with K SNPs considered for interaction with three genotypes each, 3^K genotype classes are possible. MDR uses constructive induction, a machine learning technique that constructs a new variable from available data, and creates a new binary variable from these multifactor genotype classes by comparing the ratio of the number of cases to controls within each class to a specified threshold. Genotype classes with ratios above the threshold (i.e., with more cases than controls) will be assigned 'high-risk,' and classes with ratios below the threshold (i.e., with fewer cases than controls) will be assigned 'low-risk' [6]. This newly constructed high-risk/low-risk variable has reduced the 3^K genotype classes to one-dimensional space, and this new binary variable is used to classify individuals. This reduction is equivalent to constructing a Naïve-Bayes type classifier when the data is balanced [7]. A measure of the accuracy of the classifier is evaluated, and SNP variable combinations are ranked by maximizing this accuracy. To prevent over-fitting to the sample data, internal validation measures on independent test data are used to estimate prediction accuracy.

Currently software is available to implement MDR in a number of computer languages. A user-friendly GUI implementation written in JAVA is available at http://www.epistasis.org. Additionally, an implementation using graphics processing units for parallel processing, MDRGPU, designed for more computationally efficient execution on high-dimensional data is available at http://www.sourceforge.new/projects/mdr [8]. While the traditional application of MDR is nonparametric, an R package is also available for a parametric extension, model-based MDR ('mbmdr') [9, 10]. In addition to genetic analysis of binary traits, model-based MDR can also incorporate covariates under a regression framework and can be applied to quantitative traits (*see* **Note 1**). In this chapter, we will discuss the R package 'MDR' to implement the nonparametric method for variable selection of interactions in high-dimensional data for binary traits [4, 5], which is available at http://cran.r-project.org/web/packages/MDR/. We will discuss how to use R to perform MDR analysis on genome-wide data, describe the structure of the R package, and provide an example of how to carry out these calculations.

2 R Package 'MDR' Overview

The free and open-source R statistical software is one of the most widely used statistical software environments. The open-source nature of the environment is particularly attractive for quickly evolving subdisciplines, such as bioinformatics, where extensions to current methods are constantly developed and analysis strategies involve integration of a multitude of different approaches. The 'MDR' package was designed with these characteristics in mind to provide an MDR implementation for R users with great flexibility and utility for both data analysis and research (*see* **Notes 2** and **3**). The entire workflow, including data cleaning, quality control, LD pruning, and association analysis with MDR and follow-up or comparison analyses with other statistical approaches can be accomplished in one environment (*see* **Notes 2** and **4**). The package 'MDR' provides options for internal validation and functions to summarize the fit and perform post-hoc inference. Because R is open-source, the code can be modified as extensions are developed.

The structure of the 'MDR' R package includes a base 'mdr' function used to fit the MDR model, wrappers of this base function for internal validation options, and post-hoc methods to summarize and visualize the fit. The base function 'mdr' is used specifically to fit a list of MDR models, ranked with balanced accuracy (BA), which is the arithmetic mean of sensitivity and specificity (*see* **Note 5**). The package assumes binary case–control data with categorical predictor variables. The binary response variable is coded as 0 or 1, and the categorical predictors (typically SNP genotypes) are coded numerically (0/1/2, unless otherwise specified by the user) (*see* **Note 4**). Missing SNP values ('NA') are allowed, but will be excluded when evaluating relevant genotype combinations. The particular genotype encoding and the threshold ('ratio') for assigning high-risk/low-risk status can be specified by the user (*see* **Note 6**).

Over-fitting to a particular dataset is a general concern and therefore the base 'mdr' function can be implemented via a wrapper for internal validation, using either k-fold cross-validation ('mdr.cv') or three-way split internal validation ('mdr.3WS') (*see* **Note 7**). In k-fold cross-validation with 'mdr.cv,' the data are randomly split into k intervals (tuning parameter 'cv,' *see* **Note 6**), where $k-1$ intervals are used for training and one interval is used for testing [11]. The best MDR model is determined from the training set and an estimate of the model's prediction accuracy is calculated from the testing set, and this procedure is repeated for all k possible splits of the data. Three-way split internal validation with 'mdr.3WS' is less computationally intensive. The data are randomly split into three sets for training, testing, and validation (*see* **Note 8**) [12]. MDR is first implemented in the training set for all possible combinations of loci and the x models with the highest balanced accuracy are evaluated in the testing set. The predictive ability of the best model in the testing set is then estimated in the validation set. Both internal validation methods create objects of class 'mdr,' which includes a list of the final selected model loci and its prediction accuracy, the top models and their prediction accuracies, and the high-risk/low-risk characterization of the final model.

Three methods exist for objects of class 'mdr': 'summary,' 'plot,' and 'predict' (*see* **Note 9**). The 'summary' method provides a table summarizing the model fit at each degree of interaction (i.e., single locus, pair-wise, three loci, etc.). The 'plot' method provides a contingency table of bar graphs for the final model, portraying the numbers of cases and controls in each genotype combination. The 'predict' method allows the user to predict case–control status on a new dataset using a previously fit 'mdr' object. Additional functions are also available to provide inference after an MDR model has been fit. Significance testing of the prediction accuracy estimate can be assessed with permutation testing using 'permute.mdr.' Furthermore, estimates of prediction accuracy obtained from retrospective case–control data, which may not reflect the true accuracy of prospective predictions, can be adjusted with a user-specified population prevalence rate estimate using one of the two post-hoc procedures implemented in 'boot.error' and 'mdr.ca.adj' [13].

3 Using 'MDR' for Genome-Wide Data

The 'MDR' package was initially designed for candidate gene studies, but can also be used for high-dimensional and genome-wide data. With some package modifications (see **Notes 2** and **3**), larger datasets and more intensive computation can be accommodated. Genome-wide data can be analyzed with MDR by ranking all possible SNP pairs and evaluating the top pairs for statistical significance. Previously the MDR method has been executed on genome-wide data for a study of ALS using this strategy [8]. After quality control and LD pruning, the analysis included pair-wise evaluation and permutation testing of 210,382 SNPs on the entire detection dataset without internal validation, and any "significant" SNPs in this set were evaluated in an independent test dataset.

However, a combinatorial technique such as MDR will not be feasible on genome-wide data beyond pair-wise interactions; therefore it is notable to take into account the orders of magnitude of interaction you wish to examine, because higher-order terms may involve a two-stage filtering plan (*see* **Note 10**). One strategy would be to use MDR as a screening tool to rank all possible pair-wise combinations, where the top-ranking pairs would be considered in an independent analysis (either using another analysis tool or MDR for higher-order interactions) [14]. Alternatively, a univariate screening procedure could be performed prior to MDR analysis, and follow-up of the top ranking SNPs could be executed with MDR considering higher-order interactions. For the remainder of the chapter, we will consider examples of how to use the R package 'MDR' to analyze genome-wide data with some of the aforementioned strategies.

4 Example High-Dimensional MDR Analysis

To illustrate the application of the 'MDR' package, we will utilize an example data-set simulated to reflect a realistic scenario for a complex trait likely to be encountered in real data analysis. The disease of interest has estimated prevalence of 10 % and broad sense heritability of 12 % and is assumed to be due to the common disease, common variant hypothesis, controlled by many common genetic variants with weak to moderate effects. We will assume the data comes from a retrospective case–control study and has been cleaned and formatted for MDR analysis (*see* **Note 4**), and proper quality control including LD pruning (*see* **Note 11**) has already been executed. The data consists of 2,000 samples (balanced with 1,000 cases/1,000 controls) with genotypes available at 5,000 SNPs, coded 0, 1, 2 and the disease phenotype coded 0/1. To account for over-fitting, the data has already been randomly separated into a training set and testing set. This data is available from the 'MDR' package, named 'mdr.train' and 'mdr.test.'

We begin by loading 'MDR' and both the training and testing data.

```
> library(MDR)
> train <-data(mdr.train)
> test <-data(mdr.test)
```

Both 'test' and 'train' are data frames with 1,000 rows and 5,001 columns, where the first column stores the phenotypes and the remaining 5,000 columns store the genotypes. We will first focus on a filter strategy for all possible pair-wise interactions with MDR. We rank the pairs using the base 'mdr' function without internal validation, as described in [8, 14].

```
> n.snps <-dim(train)[2]-1
> pair <-t(combn(n.snps, 2)) #generate all possible pairs of SNPs
> fit.pairs.train <-mdr(split = train, comb = pair, x = dim(pair)[1], ratio = 1, equal = "HR", genotype = c(0, 1, 2))
> head(fit.pairs.train$models) #view the top-ranking SNP pairs
     [,1] [,2]
[1,]    3    4
[2,]    1    2
[3,]    5    6
[4,]    2 1178
[5,]    2    4
[6,]    2 2202
> head(fit.pairs.train$'balanced accuracy') #view top accuracies
[1] 64.85770 64.25062 61.92874 61.58958 61.54020 61.27430
```

These ranked pairs provide a potential list for screening, where SNPs in the top *x* pairs could later be considered as candidate SNPs in further and independent analyses. In the training data, SNP pair 3 and 4 and pair 1 and 2 have the highest ranking balanced accuracies, with 64.85 and 64.25, respectively. However, these are simply the top-ranking pairs, which do not automatically imply association—this ranking could be due to chance variation. To assess whether these SNPs may be truly predictive, we evaluate the performance of the top-ranking pair in the test set by using the function 'mdr', where the top combination of loci (3 and 4) is prespecified.

```
##########################################
# Evaluate the top pair in the testing set
##########################################
> top <-t(fit.pairs.train$models[1,])
> print(top)
     [,1] [,2]
[1,]    3    4
```

```
> fit.pairs.test <-mdr(split = test, comb = top, x =
dim(pair)[1], ratio = 1, equal = "HR", genotype = c(0,
1, 2))
> head(fit.test$'balanced accuracy')
[1] 65.83305
```

Notice how the balanced accuracy of SNPs 3 and 4 in the testing data is similar to that in the training data (64.9 and 65.8 %), and both seem to be much greater than the 50 % expected by chance; if there were no association between the genotype combinations and the phenotype, we could expect a classifier to correctly predict the binary phenotype one out of every two times due to random chance. To evaluate the significance of the BA estimate for a model with Loci 3 and 4, we perform a permutation test:

```
> perm <-permute.mdr.new(accuracy = fit.pairs.test$'
balanced accuracy', loci = top, N.permute = 1000, method
= "none", data = test)
> perm$'Permutation P-value'
[1] 0
> max(perm$'Permutation Distribution')
[1] 57.1863
```

The general permutation test procedure simply evaluates the joint significance of Loci 3 and 4, and does not explicitly test for interaction. We can also test explicitly for an interaction between Loci 3 and 4 based on a likelihood ratio test [15] using the 'LRT' option:

```
> perm.intx <-permute.mdr.new(accuracy = fit.pairs.
test$'balanced accuracy', loci = top, N.permute = 1000,
ethod = "none", data = test, LRT = TRUE)
> perm.intx$'LRT P-value'
[1] 0
```

Using 1,000 permutations, we have evidence that the BA estimate of 65.8 % for SNPs 3 and 4 is significantly greater than 50 %, and significant evidence for an interaction between SNPs 3 and 4 ($p < 0.001$).

Particularly because we found evidence of potential epistasis, we may want to compare the MDR results to univariate rankings with p-values from logistic regression in the training data. For all 5,000 SNPs, we fit a univariate logistic regression model and store the odds ratio and p-value results.

```
> pvalue <-or <-rep(0, n.snps)
> names(pvalue) <-names(or) <-1:n.snps
>
> for (i in 1:n.snps) {
+ y <-train[,1]
+ x <-train[,(i + 1)]
+ fit.lr <-glm(y ~ x, family = binomial(link = logit))
+ pvalue[i] <-summary(fit.lr)$coefficients[2,4]
```

```
+ or[i] <-exp(summary(fit.lr)$coefficients[2,1])
+ }
>
```

We rank the SNPs by increasing *p*-value, and display the results for the top ten SNPs, including SNP index, *p*-value estimated with logistic regression, and Bonferroni corrected *p*-value to account for multiple testing.

```
> rank.p <-sort(pvalue,index.return = TRUE)
>    top.snps <-cbind(rank.p$ix,rank.p$x,(rank.p$x)
*5000)[1:10,]
> colnames(top.snps) <-c("SNP","P-value","Corrected")
> print(top.snps)
         SNP        P-value        Corrected
2          2    5.344528e-09    2.672264e-05
4          4    5.283348e-06    2.641674e-02
4233    4233    1.750890e-04    8.754449e-01
1390    1390    1.832230e-04    9.161152e-01
782      782    2.609099e-04    1.304549e+00
2814    2814    6.192238e-04    3.096119e+00
1148    1148    6.261407e-04    3.130704e+00
4631    4631    6.422279e-04    3.211139e+00
1801    1801    8.184807e-04    4.092403e+00
4491    4491    8.483338e-04    4.241669e+00
```

Notice that SNPs 2 and 4 are ranked highly, which were also ranked highly in the training data by pair-wise MDR.

Results from MDR analysis can be easily integrated with more traditional statistical analyses using additional functions already available in R, such as 'glm.' For example, after fitting both logistic regression and pair-wise MDR in the training data, we may be interested in post-hoc analyses using logistic regression to better characterize some of these results and provide odds ratio estimates in the testing data. Notice that it is important that we consider these post-hoc analyses in independent test data, because post-hoc estimates in the training data may be biased due to the selection/ranking process. First we obtain univariate odds ratio estimates for SNPs 1, 2, 3, and 4, the top ranking SNPs with MDR:

```
> y <-test[,1] #define the response vector
> x <-test[,-1] #define the predictor matrix
> fit.1 <-glm(y ~ x[,1],family = binomial(link = logit))
> fit.2 <-glm(y ~ x[,2],family = binomial(link = logit))
> fit.3 <-glm(y ~ x[,3],family = binomial(link = logit))
> fit.4 <-glm(y ~ x[,4],family = binomial(link = logit))
```

We summarize each of these univariate 'glm' fits, and observe that in the test data SNPs 1, 2, 3, and 4 all have estimated log odds that are significantly less than 0. For SNP 2 this is particularly interesting, because the pair-wise MDR results indicated many high ranking two-way models including SNP 2, which may indicate a potential marginal effect (*see* **Note 12**).

```
> summary(fit.1)

Coefficients:
             Estimate Std. Error z value Pr(>|z|)
(Intercept)   0.08694    0.09719   0.895    0.371
x[, 1]       -0.40941    0.09489  -4.315  1.6e-05 ***
---
Signif. codes: 0 *** 0.001 ** 0.01 * 0.05 . 0.1 1

> summary(fit.2)

Coefficients:
             Estimate Std. Error z value  Pr(>|z|)
(Intercept)   0.05479    0.09858   0.556   0.57834
x[, 2]       -0.35349    0.09324  -3.791   0.00015 ***
---
Signif. codes: 0 *** 0.001 ** 0.01 * 0.05 . 0.1 1

> summary(fit.3)

Coefficients:
             Estimate Std. Error  z value  Pr(>|z|)
(Intercept) -0.01023    0.09632   -0.106   0.91544
x[, 3]      -0.29916    0.09782   -3.058   0.00223 **
---
Signif. codes: 0 *** 0.001 ** 0.01 * 0.05 . 0.1 1

> summary(fit.4)

Coefficients:
             Estimate Std. Error  z value  Pr(>|z|)
(Intercept) -0.02952    0.09955   -0.297   0.76683
x[, 4]      -0.24323    0.09205   -2.642   0.00824 **
```

Although we observe interesting univariate results, recall that the top-ranking MDR results implicated the combinations of SNPs 1/2 and 3/4. We can also examine these pairs in the test data, and we find evidence of statistically significant interactions (consistent with the results from the permutation test):

```
> fit.12 <- glm(y ~ x[,1] + x[,2] + x[,1]*x[,2], family
= binomial(link = logit))
> summary(fit.12)

Coefficients:
Estimate Std. Error z value Pr(>|z|)
(Intercept)   -0.4470   0.1565  -2.856   0.00429 **
x[, 1]         0.6967   0.1542   4.517  6.26e-06 ***
x[, 2]         0.7391   0.1533   4.821  1.43e-06 ***
x[, 1]:x[, 2] -1.5976   0.1687  -9.467   < 2e-16 ***
```

```
---
Signif. codes: 0 *** 0.001 ** 0.01 * 0.05 . 0.1   1
> fit.34 <-glm(y ~ x[,3] + x[,4] + x[,3]*x[,4],family
  = binomial(link = logit))
> summary(fit.34)

Coefficients:
Estimate Std. Error z value Pr(>|z|)
(Intercept)   -0.7575    0.1619   -4.678    2.89e-06 ***
x[,3]          1.0903    0.1764    6.179    6.44e-10 ***
x[,4]          0.9165    0.1532    5.983    2.20e-09 ***
x[,3]:x[,4]   -1.7690    0.1793   -9.868    < 2e-16 ***
---
Signif. codes: 0 *** 0.001 ** 0.01 * 0.05 . 0.1   1
```

If obtaining a ranked list for filtering is our objective, internal validation may not be necessary. However, if additional inference is desired, we may be more concerned about over-fitting (*see* **Note 7**). In this case, we want to perform our pair-wise analysis with an internal validation measure such as 3WS. First we run 3WS, considering models of size $K \leq 2$ in dataset 'train.'

```
> fit3WS.train <-mdr.3WS(data = train, K = 2)
> summary(fit3WS.train)
Level   Best      Training   Testing    Validation
        Models    Accuracy   Accuracy   Accuracy
1       146       51.81      61.18      52.75
*2      12        64.04      64.16      65.91
'*' indicates overall best model
```

The top model includes SNPs 1 and 2, with estimated BA of 65.91 % (from the Validation Accuracy). We can better visualize the data by plotting the estimated MDR fit.

```
> plot(fit.3WStrain,train)
```

Figure 1 indicates an interaction between SNP 1 and SNP 2, where marginal components are present. For both SNPs, genotype 0 is a risk genotype, but a modifying effect is observed for the multi-locus genotype 0–0. We may be curious in how this fit might compare to the independent dataset 'test.'

```
> plot(fit3WS.train,test)
```

In the independent test data, the plot is similar (Fig. 2), with a similar pattern of high- and low-risk genotype combinations, giving stronger evidence for plausibility of the estimated model.

Because the sample data is retrospective, we may wish to report prospectively adjusted estimates of accuracy to appropriately describe the unbiased predictive ability of the selected model, assuming a disease prevalence of 0.10.

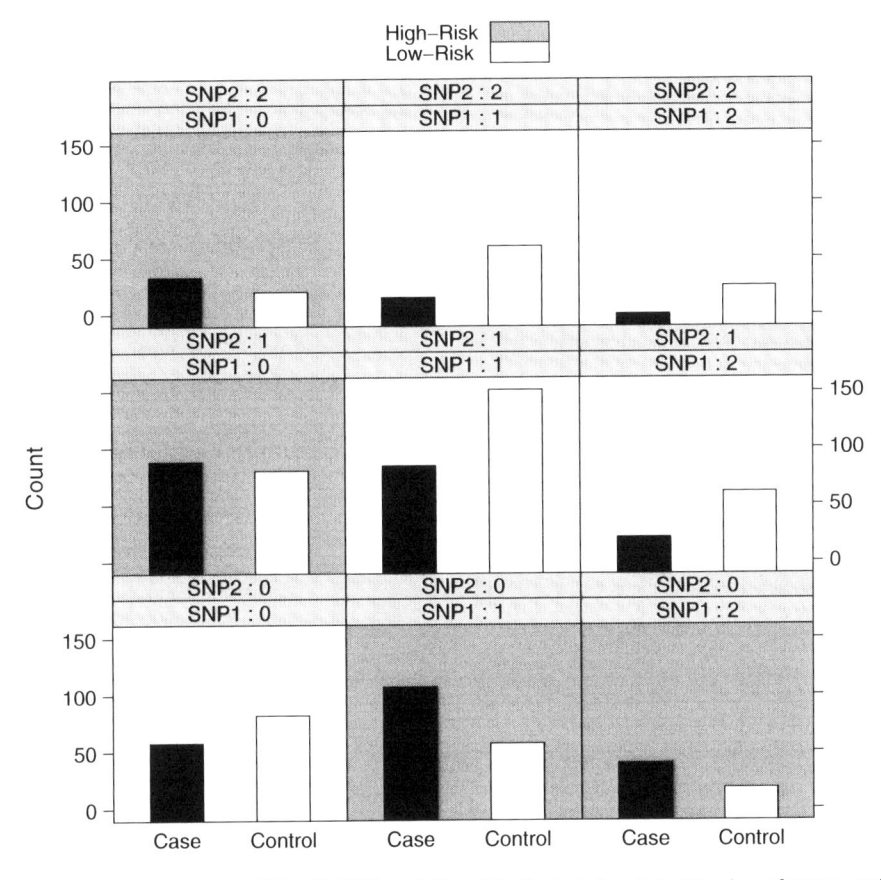

Fig. 1 The top model fit based on MDR with 3WS and $K \leq 2$ in the training data. Number of cases and controls are plotted for each genotype combination of SNPs 1 and 2

```
> mdr.ca.adj(test,prev = 0.10, model = fit3WS.train$'
finalmodel', hr = fit3WS.train$'high-risk/low-risk')
$'adjusted classification accuracy'
[1] 68.08349
$'adjusted classification error'
[1] 31.91651

> boot.error(test,prev = 0.10, model = fit3WS.train
$'final model', hr = fit3WS.train$'high-risk/low-
risk', b = 1000)
$'classification error estimate'
[1] 31.9258
$'classification accuracy estimate'
[1] 68.0742
```

With both approaches, prediction accuracy is estimated to be 68 %, indicating that the retrospective estimate of 65.9 % may be biased low.

In addition to using MDR to generate a ranked list of potential models for follow-up investigation, MDR can also be used for

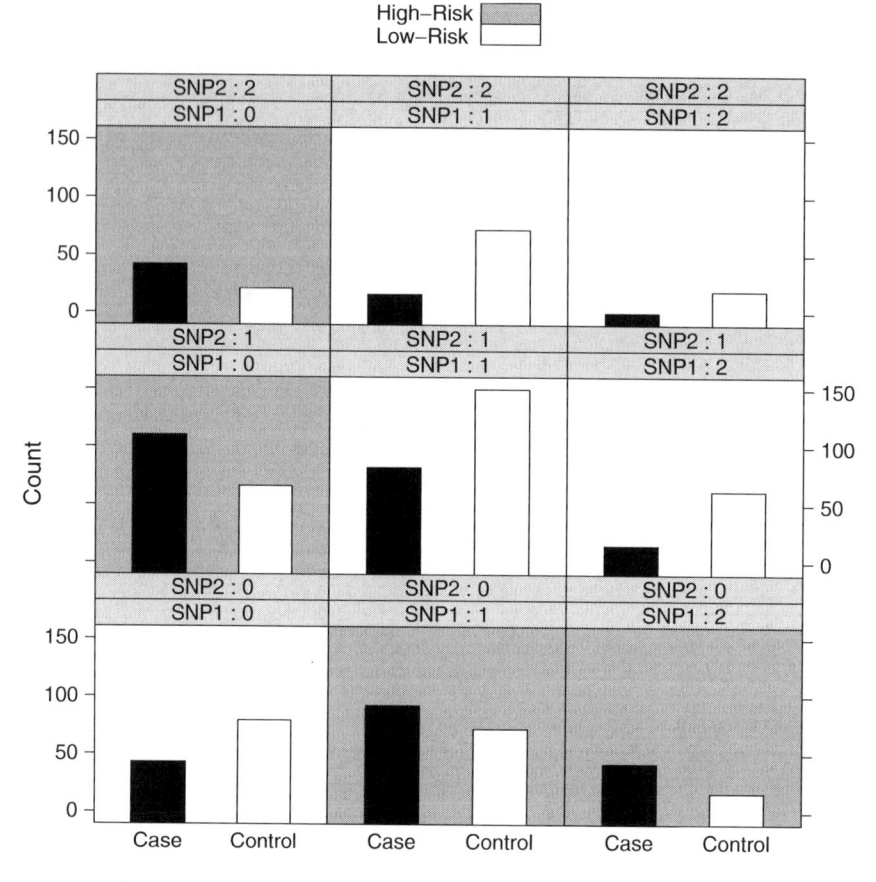

Fig. 2 The top model fit based on MDR with 3WS and $K \leq 2$ in the testing data. Number of cases and controls are plotted for each genotype combination of SNPs 1 and 2

analysis after application of a filtering technique. For instance, we may be interested in higher-order interactions, and we could use our filtered SNP list from the training data to provide a list of candidates to consider when investigating higher-order interactions in the testing data. In the filtering stage, no internal validation was necessary, since the models would be investigated elsewhere and over-fitting was not of great importance (particularly when only pair-wise models are considered). However, over-fitting may be more of a concern for higher-order interactions, and internal validation becomes more critical.

We use the top 100 pairs of SNPs in the 'train' data as a filter for four-way interaction analysis with MDR-3WS.

```
> snps <-fit.pairs.train$models[1:100,]   #Define
list of top pairs
> uniq.snps <-unique(as.vector(snps))
> #### subset training data to the filtered SNPs...
> filter.train <-train[,c(1,uniq.snps + 1)] #add one
for phenotype column
>
```

```
> ####################################################
> # Run MDR with 3WS in the training set, top 100 pairs
> ####################################################
> fit3WS.filter.train <-mdr.3WS(data = filter.train,
K = 4)
```

We determine the identified model and estimated accuracy.

```
> head(fit3WS.filter.train$'final model')
[1] 1 2 4 82
> head(fit3WS.filter.train$'final model accuracy')
[1] 74.14322
```

MDR-3WS identified a four-locus model with very high prediction accuracy; however, because this model was identified in the same dataset from which the filter was estimated, over-fitting is possible. Although not always available in every analysis, we can address this concern by refitting the model in the independent 'test' data.

```
> top <-t(fit.filter.train$'final model')
>
> fit3WS.test <-mdr(split = test,comb = top,x = 1,ratio=
1,equal= "HR",genotype = c(0,1,2))
> fit3WS.test$'models'
[1] 1 2 4 82
> fit3WS.test$'balanced accuracy'
[1] 66.75708
```

The estimated BA in 'test' is much lower, indicating possible over-fitting. In fact, the estimate of 66.8 % is similar to the BA for SNPs 1 and 2 alone. To determine whether SNPs 4 and 82 have additional predictive value, we could obtain a bootstrap confidence interval to assess the significance of this difference, which can easily be integrated into the R analysis. However, other alternatives are possible to determine whether all four SNPs are necessary. It has previously been shown that pruning with step-wise logistic regression with backward selection can improve the performance of MDR-3WS [12], which can also be easily integrated in the R analysis.

First we construct a data-frame from 'test' for regression analysis.

```
> p <-5000
> n <-1000
> data.names <-c("y",paste("snp",1:p,sep = ""))
> test.df <-data.frame(test)
> colnames(test.df) <-data.names
```

Next we define and fit the full logistic regression model with all four loci (*see* **Note 13**).

```
> fit <-
glm(y ~ (snp1 + snp2 + snp4 + snp82)^4, family = binomial(link = logit),
data = test.df)
> summary(fit)
Coefficients: (3 not defined because of singularities)
                  Estimate Std. Error z value Pr(>|z|)
(Intercept)        -0.5707     0.2475   -2.306  0.0211 *
snp1                1.0915     0.2587    4.219  2.46e-05 ***
snp2                1.0724     0.2521    4.254  2.10e-05 ***
snp4                0.1197     0.2190    0.547  0.5847
snp82              14.7102  1528.9564    0.010  0.9923
snp1:snp2          -1.9389     0.2808   -6.905  5.02e-12 ***
snp1:snp4          -0.4105     0.2217   -1.852  0.0640 .
snp1:snp82         29.7970  1765.4868   -0.017  0.9865
snp2:snp4          -0.3727     0.2187   -1.704  0.0884 .
snp2:snp82         -0.8592   882.7434   -0.001  0.9992
snp4:snp82          0.3069  1248.3877    0.000  0.9998
snp1:snp2:snp4      0.3566     0.2467    1.445  0.1484
snp1:snp2:snp82     1.7256  2791.4797    0.001  0.9995
snp1:snp4:snp82        NA        NA       NA      NA
snp2:snp4:snp82        NA        NA       NA      NA
snp1:snp2:snp4:snp82   NA        NA       NA      NA
---
Signif. codes: 0 *** 0.001 ** 0.01 * 0.05 . 0.1   1
```

The data for SNP 82 seems to be quite sparse, based on the large standard error estimates. In fact, higher-order interactions involving SNP 82 are not estimable. Therefore, even if a four-way interaction exists, there is not sufficient available data for proper evaluation, and this indicates a problem with over-fitting (*see* **Note 14**). To determine whether this SNP (or any of the SNPs) can be pruned back, we use backward selection with BIC.

```
> output <- step(fit, direction="backward", k=log(n))

... [Previous steps eliminated]

Step: AIC=1259.91
y ~ snp1 + snp2 + snp4 + snp1:snp2

             Df Deviance    AIC
<none>           1225.4  1259.9
- snp4        1  1233.1  1260.7
- snp1:snp2   1  1330.7  1358.3

> #Extract dependent variable side of model formula
> pruned <- as.character(output$formula[3])
> pruned
[1] "snp1 + snp2 + snp4 + snp1:snp2"
```

The pruning removed SNP 82 from the model, consistent with the overfit logistic regression model Additionally the pruned model gives evidence for a marginal effect of SNP 4 and an interaction between SNPs 1 and 2. Notice the consistency with results from other approaches, which have also identified these three SNPs. An alternative to using MDR-3WS plus pruning would be to consider cross-validation, which is less prone to over-fitting and does not require pruning. However, it is considerably more time-consuming, which is a major concern in high-dimensional data. For k-fold CV, computation time will be approximately k times greater than 3WS [12].

For additional comparison, we also use the top 100 pairs of SNPs in the 'train' data as a filter for four-way interaction analysis with MDR-CV with five-fold CV, and then summarize the results.

```
> ####################################################
> # Run MDR with CV in the training set, top 100 pairs
> ####################################################
>
> fitCV.filter.train<-mdr.cv(data=filter.train,K=4,cv=5)
> summary(fit.filter.train)
  Level   Best       Classification   Prediction   Cross-Validation
          Models     Accuracy         Accuracy     Consistency
  * 1     4          58.81            58.82        5
    2     1 5        65.01            64.44        3
    3     1 4 5      67.63            64.76        4
    4     1 2 4 5    72.50            64.98        3

'*' indicates overall best model
```

Using five-fold CV, the estimated overall best model in the training data includes only SNP 4. This illustrates the affinity for parsimony in the cross-validation procedure due to the incorporation of more data splits and the maximization of cross-validation consistency for model selection. With the increased avoidance of over-fitting, notice that the prediction accuracy is lower than what is seen with both the higher-order models and the models chosen with 3WS. This provides additional evidence for the marginal effect of SNP 4. We plot the results to further investigate the marginal effect of this SNP.

```
> plot(fitCV.filter.train,data=filter.train)
```

In the training data (Fig. 3), the minor allele appears to have a dominant protective effect. We further examine whether these findings are validated in the test data.

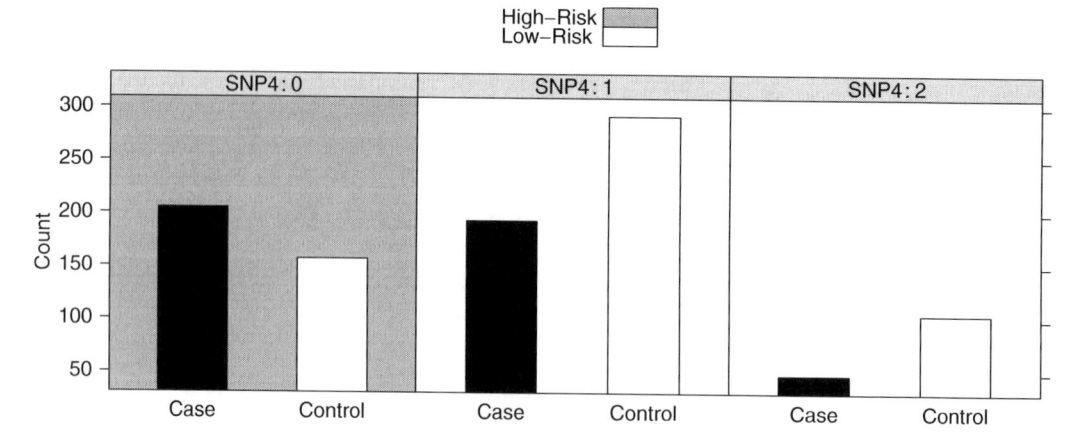

Fig. 3 The top model fit based on MDR with five-fold CV and $K \leq 4$, after performing a filter to select the top 100 SNPs, in the training data. Number of cases and controls are plotted for each genotype combination of SNP 4

```
> top <-t(fit.filter.train$'final model')
>
> fit.test <-
mdr(split = test,comb = top,x = 1,ratio =   1,equal = "
HR",genotype = c(0,1,2))
> fit.test
$models
[1] 4
$'balanced accuracy'
[1] 54.3327

$'high risk/low risk'
[1] 100
```

In the test data, the estimated balanced accuracy is much lower (54 % vs. 59 % in 'train'), which may indicate that SNP 4 is not highly predictive individually, and that additional genetic factors (such as those identified with other strategies) may be important in prediction.

5 Notes

1. Currently there is no standard approach for incorporating covariates into the MDR method. Parametric extensions based on a regression framework such as MB-MDR [10] or GMDR [16] can be utilized in this setting. Alternatively, if the covariate is discrete, stratified analyses could be performed as well as a modified analysis that considers only variable combinations including the covariate.

2. There are computational advantages and disadvantages of the 'MDR' package. Advantages include ease of implementation. Given that R is free and flexible, MDR can easily be integrated with other traditional and novel analysis approaches. Because the code is open source, it can be modified to serve your own needs and functions can be extended as methods are developed. Additionally, many steps of the statistical analysis including data-processing, quality control, and data exploration can all be executed in the same environment. However, this flexibility comes at the cost of computational efficiency. A major disadvantage of the R package 'MDR,' and of the use of R in general for high-dimensional genetic data analysis, is long computation time. R can be very slow, often significantly slower than other competitive languages. For timing experiments [5].

3. Additional R data packages are available to improve computational efficiency. Because R is open source, portions of the source code could be written in the C language. Most of the calculations are independent and could be executed in parallel, and many packages are available for parallel computing including 'foreach,' 'doMC,' and 'doSNOW' (see http://cran.r-project.org/web/views/HighPerformanceComputing.html). Memory limitations in R which may be experienced in high-dimensional datasets can be circumvented by the 'bigmemory' package which allows the user to store and analyze large datasets. These issues are discussed in more detail in [5].

4. Some pre-analysis data processing is necessary to assure that the data will be in the correct format. The binary phenotype needs to be coded as 0/1 and genotype variables need to be coded as integers. Additionally, a list of variable names and order in which they occur in the genotype matrix will need to be preserved, because the output will refer to index rather than variable name. LD pruning is necessary in addition to standard quality control procedures such as tests of HWE and verifying MAF (discussed in more detail below).

5. Let TP = number of true positives, FP = number of false positives, TN = number of true negatives, and FN = number of false negatives. Classification accuracy = (TP + FP)/N is typically used to evaluate performance in balanced data; however, if data is not balanced, classification accuracy will be biased towards the larger class. Balanced accuracy (BA) = (sensitivity + specificity)/2 = (TP/(TP + FN)) + (TN/(TN + FP))/2. Therefore errors in both classes are weighed equally, and BA is used rather than classification accuracy to account for unbalanced data. If the data is balanced, both measures are equivalent.

6. An MDR analysis involves the choice of a number of "tuning" parameters, such as *K*, *cv*, and *ratio*. Many of these have default

values suggested in the literature, but some will again need to be changed based on your analysis strategy, such as K.

7. MDR is a data-mining method and therefore there are inherent concerns with over-fitting. Depending on your specific analysis goals, this may dictate which internal validation method to use; cross-validation in conjunction with permutation testing is more conservative, but more time-consuming. A three-way split strategy or use of the base 'mdr' classifier may suffice if ranking/filtering is your primary goal. If computation time is of primary concern, it is at least recommended that you split your sample in order to refit the trained models on new independent data.

8. Results can vary by randomness (i.e., random splits of the data), so if time allows you may want to consider performing multiple runs.

9. The standard output is limited to preserve memory. Because every analysis plan will depend on the particular dataset and research goals, the standard output from the various functions might not provide everything that is desired and the source code may need to be modified. For instance, all rankings may be desired after a call to 'mdr.cv' rather than just the top ranking models for each value of K. Additionally, some of the methods such as 'summary' and 'plot' are designed only for objects of class 'mdr,' which are only created after a call to 'mdr.cv' or 'mdr.3WS' and not for the base function 'mdr.'

10. A specific analysis plan should be tailored based on research goals. For instance, are you interested in ranking/filtering to prioritize follow-up study, or is significance testing more important? This will determine if internal validation or permutation testing are used. Are you interested in only pair-wise interactions or higher-order models? This will determine the value of K to be used. The maximum K that can be investigated will be dependent on sample size, number of predictors, and amount of available computation time; generally, values $K \geq 4$ will not be feasible computationally if the number of predictors is much greater than 100, or if the sample size is less than 500 (due to sparseness of the data).

11. Pruning is necessary to deal with linkage disequilibrium in MDR to retain reasonable power. High levels of LD will confound the ability of the MDR algorithm to identify direct associations [17], and instead you may pick up the indirect association within the LD block. Therefore multiple SNPs within an associated LD block will compete for high rankings, reducing power to detect the association. Software designed to examine LD structure and select tag SNPs within haplotype blocks can be used to select a reduced set of tag SNPs prior to MDR analysis, such as Haploview [18] and Tagger [19].

12. When internal validation techniques are not used to avoid over-fitting and obtain more parsimonious models, it is still possible to identify potential marginal effects from a ranked list. Particular loci that show up multiple times in the top rankings may indicate a marginal effect (as was seen in the example), especially if there appears to be little variation in accuracy.

13. R is useful for characterizing the k-order models identified with MDR in downstream analyses. MDR analysis simply identifies combinations of loci associated with the phenotype, but does not describe the nature of the association. Formal statistical tests of main effects and lower-order interactions can be performed in R using functions such as 'glm.'

14. In this example, it is possible that the identified four-locus interaction was fit too closely to random patterns in the training data rather than true associations, and it may not generalize well to a broader population. In fact, some of the four-locus genotype combinations are so rare that they were not observed in the testing data, and the model cannot be properly evaluated in the independent set since the high-risk/low-risk status of each combination cannot be determined. The sparseness of many higher-order genotype combinations leads to unstable models, and it is likely that our model is over-fit to the training data. When fitting higher-order models for fixed sample sizes, this example underscores the need to address over-fitting when using MDR.

References

1. Moore JH (2003) The ubiquitous nature of epistasis in determining susceptibility to common human diseases. Hum Hered 56 (1–3):73–82
2. Moore JH, Williams SM (2005) Traversing the conceptual divide between biological and statistical epistasis: systems biology and a more modern synthesis. Bioessays 27(6):637–646
3. Phillips PC (2008) Epistasis—the essential role of gene interactions in the structure and evolution of genetic systems. Nat Rev Genet 9(11): 855–867
4. Ritchie MD, Hahn LW, Roodi N, Bailey LR, Dupont WD, Parl FF, Moore JH (2001) Multifactor-dimensionality reduction reveals high-order interactions among estrogen-metabolism genes in sporadic breast cancer. Am J Hum Genet 69(1):138–147
5. Winham SJ, Motsinger-Reif AA (2011) An R package implementation of multifactor dimensionality reduction. BioData Min 4(1):24
6. Moore JH, Gilbert JC, Tsai CT, Chiang FT, Holden T, Barney N, White BC (2006) A flexible computational framework for detecting, characterizing, and interpreting statistical patterns of epistasis in genetic studies of human disease susceptibility. J Theor Biol 241(2): 252–261
7. Hahn LW, Moore JH (2004) Ideal discrimination of discrete clinical endpoints using multi-locus genotypes. In Silico Biol 4(2):183–194
8. Greene CS, Sinnott-Armstrong NA, Himmelstein DS, Park PJ, Moore JH, Harris BT (2010) Multifactor dimensionality reduction for graphics processing units enables genome-wide testing of epistasis in sporadic ALS. Bioinformatics 26(5):694–695
9. Calle ML, Urrea V, Vellalta G, Malats N, Steen KV (2008) Improving strategies for detecting genetic patterns of disease susceptibility in association studies. Stat Med 27(30): 6532–6546
10. Cattaert T, Calle ML, Dudek SM, John JMM, Van Lishout F, Urrea V, Ritchie MD, Van Steen K (2011) Model-based multifactor dimensionality reduction for detecting epistasis in case-control data in the presence of noise. Ann Hum Genet 75:78–89
11. Motsinger AA, Ritchie MD (2006) The effect of reduction in cross-validation intervals on the performance of multifactor dimensionality reduction. Genet Epidemiol 30(6):546–555

12. Winham SJ, Slater AJ, Motsinger-Reif AA (2010) A comparison of internal validation techniques for multifactor dimensionality reduction. BMC Bioinformatics 11:394
13. Winham SJ, Motsinger-Reif AA (2011) The effect of retrospective sampling on estimates of prediction error for multifactor dimensionality reduction. Ann Hum Genet 75(1):46–61
14. Oki NO, Motsinger-Reif AA (2011) Multifactor dimensionality reduction as a filter-based approach for genome wide association studies. Front Genet 2:80
15. Edwards TL, Turner SD, Torstenson ES, Dudek SM, Martin ER, Ritchie MD (2010) A general framework for formal tests of interaction after exhaustive search methods with applications to MDR and MDR-PDT. PLoS One 5(2):e9363
16. Lou XY, Chen GB, Yan L, Ma JZ, Mangold JE, Zhu J, Elston RC, Li MD (2008) A combinatorial approach to detecting gene-gene and gene-environment interactions in family studies. Am J Hum Genet 83(4):457–467
17. Grady BJ, Torstenson ES, Ritchie MD (2011) The effects of linkage disequilibrium in large scale SNP datasets for MDR. BioData Min 4(1):11
18. Barrett JC, Fry B, Maller J, Daly MJ (2005) Haploview: analysis and visualization of LD and haplotype maps. Bioinformatics 21(2): 263–265
19. de Bakker PI, Yelensky R, Pe'er I, Gabriel SB, Daly MJ, Altshuler D (2005) Efficiency and power in genetic association studies. Nat Genet 37(11):1217–1223

Chapter 24

Higher Order Interactions: Detection of Epistasis Using Machine Learning and Evolutionary Computation

Ronald M. Nelson, Marcin Kierczak, and Örjan Carlborg

Abstract

Higher order interactions are known to affect many different phenotypic traits. The advent of large-scale genotyping has, however, shown that finding interactions is not a trivial task. Classical genome-wide association studies (GWAS) are a useful starting point for unraveling the genetic architecture of a phenotypic trait. However, to move beyond the additive model we need new analysis tools specifically developed to deal with high-dimensional genotypic data. Here we show that evolutionary algorithms are a useful tool in high-dimensional analyses designed to identify gene–gene interactions in current large-scale genotypic data.

Key words Evolutionary algorithms, Decision trees, Machine learning, Genetic algorithm, Epistasis, Nonadditive interactions

1 Introduction

In the last few years, the genomic revolution has provided researchers with an unprecedented amount of data. These include whole genome sequences or genotypes from many individuals representing multiple species, each with tens of thousands to millions of SNP markers, a multitude of phenotypes from direct measurements, or microarray studies. From this magnitude of data, a very small number of studies have led to biological explanations of how the genetic architecture determines complex phenotypes [1–5]. The challenge to develop methods that will enable us to better understand how multiple genes interact with each other and the environment to shape an individual phenotype thus still remains. This chapter will deal with methods that utilize evolutionary algorithms (EAs) for exploring the contribution of epistasis and gene–environment interactions to complex trait expression. These methods may hold the key to help better understand phenotypic variation from a holistic-genetic perspective, or at least support currently available methods.

EAs are meant to supplement, not to replace, current methods used to analyze genomic data. They will be particularly useful in multidimensional datasets when aiming to detect epistasis (*see* also ref. 3]. The main advantage of EAs is the ability to, more effectively than most other methods, deal with the computational complexity that arises in analyses aiming to detect epistasis in genomic size datasets. The term "evolutionary algorithm" refers to a general class of global optimization methods that earlier have been shown to be useful for optimizing models describing the link between genetic and phenotypic variations [3, 6–8]. Machine learning and data mining methods are well suited for model-free explorations of the multidimensional genotype–phenotype map underlying complex traits, and combining these with optimization by an EA is a promising approach. Since no complete synthesis of how these methods can be combined in the analysis of genome-wide association studies (GWAS) exists, we shall here focus on describing and discussing the gaps that can be filled by EAs in current GWAS. We will describe machine learning methods that are suitable for optimization with EAs, how a basic genetic algorithm (GA) can be applied to genomic data, which methods are likely to make efficient use of some particular features of EAs, or are suitable for optimization by it, and conclude by providing some more insight into the technical advantages and disadvantages of using a combination of EAs and machine learning methods in GWAS.

2 Nonadditive Interactions and Dimensionality

Since 2007, GWAS have successfully associated a number of genes to a disease and complex traits in humans and other organisms [1, 4, 5]. The modeling paradigm used in these studies originates from the idea that most complex phenotypes are likely to be affected by a few common variants with large additive effects. However, while the GWAS are powerful for finding a few additively acting genes among thousands of markers, they have little power, given current sample sizes to find even the simplest form of interactions including two loci. Much of the genetic contribution to the phenotypic variation is thus likely to remain unexplained and we need to develop methods that are able to deal with more intricate inheritance modes including high-dimensional gene–gene interactions [7, 9, 10].

When developing methods for dissecting the genetics underlying a complex trait, we first need to consider how the genome can affect a given trait. It is well known that sometimes a single gene or a few additive genes can make a very large contribution to the observed variation in a complex trait. For other traits, however, it has been shown that the contribution of gene–gene interactions (epistasis) [11–13] and gene–environment interactions [14] could be of major importance. There are often, before the start of an analysis, prior knowledge of a number of environmental factors or even candidate genes, which are expected to have an influence on the trait under investigation. These expectations may originate from earlier studies of the same species, population, or trait, or from our understanding of genetics, but such knowledge is currently often not used, due to model restrictions. Optimally, analysis methods should allow for the inclusion of any number of genes and environmental factors and allow them to interact in any conceivable way. Although the additive model is a useful tool for initial analyses aiming to detect main loci, it is too restrictive to form the basis for a method designed to provide a more complete understanding of genotype to phenotype relationship. As it is not likely that a single method will be able to explore all aspects of genetic inheritance in a population, a realistic scenario is that the results of multiple analysis methods will be combined to identify the genes that affect a trait either as single additive genes or as interacting gene–gene or gene–environment complexes. Since some genes or interactions will affect traits to a greater or lesser extent, it is important for the interpretation of the results that their effect sizes can be ranked. Only after combining all extractable information from current and future datasets, our understanding of the path from genotype to phenotype can substantially increase.

When designing a method with the objective to analyze data in order to explore the potentially complex interactions underlying trait variation, it is necessary to consider how to develop the methods that will lead us beyond the traditional GWAS approach and into the high-dimensional interaction space. Hundreds-of-thousands, or even millions, of SNP markers per individual are common in GWAS (*see* **Note 1**). This implies at least hundreds-of-thousands of tests to find single genes with main effects on the phenotype and including even the simplest form of interactions, two-way epistasis, the number of possible tests increase to $n(n-1)/2$, where n is the number of markers. As the number of test grows exponentially with every additional interaction dimension evaluated, the massive computational challenge, commonly known as the *curse of dimensionality*, leads to two issues that need to be addressed when aiming to explore higher order interactions. The first is obvious: how to explore the search space containing an astronomical number of interactions that need to be evaluated. The second problem is that there are many more variables than samples, i.e., the *small n, large p*, or *ill-defined*, problem (*see* **Note 2**).

When using an additive model-based GWAS paradigm, it is computationally feasible to test all possible associations between single marker genotypes and the phenotype using currently available hardware. Although the computational demand increases considerably, the use of large-scale computing makes it possible to extend the approach to evaluate all possible two-way interactions. Such analyses are, however, not a realistic standard analysis approach due to the high cost for the necessary computing resources. Furthermore, anything beyond first order interactions reach beyond the limit of the computing power available to most users today. Contemporary computers cannot traverse the whole search space in sensible time—if we, e.g., would like to test for all possible associations in a dataset of 50,000 SNP markers, starting with single gene association and subsequently include all higher order interactions, the total number of possible tests will be 2^{50000}. Although this is a rather unrealistic example, it illustrates that exploration of high-order interactions is a challenge where more efficient algorithms are needed to efficiently explore the search space without performing unnecessary evaluations. For this problem, EAs are one of the promising tools as they have the potential to efficiently explore the extremely large number of potential associations to find the ones with the strongest influence on the phenotype. We will later discuss a number of data mining and dimension reduction techniques that can be used to evaluate the association between multi-locus genotypes and the phenotype and how these combined or optimized with EAs can be used to address this challenge.

In classical GWAS, tests for association between the marker genotype and phenotype are performed only for one or a few markers at a time. This means that at no stage the full model, including all markers associated with the phenotype, is tested simultaneously (i.e., not even all the main effects are tested jointly in a single model). This strategy avoids the problem of having more observations (markers) than samples (individuals). The disadvantage is that it instead results in a multiple testing problem with spurious associations and correlated tests. Multiple testing correction is therefore needed and, e.g., Bonferroni correction or permutation testing is routinely used to counteract the effect of spurious associations. As mentioned above, the tests are also not independent (not orthogonal) due to, for example, the physical linkage of the SNP markers (i.e., located close together on the same chromosome), genes affecting the trait via a common pathway, interactions of the gene products, or a myriad of other ways that confound the data. Such effects are much harder to handle and they are today most often ignored. All the issues encountered in a GWAS for single genes will be amplified when higher order interactions are included in the models (i.e., the number of variables when considering interactions increase by several orders of magnitude). In other words, if no multiple testing correction is applied, the number of false positives increases exponentially. By default, tests for interactions are non-independent and consequently the complex layers of non-independence make it difficult to apply the correct adjustments for multiple testing which will often reduce the power to find epistasis. The standard GWAS paradigm is therefore not suited for finding higher order epistasis. It should also be noted that merely increasing the number of samples will not solve the inherent problems with genomic data, as it will not be possible to get more samples than observations when interactions are considered. In the next section we describe how machine learning can be used to analyze and interpret genomic data and then show how these methods benefit from using evolutionary algorithms to optimize them. When combined, these analysis methods will circumvent, or deal with at least some of the problems of searching for higher order interactions in genomic data.

3 The GWAS as a Machine Learning Problem

The main purpose of machine learning is to use computers to learn general concepts about a hypothesized problem from examples provided in the so-called training set—e.g., to provide a genetic predictive model for a complex disease in a population from a training dataset of marker genotypes and phenotypes from a case–control sample. The first definition of machine learning was given in 1959 by Arthur Samuel [15]: *Machine learning is a field of study that gives computers the ability to learn without being explicitly programmed*. Samuel designed and implemented an algorithm that gave computers the ability to learn how to play checkers and, with experience, to outcompete human players. Until now, several machine learning algorithms have been developed and successfully used in many fields. The most popular algorithms include: decision trees, random forests, support-vector machines, rough sets, neural networks, and fuzzy sets. The border between statistics and machine learning is not clearly defined and, together with more "traditional" methods such as linear models and Bayesian classifiers, the machine learning methods can be referred to as the *statistical learning* methods.

Regardless of the learning algorithm applied, the general scheme of constructing a predictive model is common to all statistical learning methods. The workflow involves: (1) construction of a proper training set (including data transformations, if necessary); (2) feature extraction and feature selection; (3) model construction; and (4) model validation. These steps can be reiterated several times to obtain the best model that can be used to: (1) make predictions for previously unseen cases; and (2) increase the general understanding of the modeled phenomenon.

A GWAS can be described as a machine learning problem (*see* Table 1). The measured genotypes at the examined markers, together with other descriptors such as sex or subpopulation, constitute the so-called *condition attributes* (*attributes* or *features* for short). The measured value for the studied trait(s) is the so-called *decision attribute*. Each individual is described by its own, possibly unique, vector of *condition-* and *decision attributes*, called *an instance*. A collection of such represented individuals, *instances*, constitute a *training set*. The goal is to use the training set to

Table 1
Translation between genome-wide association studies (GWAS) and machine learning terminology

Machine learning	GWAS
(Condition) attribute/feature	Genotype
(Condition) attribute/feature	Covariate
Decision attribute	Phenotype
Instance	Individual from the population described by its condition and decision attributes
Training set (in particular training set or test set)	Population sample

construct a model that will relate the observed patterns of attribute values to the decision attributes. In short, the model aims to predict the phenotype from the genetic (genotype) and nongenetic (covariates, e.g., environmental variables) information measured on the individuals in the training set.

3.1 Feature Extraction and Feature Selection in GWAS

From a machine learning perspective, the choice of appropriate covariates and detection of associations is a feature extraction and feature selection task. Feature extraction is an initial step, where dimensionality of the problem is reduced manually using the available field-specific knowledge (e.g., using information about correlation between features). In a GWAS, feature extraction includes the choice of potentially valuable covariates based on biological knowledge or a subset of markers based on the known genetic architecture (e.g., using only information from exons). The extracted features are included in the training set that will subsequently be used in the feature selection step.

The main aim of feature selection (and feature extraction) is to reduce feature-space to, in the end, include a minimal subset of only those features that are relevant for building the predictive model. In a GWAS, feature extraction and feature selection thus aim at finding only the covariates and markers that are associated with (or help to explain) the observed phenotypic values. All spurious associations should, ideally, be removed from the final model. Although feature selection and extraction can be clearly defined (Box 1), accomplishing such a task is not easy in practice. The algorithm needs to be implemented to efficiently explore the feature-space and, as we describe later, can be useful for finding the most predictive final model. Also, as in all other GWAS analyses, several biological factors can complicate the analyses and influence the results including linkage disequilibrium between markers, allele frequencies,

> **Box 1**
> **A typical feature selection scheme used in machine learning**
>
> 1. Create training set(s) that contain a subset of (usually randomly selected) features.
> 2. For each set, build a predictive model using the machine learning method of choice.
> 3. For each feature, record its predictive value taking into account the predictive quality of the model(s) where the feature was used.
> 4. If an optimal subset of features has not been found—repeat steps 1–3 using only the features with the highest predictive power so far.

population structure, or cryptic relatedness between individuals, just to mention some of the most common ones.

Several feature selection algorithms exist within the machine learning framework. The most common approach relies on examining quality of models based on various subsets of available features and then choosing the features that, on average, contribute the most to the classification process. This type of approach is often implemented in an iterative manner (Box 1).

3.2 Building a Predictive Model Using an Evolutionary Algorithm

Many optimization algorithms can be used in the feature selection step, including simple greedy search algorithms such as *forward selection* and *backward elimination*, which guarantee to find local optima but not necessarily the global ones. More advanced algorithms include *attribute space* search algorithms that combine forward selection and backward elimination into, e.g., the *bidirectional best-first search* approach where the selection procedure does not stop when there is no further improvement in predictive power, but keeps record on the performance of all previously evaluated feature subsets and can revisit earlier configurations if necessary. Both these approaches do, in addition to problems of finding global optima, have one more disadvantage—their computational complexity grows fast with the number of features to be examined, making them practically inapplicable to GWAS where number of markers (features) often exceeds hundreds-of-thousands or even millions and the number of individuals (instances) remains within the order of thousands. Many alternative machine learning methods, including feature selection algorithms, have been designed to deal specifically with this small n, large p problem. Some algorithms, including the EAs, that are particularly suitable for analyzing ill-defined problems, belong to the family of the so-called Monte Carlo methods. They are efficient global optimization algorithms that also cover the search space in a reasonable time. We will now discuss the principles of EAs in more detail with an emphasis on how they can be used to optimize feature selection in machine learning when applied to genomic data. This will be followed by a short description of methods that are frequently used to build predictive models in GWAS utilizing machine learning.

4 Evolutionary Algorithms and Genomic Data

Genetic programming and GAs are automated computational discovery tools that utilize an algorithm inspired by the biological processes of evolution via selection and reproduction [16, 17], both falling under the general class of EAs. It is a promising method for unraveling the complex genetic architecture in genomic data when combined with machine learning techniques. All GAs include a number of steps executed iteratively (Fig. 1). Some slight variations of this procedure exist, and are mainly used to tailor the GA to a specific dataset or function being optimized. We will here provide a brief description of a generic GA and then outline how a GA can be used to optimize decision trees, a machine learning technique, applied to genomic data.

4.1 The Genetic Algorithm

GAs are global optimization algorithms capable of solving combinatorial problems. They are a very efficient way to traverse through large datasets especially when the search space is increased due to high dimensionality [16]. The advantages of GAs have been exploited in various fields, including biology, for example, the identification of tumor cells from microarray data [18], prediction of secondary structure in RNA [19], protein-based mass spectrometry [20, 21] as well as regression models to map quantitative trait

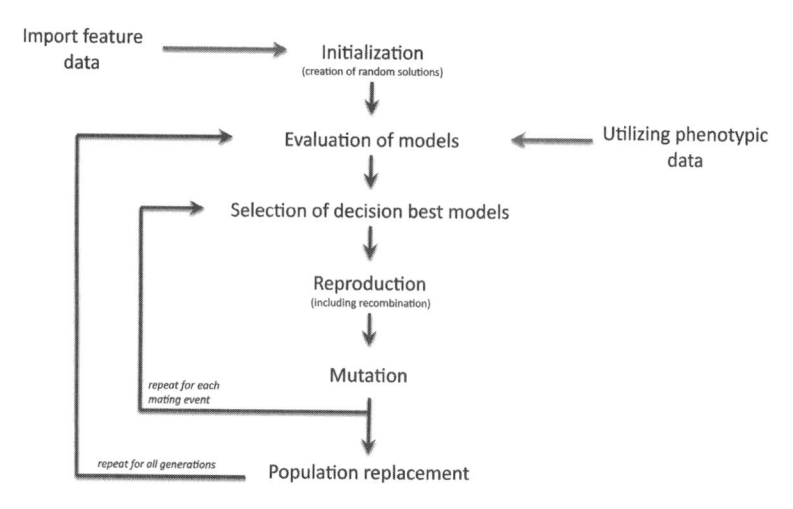

Fig. 1 Basic steps in a genetic algorithm

loci (QTL) in genomic data [22], including epistatic models [6] and environmental factors [23]. Large genomic datasets have been analyzed with GAs to a limited extent [24, 25]. It is, however, important to select an appropriate model to optimize. A number of machine learning and feature selection techniques have been applied to genomic data and hold promise to be a good starting point when developing methods incorporating EAs (as further described in Subheading 5).

A basic GA is an optimization algorithm that uses a set of functions resembling those in genetics and evolution to find a solution to a high-dimensional problem where it is not feasible to explore all possibilities. Key steps in this procedure are selection and reproduction, where all the solutions are evaluated and combinations of the best solutions passed to the next generation of solutions (using recombination and mutation operators). This enables the solutions to evolve via the power of selection, mutation, and recombination. Box 2 illustrates the basic steps in an evolutionary algorithm.

Box 2
Outline of a generic GA

All GAs have the basic outline shown in Fig. 1. Following is a short description of the steps required:

- Initialization: A population of random solutions is generated. Each solution is encoded in a modular fashion (the subcomponents are often called chromosomes). Solutions can be represented as an array of bits, mathematical functions, subprograms, or tree-like structures.

- Evaluation: Every solution is evaluated using a fitness function. The fitness function serves as the environment to which the solutions should adapt.

- Selection: The solutions that are best adapted to the environment are selected for reproduction. Various selection procedures can be used. However, it has been shown to be preferential to use a selection method that enables less-fit solutions to sometimes be selected. This reduces the risk of the solutions to become stuck on local optima.

- Reproduction: The selected solutions can be cloned or mated with each other. The latter entails that some parts of two solutions recombine to create a novel solution.

- Mutation: Some of the novel, or cloned, solutions are selected to undergo mutation to create novel variation at some of the modules.

(continued)

> **Box 2**
> **(continued)**
>
> – Population replacement: The newly created, or cloned, solutions replace the old population of solutions. This can be done at once resulting in discrete generations or be continuous resulting in overlapping generations. Note that, when using an inner loop such as in Fig. 1, the generations become discretized.
> – Termination: All the steps following initialization are repeated until a termination condition is fulfilled. Common conditions are, e.g., when the solutions do not increase in fitness after a specified number of iterations, or when a maximal and predefined number of repeats have been completed.

5 Machine Learning and GA Methods Applied to Genetic Data

5.1 Monte Carlo Methods

The term Monte Carlo methods was coined in the 1940s by three researchers involved in the Manhattan project, John von Neumann, Stanisław Ulam, and Nicholas Metropolis, to describe a family of algorithms that rely on repeated random sampling to arrive at final results [26, 27]. Named after the famous Monte Carlo Casino, the methods are nondeterministic, i.e., when run two or more times they are not guaranteed to arrive at exactly the same results. Nevertheless, they proved to be suitable for solving numerous problems. *Random forests* is one of the Monte Carlo methods that can be used for feature selection in GWAS.

5.2 Random Forests

Random forests were developed by Leo Breiman [28] in the second-half of the 1990s and are based upon construction of a number of decision-tree classifiers. The method has primarily been developed as a predictive model-construction algorithm (classification method). In addition to this, it also produces a useful ranking of features sorted by their predictive value and shows the joint predictive power of any pair of features. The prediction based on combinations of features is particularly useful in genetics as it allows us to study marker–marker interactions, i.e., epistasis (*see* **Note 3**).

We will first introduce the concept of a decision tree, as this is necessary to understand random forests, and the methods based on this concept are described in the coming sections. Formally, a decision tree is a directed acyclic graph and in Fig. 2, we present the tree-specific terminology. The initial node of a tree is called the *root* (Locus A). The root is connected with its *children nodes*

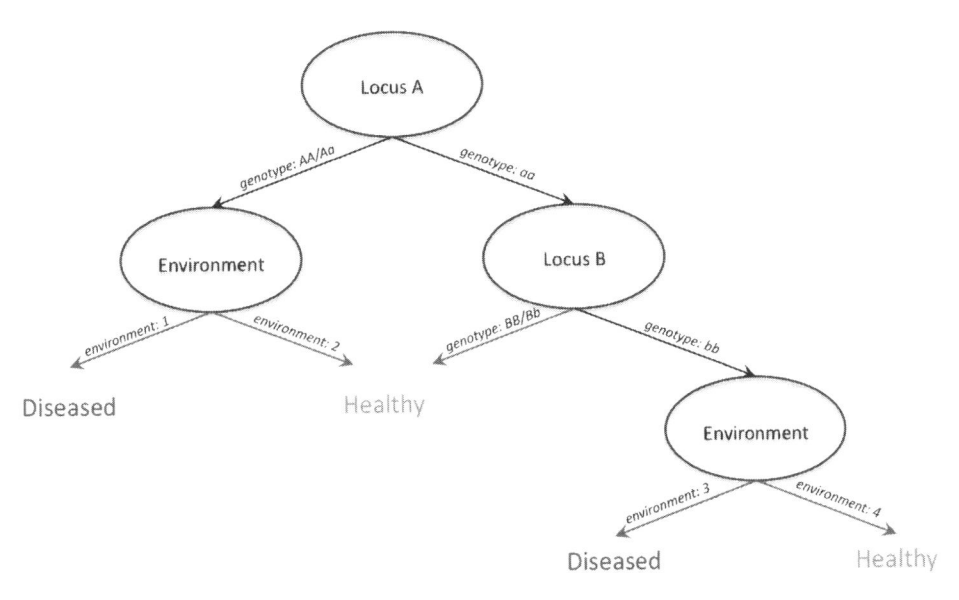

Fig. 2 A simple decision tree. Disease can be predicted from the genotype and environment, e.g., an individual in environment 4 with genotype aabb will be diseased

(here Environment and Locus B) by directed *edges* and is their *parent node*. The last-level nodes (Diseased and Healthy), that do not have any children, are called *leaves*. A decision tree can be seen as a tree where the whole population (training set) is contained within the root. Edges represent the rules of directing instances from the parent node to the specific child node based on the value of a particular feature. Finally, leaves represent decision attributes. Ideally, all the instances that reached a particular leaf should have the same value of a decision attribute. The aim of the training phase is to use the training set to assign an appropriate redirection rule to each edge in the tree. When a new instance needs to be classified, it is placed in the root and directed along a *path* determined by the redirection rules until it reaches a leaf. The leaf determines the predicted value of the decision attribute for this instance.

To give an example of how the random forest algorithm can be used for analyses of GWAS data, let us consider a population of 100 individuals (50 cases and 50 controls). Each individual is described by a vector of ten features, where each feature is the genotype at one of the ten SNP markers. For each individual, its phenotypic value for a binary trait is known: case and control encoded as 0 and 1, respectively. Tree construction begins by placing all the individuals in the root. The next step is to split the population into three subpopulations according to their value (genotype) of one of the ten markers. There will be three edges representing three possible redirection rules: (i) If genotype at marker X is AA then go to node 1; (ii) If genotype at marker X is AB then go to node 2; and (iii) If genotype at marker X is BB then go to node 3. Trees can now

Table 2
Illustration of the purity for markers P and Q

Marker	Node 1 (AA)	Node 2 (AB)	Node 3 (BB)
P	0 controls 50 cases	20 controls 0 cases	30 controls 0 cases
Q	10 controls 40 cases	15 controls 5 cases	25 controls 5 cases

be constructed using any of the available markers, but the crucial point of the algorithm is how to select the marker, X, that produces the best prediction. The answer is that X is the marker producing the purest nodes in terms of the decision attribute values (see **Note 4**). To illustrate what is meant by *pure*, let us consider an example including two markers, P and Q (Table 2). Here, splitting the initial population on the genotype at marker P results in three nodes that are *pure* in terms of the value of decision attribute (i.e., perfectly separates cases from controls). This finishes tree construction, since the perfect classifier has been created. In contrast, splitting on genotype at marker Q will yield a tree with impure nodes and thus it is not as good as when splitting on marker P. In practice, it is rarely possible to achieve pure nodes and the feature that produces the purest possible nodes has to be used. Various measures of node purity are used and they are often based on entropy. We will not discuss them here and recommend the reader to consult, e.g., Witten et al. [29] for details.

As mentioned before, to achieve an accurate classifier/predictive model, random forests rely on the construction of a large number of decision trees, each based on a subset of features (*see* **Note 5**). When a previously unseen instance has to be classified, these trees are used to predict the decision attribute. The final decision is a result of voting in which each tree participates with the number of votes proportional to the predictive power of this particular tree. In other words, the better the tree is in classifying, the more votes it casts. The original random forest algorithm is summarized in Box 3.

Let us note that sampling of instances is done with replacement. Thus, the chance of an instance being used in the construction a particular training set is about 0.66. The remaining of the instances constitute the test set and can be used to evaluate the tree-based classifier giving the so-called OOB (Out Of Bag) unbiased error estimate. In order to evaluate the importance of a single feature f, the feature is permuted in the test set and the drop in prediction quality (compared to the original OOB error) is recorded. The drop, averaged over the entire forest, gives a reliable

> **Box 3**
> **An overview of the classical random forest construction algorithm proposed by Breiman**
>
> - Training set creation. From the original dataset with M attributes and N instances, randomly without replacement sample the predefined number m of attributes ($m << M$). Next, randomly with replacement sample N instances. This is the training set. The ~1/3 of the instances that have not been selected to the training set constitute the test set.
> - Forest building. Repeat the training set creation step several (typically several thousand) times. This step results in a number of training set–test set pairs.
> - Tree growing. For each pair, use the training set to construct a decision tree without tree pruning. Evaluate the tree using the test set. Store the tree and its classification quality measures.
> - Classification. Given an instance to classify, each tree from the forests votes for the final decision. The number of votes from each tree is proportional to the classification performance of this tree. Final decision is the result of voting.

estimate of the importance of the f feature. However, when there is a strong correlation between any two (or more) features, a masking effect can be observed, i.e., only one of the correlated features will be significant while the other will be considered redundant (e.g., ref. 30). The Monte Carlo Feature Selection and Interdependency Discovery (MCFS-ID) algorithm is an alternative, random forest-inspired approach that overcomes this limitation [31] and facilitates the detection of higher order interactions between features. The algorithm has been successfully applied to gene expression datasets and in proteomics [31, 32].

6 GA Applied to Genomic Data

EAs and, in particular, GAs are well suited to analyze higher order interactions in genomic data. It is, however, important to select a suitable way to represent the genetic model. To move away from the additive GWAS paradigm and instead focus on a model-free approach including nongenetic data (i.e., population structure and environmental data), we suggest exploring alternatives to traditional statistical models.

Genetic algorithms have been used in combination with neural and other network structures and successfully applied to genetic data [7, 25, 33, 34]. Similar to a decision tree, a neural network is composed of connected nodes (or neurons). Flow of information through this network allows the phenotype to be predicted. It has been demonstrated that neural networks that have been optimized by GAs have better predictive power than neural networks optimized by other methods [33].

Feature selection methods used in machine learning provide fitting alternatives, especially since they are often capable of handling large datasets (for an example on how machine learning methods have been applied to genomic data see the following sections). Two methods that combine well with GAs are decision trees and neural networks. Here, we will outline how one of these methods, decision trees, can be paired with a genetic algorithm and applied to genomic data [24]. Note that GAs could be used as an optimization algorithm also in other machine learning techniques or in more traditional statistical testing-based methods; the implementation with decision trees described here is only one example.

Decision trees are well suited for optimization by GAs and provide a very nice representation of features (i.e., genes and environmental factors) and their interactions (Fig. 2). The biological interpretation of the selected features in a decision tree is intuitive and the subsequent construction of a genotype–phenotype map is uncomplicated. In addition, the relative importance of different features is easy to infer from the tree structure.

Following the outline in Box 24.2, a GA can be implemented with decision trees to analyze genomic data as follows.

First, a population of decision trees is constructed. Each tree has a random number of nodes. Each node represents an SNP or environmental feature chosen at random from all the SNPs and environmental factors in the dataset.

Second, the trees are evaluated. This is a simple procedure where every tree uses the genotypes of all the individuals in the dataset to predict the phenotype. The fitness value is related to the number of accurate predictions. To avoid unnecessary nodes with low explanatory power, a penalty is included for each additional node in the tree (*see* **Note 6**).

Third, trees with high fitness are selected more often than trees with low fitness. This means that trees that predict the phenotype better, i.e., SNP combinations that have an effect on the phenotype, have a high chance to be part of the solution during the next round of optimization. However, we suggest not to select the best trees only, as this could lead to the solutions becoming stuck on a local optima (this is an important consideration in all EAs).

Fourth, during reproduction the selected trees are either cloned or mated with each other. The cloned offspring are exact copies of the parental trees. Trees created during recombination (i.e., the mating of two trees) contain branches from different parental trees. To ensure that the search space is efficiently covered, tree topologies are not limited to the initial population's tree structures, and new variation is released for optimization in later generations by mutating some of the offspring. The mutations should include random mutations of nodes where one SNP is replaced by any other SNP in the data, and stepwise mutations where the SNPs at the node are replaced with adjacent SNPs on the genome. In the latter case the LD is utilized when optimizing the solutions without any additional computational cost. Also, if a cost is included for large trees, the excess of redundant nodes will be reduced. In addition to mutations of the nodes, the trees can also be mutated to grow or shrink by one node at a time.

The final step is population replacement. It is often easier to keep track of the performance of the solutions when the generations are discrete. This means that all the trees are evaluated simultaneously followed by selection, reproduction, and mutation after which the whole population is replaced by the offspring generation.

The GA should be terminated when the combination of SNPs and environmental factors accurately predicts the phenotype. To specify such a termination condition is not trivial and often the change in average or maximum fitness is used as an indicator. Once there is no increase in the fitness, an optimum is found. It is, however, difficult to determine whether this is a local or global solution to the problem. Repeating the whole GA with different random trees in the initial population is one way to evaluate the robustness of the decision trees in the final populations.

The method outlined above is expected to provide a thorough exploration of the genetic architecture underpinning the phenotype. To ensure that the high-dimensional search space is adequately traversed, both a large number of trees and many generations of simulation in the GA are recommended. It is possible to balance these two factors to some extent, but an increase in either has the disadvantage of making the analysis time longer (this is a drawback of EAs in general).

Decision trees indicate the importance of a feature, depending on the position in the tree (i.e., the effect size of an allele is larger when the node with the SNP is closer to the top). However, if the effect of two SNPs is additive and equal it may not be obvious from looking at the tree with the highest fitness (*see* **Note** 7). The entire population should therefore be inspected as well as the top trees in the final population of a separate analysis run. Variation in the SNP locations across trees could indicate additive and equal effects. Conversely, if repeated analyses yield the same tree topology with the same hierarchy in SNP distribution, it indicates a stable relation between the effects of the SNPs and potentially also epistasis. Environmental effects and population structure can be included in this type of analysis, where each such factor can be included as a node in the decision tree. This means that the environmental factors and population structure will also be used for the prediction of the phenotype. Lastly, the penalty parameter, for including additional SNPs can be varied, where using a lower penalty for more nodes will lead to SNPs with smaller effects to be included in the tree. In this way, a more complete prediction model can be obtained using a stepwise model building procedure where the stringency for SNP inclusion is successively decreased.

Although this method can be used to build a predictive model for the phenotype, we do not think that it should be the endpoint in the analysis of genomic data but rather a complementary approach to the standard GWAS. Also, once the candidate SNPs are identified from the GA, it is possible to do regression model-based analyses on these only (including the identified interactions). This will allow estimation of effect sizes and comparisons between those and the relative effect sizes indicated by the decision tree.

The R package R/Freak provides a framework for evolutionary computation (*see* **Note 8**) and is available from CRAN (*see* ref. 24]. This package is a procedure called Genetic Programming for Association Studies (GPAS) that enables the user to use genomic data in a decision tree and evolutionary selection framework (*see* **Note 9**), similar to the method outlined above.

7 Conclusion

The most promising approach for future analyses of genomic data will most likely be based on a combination of complementary analysis strategies. No single analysis method will be able to completely unravel the genetic architecture of a complex trait and methods that are powerful in detecting single or additive genes with main effects are unlikely to be those that properly model and find the epistatic interactions. This is where EAs can play an important role in exploring the contribution of high-dimensional interactions, for example, when combined with model-free machine learning methods.

8 Advantages and Disadvantages

The method outlined in this chapter is suitable for analyzing genomic data and is especially suitable for investigation of interactions. We have discussed the main reasons for using EA on genomic data and will here conclude by providing a list of more technical advantages and disadvantages (and some solutions to these) that those interested in implementing EA for analyses of genomic data need to be aware of.

8.1 Advantages

EAs are fast and efficient in searching through large genomic datasets for interacting genes when compared to exhaustive algorithms.

In particular the speed of the analysis scales well with the increase in markers or other features added to the dataset or the solution.

EAs are better at exploring the global search space than naive greedy algorithms and are therefore less prone to be stuck at local optima.

EAs can incorporate other important data in the optimization. These include environmental data, population structure, pedigree data, or data on any other factor that may have an effect on the phenotype.

EAs are a general class of global optimization methods. Optimization is based on an external fitness function that can be either model-based (i.e., regression modeling) or model-free (i.e., random forests). Here, we propose a model free approach that takes full advantage of this feature of the EA.

No priors are required for an EA and there is thus no bias at the start of the analysis (*see* **Note 10**).

8.2 Disadvantages

Doing a genome-wide, stepwise regression (i.e., an exhaustive search) to evaluate the associations between each of the markers individually and the trait is faster than applying a global optimization algorithm (such as a GA) to the problem when the aim is to find the marker with the biggest main effects on the phenotype. Once multidimensional scans involving interactions are to be investigated, a properly implemented GA will be more efficient.

An EA will, like any other Monte Carlo-based optimization method, not guarantee that the global optimum will always be found. Consequently, it may arrive at a number of different final solutions within an analysis of the same data. EAs are good at global optimization and are therefore likely to detect global optima that are far superior to other solutions. They are, however, less effective at local optimization and consequently they are challenged when there exist multiple local optima in the search space (i.e., if the landscape is very rugged with many local optima at the same level). If differentiation between multiple roughly equal optima is required in the analyses, a hybrid-EA is recommended where the multiple optima indicated by the GA are explored in detail using a more efficient local optimization method (*see* **Note 7**).

The method discussed here that combines random forests with a GA is challenged when multiple local optima or features with similar effects are encountered, as it is not efficient in ranking such solutions. It is therefore difficult for the researcher to evaluate how good a solution is in the general sense by just studying the results from this method. In such cases, it is recommended that the features identified in the analyses are evaluated further using, e.g., model-based regression approaches that will provide complementary information needed for a better understanding of the relationship between the effects of important selected features (*see* **Note 7**).

Over fitting is a potential problem when constructing random forests in GWAS data. It is easy to add more features, which will allow us to predict the data in the training set better. It is, however, not certain that this will improve the prediction in the general population. A way to reduce this problem is by using stronger penalties for solutions with more features. However, selecting this penalty remains a problem since a large penalty will reduce the chance of finding complex interactions, while a small penalty will result in a solution with many features having very small effects (*see* **Note 6**).

It might be difficult to interpret the solution from a tree from a biological perspective if there are many SNPs included and thus many interactions. However, weighting the influence of each SNP and building a stepwise genetic model will help to interpret the results in a biologically more meaningful way. Alternatively, repeated construction of trees, where the penalty is successively reduced, will provide an indication of the importance of individual markers as well as marker complexes that will help in constructing a biologically relevant model.

9 Notes

1. Around ten million SNPs have been identified in the human genome. Current commercial SNP-arrays cover approximately two million of these SNPs.
2. In current and future genetics studies, the number of samples will almost always be small compared to the amount of genetic data available. The *ill-defined* problem has to be addressed in analysis methods since good sampling strategies will not eliminate this problem.
3. Random forests can be easily run using either Java implementation in WEKA software [29] or the randomForest R package [35]. randomForest is an R wrapper over the original Fortran implementation by Breiman and Cutler [28].
4. When setting parameters for running random forest, as a rule of thumb, it is good to start with the *m* number of attributes equal to the square root of the total number of attributes.
5. Construction of a random forest with too few trees is a common mistake. It is good practice to begin with at least 1,000 trees and after having the model optimized it is good to run it with 5–10 k trees.
6. Changing the penalty parameter will influence the maximum number of nodes in the tree. Lowering it will lead to selection of more nodes, but although the actual number of interactions can be larger, this might not necessarily make the results more interpretable.

7. If the trees at the end of the run (or at the end of multiple runs) include the same features, but at different locations in each tree, it indicates that a number of additive genes affect the phenotype. However, if the features are structures in the same relative positions in different trees, it indicates epistasis.

8. The R package R/Freak has the function *GPASInteractions()* that allows the user to search for interaction in genetic data using an evolutionary algorithm. This function, with the accompanying documentation and data, is a good place to start investigating EAs.

9. Parts of the evolutionary algorithm in the R/Freak package are written in Java but wrapped in the R-functions. The R-environment does not handle looping functions well and is not well suited for most genetic algorithms.

10. It is possible to create the first population using preexisting information about the system. In general it is not recommended since it may limit the initial space that can be searched and bias the analysis. However, in some circumstances, it may be useful in a local optimization for elucidating more accurate final models around global maxima.

References

1. McCarthy MI et al (2008) Genome-wide association studies for complex traits: consensus, uncertainty and challenges. Nat Rev Genet 9:356–369
2. Manolio TA (2009) Finding the missing heritability of complex diseases. Nature 461:747–753
3. Moore JH, Asselbergs FW, Williams SM (2010) Bioinformatics challenges for genome-wide association studies. Bioinformatics 26:445–455
4. Visscher PM et al (2012) Five years of GWAS discovery. Am J Hum Genet 90:7–24
5. Hindorff LA et al. Catalogue of published genome-wide association studies. www.genome.gov/gwastudies. Accessed 2012
6. Carlborg Ö, Andersson L, Kinghorn BP (2000) The use of a genetic algorithm for simultaneous mapping of multiple interacting quantitative trait loci. Genetics 155:2003–2010
7. McKinney BA et al (2006) Machine learning for detecting gene-gene interactions. Appl Bioinformatics 5:77–88
8. Moore JH, White B (2007) Genome-wide genetic analysis using genetic programming: the critical need for expert knowledge. In: Riolo R, Soule T, Worzel B (eds) Genetic programming theory and practice IV. Springer, New York, pp 11–28
9. Frazer KA et al (2009) Human genetic variation and its contribution to complex traits. Nat Rev Genet 10:241–251
10. Eichler EE et al (2010) Missing heritability and strategies for finding the underlying causes of complex disease. Nat Rev Genet 11:446–450
11. Carlborg Ö, Haley CS (2004) Epistasis: too often neglected in complex trait studies? Nat Rev Genet 5:618–625
12. Phillips P (2008) Epistasis—the essential role of gene interactions in the structure and evolution of genetic systems. Nat Rev Genet 9:855–867
13. Cordell HJ (2009) Detecting gene-gene interactions that underlie human diseases. Nat Rev Genet 10:392–404
14. Thomas D (2010) Gene-environment-wide association studies: emerging approaches. Nat Rev Genet 11:259–272
15. Samuel AL (1959) Some studies in machine learning using the game of checkers. IBM J Res Dev 3:210–229
16. Goldberg D (1989) Genetic algorithms in search, optimization, and machine learning. Addison-Wesley Publishing Company, New York
17. Koza JR (1992) Genetic programming as a means for programming computers by natural selection. Statistics Computing 4:87–112
18. Li L et al (2001) Gene selection for sample classification based on gene expression data: study of sensitivity to choice of parameters of the GA/KNN method. Bioinformatics 17:1131–1142

19. van Batenburg FH, Gultyaev AP, Pleij CW (1995) An APL-programmed genetic algorithm for the prediction of RNA secondary structure. J Theor Biol 174:269–280
20. Petricoin EF et al (2002) Use of proteomic patterns in serum to identify ovarian cancer. Lancet 359:572–577
21. Li L et al (2004) Application of the GA/KNN method to SELDI proteomics data. Bioinformatics 20:1638–1640
22. Zhang B, Horvath S (2005) Ridge regression based hybrid genetic algorithms for multi-locus quantitative trait mapping. Int J Bioinform Res Appl 3:261–272
23. Mukhopadhyay S, George V, Xu H (2010) Variable selection method for quantitative trait analysis based on parallel genetic algorithm. Ann Hum Genet 74:88–96
24. Nunkesser R et al (2007) Detecting high-order interactions of single nucleotide polymorphisms using genetic programming. Bioinformatics 23:3280–3288
25. Mooney M, Wilmot B, McWeeney (2012) The GA and the GWAS: using genetic algorithms to search for multi-locus associations. IEEE/ACM transactions on computational biology and bioinformatics 9:899–910
26. Metropolis N, Ulam S (1949) The Monte Carlo Method. J Am Stat Assoc 44:335–341
27. Metropolis N (1987) The beginning of the Monte Carlo method. Los Alamos Sci (Special Issue dedicated to Stanisław Ulam) 15:125–130
28. Breiman L (2001) Random forest. Machine Learning 45:5–32
29. Witten I, Frank E, Hall MA (2011) Data mining. Practical machine learning tools and techniques, 3rd edn. Morgan Kaufmann, Burlington, MA
30. Tuv E et al (2009) Feature selection with ensembles, artificial variables and redundancy elimination. J Machine Learning Res 10:1341–1366
31. Dramiński M et al (2008) Monte Carlo feature selection for supervised classification. Bioinformatics 24:110–117
32. Kierczak M et al (2010) Computational analysis of molecular interaction networks underlying change of HIV-1 resistance to selected reverse transcriptase inhibitors. Bioinform Biol Insights 4:137–145
33. Ritchie MD et al (2003) Optimization of neural network architecture using genetic programming improves detection and modelling of gene-gene interactions in studies of human diseases. BMC Bioinformatics 4:28
34. Bush WS et al (2005) Can neural network constraints in GP provide power to detect genes associated with human disease? Lecture Notes Computer Sci 3449:44–53
35. Liaw A, Wiener M (2002) Classification and regression by randomForest. R News 2:18–22

Chapter 25

Incorporating Prior Knowledge to Increase the Power of Genome-Wide Association Studies

Ashley Petersen, Justin Spratt, and Nathan L. Tintle

Abstract

Typical methods of analyzing genome-wide single nucleotide variant (SNV) data in cases and controls involve testing each variant's genotypes separately for phenotype association, and then using a substantial multiple-testing penalty to minimize the rate of false positives. This approach, however, can result in low power for modestly associated SNVs. Furthermore, simply looking at the most associated SNVs may not directly yield biological insights about disease etiology. SNVset methods attempt to address both limitations of the traditional approach by testing biologically meaningful sets of SNVs (e.g., genes or pathways). The number of tests run in a SNVset analysis is typically much lower (hundreds or thousands instead of millions) than in a traditional analysis, so the false-positive rate is lower. Additionally, by testing SNVsets that are biologically meaningful finding a significant set may more quickly yield insights into disease etiology.

In this chapter we summarize the short history of SNVset testing and provide an overview of the many recently proposed methods. Furthermore, we provide detailed step-by-step instructions on how to perform a SNVset analysis, including a substantial number of practical tips and questions that researchers should consider before undertaking a SNVset analysis. Lastly, we describe a companion R package (*snvset*) that implements recently proposed SNVset methods. While SNVset testing is a new approach, with many new methods still being developed and many open questions, the promise of the approach is worth serious consideration when considering analytic methods for GWAS.

Key words SNVset, SNP-set, Gene set, Pathway, GWAS

1 Introduction

Within the last decade, the technological and statistical framework for the design and testing of genotype–phenotype associations in genome-wide studies (GWAS) has matured significantly. In particular, the use of two-stage study designs, methods of sample preparation and quality control, and guidelines for missing data and error handling have all been generally accepted by the community, with detailed guidance on these issues provided by others in this tome.

Once a set of accurate genotypes have been generated on a sample of individuals of known phenotypes, the most common analytic approach is to use a general linear modeling (regression) framework to evaluate genotype–phenotype association. Typically, the model is framed as Phenotype = Genotype + Covariates. In particular, a logistic regression model is used for dichotomous phenotypes (e.g., case–control studies) and linear regression model for quantitative phenotypes. The genotype is typically coded 0, 1, or 2 at the single nucleotide variant (SNV) locus of interest, in order to represent an additive disease mode of inheritance, where the value of the genotype represents the number of minor alleles (other coding schemes are possible to represent other modes of inheritance). Covariates typically represent various demographic (e.g., population stratification) or disease related variables which are potentially associated with both the genotype and phenotype, are typically not of interest to the researcher, but can bias the analysis if not properly considered.

In a traditional analysis, the regression model described above is applied at each of the measured SNVs that passes strict quality control criteria. For large SNV microarrays (e.g., Illumina Omni5®, Affymetrix myDesign Genotyping arrays which now feature 5,00,000 to well over six million plus SNVs), the traditional analysis strategy involves running a separate regression model for each measured SNV, with additional models for SNVs measured in silico via imputation strategies using appropriate haplotype reference panels.

Thus, a researcher will ultimately conduct at least hundreds of thousands (if not multiple millions) of tests of significance. Strict multiple-testing penalties are traditionally applied to minimize the proportion of false positives (e.g., significance level of $p < 1 \times 10^{-8}$ or lower). Even with rapidly growing sample sizes, the power to find causal variants with only modest effects (Odds Ratios of 1.1–1.3) remains low, with increasing evidence suggesting that most causal variants for common diseases have only modest effects [1]. Compounding matters even further is the fact that the resulting list of "significant" SNVs may or may not actually represent functional risk-inducing (or risk-reducing) variants. Instead, these SNVs may only be in linkage disequilibrium (LD; high correlation) with the causal variants. Even in cases where the SNVs are causal, the list of significant SNVs may yield little direct understanding of disease etiology when interpreted in isolation. These challenges are a substantial part of why some have argued that GWAS have not yet achieved their full potential [2]. In response, numerous alternative design and analytic approaches have been proposed to address the power problems of GWAS including alternative multiple-testing strategies, alternative sample designs, and alternatives to the traditional single SNV regression model approach; the latter of which is our focus here.

Many of the alternative analysis approaches for GWAS find their roots in the analysis of gene expression data, which consists of more mature technology and analysis methods. And so, it is worth a brief historical detour on gene expression data analysis. Gene expression data analysis is traditionally conducted by evaluating the differential expression of two genes across two different phenotypes. The genes showing the most differential expression from a rank-ordered list of genes are examined. Historically, biologists look at these "most differentially expressed genes" and attempt to give a biological explanation for their differential expression. This approach, however, suffered from some of the same limitations as GWAS in that (a) multiple-testing penalties and a strict, arbitrary cutoff limited power and (b) identifying the significantly differentially expressed genes was of little interest compared to identifying the biological explanation which caused the differential expression.

Recognizing these limitations, analytic methods for gene expression data expanded to use Fisher's exact test as a post hoc approach to increase power and biological interpretability. In particular, Fisher's exact test can be used to compare the proportion of differentially expressed genes within an a priori defined biological set of interest to the proportion of differentially expressed genes not in the biological set of interest. For example, if there are 15 genes in a particular biological pathway and 10 of those genes show significant differential expression and there are 19,000 genes not in the pathway, of which 500 show differential expression, using Fisher's exact test to compare $10/15 = 66.7\ \%$ to $500/19{,}000 = 2.6\ \%$ shows strong statistical significance ($p = 3 \times 10^{-5}$) that the genes in the pathway of interest show more evidence of differential expression than the genes not in the pathway. Using Fisher's exact test as a post hoc method for gene expression data analysis remains popular today, though numerous alternatives have been proposed. In particular, many methods have attempted to address the limitation that Fisher's exact test requires a priori classification of genes as differentially expressed or not differentially expressed (e.g., see refs. [3–7]). This class of methods has become known as gene set or pathway testing.

In a landmark paper in 2007, Wang et al. [8] described how to apply gene set (pathway) methods to GWAS data. The approach proposed by Wang et al. requires (1) assigning SNVs to genes and (2) assigning genes to sets (e.g., pathways) of interest. Much as was the case in gene expression data analysis, the approach by Wang et al., and the idea of gene set (pathway) analysis for GWAS caused an explosion of sorts in the number of methods that can be used for gene set analysis. However, these approaches typically seek the same goals: (1) reduce multiple testing by running fewer tests, since individual SNV–phenotype associations are aggregated into larger sets, (2) improve power through combining multiple weak, but true, signals, and (3) enhance biological interpretability by obtaining p-values for sets of SNVs (genes) that have biological relevance. In general, these methods can be viewed as part of a movement toward so-called "knowledge-based approaches" to genomic data analysis, and away from "agnostic" approaches (e.g., see ref. [9]). For example, running 5,00,000+ tests is viewed as agnostic because we do not, a priori, condition on any biological information about the SNVs (e.g., which SNVs may be functional, which genes they are in, which pathways those genes are part of, etc.). Gene set (pathway) approaches are considered "knowledge-based" as they directly incorporate preexisting biological knowledge into the tests of association. We direct interested readers to a comprehensive high-level review of such methods [10], which includes description of some successful implementations of gene set methods for GWAS data not described here. Though we caution the reader that, as described later in this chapter, important methods have been developed since [10] was published.

Currently, numerous methods of pathway/gene set analysis have been proposed for GWAS. However, to date, no clear optimal choices for use in practice have emerged. The lack of a clear, universal "best-approach" can be explained for a number of reasons: (1) The development of the methods is still rather recent (the first attempt was only published in 2007, with an exponential growth in methods papers on this topic since that time), (2) The strengths and weaknesses of these approaches have received little systematic comparative analyses in the literature, and (3) When direct comparisons have been made, researchers have typically found that optimal methods are highly dependent upon the genetic architecture of disease, correlation patterns (linkage disequilibrium) within the SNVs/genes in the set of interest, and the number of noncausal variants that are included in the pathway of interest.

There is one remaining important point worth noting. Many gene set/pathway approaches for GWAS may be better labeled as "SNVset" approaches because they are truly creating sets of SNVs, not genes—the name gene set/pathway approach is a relic from gene expression data analysis. Indeed, many of these SNVset methods focus on the gene as the "set" of interest. We will use the term SNVset from here forward when describing these approaches.

In this chapter we will attempt to provide some clarity to the reader on practical considerations for the use of SNVset analysis methods for GWAS data, providing a general framework for the analytic workflow, providing links to relevant software, and specific implementation tips and guidance. Additionally, we will describe specific open questions about best-practices.

2 Materials

In order to run a SNVset analysis there are numerous initial questions that should be answered which will help to determine the methods that should be used and the source data that should be obtained.

We summarize the source data questions in the following three subsections. see Subheadings 3 and 4 for more explanation of methods choices and implementation details.

2.1 Source Data #1: Genotype–Phenotype Data

SNVset analysis methods can be grouped into two large categories: (1) methods that take as input genotype–phenotype level data files (raw genotype methods) and (2) methods that take as input genotype–phenotype associations at each SNV of interest (p-value methods). The first set of methods requires the genotype calls on each individual at each SNV of interest, along with all phenotypes (and covariates, where appropriate) of interest for the analysis. These data will be readily available to the primary researchers involved in the study and are what is used in traditional analysis of GWAS data.

The second set of methods takes as input measures of the genotype–phenotype association (e.g., p-values) at each SNV of interest. For example, the input data may be a two column matrix consisting of rs# (SNV ID) and p-value from the individual SNV regression model. This second set of methods is popular because the computational demands of these methods are generally lower and researchers can rely on the high quality of information received from the mature field of individual SNV methods. This class of methods also allows researchers who do not have access to the "raw" data to apply these methods to publicly available datasets. For more details *see* **Note 6**.

2.2 Source Data #2: Functional (Set) Information

A critical step in the application of set methods is to consider the source of functional/set information that will be incorporated in the analysis. In short, any biological information can be used to group SNVs into sets. In general, the goal is to minimize the number of noncausal SNVs in the set while maximizing the number of causal SNVs in the set—though methods differ in their robustness to increased numbers of noncausal SNVs and different causal SNV effects, as we will describe later (*see* **Note 3**). Of course, optimizing the proportion of causal SNVs is challenging—if disease architecture was fully understood, you would not need to do the analysis! Here are some of the most common sources of functional/set definitions.

1. *Positional (region) based sets*. Positional based sets can include SNVs directly within a particular region of interest (e.g., promoter, intron, exon, gene, recombination hotspot, cytogenetic band/subband, arm (p/q), chromosome), those within a fixed distance of a region of interest (e.g., gene start/stop ±50 kb), those in LD with a region of interest (e.g., all SNVs within the gene plus SNVs in an LD-block overlapping the region of interest), or a combination of distance and LD.

    Positional sets can be obtained from any number of sources of biological information including MSigDB [11], ENSEMBL [12], dbSNP [13], HapMap [14], and 1,000 Genomes [15].

2. *Nonpositional based sets.* Nonpositional based sets can also arise from numerous sources. Common choices are Gene Ontology [16], Biocarta [17], KEGG [18], Bioconductor [19] (via links to other databases) and MSigDB [11] and other sources of curated and noncurated sets of genes. However, for any particular disease there may be novel, hypothesized disease-causing pathways that are of interest (e.g., novel cancer pathways). For example, prior GWAS, other -omics data, or other biological research may suggest a particular set of genes that may be related to the disease of interest. Another common choice of nonpositional based sets of SNVs is to consider only those SNVs with known (or predicted) functional impact on protein creation (e.g., synonymous vs. nonsynonymous variants; e.g., PolyPhen score [20], SIFT score [21]).

2.3 Source Data #3: Weights

Many, though not all, set-based analysis strategies can incorporate SNV weights. Use of weights can be considered a generalization of the traditional set-based framework. When defining sets, we typically are inferring that a 0/1 indicator variable indicates whether or not a particular SNV is in a set of interest. For example, we could define a set as all (predicted) nonsynonymous SNVs within a gene of interest. However, weighting all of the SNVs in a gene can reflect the potential uncertainty in predicting the function of a particular SNV. For example, we might wish to define the set as all SNVs within the gene, but then assign a weight to each SNV within the gene reflecting some measure of prior information about potential deleterious impact of the SNV (e.g., CAROL score which integrates PolyPhen and SIFT scores; [22]). Prior analytic work can also inform weighting—each SNV can be weighted based on prior evidence from other GWAS or -omics studies instead of applying a strict threshold approach (e.g., gene expression analysis shows potential impact of certain genes within a pathway of interest that may be related to cholesterol level, so SNVs in those genes are upweighted relative to other genes in the pathway). These weighting strategies, while not explicitly Bayesian, reflect a Bayesian-like approach of updating prior evidence of phenotypic association with new evidence from the current GWAS study.

3 Methods

We will now provide both general and specific instructions for readers as to how to specifically implement SNVset analysis strategies on their own GWAS data. We start by discussing the choice of a SNVset statistic, and then give detailed instructions about implementing a SNVset strategy for the SNVset statistic of choice.

3.1 Methodological Decision: Choosing a SNVset Test Statistic

While choosing a weighting scheme and source of functional (set) information is nontrivial due to the number of choices, the choice of SNVset test statistic may be even more challenging. There are numerous options available, and to date, little comparative data are available on these methods to help understand when particular methods are best and why. As noted earlier, no decisive choices have emerged to date [10]. We will summarize some of the options here, along with guidance about which test statistic should be used. However, we caution the reader to carefully evaluate the options by reading the papers proposing the methods and comparative papers which will be published soon (e.g., see ref. [23]), which further illuminate the pros and cons of various methods. As noted earlier, methods can be divided into two main categories: methods that use raw genotype–phenotype data and methods that use aggregate genotype–phenotype association data for each SNV (SNV *p*-value approaches). We will discuss these methods separately. Additionally, we focus only on methods for which publicly available software is available in order to aid in the implementation of the methods. We reserve a separate section (see 3.1.3 below) to briefly list some published methods for which no publicly available software is available.

3.1.1 SNV p-Value Approaches

SNV *p*-value methods assume that the researcher has access to *p*-values from single SNV association analyses with the phenotype of interest. We can further break down SNV *p*-value methods into two subtypes: self-contained tests and competitive tests. Self-contained tests use only the *p*-values from SNVs in the set of interest, while competitive tests also consider *p*-values from SNVs outside of the set of interest.

Competitive Tests

Most of the initial SNVset methods proposed, as well as many recently proposed tests, can be considered "competitive tests" in that they evaluate the evidence of statistical significance between the set of the interest and the phenotype in comparison to the evidence of statistical significance for other SNVs not in the set, or in random sets. For example, the use of Fisher's exact test [24, 25], or derivations of this hypergeometric approach [26, 27], is a common competitive approach. In essence these methods compare the proportion of significant SNVs in the set of interest to the proportion of significant SNVs not in the set. An early alternative to Fisher's exact test that attempts to address the potential limitation of the dichotomous nature of Fisher's exact test (classifying each SNV as "significant" or "not significant") used a weighted Kolmogorov–Smirnov test to compare the (weighted) distribution of *p*-values for SNVs in the set to SNVs not in the set [8, 28]. Recently, modifications to this approach have been proposed [29, 30]. Even more recently, Yaspan et al. [31] proposed a unique method to combine *p*-values from multiple SNVs accounting for LD structure, by comparing evidence for association in the set of interest to

Table 1
SNV p-value approaches to SNVset analysis

Name	Summary of analysis strategy	Link to software
ALIGATOR [27][a]	Define significant p-values by prespecified p-value cutoff; count significant genes in each pathway	http://x004.psycm.uwcm.ack.uk/~peter
GATES [33][b]	Extended Simes-like approach to combining p-values; SNV weights are possible	http://bioinfo.hku.hk/kggweb/ or SNVset R package (http://www.dordt.edu/statgen)
GESBAP [26][a]	Fisher's exact test-like approach; most-significant SNV per gene	http://bioinfo.cipf.es/gesbap/
GSA-SNP [29][a]	Average and max approaches; significance through permutation; GSEA also available	http://gsa.muldas.org
GSEA-SNP [28][a]	Extension of GSEA method	http://www.nr.no/pages/samba/area_emr_smbi_gseasnp
HGLMM [36][b]	Hierarchical generalized linear mixed model	http://biostat.mc.vanderbilt.edu/wiki/Main/LilyWang
iGSEA4GWAS [30][a]	GSEA-like approach; significance by permuting SNV labels	http://gsea4gwas.psych.ac.cn/
LCT, QT and DT [35][b]	Three separate tests to combine dependent p-values	https://sph.uth.tmc.edu/hgc/faculty/xiong/software-A.html
PARIS [31][a]	Pathway analysis through randomization by incorporating topological structure	http://ritchielab.psu.edu/ritchielab/software
VEGAS [34][b]	Combine p-values using average or max p-value (other options are possible); adjust for LD using simulation approach	http://genepi.qimr.edu.au/general/softwaretools.cgi or SNVset R package (http://www.dordt.edu/statgen)

[a]Competitive
[b]Self-contained

evidence in randomly chosen sets of SNVs. Table 1 provides a list of related software packages for these methods. Jia et al. [32] provided some head-to-head comparisons of these methods.

Self-Contained Tests

Self-contained tests only use the p-values (or other association measure) for the SNVs in the set itself, ignoring all SNVs outside of the test. In traditional statistical applications, Fisher's combination test, a Simes test, or other similar options would be used; however, these tests generally assume independence of the tests being conducted—which is generally not the case when combining SNVs, especially within a gene [33]. Recently, a few methods have been proposed to combine p-values while accounting for correlation between the SNVs. Both GATES [33] and VEGAS [34] attempt to correct for LD structure through low computational overhead approaches. A head-to-head comparison of the two methods is available in Li et al. [33]. In general, the two methods were similar. Though, we note that VEGAS has the flexibility of using different statistics (e.g., sum of all p-values, max of all p-values, or other choices, *see* **Note 13**). Other recently proposed options are the three methods of Luo et al. [35] and HGLLM [36], though these methods have not benefitted from direct comparison with other self-contained p-value tests.

3.1.2 Raw Genotypes Approaches

When raw genotype data are available, many additional methods become available to the researcher, though these methods are generally more complex and have a higher computational burden. A number of the methods involve regression-based approaches. For example, Gauderman et al. [37] propose regressing all SNVs in the set onto the phenotype or, more efficiently, regressing principal components representing the majority of the variability explained by the SNVs in the set onto the phenotype. More recently, Wu et al. [38] proposed a kernel machine based logistic regression analysis approach, Chen et al. [39] proposed a group ridge regression approach, and Schwender and Ruczinski [40] proposed a logic regression approach. In general, these approaches attempt to reduce the dimensionality of the problem, and then regress a reduced genotype dimension matrix onto the phenotype, with the potential for modeling complex interaction and epistatic relationships, while adjusting for covariates.

Other raw genotype approaches involve methods similar to the competitive tests using only p-values, but which utilize the raw data [41–43], Bayesian model fitting [44], multilocus tests for quantitative traits [45], and other recent novel approaches [46, 47]. Lastly, PLINK [48] offers a SNVset test that uses permutation to assess significance, but otherwise is very similar to [34].

In general, little comparative data are available on these approaches to establish when particular methods are optimal or more robust than other approaches. Table 2 provides an overview of the methods for which software is available.

3.1.3 Methods with No Software

Numerous other methods have been recently proposed but without software readily available. Table 3 provides a list of these methods and a brief description for the interested reader. It is possible that some of these methods are more optimal than methods in Tables 1 and 2; however, without readily available software they require custom software creation by the user. With a lack of comparative data for most methods arguing their general reliability, validity and power, undertaking the software development effort may be prohibitive.

Table 2
Raw genotype–phenotype approaches to SNVset analysis

Name	Summary of analysis strategy	Link to software
CCA [47]	Canonical correlation analysis	http://genepi.qimr.edu.au/staff/manuelF/gene/main.html
GRASS [37]	Group ridge regression, (also implements methods of [8, 27, 46])	http://linchen.fhcrc.org/grass.html
GWiS [44]	Greedy Bayesian model selection to identify independent effects within a gene; combined to generate single signal	www.baderzone.org
Kernel regression [38]	Kernel machine based logistic regression set	http://www.bios.unc.edu/~mwu/software/
Logic Regression [40]	Adaptive discrimination and regression procedure	logicFS R package (http://www.bioconductor.org)
LSKM [45]	Multilocus association test for quantitative traits	http://genetics.emory.edu/labs/epstein/software
Pathways of Distinction [46]	Evaluate the similarities of cases to cases and controls to controls using vectors representing the genotypes in the set	http://braun.tx0.org/PoDA
PLINK: *set-based test* [48]	Combines individual SNV statistics into a single score; permutation to assess significance	http://pngu.mgh.harvard.edu/~purcell/plink/
RS-SNP [42]	Evaluate whether proportion of significant SNVs in the set is more than expected by chance	http://www.biomedcentral.com/1471-2164/12/166/additional
SNP-Ratio Test [41]	Evaluates chance of observed percent of significant SNVs in set of interest	https://sourceforge.net/projects/snpratiotest/
SSEA [43]	Extension of GSEA	http://cbcl.ics.uci.edu/SSEA.

3.2 Implementing a SNVset Analysis

Because of the diversity of SNVset analysis strategies and software packages available, we cannot provide a detailed description of an analysis strategy that will work in all situations. Instead we will provide a broad overview of the procedure, followed by a specific example using a flexible, companion software package developed by our group [70].

3.2.1 Overview of Implementing a SNVset Analysis

Step 1. Select a SNVset method. While numerous options exist, self-contained *p*-value only tests are convenient, computationally simple, and theoretically less complicated than competitive and raw genotype tests. A companion R package implementing two recent, popular SNVset methods (GATES [33] and VEGAS [34]) is described in the following section (Subheading 3.2.2, [70]). If a *p*-value only test is being used, then all QC procedures (e.g., HWE)

Table 3
SNVset methods for which no software is available

Paper	Brief description of approach
Ballard et al. [49–51]	Regression with/without PCA, Summation and Binomial approximation
Chai et al. [52]	Extension of Fisher's method which accounts for correlation
Chasman et al. [53]	GSEA and K–S test, and hypergeometric variations for quantitative traits
Chen et al. [54]	Risk statistic for pathways
Chen et al. [55]	Markov random field that incorporates pathway topology
De la cruz et al. [56]	Combining p-values and adjusting for LD
Gao et al. [57]	Extension of regression approaches to kernel PCA
Guo et al. [58]	Extension of GSEA using SNV permutation
Hong et al. [59]	GSEA and Fisher's exact test extensions
Lebrec et al. [60]	Hierarchical Bayes model
Lee et al. [61]	Parametric gene set analysis approach
Li et al. [62]	Weighted LD based test
Menashe et al. [63]	Implementation of GSEA on GWAS data
Peng et al. [64]	Simes', Sidak's, Fisher's combination test, etc.
Shahbaba et al. [65]	Bayesian multinomial logit model
Sohns et al. [66]	Hierarchical Bayes model
Tintle et al. [67]	Summation and weighted summation methods on gene-level statistics
Wang and Elston [68]	Weighted score test
Wang et al. [8]	The original extension of GSEA to GWAS paper
Yu et al. [69]	Adaptive rank truncated product statistic

should be applied before running the test, and covariates should be accounted for in the SNV p-values. We also recommend running a few different tests as this can help to alleviate concerns about optimizing the methods choice for only certain disease architectures (*see* **Notes 1** and **13**), something that may be difficult to specify a priori. If users are already using PLINK to conduct their GWAS analysis, running the PLINK SNVset method is also a convenient option; however, we note that the methods implemented in our R package (GATES [33] and VEGAS [34]) are computationally simpler approximations of the method in PLINK [48]. Careful reading of Subheadings 3 and 4 (*see* **Notes 1–3** and **6–17**) is necessary to guide in making a decision on SNVset method. Of course, other methods in Tables 1 and 2 may also be appealing to researchers depending on their context.

Step 2. Identify functional sets of interest and create data files representing the functional sets which can be incorporated into the SNVset analysis software. Common choices are either gene-based tests or pathway-based tests. For gene-based tests, all SNVs within (or near) a gene are included in the SNVset for the gene. We recommend using all intragenic SNVs, and SNVs nearby to the gene (e.g., within 5–10 kb or in high LD with portions of the gene [23]). For pathway-based approaches more uncertainty exists. While the Gene Ontology and KEGG are popular choices, these databases generate many large sets of SNVs which may yield underpowered tests [71], and so smaller sets (like genes) may be more appropriate. Another option is the use of two-level analysis (*see* **Note 2**). If a weighting strategy is to be used that should also be developed at this stage (*see* **Note 4**).

Step 3. Run the software on the sets of interest and obtain *p*-values for all SNVsets tested. Multiple testing should be controlled across the number of sets tested in this stage. Post hoc analyses are available for some software packages, but no particular standards exist (*see* **Note 5**). A key question is whether the observed significant sets are the result of a single significant SNV in the set or the result of multiple significant SNV associations. Typically, individual SNV results should be viewed as a form of post hoc analysis to a SNVset analysis. In the case of a two-step SNVset analysis (*see* **Note 2**) the post hoc analysis may look at gene-based significance instead.

3.2.2 Example Implementing a SNVset Analysis

Here we discuss how an R package created by our group [70] which currently implements two self-contained, *p*-value tests (GATES and two different versions of VEGAS) can be used to conduct a SNVset analysis.

Simulation

Our example analysis consists of simulated data. We will provide a brief description of the simulation parameters here, a more detailed description is available in [23]. We simulated data for 5,000 cases and 5,000 controls, and consider data for 18 SNVs. Nine of the SNVs have minor allele frequency, MAF $= 0.10$ and nine of SNVs have MAF $= 0.4$. The nine low MAF SNVs are broken into three groups: a high LD-block (three SNVs, pairwise correlation, $r = 0.9$ between all three SNVs), a low LD-block (three SNVs, pairwise correlation, $r = 0.5$ between all SNVs), and a no LD group (pairwise correlation $= 0$ between all three SNVs); there is no correlation between any of the SNVs across the groups. The nine high MAF SNVs have similar LD structure, and there is no correlation between any of the high and low MAF SNVs. Of the 18 SNVs, 6 are causal: one SNV in each of the two high LD-blocks, one in each of the two low LD-blocks, and one low and one high MAF SNV that is not in LD with any other SNV. We note that while, realistically, it is possible to have sets of SNVs where none are causal, but some are in LD with causal SNVs we do not consider that case here—for most methods the impact will simply be a loss in power (see, e.g., ref. [23]).

In all cases the relative risk of disease from the causal SNV is 1.50, following an additive mode of inheritance. The disease prevalence is set to 10 %, with each of the six causal SNVs contributing 10 %/6 = 1.67 % of disease prevalence. Genotype and phenotype data were simulated and raw genotypes and phenotypes are provided in the files *genos2.txt* and *phenos2.txt*, which are part of the example files that go with [70].

Running the Analysis

Use of the SNVset package is explained in more detail in the package's help files [70], but we provide a brief description here. Two options are available: (1) The user generates *p*-values for each SNV and creates the SNV LD-matrix using other software. (2) The SNVset package generates *p*-values for each SNV and the SNV LD-matrix. In the case of option (2) the *raw_genotypes()* function is called passing in the following two files: *genos2.txt* (matrix of raw genotypes) and *phenos2.txt* (categorical or quantitative disease variable and relevant covariates). A general linear model (logistic or linear) is run generating a *p*-value for each SNV, after controlling for covariates. An LD-matrix is generated based on the observed correlation between the genotypes. Use of option (1) allows the researcher to bring in *p*-values generated by any method, along with an LD-matrix that could be based on the sample or haplotype databases (e.g., HapMap [14] and 1,000 Genomes [15]).

The *snvset_analyze* function is then run passing in the LD-matrix (*ld_matrix_sparse2.txt*) and the individual SNV *p*-values (*snv_assoc2.txt*). Additionally, SNV sets are passed using an indicator matrix format (*snv_set.txt*). For this example, we created four SNV sets: (1) all 18 SNVs, (2) all 6 causal SNVs, (3) all 4 noncausal SNVs that are not in LD, and (4) same as set #3, plus the MAF = 0.4 causal SNV that is not in LD. Running the *snvset_analyze* function with default settings generates three *p*-values for each set: the GATES *p*-value [33] and two VEGAS *p*-values: one using the SUM statistic and one using the MAX statistic [34]. Table 4 gives

Table 4
***p*-Values from example analysis using the SNVset R package**

Set	GATES	VEGAS-SUM[a]	VEGAS-MAX[a]
1 (18 SNVs: 6 causal, 12 noncausal)	0.004	<0.001	0.083
2 (6 causal SNVs)	0.002	<0.001	0.005
3 (4 noncausal SNVs)	0.249	0.195	0.228
4 (4 noncausal and 1 causal SNV)	0.003	0.002	0.003

[a]Because *p*-values for VEGAS are estimated by simulation results may differ slightly each time the function is called

Table 5
Post hoc analysis examining individual SNV *p*-values

SNV	Description	*p*-Value
1	Causal, MAF = 0.1, correlated with SNVs 2,3 ($r = 0.9$)	0.0006
2	Noncausal, MAF = 0.1, correlated with SNVs 1,3 ($r = 0.9$)	0.029
3	Noncausal, MAF = 0.1, correlated with SNVs 1,2 ($r = 0.9$)	0.03
4	Causal, MAF = 0.1, correlated with SNVs 5,6 ($r = 0.5$)	0.04
5	Noncausal, MAF = 0.1, correlated with SNVs 4,6 ($r = 0.5$)	0.16
6	Noncausal, MAF = 0.1, correlated with SNVs 4,5 ($r = 0.5$)	0.59
7	Causal, MAF = 0.1, not correlated with any other SNV	0.0006
8	Noncausal, MAF = 0.1, not correlated with any other SNV	0.34
9	Noncausal, MAF = 0.1, not correlated with any other SNV	0.40
10	Causal, MAF = 0.4, correlated with SNVs 11,12 ($r = 0.9$)	0.002
11	Noncausal, MAF = 0.4, correlated with SNVs 10,12 ($r = 0.9$)	0.023
12	Noncausal, MAF = 0.4, correlated with SNVs 10,11 ($r = 0.9$)	0.025
13	Causal, MAF = 0.4, correlated with SNVs 14,15 ($r = 0.5$)	0.13
14	Noncausal, MAF = 0.4, correlated with SNVs 13,15 ($r = 0.5$)	1.0
15	Noncausal, MAF = 0.4, correlated with SNVs 13,14 ($r = 0.5$)	0.83
16	Causal, MAF = 0.4, not correlated with any other SNV	0.0009
17	Noncausal, MAF = 0.4, not correlated with any other SNV	0.06
18	Noncausal, MAF = 0.4, not correlated with any other SNV	0.36

the *p*-values for the three methods and each of the four sets. A post hoc analysis looking at the individual SNV *p*-values (see Table 5) gives a fairly clear picture of which SNVs are driving the association within each set. Notably, none of the 18 SNVs analyzed would have met genome-wide significance (e.g., $p < 1 \times 10^{-8}$).

4 Notes

1. *Impact of disease architecture*. Simulation results [23, 33, 34] suggest a strong connection between identifying which SNVset methods are optimal and the underlying disease

architecture. However, little is known about which disease architectures are most reasonable and common. Furthermore, despite these recent results, no method has been deemed robust and powerful across a variety of disease architectures. Despite all of the general uncertainty around this issue there are two specific disease architectures worth considering: (1) The case where all causal SNVs in the set have similar weak to modest effects and (2) The case where one SNV has a strong effect and the remaining causal SNVs have relatively weak effects. In the first case, averaging the effects will generally prove most powerful, whereas in the second case taking the maximum observed effect will generally prove most powerful. While the first case is the one that motivates SNVset tests, and the second case motivates single SNV analyses, it is unknown how common these different disease architectures are biologically. A related methodological issue is described in the following note.

2. *Two-level vs. direct aggregation.* Related to the impact of disease architecture on choice of SNVset method, SNVset methods can be grouped as those methods that directly aggregate vs. those methods that employ a two-level aggregation strategy. Most gene-based tests employ direct aggregation. That is, they seek to directly combine the individual association signals from the SNVs in the set. On the other hand many "pathway" or "gene set" SNVset methods utilize a two-level aggregation strategy. For example, they advocate first combining SNV signals within genes, then combining gene-level signals into pathways. One of the most common choices for a gene-level aggregation statistic in a two-level aggregation strategy is to use the strongest observed SNV association within the gene, reflecting a general disease architecture of a single causal SNV within the gene (disease architecture #2 in Note 1). Having obtained a statistic for each gene, the gene-level statistics are often combined in a manner more consistent with a disease architecture reflecting many causal genes (SNVs) with similar effects (disease architecture #1 in Note 1). As noted in Note 1, the choice of strategy is dependent upon the beliefs the researcher has about the underlying disease architecture. Of particular note, however, is that two-level aggregation methods will generally have lower power than direct aggregation methods if the genes in the pathway contain multiple causal variants (see, for example, ref. [72]). The tradeoff, however, is the number of noncausal variants that are included in the set, as is discussed in the next note.

3. *The impact of noncausal variants.* In general, the more noncausal variants included in the set, the lower the power will be [23]. That said, some methods are more robust to the inclusion of noncausal variants than others. This robustness, however, comes at the expense of sensitivity to causal variants with weak effects. For example, implicit in the use of a two-level aggregation strategy as described in Note 2 is that direct aggregation of all SNVs in a pathway will introduce a very high percentage of noncausal variants to the set and limit power. To combat this potential power loss, SNVs are first aggregated to genes using the maximum observed effect within the gene—an approach that is the most robust to the inclusion of noncausal variants but the least powerful for situations with multiple weaker signals within the gene. When the pathway is finally tested by combining results across genes, the hope is that the effects of noncausal variants have been minimized. As we have noted already, selecting an optimal strategy is highly dependent on disease architecture. One way to more finely tune strategies and reduce the impact of noncausal variants is through the use of weighting, as discussed next.

4. *Weighting.* Traditionally, SNVset methods require rigid set definitions—the SNV is in the set, or it is not. The use of SNV weights, however, help the researcher to better reflect prior knowledge and biological insights. For more information about the many and varying sources of weights see Subheading 3.2. The goal of weighting is to accurately reflect a priori information about the SNVs in a way that reflects true disease architecture, resulting in increased power to identify true causal variation. This is a difficult task but a promising prospect as we learn more about disease etiology. That said, the highly capricious nature of weighting strategies is certainly worth careful consideration, and we recommend directly assessing the impact of weighting strategies by comparing unweighted and weighted results. Additionally, we recommend utilizing a post hoc analysis strategy to help further discern why any significant sets were significant, which we describe in more detail next.

5. *Post hoc analyses.* There is a dearth of research as to how best to follow-up a SNVset analysis that identifies a significant set. Clearly, some effort is needed to identify causal SNVs within the set when an entire set is identified as significant. The most straightforward option is to examine the individual SNV association results for all SNVs within the significant set, or for each gene when using a two-level aggregation strategy (*see* **Note 2**), with the application of an appropriate false discovery correction. However, this simplistic approach may be substantially hindered by low statistical power for such an analysis. Little is available in the literature as to what an appropriate false

discovery correction might be and if other more sophisticated post hoc testing methods may be more optimal.

6. *Individual level data*: *Use it or not?* Wang et al. [10] suggests that individual level (raw genotypes) data should be used where available. Undoubtedly, using raw genotypes increases the number of methodological choices available. However, we note that at the time Wang et al. [10] was written four of the five self-contained, *p*-value only methods (see Table 1) were not available, and the methods of Luo et al. [35] had only been very recently published. These methods show promise in head-to-head comparisons [23, 33]. We remind the reader that *p*-value only methods allow for substantially reduced computational overhead, and a substantially reduced methodological complexity. The balance of computational efficiency and methodological flexibility suggests that the use of raw genotypes where available is no longer an unambiguous decision.

7. *Self-contained vs. competitive null hypothesis.* In Subheading 3 we identified and discussed the difference between self-contained and competitive null hypothesis tests. A self-contained null hypothesis states that the SNVs in the set show no association with the phenotype, while a competitive null hypothesis states that the SNVs in the set show no more association with the phenotype than the SNVs not in the set. Wang et al. [10] are of the opinion that self-contained tests are better, and, in principle, we agree. Historically, it seems that competitive tests attempted to alleviate the deficiency of some methodological approaches in finding an appropriate null distribution with which to evaluate the observed test statistic. However, some of the newer classes of tests (self-contained, *p*-value only methods, *see* **Note 6**) seem to have managed to have a self-contained null hypothesis while still controlling the type I error rate.

8. *Impact of LD structure.* The impact of the linkage disequilibrium (LD; correlation) structure among the SNVs in the set on SNVset methodologies is varied and complex [23]. At the most basic level, when we aggregate individual SNV–phenotype associations we must account for the fact that some of the association signal may be inflated due to correlation structure between the SNVs in the set. This is, for example, why methods like GATES and VEGAS arose as alternatives to meta-analytic methods (e.g., Fisher's Method) that assume independent *p*-values. Instead, we must account for the correlation between SNVs in some way. Much, however, remains unknown about the impact of LD structure on SNVset tests, as its impact is specific to the various analytic approaches.

9. *Permutation methods.* Most methods that have been proposed to assess significance for SNVset tests involve permutation or

resampling to assess statistical significance. There are two main points worth mentioning about this practice. First, arguably, one of the things keeping people from widespread utilization of SNVset testing methods is the computational overhead required for permutation test approaches. Secondly, and importantly, caution is needed when considering the permutation approach for any specific method. Wang et al. [10] point out that permutations at the SNV level can disrupt LD patterns which, in turn, can inflate the type I error rate. The plethora of methods available and dearth of comparative evaluations of these methods preclude our ability to pinpoint specific problematic methods here.

10. *Covariates*. Adjusting for covariates is a critical component of most genetic association analyses, and should be strongly considered as part of a SNVset analysis. However, the way covariates are considered varies widely. The simplest way to handle covariates is to use a SNV *p*-value method. In this case, the SNV *p*-value used should reflect the evidence that the SNV is associated with the phenotype, after covariate adjustment. When using raw genotype approaches some methods can, and others cannot, incorporate a covariate adjustment. Regression-based approaches simply use the covariates as predictors in the model equation, while other more complex methods are used in other cases.

11. *Pathway topology*. Recently, some SNVset approaches have attempted to incorporate SNVset topology into SNVset methods (e.g., see ref. [55]). For example, imagine that there is a metabolic pathway set of ten genes. A database like KEGG [18] will indicate that certain genes are directly connected in the metabolic pathway, while others are not. In a traditional SNVset analysis all SNVs in the set are treated equally; however, it may be possible to leverage set topology (connections between genes) to consider subsets of the larger set that are coassociated with disease risk, while ignoring subsets that make less biological sense to be coassociated with disease risk. In essence, these approaches more realistically integrate biological knowledge in the statistical analysis. This is sure to be an area of additional work. For more comments on set choice *see* **Note 15**.

12. *Rare variants*. While methods for SNVset testing for common variants have been our focus here, we must acknowledge that a parallel research track is in the area of rare variant testing methods for sequence data. Essentially, researchers use SNVset methods, with typical sets focusing on genes as the unit of analysis, or even just the exonic region of genes. Initial methods developed for rare variants focused exclusively on rare variants (e.g., see refs. [72–76]) and ignored the SNVset methods described here, which focus on common variants. More

recently, and a trend that we see continuing, rare variant and common variant SNVset methods are being considered simultaneously (e.g., see ref. [77]).

13. *Replication of results.* As noted by Wang et al. [10], and not surprising based on Note 1, different SNVset methods may give different results about whether a set is significant or not. Thus, the issue of replication is important. Applying a few dramatically different SNVset approaches may be advantageous in identifying sets that are significantly associated with the phenotype. For example, application of the VEGAS method [34] using both the maximum statistic and the sum statistic may identify two different types of sets as statistically significant and increase the sensitivity of the SNVset approach. However, if this approach is taken, caution is needed with regards to both multiple-testing issues and post hoc analyses (*see* **Note 5**).

14. *Informal pathway analysis.* Given all of the open questions and unknowns surrounding the current state of SNVset methods and testing, taking the traditional ad hoc, informal approach to interpretation is reasonable [10]. For example, if a few SNVs in a pathway of interest meet genome-wide significance when analyzed separately while a few others in the pathway are borderline significant, that is fairly strong evidence that the pathway is, overall, associated with the phenotype. *see* **Notes 15** and **16** and **17** for more thoughts.

15. *Method and set choice.* In the end, how should researchers go about choosing a method? Careful thought needs to go into what hypotheses are being tested, to what extent prior biological hypotheses exist about the disease process, and whether complementary data will be included via weighting. Decisions must be made about which SNVsets will be tested. While it is tempting to entertain the possibility of testing every set possible, the analysis will then become much like a typical single SNV analysis and many of the advantages of the SNVset approach are lost. A critical question is deciding whether to focus on gene-based sets or pathway-based sets (though, other options are also available, and doing both is also reasonable). It is clear that set choice has a critical impact on power, but little is known on a broad scale about which sources of sets are generally the best. One exception is our recent work showing that GO and KEGG based sets were nonoptimal across a variety of statistical methods, compared to smaller more refined sets—however, our analysis was restricted to prokaryotic gene expression data [71]. Further work is necessary to extend these results to other DBs, -omics data, and eukaryotes.

We also see recent self-contained, *p*-value methods (*see* Table 1) as potentially setting a new industry standard for ease of use/implementation, power and maintenance of the type I error rate. While we are optimistic about these methods, the next few years will dictate whether these approaches live up to their early promise. We have implemented two of these methods in a companion R package for this chapter, as described earlier [70].

16. *Further reading.* For the interested reader, additional recent reviews are available in Cantor et al. [78], Tintle et al. [79], Wang et al. [10], and Jia et al. [32].

17. *Final remarks.* A fair question is whether it is wise to consider using SNVset methods at all, given the general uncertainty around key methodological questions. Stated fairly, to date, SNVset methods have yet to deliver on their anticipated promise and have not yet yielded many meaningful biological insights. However, this lack of success in the application of SNVset methods may be due, in large part, to the variety of methodological choices and challenges posed by the unknown impact of underlying disease architecture. In the early days of single SNV analysis, numerous analysis methods were available until strong arguments were made that using an additive mode of inheritance model (e.g., coding genotypes as 0, 1, 2) was a broadly robust approach regardless of true disease architecture. We anticipate similar clarity emerging in the field of SNVset testing as methods mature and increasing knowledge of disease architecture emerges. We are particularly optimistic given recent methods implementing self-contained, p-value only approaches that are computationally simple. However, until these methods are fully vetted, we see value in using SNVset methods as a complementary analysis, which may give additional insights into the biological underpinnings of complex disease, to a (traditional) single SNV analysis. Importantly, continued attention should be paid to recently published research in this emerging area of genetic data analysis.

References

1. Hindorff LA, Sethupathy P, Junkins HA et al (2009) Potential etiologic and functional implications of genome-wide association loci for human diseases and traits. Proc Natl Acad Sci 106:9362–9367
2. Visscher P, Brown MA, McCarthy M et al (2012) Five years of GWAS discovery. Am J Hum Genet 90:7–24
3. Subramanian A, Tamayo P, Mootha VK et al (2005) Gene set enrichment analysis: a knowledge-based approach for interpreting genome-wide expression profiles. Proc Natl Acad Sci 102:15545–15550
4. Dinu I, Potter JD, Mueller T et al (2007) Improving gene set analysis of microarray data by SAM-GS. BMC Bioinformatics 8:242
5. Tian L, Greenberg SA, Kong SW et al (2005) Discovering statistically significant pathways in expression profile studies. Proc Natl Acad Sci 102:13544–13549
6. Efron B, Tibshirani R (2007) On testing the significance of sets of genes. Ann Appl Stat 1:107–129

7. Tintle NL, Best AA, DeJongh M et al (2008) Gene set analyses for interpreting microarray experiments on prokaryotic organisms. BMC Bioinformatics 9:469
8. Wang K, Li M, Bucan M (2007) Pathway-based approaches for analysis of genomewide association studies. Am J Hum Genet 81:1278–1283
9. Ala-Korpela M, Kangas AJ, Inouye M (2011) Genome-wide association studies and systems biology: together at last. Trends Genet 27(12): 493–498
10. Wang K, Li M, Hakonarson H (2010) Analysing biological pathways in genome-wide association studies. Nat Rev Genet 11(12): 843–854
11. http://www.broadinstitute.org/gsea/msigdb
12. http://www.ensembl.org
13. http://www.ncbi.nlm.nih.gov/projects/SNP/
14. http://www.hapmap.org
15. http://www.1000genomes.org
16. http://www.geneontology.org
17. http://www.biocarta.com
18. http://www.genome.jp/kegg/
19. http://www.bioconductor.org
20. http://genetics.bwh.harvard.edu/pph/
21. http://sift.jcvi.org/
22. Lopes MC, Joyce C, Ritchie GRS et al (2011) A combined functional annotations score of non-synonymous variants. Hum Hered 73:47–51
23. Petersen A, Alvarez C, DeClaire S, Tintle NL (2013) Assessing methods for assigning SNPs to genes in gene-based testes of association using common variants. PLoS One. In press
24. Elbers CC, van Eijk KR, Franke L (2009) Using genome-wide pathway analysis to unravel the etiology of complex diseases. Genet Epidemiol 33:419–431
25. Torkamani A, Topol E, Schork N (2008) Pathway analysis of seven common diseases assessed by genome-wide association. Genomics 92(5): 265–272
26. Medina I, Motaner D, Bonifaci N et al (2009) Gene set-based analysis of polymorphisms: finding pathways or biological processes associated to traits in genome-wide association studies. Nucleic Acids Res 37:W340–W344
27. Holmans P, Green E, Pahwa JS et al (2009) Gene ontology analysis of GWA data sets provides insights into the biology of bipolar disorder. Am J Hum Genet 85:13–24
28. Holden M, Deng S, Wojnowski L et al (2008) GSEA-SNP: applying gene set enrichment analysis to SNP data from genome-wide association studies. Bioinformatics 24(23):2784–2785
29. Nam D, Kim J, Kim S et al (2010) GSA-SNP: a general approach for gene set analysis of polymorphisms. Nucleic Acids Res 38: W749–W754
30. Zhang K, Cui S, Chang S et al (2010) i-GSEA4GWAS: a web server for identification of pathways/gene sets associated with traits by applying an improved gene set enrichment analysis to genome-wide association study. Nucleic Acids Res 38:W90–W95
31. Yaspan BL, Bush WS, Torstenson ES et al (2011) Genetic analysis of biological pathway data through genomic randomization. Hum Genet 129:563–571
32. Jia P, Wang L, Meltzer HY et al (2011) Pathway-based analysis of GWAS datasets: effective but caution required. Int J Neuropsychopharmacol 14:567–572
33. Li M, Gui H, Kwan J et al (2011) GATES: a rapid and powerful gene-based association test using extended simes procedure. Am J Hum Genet 88:283–293
34. Liu JZ, Mcrae AF, Nyholt DR et al (2010) A versatile gene-based test for genome-wide association studies. Am J Hum Genet 87:139–145
35. Luo L, Peng G, Zhu Y et al (2010) Genome-wide gene and pathway analysis. Eur J Hum Genet 18:1045–1053
36. Wang L, Jia P, Wolfinger RD et al (2011) An efficient hierarchical generalized linear model for pathway analysis of genome-wide association studies. Bioinformatics 27(5):686–692
37. Gauderman WJ, Murcray C, Gilliland F et al (2007) Testing association between disease and multiple SNPs in a candidate gene. Genet Epidemiol 31:383–395
38. Wu MC, Kraft P, Epstein MP et al (2010) Powerful SNP-set analysis for case-control genome-wide association studies. Am J Hum Genet 86:929–942
39. Chen LS, Hutter CM, Potter JD et al (2010) Insights into colon cancer etiology using a regularized approach to gene set analysis of GWAS data. Am J Hum Genet 86:860–871
40. Schwender H, Ruczinski I (2011) Testing SNPs and sets of SNPs for importance in association studies. Biostatistics 12:18–32
41. O'Dushlaine C, Kenny E, Heron E et al (2009) The SNP ratio test: pathway analysis of genome-wide association datasets. Bioinformatics 25(20):2762–2763
42. D'Addabbo A, Palmieri O, Latiano A et al (2011) RS-SNP: a random-set method for genome-wide association studies. BMC Genet 12:166

43. Weng L, Macciardi F, Subramanian A et al (2011) SNP-based pathway enrichment analysis for genome-wide association studies. BMC Bioinformatics 12:99
44. Huang H, Chanda P, Alonso A et al (2011) Gene-based tests of association. PLoS Genet 7(7):e1002177
45. Kwee L, Liu D, Lin X et al (2008) A powerful and flexible multilocus association test of quantitative traits. Am J Hum Genet 82:386–397
46. Braun R, Buetow K (2011) Pathways of distinction analysis: a new technique for multi-SNP analysis of GWAS data. PLoS Genet 7: e1002101
47. Tang CS, Ferreira MAR (2012) A gene-based test of association using canonical correlation analysis. Bioinformatics 28(6):845–850
48. Purcell S, Neale B, Todd-Brown K et al (2007) PLINK: a toolset for whole-genome association and population-based linkage analysis. Am J Hum Genet 81:559–575
49. Ballard D, Abraham C, Cho J et al (2010) Pathway analysis comparison using Crohn's disease genome wide association studies. BMC Med Genet 3:25
50. Ballard DH, Aporntewan C, Lee JY et al (2009) A pathway analysis to genetic analysis workshop 16 genome-wide rheumatoid arthritis data. BMC Proc 3(Suppl 7):S91
51. Ballard DH, Cho J, Zhao H (2010) Comparisons of multi-marker association methods to detect association between a candidate region and disease. Genet Epidemiol 34:201–212
52. Chai HS, Sicotte H, Bailey KR et al (2009) GLOSSI: a method to assess the association of genetic loci-sets with complex diseases. BMC Bioinformatics 10:102
53. Chasman DI (2008) On the utility of gene set methods in genome wide association studies of quantitative traits. Genet Epidemiol 32:658–668
54. Chen L, Zhang L, Zhao Y et al (2009) Prioritizing risk pathways: a novel association approach to searching for disease pathways fusing SNPs and pathways. Bioinformatics 25(2): 237–242
55. Chen M, Cho J, Zhao H (2011) Incorporating biological pathways via a markov random field model in genome-wide association studies. PLoS Genet 7(4):e1001353
56. De la Cruz O, Wen X, Ke B et al (2010) Gene, region and pathway level analyses in whole-genome studies. Genet Epidemiol 34:222–231
57. Gao Q, He Y, Yuan Z et al (2011) Gene- or region-based association study via kernel principal component analysis. BMC Genet 12:75
58. Guo Y, Li J, Chen Y et al (2009) A new permutation strategy of pathway-based approach for genome-wide association study. BMC Bioinformatics 10:429
59. Hong M, Pawitan Y, Magnusson PKE et al (2009) Strategies and issues in the detection of pathway enrichment in genome-wide association studies. Hum Genet 126:289–301
60. Lebrec JJ, Huizinga TW, Toes RE et al (2009) Integration of gene ontology pathways with north American rheumatoid arthritis consortium genome-wide association data via linear modeling. BMC Proc 3(Suppl 7): S94
61. Lee J, Ahn S, Oh S et al (2011) SNP-PRAGE: SNP-based parametric robust analysis of gene enrichment. BMC Syst Biol 5(Suppl 2):S11
62. Li M, Wang K, Grant SFA et al (2008) ATOM: a powerful gene-based association test by combining optimally weighted markers. Bioinformatics 25(4):297–503
63. Menashe I, Maeder D, Garcia-Closas M et al (2010) Pathway analysis of breast cancer genome-wide association study highlights three pathways and one canonical signaling cascade. Cancer Res 70(11):4453–4459
64. Peng G, Luo L, Siu H et al (2010) Gene and pathway-based second-wave analysis of genome-wide association studies. Eur J Hum Genet 18:111–117
65. Shahbaba B, Shachaf CM, Yu Z (2012) A pathway analysis method for genome-wide association studies. Stat Med 31:988–1000. doi:10.1002/sim.4477
66. Sohns M, Rosenberger A, Bickeboller H (2009) Integration of a priori gene set information into genome-wide association studies. BMC Proc 3:S95
67. Tintle N, Borchers B, Brown M et al (2009) Comparing gene set analysis methods on single-nucleotide polymorphism data from genetic analysis workshop 16. BMC Proc 3:S96
68. Wang T, Elston RC (2007) Improved power by use of a weighted score test for linkage disequilibrium mapping. Am J Hum Genet 80:353–360
69. Yu K, Li Q, Bergen AW et al (2009) Pathway analysis by adaptive combination of p-values. Genet Epidemiol 33(8):700–709
70. SNVset, R package. http://www.dordt.edu/statgen
71. Tintle NL, Sitarik A, Boerema B et al (2012) Evaluating the quality of gene sets used in the analysis of bacterial gene expression data. BMC Bioinformatics 13:193

72. Madsen BE, Browning SR (2009) A groupwise association test for rare mutations using a weighted sum statistic. PLoS Genet 5:e10000384
73. Li B, Leal S (2008) Methods for detecting associations with rare variants for common diseases: applications to analysis of sequence data. Am J Hum Genet 83:311–321
74. Morris AP, Zeggini E (2010) An evaluation of statistical approaches to rare variant analysis in genetic association studies. Genet Epidemiol 34:188–193
75. Zawistowski M, Gopalakrishnan S, Ding J et al (2010) Extending rare-variant testing strategies: analysis of noncoding sequence and imputed genotypes. Am J Hum Genet 87:604–617
76. Wu MC, Lee S, Cai T et al (2011) Rare variant association testing for sequencing data with the sequence kernel association test (SKAT). Am J Hum Genet 89:82–93
77. Dai Y, Jiang R, Dong J (2012) Weighted selective collapsing strategy for detecting rare and common variants in genetic association study. BMC Genet 13:7
78. Cantor RM, Lange K, Sinsheimer JS (2010) Prioritizing GWAS results: a review of statistical methods and recommendations for their application. Am J Hum Genet 86:6–22
79. Tintle N, Lantieri F, Lebrec J et al (2009) Inclusion of a priori information in genome-wide association analysis. Genet Epidemiol 33:S74–S80

Chapter 26

Genomic Selection in Animal Breeding Programs

Julius van der Werf

Abstract

Genomic selection can have a major impact on animal breeding programs, especially where traits that are important in the breeding objective are hard to select for otherwise. Genomic selection provides more accurate estimates for breeding value earlier in the life of breeding animals, giving more selection accuracy and allowing lower generation intervals. From sheep to dairy cattle, the rates of genetic improvement could increase from 20 to 100 % and hard-to-measure traits can be improved more effectively.

Reference populations for genomic selection need to be large, with thousands of animals measured for phenotype and genotype. The smaller the effective size of the breeding population, the larger the DNA segments they potentially share and the more accurate genomic prediction will be. The relative contribution of information from relatives in the reference population will be larger if the baseline accuracy is low, but such information is limited to closely related individuals and does not last over generations.

Key words Genomic selection, Breeding programs, Reference population, Livestock genetics

1 Introduction

Breeding programs are mainly driven by the choice of traits in the breeding objective, and their relative importance, the investment in trait measurement, and decisions about selection and mating based on estimated breeding value (EBV). Currently, the main tools available to breeders are EBVs and indices, and the information used herein are phenotypes measured on selection candidates and/or their relatives. EBVs are best predictions of an animal's breeding value given all data available on phenotypic measurement and pedigree, and this can be enhanced by genomic information. One of the key questions for individual breeders is what information should be collected to drive breeding programs. With the advent of genomic selection, a typical question that arises is "should I invest in genotyping and which animals should I DNA test?"

Genomic selection can have a major impact on animal breeding programs, especially where traits that are important in the breeding objective are hard to select for otherwise. Genomic selection provides more accurate estimates for breeding value earlier in the life of breeding animals, giving more selection accuracy and lower generation intervals. Especially in dairy cattle, where the important breeding objective traits can only be measured on females and only after they have reproduced, genomic selection can potentially nearly double the rate of gain. In beef cattle, pigs, and sheep, growth is an important trait and this can be measured before reproduction. However, further improvements of breeding objectives call for breeding value information on traits that may be difficult or expensive to measure on-farm. Breeding programs that have objectives including the so-called hard-to-measure traits such as meat quality, feed efficiency, reproduction, and disease resistance will benefit more from genomic selection as these traits can be improved more effectively. Overall, rates of genetic gain could improve 20–100 %, somewhat depending on the breeding objective and the size of the reference population. The effect of genomic selection on the rates of gain can be predicted by selection index theory combined with an algorithm to optimize age structure. Further detail on these methods will be presented in this chapter.

To predict breeding value based on genomic information does require a reference population that needs to be large (thousands of animals measured) and to some extent represent the lineages and breeds found in the commercial breeding population. It is important to determine the required size and the genetic constitution of a reference population as this will have an important impact of the effective implementation of genomic selection. The accuracy of genomic predictions of breeding value (GBV) depends on the size of the reference population, how related it is to the animals to be predicted, the effective population size of the breed, and the heritability of the trait. The effective population size (N_e) is mainly related to the number of sires used in the breeding population and it determines how much individuals within a population have on average in common genetically if they are not direct relatives. The smaller the effective population size, the larger the DNA segments they potentially share and the more accurate genomic prediction will be.

Simulation modeling as well as analysis of real data shows that genomic predictions are weaker when the animals to be predicted are more distantly related from the reference population. If an animal has many closely related animals in the reference population, then genomic prediction accuracy is high, but a pedigree-based BLUP prediction of breeding value will be reasonably accurate. If animals have no direct relatives in the reference set, their prediction will depend on "average relatedness" with an accuracy that can be predicted from the effective size of the breeding population [1]. The latter accuracy can be called "baseline accuracy," which is an accuracy to be expected for a member of a breed without direct relatives in the reference. High baseline accuracy can only be achieved with large reference populations, and predictions based on this information will last for a few generations. The contribution of relatives' information will be larger if the baseline accuracy is low, but information from relatives is limited to closely related individuals and does not last over generations. Genomic relationships across breeds would be much smaller than within breeds. Theoretically they could be determined with a very high density of markers information. However, to date there is limited evidence that genomic prediction can be accurate across breeds. This chapter will present methods and a discussion on the design of reference populations for genomic selection. The chapter will mainly discuss livestock examples, but most of the principles discussed are general and are also relevant for plant breeding programs.

2 Prediction of Increased Rates of Genetic Gain in Breeding Programs

2.1 Selection Index Theory

Multiple-trait selection index theory can be used to predict the potential benefit of including DNA information in selection decisions. The genetic basis for constructing selection indexes was developed more than 60 years ago [2]. The selection index methodology determines the selection criterion coefficients that maximize the response in a given breeding objective on the basis of available information. The phenotypic records available for use as selection criteria in indexes can be assumed to be those that are routinely recorded by breeders. Hence, a list of traits needs to be compiled with information on whether they are measured, at what age and on what type of relatives, and estimates of genetic and phenotypic parameters (variances, heritabilities, and correlations) and, finally, their relative economic value. To model the value of genomic information, one additional parameter is required: the proportion of additive genetic variance explained by the genomic information.

The information used for constructing a selection index can be summarized in a vector **x** where element x_i refers to a phenotype, or an average phenotype, on the individual, its sire, its dam, its full sibs, its half sibs, and/or its progeny for each of the measured traits. We define the matrix **P** = Var(**x**) as the matrix with variances and covariances among all information sources and the matrix **G** as Cov(**x**, **a**), where **a** is the vector with breeding values for all traits. The construction of **P** and **G** matrices for standard selection index theory is summarized below and requires knowledge of genetic and phenotypic parameters and of the type of information used in x_i (which relative, how many records).

Summarizing the rules for setting up selection index equations (single trait):

*For the **P**-matrix*:

$\text{Var}(x_i) = \sigma_P^2$ for the variance of any single measurement and σ_P^2 is the phenotypic variance

$\text{Var}(\bar{x}_i) = r\sigma_P^2 + (1-r)/n\sigma_P^2$ for the variance of a mean of n individuals, where r is the intra-class correlation between the measurements in the group, e.g., $r =$ repeatability for the mean of repeated measurements on the same animal, and $r = \frac{1}{4} h^2$ for a half sib or progeny mean

$\text{Cov}(x_i, x_j) = a_{ij}\sigma_A^2$ for the covariance between two single measurements on two different individuals, a_{ij} is the additive genetic relationship between these individuals. In some cases the individuals could share a common environment, in which case we add a term $c^2\sigma_P^2$, where c^2 is the proportion of variation due to common environment

$\text{Cov}(x_i, \bar{x}_j) = a_{ij}\sigma_A^2$ for the covariance between a single measurement and the mean of measurements, a_{ij} is the additive genetic relationship between one individual and each of the members that make up the mean

In working out selection index equations, it is convenient to take a mean of a group of individuals, which is allowed if all members of that group have the same relationship with each other, as well as the same relationship with the animal for which we calculate an EBV. Hence, the group needs to be "homogeneous."

*And for the **G**-vector*:

$\text{Cov}(x_i, a) = a_{ij}\sigma_A^2$ for the covariance between a single measurement and the breeding value of the EBV-animal

$\text{Cov}(\bar{x}_i, a) = a_{ij}\sigma_A^2$ for the covariance between a mean of measurements and the breeding value of the EBV-animal, a_{ij} is the additive genetic relationship between the EBV-animal and each of the members that make up the mean

For multi-trait selection indices, the **P** matrix will have diagonal blocks for the information about each trait, and off-diagonal blocks to indicate covariances among these information sources between traits. These elements are essentially the same as above, except that genetic and phenotypic variance is replaced by genetic/phenotypic covariance. For information sources based on average performances, it is possible that there are unequal numbers used for different traits and an assumption has to be made about the number (e.g., repeated records, or full sib, half sib, or progeny records) in common between the two traits. For example, if a selection candidate has five half sibs measured for one trait and ten for another trait, it is relevant to know how many half sibs have both traits recorded.

The selection index weights are calculated as $\mathbf{b} = \mathbf{P}^{-1}\mathbf{G}\mathbf{v}$, where $\mathbf{v}$ is a vector with economic values and $\mathbf{b}$ are weights such that the index $I = \mathbf{b}'x$ provides a ranking among selection candidates that gives the highest expected correlation with their ranking based on true genetic merit according to the breeding objective. The predicted response to selection (per unit of index SD) based on the index is $\Delta \mathbf{G} = \mathbf{b}'\mathbf{G}/\sigma_I$ where $\sigma_I = \sqrt{\mathbf{b}'P\mathbf{b}}$ is the standard deviation of the index. The accuracy of predicting the breeding objective is calculated as σ_I/σ_H where $\sigma_H = \sqrt{\mathbf{v}'C\mathbf{v}}$ is the standard deviation of the aggregate genotype of the breeding objective, and $\mathbf{C} = \mathrm{Var}(\mathbf{a})$. Response for each trait is determined with the regression of each trait on the index. The response for trait i is $\delta g_i = \mathbf{b}'G_i/\sigma_I$, where G_i is the ith column of $\mathbf{G}$ (*see* **Note 1**).

Under directional selection, the additive genetic variance in a population decreases and establishes at an equilibrium but lower level than the unselected variance. This phenomenon has been described as the Bulmer effect, and overall responses will be about 20 % lower after the Bulmer correction. Bijma [3] showed that this also affects the relative gains of genomic selection over non-genomic selection. It will make the advantage of genomic selection relatively higher, especially when non-genomic selection is strongly dominated by family information (e.g., when the selection criterion is parent average). If non-genomic selection contains information from own performance, the Bulmer effect is likely to have less influence on the comparison.

2.2 Including Genomic Prediction in a Selection Index Framework

Information from DNA test information can be included in $\mathbf{x}$ as a molecular breeding value (q_i) explaining a proportion (ρ) of the additive genetic variance (σ^2_{ai}) in trait i; $V_{qi} = \rho\sigma^2_{ai}$, as described by Lande and Thompson [4]. Hence, the diagonal in $\mathbf{P}$ for this information is V_{qi} and off diagonals in $\mathbf{P}$ are covariances between q_i and other information sources is $\mathrm{Cov}(q_i, x_i) = a_{ij}\rho\sigma^2_{ai}$, where a_{ij} is the additive genetic relationship between individuals in x_i and the selection candidate.

As an example, assume a trait with $\sigma^2_P = 1$, $\sigma^2_A = 0.25$, and $\rho = 0.3$, i.e., 30 % of the additive genetic variance is explained by the genomic test and the accuracy of the genomic test is $\sqrt{0.3}$. The variance of the genomic test is $\sigma^2_q = 0.075$. A selection index for a single own phenotype can be considered. With no genomic selection, we simply have $\mathbf{P} = \sigma^2_P$, $\mathbf{G} = \sigma^2_A$, and $\mathbf{b} = \sigma^2_A/\sigma^2_P = 0.25$ (is equal to the heritability). The accuracy is $\sqrt{[\mathbf{b}'P b/\sigma^2_A]} = 0.5$. With genomic selection we can write the **P**-matrix as

$$\mathbf{P} = \begin{bmatrix} 1 & 0.075 \\ 0.075 & 0.075 \end{bmatrix} \text{ and } \mathbf{G} = \begin{bmatrix} 0.25 \\ 0.075 \end{bmatrix} \text{ such that } \mathbf{b} = \begin{bmatrix} 0.189 \\ 0.811 \end{bmatrix}$$

and the accuracy is 0.658

An alternative model can be written where the breeding value is separated into a QTL and a polygenic component (*see* **Note 2**).

The covariances between variables for different traits (traits i and i') are: $\mathrm{Cov}(q_i, q_{i'}) = \rho_i\rho_i\sigma_{aii'}$, where $\sigma_{aii'}$ is the genetic covariance and $\text{Cov}(q_i, x_{i'}) = a_{ii'}\rho_i\sigma_{aii'}$. In **G** we have to define the covariance between q_i and the true breeding value (a) is $\text{Cov}(q_i, a) = \rho\sigma_{ai}^2$ and $\text{Cov}(q_i, a_{i'}) = \rho_i\sigma_{aii'}^2$ (*see* **Note 3**).

The value for ρ_i, the reliability of the genomic test for trait i, could be determined based on data analysis, but in modeling of genomic selection scenarios, prior assumptions can be made. In general, this reliability will be dependent on the size of the reference population (see Subheading 2.2) and for a given size of reference population; it will also depend on the heritability of the trait. It may be convenient to model these values as a proportion of heritability, assuming that the genomic test explains a percentage of additive genetic variance. For example, to mimic scenarios that vary in the size of the reference population one can assume that the reliability of the genomic test is equal to either one quarter ($\rho_i = \frac{1}{4}h_i^2$; low accuracy scenario), one half ($\rho_i = \frac{1}{2}h_i^2$; intermediate accuracy scenario), or the full heritability ($\rho_i = h_i^2$; high accuracy scenario). Alternatively, formula's by Goddard [1] or Daetwyler et al. [10] or Goddard et al. [8] could be used to predict genomic selection accuracy for a given reference population size (see Subheading 3).

2.3 Optimizing Age Structure in Breeding Programs

The rate of gain in a breeding program is predicted by the breeder's equation: $R = ir\sigma_A/L$, where i = selection intensity, r is accuracy and L is generation interval. These components are not independent. Usually, when selection candidates become older, more information can be collected about their breeding value, e.g., own performance record(s), sib, and progeny information. Hence, EBVs of older animals are generally more accurate. In optimizing selection strategies, one can choose between selecting young breeding animals (low r, low L) or older breeding animals (high r, high L). An optimal strategy is usually to select the very best of both. This can be achieved by truncation selection across age classes [5] as illustrated in Fig. 1.

Figure 1 shows overlapping distributions, representing values of animals from two different age classes. These values could be phenotypes, EBVs or index values, and assuming selection is for such a single criterion. The distributions to the left could represent the oldest animals, as they will have the lowest mean if there is a positive genetic trend.

The objective is to find a common truncation point x where selection of all animals across all available distributions leads to a total proportion selected **P**. There will be one truncation point across all distributions. This truncation point has a single value, but relative to the mean of each distribution, the truncation points will differ. Therefore, selected proportion within each cohort as well as selection intensities will differ.

There is no algebraic-closed solution to finding the optimal truncation point but the solution can be found iteratively. If the

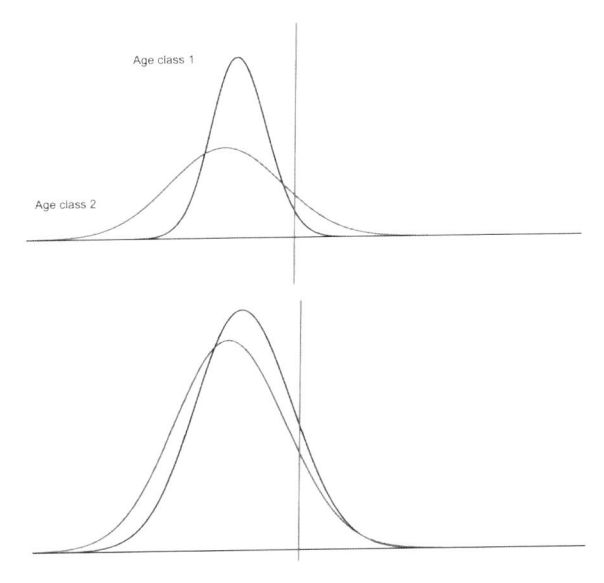

Fig. 1 Example of truncation selection across age class: the distributions represent those of selection criteria (Estimated Breeding Values—EBVs) for two age classes, with the older animals having more variation (due to higher EBV-accuracy) but slightly lower mean. The *top* figure is with no genomic selection (accuracies 0.4 and 0.8, respectively), the *bottom* figure is with genomic selection (accuracies 0.7 and 0.8, respectively). Without genomic selection, only 27 % of selected individuals will be from age class 1 whereas this is 54 % in the case with genomic selection

truncation point is too far to the right, the overall selected proportion will be too small, and if too far to the left, too many will be selected. Therefore, an iterative algorithm [6] can work as follows:

1. Choose starting values to the left and far (e.g., six standard deviations) to the right: xL and xU for lower and upper threshold.
2. Choose a truncation point in the middle, xm = (xL + xU)/2.
3. For a given xm, determine the total number selected.
   This can be done by working out for each distribution:
   (a) xi, truncation point in standard units relative to that distribution.
   (b) From xi, determine pi: proportion selected from that cohort.
   (c) Total number selected N_t is the sum of the number selected from each cohort.
4. If the N_t is close to the number that needs to be selected (N), stop the iteration.
5. If $N_t < N$, then xU = xm.
6. If $N_t > N$, then xL = xm.
7. Return to step 2.

In determining the outcomes of various scenarios with and without genomic selection (with various accuracies), we can (1) determine the selection accuracy of each age class and for each sex, (2) optimize selected proportion from each age class, and (3) calculate the overall selection accuracy and overall response as well as response per trait (*see* **Note 4**).

3 Theoretical Prediction of Genomic Prediction Accuracy

3.1 Approaches to Predict Genomic Selection Accuracy

Genomic prediction can be seen as predicting the effects of polymorphisms in many segments of the genome on phenotype. The phenotype (or breeding value) of young animals can be predicted by comparing their segments with those that have been both genotyped and phenotyped, the latter being referred to as the reference population. Another way of looking at it is to compare the similarity of genome segments among individuals, where a selection candidate is predicted based on covariances with other individuals that have been measured. These covariances are based on genomic relationships and they are derived from genomic information. The two approaches are represented by two methods termed Ridge Regression Best Linear Unbiased Prediction (RR-BLUP) and Genomic Best Linear Unbiased Prediction (GBLUP) and they are equivalent, i.e., they provide the same ranking of animals [7]. When predicting the accuracy of genomic prediction, both approaches can be used: how accurate are the predicted effects of the various genome segments effects and how strongly a selection candidate is related (genomically) to other animals with phenotype. The first approach was mainly followed by Goddard [1] but the second approach is followed by Goddard et al. [8] and this one becomes useful when we consider the effect of having more direct relatives in the reference population.

3.2 Predicting the Baseline Accuracy

The baseline accuracy of genomic prediction is predicted by the work described in Goddard [1]. He did actually not use this term, rather it was defined by Clark et al. [9] as the accuracy of genomic prediction for a member of a population that has no direct relatives in the reference population. In such cases the accuracy of GBV depends on the size of the reference population, the effective population size, and the heritability of the trait. It also depends on the genetic model underlying the trait. If phenotypic variation is due to polymorphisms in only a few major genes with large effects, the prediction accuracy can be higher. However, evidence so far points to the fact that many quantitative traits are being affected by very many genes, each with very small effects on trait variation, and that these genes are generally scattered across the whole genome. We will mostly assume the latter model.

Goddard [1] and Daetwyler et al. [10] presented formulas that could be used to predict the reliability of genomic prediction. These predictions are dependent on the size of the reference population, the effective population size of the breed, the heritability of the trait, and the length of the genome. The method is based on first estimating the number of independent segregating segments in a genome, and second on the accuracy of estimating the effects of each of these segments. The number of independent segments is dependent not only on the size of the genome but also on the effective population size. The effective population size (N_e) is formally measured via the rate of inbreeding and is a reflection of how related animals within a population are to each other. It is totally unrelated to true population size, and mainly depends on the genetic diversity of sires used in breeding programs. For example, the Holstein Friesian population has a population size of millions of dairy cattle, but due to a narrow genetic base (most sires used worldwide are descending from a handful of grandsires), the effective population size is less than 100. This means that the rate of inbreeding is the same as in a population of 100 individuals with equal number of males and females. Various methods exist to approximate a value for N_e, either via pedigree information [11] or via molecular marker information [12] (see **Note 5**).

The effective size is strongly related to genomic selection accuracy as it affects the size of the genomic segments that animals in the population share. In populations with small effective size, such as Holstein Friesian (N_e is smaller than 100), these shared segments are large whereas they are small in the merino breed which is more diverse and has a much larger effective size (N_e is around 1,000). In populations with small N_e there are larger segments, hence there are fewer of them (there are fewer "independent loci") and their effects are larger and easier to estimate.

The overall N_e will govern the effective number and size of chromosome segments (M_e) that are segregating in the population. If the effective population size is small, it is expected that animals will share larger chromosome segments and the genomic predictions will be more accurate. The accuracy for an individual with no phenotype as described by [1] is then predicted as:

$$\text{Accuracy} = \sqrt{1 - \lambda/(2N\sqrt{a}) \cdot \log((1+a+2\sqrt{a})/(1+a-2\sqrt{a}))}$$

where $a = 1 + 2 \times \lambda/N$, N is the number of animals in the reference,

$$\lambda = \sigma_e^2/\sigma_u^2$$

where σ_e^2 is the residual variance and σ_u^2 is the genetic variance at a single locus. σ_u^2 is estimated by

$$\sigma_u^2 = h^2/M_e k$$

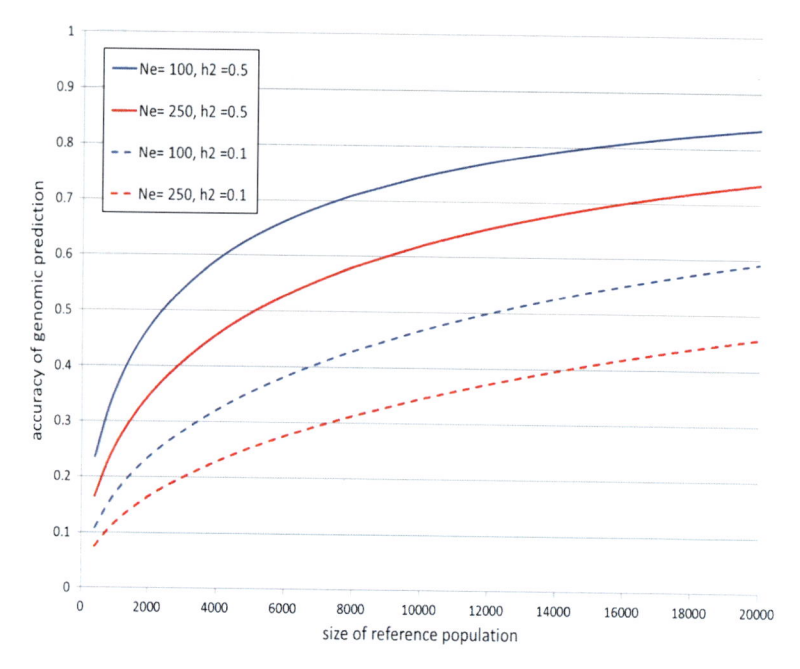

Fig. 2 Accuracy of genomic prediction depending on size of reference populations for two heritabilities (h^2) and effective population sizes (N_e)

where $M_e = 2 N_e L$ is the effective number of chromosome segments, L is the genome length in Morgans, h^2 is the heritability (*see* **Note 6**), and $k = 1/\log(2 N_e)$.

For a further discussion on M_e, *see* **Note 7**.

The other important parameter determining accuracy is heritability. With lower heritability, there is more error variance, and it is more difficult to estimate the effect of genome segments accurately. More records are needed for accurate prediction. Goddard's formula allows to show the relationship between number of records used (size of the reference population) and genomic prediction accuracy. This is illustrated in Fig. 2 for two levels of heritability and for two values for effective population size. The figure is derived using the formula of Goddard [1] and using $2 N_e L$ as an approximation for the number of effective loci (M_e) and L is the genome length.

3.3 Effect of Having Relatives in the Reference Population

Hayes et al. [13] presented formulas for genetic prediction accuracy with relatives. They assumed genomic prediction was based on gBLUP, using the genomic relationship matrix to replace the pedigree-based numerator relationships as used in traditional BLUP. Table 1 shows how the genomic predictions can gain in accuracy with relatives in the reference. The additional gains compared to current BLUP are not large, but the accuracy from using information on a few relatives is large compared to the accuracy obtained from information on "non-relatives." For example, information on 100 half sibs gives an accuracy of 0.44 for a low

Table 1
Accuracy based on information on *N* half sibs, using either pedigree-based relationships (*A*) or genomic relationships (*G*), [12]

	N	Accuracy A	G
$h^2 = 0.5$	10	0.37	0.39
	100	0.48	0.52
	1,000	0.50	0.63
$h^2 = 0.1$	10	0.22	0.22
	100	0.42	0.44
	1,000	0.49	0.56

heritable trait (Table 1) whereas this would require many thousands of animals in a reference population if only unrelated individuals were used.

Whereas pedigree relationships predict that full sibs have 50 % and half sibs have 25 % of genetic material in common (by descent) the genomic information will give a better prediction of actual gene sharing. The actual relationship among half sibs can vary from 0.2 to 0.3, with 0.25 only being the average. Therefore, genomic relationships give more accurate predictions than pedigree relationships.

3.4 Relationships or LD?

Results in Table 1 show that, when using genomic relationship with relatives, the increase in accuracy relative to pedigree-based prediction is relatively small unless there are extremely large family sizes. However, in genomic prediction we can also use information on less related individuals, and there can be very many of those. This information is not used in pedigree-based prediction. The baseline accuracy of genomic prediction is in effect information on "unrelated members of the same population" and if there are very many of these, accuracy can be high, as illustrated in Fig. 2. In the context of a breeding program, it may be difficult to form reference populations with direct relatives to all future selection candidates. Furthermore, predictions based on direct relatedness only last for one to two generations as relationships become smaller. Predictions based on many "unrelated" individuals in the reference population will wear out less quickly, and persist over more generations.

Genomic predictions are often discussed as based on linkage disequilibrium (LD) versus based on (genomic) relationships. LD refers to the marker alleles (SNPs) being in linkage disequilibrium with QTL (quantitative trait locus) alleles such that individuals with the same SNP allele (in a given region) are likely to have the same gene effects in that region. Within families such predictions are possible over large segments of chromosomes. Marker-based predictions are possible within families even if marker and QTL are not close; this is termed prediction based on "linkage." Within breeds,

Table 2
Comparison of prediction accuracy between groups differing in relatedness with the reference population for pedigree-based BLUP prediction based on a shallow pedigree (BLUP-S), a deep pedigree (BLUP-D), and genomic relationship matrix (GBLUP)

Method	Close	Distant	Unrelated
Relatedness with reference	0–0.25	0–0.125	0–0.05
Method		Accuracy	
BLUP-S	0.39	0.00	0.00
BLUP-D	0.42	0.21	0.04
GBLUP	0.57	0.41	0.34

there will be LD between SNPs and QTL over small distances, e.g., <50 kb in sheep and cattle. With dense markers, we can assume that all QTL are at least in LD with one marker, making predictions based on SNP genotypes possible within breed. The amount of LD (i.e., the distance on the chromosome over which loci are in LD) is smaller for populations with large N_e. LD across breeds exists over even shorter distances and the 50k chip is not expected to predict breeding value across breed as there is not sufficient LD.

LD and relationships are to some extent highly related. One could say that there is a lot of LD within families, and also that all individuals within a breed are somewhat more related relative to relatedness across breed. It has been shown that genomic prediction accuracy declines when an animal is less related to the reference population. The accuracy of "unrelated" animals falls back to the baseline accuracy, which is dependent on the number of individuals in the reference population. Prediction across breed is not possible if the marker density is insufficient. Whether a denser SNP chip will overcome this problem will also depend on other considerations, e.g., whether SNP effects are somewhat consistent across a wide variety of genetic backgrounds (i.e., across breeds).

In a simulation study Clark et al. [9] illustrate how accuracy is affected by relatedness (Table 2). They predicted a test set of animals based on a reference population that was either highly or moderately related, or unrelated to the test set. The first group had at least 20 half sibs in the reference set, those with moderate relationships had cousins and the unrelated reference set was up to ten generations removed from the test set. The test set consisted of 750 animals, the reference set of 1,750 animals, the heritability was 0.3, and the effective population size was 100.

For closely related animals, the accuracy of pedigree-based BLUP predictions was high and the accuracy of GBLUP was somewhat higher. For moderately related animals, genomic prediction accuracy was lower, but pedigree-based BLUP was much lower and even zero if only a one-generation pedigree was used. For the "unrelated" test set, pedigree-based BLUP had virtually zero accuracy but GBLUP still gives a decent accuracy. As it turns out, the value of 0.38 is close to the theoretical value predicted by the Goddard equation, which gives 0.36 for "baseline accuracy."

This illustrates that the accuracy predicted from the Goddard equation (Fig. 1) can be considered as accuracies for individuals that are not directly related to the reference, but are members of the same breed (breeding population). Hence, they may not be related within two generations, but they are related at least within the last ten generations. The average prediction accuracy of selection candidates will be somewhat higher because many will have more direct relatives in the reference population.

4 Empirical Evidence of Genomic Prediction Accuracy

Real data can be used to check whether the reliability of genomic selection can be accurately predicted using this theory. I am not going to give an overview of results published so far (and the number of this is rapidly growing these days) but rather provide some general comments. Firstly, there are two main methods to determine prediction accuracy. A third method could be based on realized selection response, but this is more time-consuming. The two main methods are

1. *Cross validation*. The best method is based on cross validation, where the whole dataset is divided into n subsets, and $n-1$ of these subsets are used as a reference set to predict the values in the nth set, which is the validation set or the test set. In the test set, these predictions can be correlated with the observed values. This process can be repeated n times, such that there are n independent samples of accuracy (measured as correlation). There is some debate about the value of n. Large values for n give many samples but a small test set. A value of $n = 10$ seems the maximum but this depends somewhat on the size of the total dataset.

2. *One validation set*. A set of animals exists with a very high accuracy of EBV, preferably based on a large progeny test. The EBVs of these animals can be used as a good approximation of true breeding value. Correlating GBV with EBV gives then a good indication of realized GBV accuracy. Note that the reference set does not necessarily need much progeny test data, these could even have just own performance data. This could be a powerful validation if the test set is large. However, a disadvantage is that there is only one test set, so the result can easily be affected by some outliers. Furthermore, we have no measure of variation in the correlation values, as we have in cross validation.

It is useful here to point out three potential problems:

1. Method 1 (and sometimes method 2) is often inaccurate if the GBV is not correlated with very highly accurate EBVs. In method 1, phenotypes are often used and these are (after correction for fixed effects) equivalent to EBVs with a reliability equal to heritability. The correlation between EBV and GBV would then underestimate the actual accuracy of GBV predicting true GBV. Dividing the correlation with the EBV accuracy would give an unbiased expectation of genomic prediction accuracy, but this also inflates the standard error of the estimated correlation by that amount.

2. If the validation set is heterogeneous, e.g., animals are from different genetic groups (i.e., have a different expected mean EBV, due to a subpopulation structure or animals being from different birth years) the correlation would be an overestimate of the prediction accuracy. Presumably, the usual genetic evaluation can take into account genetic differences in genetic group means, and we are interested in the accuracy of genomic prediction within groups. Hence, accuracy should in such cases be estimated as a correlation between residuals in a bivariate analysis of GBV and EBV, where the genetic groups are fitted as (fixed) effect, e.g., as birth year or subgroup.

5 Deciding on the Required Size of a Reference Population

Often, reference populations are simply formed by the previous generations of breeding animals. However, when genomic selection is aimed at selecting for new traits that are only measured in a reference population, a new group of animals needs to be selected for phenotypic measurement and genotyping. This could be the case when aiming improvement of meat quality and slaughter characteristics, or feed efficiency.

The amount of resources that one can afford to invest in reference populations will depend on the additional gains that are expected from genomic selection. A larger reference population will deliver more accuracy of genomic prediction, but there is a diminishing return. The additional genetic gain that can be achieved through genomic selection depends on the breeding objective and varies between species and between breeds and objectives. It is mainly driven by how well traits can be selected for with current methods. The potential gain of genomic selection will depend on the "measurability" of the main traits in the breeding objective.

More records are needed for low heritable traits to achieve the same accuracy. However, typically in breeding programs, it is not possible or efficient to obtain an equally high accuracy for all traits,

Table 3
Accuracy (lower limit) of genomic prediction depending on size of reference population, for various values of heritability and effective size and predicted additional genetic gain in sheep breeding programs [13]

N_e	1,000	250	100
Size of reference population	30,000	10,000	5,000
Progeny measured per year[1]	3,750	1,250	625
$h^2 = 0.1$	0.33	0.34	0.35
$h^2 = 0.3$	0.51	0.53	0.54
$h^2 = 0.5$	0.60	0.62	0.63
Predicted additional genetic gain	40 %	20 %	15 %
Size of reference population	12,000	4,000	2,000
Progeny measured per year[1]	1,500	500	250
$h^2 = 0.1$	0.22	0.23	0.23
$h^2 = 0.3$	0.36	0.37	0.38
$h^2 = 0.5$	0.44	0.46	0.47
Predicted additional genetic gain	20 %	10 %	8 %

[1]assuming the reference population is "refreshed" every 8 years

but rather the accuracy is increased proportionally to the heritability of the trait. The squared value of accuracy of an EBV is a measure of how much variance of the true breeding value is captured with the EBV. The additional genetic gain is more or less proportional to the explained variance [14].

The accuracies predicted by Goddard's formula are expected to last over several generations, as they are not based on direct relationships. Therefore, such large reference populations "last" at least for several generations. The number required to be measured each year can be a proportion of the total such that the reference is gradually refreshed and kept up to date. For example, if the reference population was "refreshed" every 8 years, then only 12.5 % would have to be measured each year. Table 3 gives an example of numbers needed to be phenotyped, for populations of various effective sizes.

6 Design of the Reference Population

Animals to be tested in the reference populations should be selected not only from a diverse genetic background within the breed but also from family lines that can be expected to contribute to the future gene pool in that breed. So there needs to be a balance between merit and diversity. A good strategy is to select progeny from young sires that have a high genetic merit, yet that are relatively unrelated to each other. The number of progeny tested per sire should be small, about ten progeny per sire, but accuracy is not highly dependent on that number. However, smaller progeny groups allow testing more sires which is desirable from an "industry engagement" point of view. All progeny should be genotyped.

Genotyping sires only gives no information about segregating alleles within the sire families. A strategy where only sires are genotyped requires a lot more progeny measurement. For example, for $N_e = 250$, $h^2 = 0.3$ and 2,000 individuals measured and genotyped gives an accuracy of 0.27. The same accuracy would be achieved if 1,300 sires were genotyped, each with ten progeny, where "heritability of progeny mean" is equal to 0.45. Hence, although fewer animals need to be genotyped, a lot more will require a phenotype.

7 Notes

1. The response presented here is per unit of index standard deviation. The actual response per year is calculated as $(i_m + i_f)/(L_m + L_f)\sigma_I$. Hence the response per trait can be multiplied by $(i_m + i_f)/(L_m + L_f)$ to obtain response per year. If males and females have different indices because they have different sets of information measured, then the male and female index results are weighted by $i_m/(L_m + L_f)$ and $i_f/(L_m + L_f)$, respectively.

2. The breeding value of an individual can be separated into a component predicted from the genomic test (q_i), and a residual additive genetic component that requires prediction accuracy based on the phenotypic information. This term can be referred to as "polygenic," with a variance in this example equal to $(1 - \rho)\sigma_a^2 = 0.175$ and the polygenic part and the QTL part are uncorrelated and only the polygenic part benefits from more phenotypic information. The phenotypic information is corrected for the genomic test and its variance is $\sigma_p^2 - \sigma_q^2$. In the example,

$$\mathbf{P} = \begin{bmatrix} 0.925 & 0 \\ 0 & 0.075 \end{bmatrix} \text{ and } \mathbf{G} = \begin{bmatrix} 0.175 \\ 0.075 \end{bmatrix} \text{ such that } \mathbf{b} = \begin{bmatrix} 0.189 \\ 1.00 \end{bmatrix}$$

and the accuracy is 0.658.

Note that the accuracy is increased, the weighting of the genomic information is 1 and the weight of the phenotypic information is lower than heritability as it only refers to the polygenic term: the weight $= 0.189 = (\sigma_a^2 - \sigma_q^2)/(\sigma_p^2 - \sigma_q^2)$.

The accuracy is the same as obtained with the Lande model (*see* **Text**), hence this approach is equivalent to this case. However, if information on relatives is included, a distinction can be made between a case where relatives are also genotyped and one where they are not. For relatives that have been genotyped, the phenotypic data needs to be included after correcting it for the animals GBV. Otherwise, the phenotypic data will be included without such a correction. The genomic breeding value on relatives will not provide any information about an animal that is genotyped itself.

For example, if we include information on a genotyped sire, we get

	P			G	b	VarIndex	Accuracy
ownPoly	0.925	0	0.0875	0.175	0.1819	0.1136	0.6741
ownGBV	0	0.075	0	0.075	1.0000		
sirePoly	0.0875	0	0.925	0.0875	0.0774		

and note that this is slightly different from including the sires' phenotype, uncorrected for the GBV:

	P			G	b	VarIndex	Accuracy
ownPoly	0.925	0	0.0875	0.175	0.1823	0.1133	0.6732
ownGBV	0	0.075	0.0375	0.075	0.9635		
Sirepheno	0.0875	0.0375	1	0.125	0.0729		

3. These covariances are based on assumptions and these could be reasonable approximations if we assume that the genomic term is also representing the effect of many loci. In that case the correlation between genomic tests is the same as the correlation between breeding values for two traits. However, if the genomic test is based on just a few QTL, it becomes more likely that the test does not explain the same proportion of the variances and the covariances of two traits. For example, the QTL could have an effect on one, but not the other trait. With sufficient data, the correlation between genetic predictions for various traits could be calculated and such assumptions could be verified.

4. Note here that the relative response per trait can differ between age classes because animals from different age classes have different accuracies for each trait. It is therefore important to average the response calculated from selection index response per trait for each age class, depending on the accuracy and the selection intensity in that age class.

5. Most breeding populations have seen a large reduction in effective size due to more targeted selection and more use of

common sires (hence, more inbreeding). Moreover, populations are often not fully closed, there is breed admixture, and they are not always homogeneous, e.g., some breeds really consist of subpopulations. Therefore, in modeling accuracy of genomic prediction it is a challenge to find the correct parameter for effective size.

6. Note that sometimes in a reference population (esp. in dairy), means of progeny are used as phenotypes rather than the phenotypes on the genotyped individual themselves. In that case one should not use the trait heritability, the "heritability of the progeny mean," which is equal to nprog/(npog + [(4 − h^2)/h^2]), where nprog is the number of progeny per animal genotyped.

7. In the theoretical prediction methods there is some ambiguity about the approximation of M_e with proposed values equal to $2N_eL/\ln(4N_eL)$ or $4N_eL$ [1] or $2N_eL$ [13]. A simulation study [9] using these approximations resulted in very different predicted accuracies of (a) 0.74, (b) 0.27, and (c) 0.36 whereas an empirical value of 0.38 was found. Consequently $2N_eL$ appears to be the most appropriate variable for baseline accuracy, at least in this simulation example. This is supported by other evidence (Ben Hayes, personal communication).

Acknowledgements

The chapter is based on ideas that were developed mainly during work associated with the Australian sheep CRC data and discussion with the team and other colleagues. For this, Hans Daetwyler, Sam Clark, Andrew Swan, Nasir Moghaddar, Ben Hayes, John Henshall, Brian Kinghorn, John Hickey, and Rob Banks are acknowledged.

References

1. Goddard ME (2009) Genomic selection: prediction of accuracy and maximisation of long term response. Genetica 136:245–257
2. Hazel LN (1943) The genetic basis for constructing selection indexes. Genetics 28:476–490
3. Bijma P (2012) Accuracies of estimated breeding values from ordinary genetic evaluations do not reflect the correlation between true and estimated breeding values in selected populations. J Anim Breed Genet 129:1–14
4. Lande R, Thompson R (1990) Efficiency of marker-assisted selection in the improvement of quantitative traits. Genetics 124:743–756
5. James JW (1987) Determination of optimal selection policies. J Anim Breed Genet 104:23–27
6. Ducrocq V, Quaas RL (1988) Prediction of genetic response to truncation selection across generations. J Dairy Sci 71:2543–2553
7. Habier D, Fernando RL, Dekkers JCM (2007) The impact of genetic relationship information on genome-assisted breeding values. Genetics 177:2389–2397
8. Goddard ME, Hayes BJ, Meuwissen TH (2011) Using the genomic relationship matrix to predict the accuracy of genomic selection. J Anim Breed Genet 128:409–421

9. Clark SA, Hickey JM, Daetwyler HD, van der Werf JHJ (2012) The importance of information on relatives for the prediction of genomic breeding values and the implications for the makeup of reference data sets in livestock breeding schemes. Genet Sel Evol 44:4
10. Daetwyler H, Villanueva B, Woolliams JA, Weedon N (2008) Accuracy of predicting the genetic risk of disease using a genome wide approach. PLoS One 3:e3395. doi: Doi:10.137
11. Cervantes I, Goyache F, Molina A, Valera M, Gutierrez JP (2011) Estimation of effective population size from the rate of co-ancestry in pedigreed populations. J Anim Breed Genet 128:56–63
12. Kijas JW, Lenstra JA, Hayes BJ et al (2012) Genome-wide analysis of the world's sheep breeds reveals high of historic mixture and strong recent selection. PLoS Biol 10: e1001258
13. Hayes BJ, Visscher PM, Goddard ME (2009) Increased accuracy of artificial selection by using the realized relationship matrix. Genet Res 91:47–60
14. Van der Werf JHJ (2009) Potential benefit of genomic selection in sheep. Proc Assoc Advmt Anim Breed Genet 18:38

INDEX

A

Accuracy
    of genomic prediction 296, 544, 545, 550–557, 560
    of imputation 364, 396–398, 401, 408–409
Additive relationship matrix 183, 184, 186, 366
Additive 7, 59, 60, 65–68, 70, 150, 161, 163, 173, 174, 180, 183, 184, 186, 187, 207, 209, 215, 217, 218, 225, 226, 238, 254, 280, 288, 317, 323–325, 366, 372, 438, 439, 440, 443, 450, 453, 454, 460, 465, 479, 500, 501, 511, 513, 514, 517, 520, 531, 538, 545–548, 558
Affymetrix 100, 102–107, 109, 110, 113, 114, 115, 119, 120, 520
Age structure 544, 548–550
AlphaImpute 396, 398–409
AlphaPhase 382–393, 399–402, 404, 405, 408
Analysis of variance 25, 54, 424
Ancestral haplotypes 347–379, 399
Animal breeding 280, 453, 543–560
Approximate Bayes factor 43, 55, 57–59, 65, 67, 85
ASREML 156, 160, 163, 185, 280, 326, 329
Association weight matrix (AWM) 437–446

B

Bayes factor 39, 41–43, 47–51, 53–60, 63–65, 67, 74–88, 91, 95, 171, 173, 174, 176, 182, 186, 187, 190
BayesA 165, 238–241, 244, 245, 252–253, 283–285, 293, 294, 295
BayesB 238–241, 243, 245–246, 250, 252–253, 256, 257, 283–285, 289, 293, 294, 295, 297, 317
BayesC 238–241, 246–248, 253, 257, 283–286, 293–297, 317
BayesCπ 238, 240–241, 246–248, 253, 283, 293, 294, 296
Bayesian LASSO (BL) 283, 301, 302, 305–309, 316
Bayesian linear regression (BLR) 283, 299–320
Bayesian ridge regression (BRR) 301, 302, 305, 306, 316
Bayesian statistics 39–41
BEAGLE 196, 371, 381, 396

Best linear unbiased prediction (BLUP) 158, 165, 237–239, 256, 283, 322–324, 326–328, 414, 544, 552, 554, 555
BIC criterion 59–60, 173, 451
Binary data 2, 27, 33, 197, 481
Biological filter 468–470
BL. *See* Bayesian LASSO (BL)
BLR. *See* Bayesian linear regression (BLR)
BLUP. *See* Best linear unbiased prediction (BLUP)
Box-Cox transformation 30, 32–33
BRR. *See* Bayesian ridge regression (BRR)

C

Call rates 141–142, 177, 201, 277
Case–control studies 38, 53, 57, 60, 65–71, 85, 141, 146, 195, 222, 348, 439, 483, 520
Causal variants 177, 217–221, 233, 373, 375, 376, 520, 533, 534
Centrality 22–23
Confidence interval 42, 44, 156–157, 205, 231, 366, 368, 375, 377, 379, 491
Continuous data 25, 282
Covariance matrix ... 183, 185, 226, 227, 238, 239, 242, 247, 249, 255, 256, 300, 301, 323, 326, 450, 452–455, 459, 460, 545
Covariates 177–179, 201, 203, 205, 207–209, 213, 217–220, 232, 238, 248, 251, 254, 279, 282, 285, 293, 296, 300, 366, 367, 446, 453, 460, 467, 481, 494, 504, 520, 523, 527, 529, 531, 536
Cross-validation 156, 165, 293, 302, 312, 313, 316, 361, 412, 418, 419, 482, 493, 494, 496, 555
Cryptic relatedness 177, 219, 224, 225, 505

D

Data transformation 23, 30, 32, 101, 503
Database 1, 100–104, 108–111, 117–126, 130–133, 135, 146, 147, 179–181, 208, 211, 212, 278, 445, 468, 470, 524, 530, 531, 536
Database schemas 103, 104, 109–111, 117, 119, 124, 132, 133, 147
Decision trees 466, 467, 503, 506, 508–514
Descriptive statistics 19–34, 424

Index

Design matrix 150, 158, 163, 226, 323, 324, 365, 366
Dispersion 22, 23, 451, 459, 460, 461
Divergent 426, 429, 431

E

Epistasis 465–474, 479, 485, 499–517
Evolution 331, 375, 473, 499, 506, 507
Evolutionary computation 499–517
Exhaustive search 467, 469, 471, 515
Explanatory variables 21, 22, 25–28, 30, 31, 33
Extended haplotype homozygosity 425

F

Family structure 27–28, 275, 372, 381, 387
fastPhase 361, 381, 382, 396
Feature selection 471, 503–508, 511, 512
Filter and wrapper approaches 471
Fixed effects 25–27, 150, 158, 219, 227, 238, 241, 243, 249, 251, 252, 275, 279–283, 292, 293, 295, 303–305, 318, 319, 323, 324, 326, 367, 412, 446, 450, 454, 459, 461, 556
Frequentist statistics 39, 42, 46
Functional set analysis 523

G

gBLUP. *See* Genomic best linear unbiased prediction (gBLUP)
GCTA. *See* Genome-wide complex trait analysis (GCTA)
Gene network 433, 437–440, 443–446, 466
Gene-gene interactions 466, 470, 479, 480, 500
Gene-set analysis 521, 522, 529
Genetic algorithms 500, 506–515, 517
Genome map 441
Genome-wide association studies (GWAS) 1–16, 29, 37, 45, 99, 129–147, 149–167, 193–213, 215, 217, 237, 275–296, 322, 347–379, 381, 395, 411–419, 437–446, 465, 479, 500, 504, 519–538
Genome-wide complex trait analysis (GCTA) 215–235
Genomic best linear unbiased prediction (gBLUP) 321–329, 550, 552, 554, 555
Genomic prediction 8, 164, 249, 275–296, 322, 326, 413, 415, 416, 544, 545, 547–548, 550–557, 560
Genomic relationship matrix (GRM) 217–221, 223–225, 232, 322–326, 328, 329, 372, 378, 391, 430–432, 552, 554
Genomic selection 163, 238, 248, 309, 361, 364, 367, 379, 411–418, 543–560
Genotype calling 136–138, 400, 438
Genotype-environment interaction 225–226

GenSel 249, 257, 276, 279–286, 289, 291, 294, 295, 296, 297
GRM. *See* Genomic relationship matrix (GRM)
GWAS. *See* Genome-wide association studies (GWAS)

H

Haplotype 149, 172, 207, 276, 332, 347, 381, 396, 417, 424, 496, 520
Haplotype score 372
Haplotyping 376, 404
Hardy-Weinberg equilibrium (HWE) 45, 138–141, 228, 277, 278, 438, 495, 528
Heritability 60, 64, 85, 86, 215–217, 220, 221, 232, 280, 288, 292, 294, 306, 307, 316, 321, 322, 327, 411, 413, 415, 465, 466, 483, 544, 545, 547, 548, 550–552, 554, 556–558, 560
Heterozygosity 142–146
Higher order interactions 471, 483, 490, 492, 499–517
Homozygosity 159, 331–344
HWE. *See* Hardy-Weinberg equilibrium (HWE)

I

IBD. *See* Identity by descent (IBD)
IC. *See* Inductive causation (IC)
Identity by descent (IBD) 159, 161–164, 166, 183–186, 193, 332, 336, 338, 374, 375, 376, 379, 381, 384
Illumina 4, 100, 103–105, 107–109, 111, 113, 114, 115, 119–122, 130, 131, 136, 137, 146, 147, 276–278, 285, 367, 520, 525
Imputation 130, 146, 172, 228, 348, 361, 364, 377, 379, 381, 383, 385, 395–409, 520
IMPUTE 172, 196, 376, 386, 395, 398, 400, 404, 408
Inductive causation (IC) 452, 453, 455–458, 460–463

L

Likelihood ratio test (LRT) 34, 85, 163, 221, 366–368, 370, 371, 374, 485
Linear models 19, 20, 25, 29–31, 33, 54, 59, 84, 163, 171–173, 216, 220, 223, 234, 238, 283, 322, 503, 520, 531
Linkage analysis 28, 193, 375, 379
Linkage disequilibrium (LD) 46, 60, 61, 63, 65, 67, 68, 70, 72, 73, 75, 78, 80, 83, 85, 87, 88, 90, 91, 149–152, 154, 156, 157, 159, 161, 162, 164, 166, 167, 171–173, 177, 179, 183–190, 206–208, 211, 212, 216, 220–221, 227, 229–231, 239, 250, 286, 288, 294, 326, 332, 348, 361, 365, 366, 369, 371, 376, 379, 396,

397, 416, 425, 427, 441, 442, 474, 483, 495, 496, 504, 513, 520, 522, 523, 525, 526, 527, 529, 530, 531, 535, 536, 553–555
Linux 106, 117, 118, 125, 205, 207, 312–314
Logistic regression 27, 33, 209, 231, 232, 466, 467, 485, 486, 491, 493, 520, 527, 528
Long-range phasing 381–383, 385, 399, 404, 409
LRT. *See* Likelihood ratio test (LRT)

M

Machine learning................ 466–468, 473, 480, 499–517
MAF. *See* Minor allele frequency (MAF)
Marginal probabilities 174, 175
Markov chain Monte Carlo (MCMC) 41, 174, 186, 239, 241–250, 252, 285, 286, 295, 302, 316, 424, 455, 456, 460, 461
MCMC sampling 239, 241–248, 252
MDR. *See* Multifactor dimensionality reduction (MDR)
MDS. *See* Multidimensional scaling (MDS)
Memory management..8–10
Minor allele frequency (MAF) 63, 86, 88–90, 93, 138–141, 147, 154, 197, 208, 220, 221, 277, 297, 323, 325, 438, 495, 530–532
Missing genotypes...................... 143, 172, 195, 196, 206, 278–279, 377, 390, 397, 401, 406
Missing heritability.. 465, 466
Mixed model 39, 59, 156, 158, 178, 183–186, 232, 239, 243, 252, 280, 281, 288, 323, 326, 366, 367, 371, 374, 452, 526
Model comparison 25–27, 30–31, 33, 451
Model diagnostics 20, 31, 34
Multidimensional scaling (MDS).............. 198, 201, 202, 203, 207
Multifactor dimensionality reduction (MDR)...... 465–497
Multiple SNP association..............................227–228, 230
Multiple SNP regression... 227
Multiple testing 46, 141, 153, 154, 164, 205, 446, 486, 502, 520, 521, 530, 537

N

Network inference... 437
Non-additivity 226, 466, 500–503
Non-causal variants 522, 533, 534
Numerator relationship matrix 184, 300–302, 317, 323–326, 391, 452

O

Opposing homozygotes....................... 382, 385, 389, 404
Outliers 20, 29–31, 142, 144, 192, 204, 231, 555
Over-fitting 31, 178, 229, 481–483, 488, 490–493, 496, 497, 516

P

Parallel computation 11–15
Pathway analysis 521, 526, 537
Pedigree 27, 85, 154, 156, 158, 166, 178, 183–187, 197, 217, 218, 226, 232, 278, 279, 280, 293, 294, 301, 302, 308–309, 317, 318, 322, 324, 327, 329, 333, 334, 348, 349, 363–364, 366, 372, 374, 376–378, 384, 386–392, 395–409, 412, 414, 418, 419, 432, 450, 452, 515, 543, 544, 551–555
Permutation tests 154, 467, 471, 474, 482, 483, 485, 487, 496, 502, 536
Phasing 130, 179, 348, 349, 364, 377, 381–394, 396, 398–405, 408, 409, 425
Phenotype..........7, 19, 37, 99, 130, 150, 178, 193, 216, 237, 276, 299, 322, 333, 347, 411, 424, 437, 449, 468, 479, 499, 519, 543
Phenotype network 449
PLINK 101, 126, 193–213, 218, 333–344, 426, 432, 527–529
Population
    genetics 201, 284
    size 162, 184, 188, 316, 331, 360, 361, 375, 376, 386, 387, 432, 544, 548, 550–552, 554
    stratification 194, 198–203, 207–209, 224, 225, 520
    structure........... 60, 72, 73, 147, 150, 157–159, 166, 177–183, 186, 190, 198, 210, 225, 275, 276, 281, 292, 293, 396, 412, 414, 419, 453, 460, 505, 511, 513–515, 556
Posterior distribution 40, 41, 57, 79, 95, 173, 187, 190, 239, 241, 248, 249, 253, 256, 257, 276, 283, 284, 285, 288, 291, 294, 296, 424, 453, 455, 456, 461, 462
Posterior probabilities 38, 42, 53, 59, 60, 86, 87, 166, 173, 174, 186, 256, 257
PostgreSQL 104, 114, 117, 118, 123, 124, 126
Power calculations........................47–54, 60–86, 182–183
Predictive model 177, 503–506, 508, 510, 514
Principal components 159, 178–180, 182–184, 198, 219, 430, 432, 527
Prior distribution 40, 42, 44, 45, 47, 55, 58, 86–95, 165, 187, 190, 238, 241, 242, 244, 247, 276, 283–285, 323, 424, 454, 460, 461
Python 102, 108–115, 117, 120, 126, 474

Q

Quality control (QC)............1, 102, 104, 129–147, 193, 197, 198, 218, 223, 233, 277, 371, 438, 481, 483, 495, 519, 520, 528
Quantitative trait loci (QTL) 149, 150, 154, 182, 321, 507

Quantitative trait loci (QTL) .... 149 (*cont.*)
    mapping 59, 153–155, 163, 173, 174, 183,
        187–189, 365, 366, 368, 371, 374, 378, 379

R

R programming 15
R statistics 481
Random effects 39, 56, 158, 160, 163, 164, 178,
        183–186, 216, 223, 226, 227, 237, 239, 279,
        283, 292, 300, 302, 305, 308, 315, 319, 320,
        322, 453
Recessive disease 333, 338, 343, 371, 373, 374
Recombinant inbred lines 349–351
Reference population 172, 323, 328, 544, 545, 548,
        550–560
Regression best linear unbiased prediction
    (RR-BLUP) 238–240,
        243–247, 250–253, 255–272, 283, 323, 324,
        326, 328, 550
Response variable 20, 22, 25–27,
        29–30, 482
Restricted maximum likelihood (REML) 56, 160,
        217–223, 232–235, 443, 444
    analysis 217–220, 223, 232
RR-BLUP. *See* Regression best linear unbiased prediction
    (RR-BLUP)
RSQLite 131, 133
Runs of homozygosity 331–343

S

Selection index 544–548, 559
SEM. *See* Structural equation models (SEM)
Shapiro-Wilk test 29
Sib disequilibrium test (SDT) 71, 78, 83–84
Sib transmission disequilibrium test (S-TDT) 53, 57,
        58, 60, 71, 78–80
Signatures of selection 423–434
Single marker regression 149–153, 163, 322
Single nucleotide polymorphism (SNP) 2, 37, 99,
        129, 150, 178, 195, 215, 238, 276, 315, 321,
        333, 347, 382, 396, 411, 425, 438, 465, 479,
        499, 523, 553
Smoothed F_{st} 425–429, 431–434
SNP map 277, 278, 426, 428, 433
SNPpy 99–126
Stochastic search 468–471
Structural equation models (SEM) 449–463

T

Transmission disequilibrium tests (TDT test) 71–83,
        85, 157, 158, 178

V

Validation of GWAS 411–419
Variable selection 256, 299, 317, 480, 481
Vectorization 4–7, 456

Printed by Publishers' Graphics LLC
MLSI130618.15.15.66 20130618